Functional Analysis and Geometry: Selim Grigorievich Krein Centennial

CONTEMPORARY MATHEMATICS

733

Functional Analysis and Geometry: Selim Grigorievich Krein Centennial

Peter Kuchment
Evgeny Semenov
Editors

AMS AMERICAN MATHEMATICAL SOCIETY

2010 *Mathematics Subject Classification.* Primary 01Axx, 14Rxx, 32Axx, 37Bxx, 46Axx, 46Bxx, 46Exx, 47Bxx, 52Axx, 53C65.

Library of Congress Cataloging-in-Publication Data

Names: Kuchment, Peter, 1949– editor. | Semenov, E. M. (Evgeniĭ Mikhaĭlovich), editor.
Title: Functional analysis and geometry : Selim Grigorievich Krein centennial / Peter Kuchment, Evgeny Semenov, editors. Other titles: Selim Grigorievich Krein centennial
Description: Providence, Rhode Island : American Mathematical Society, [2019] | Series: Contemporary mathematics ; volume 733 | Includes bibliographical references.
Identifiers: LCCN 2019006068 | ISBN 9781470437824 (alk. paper)
Subjects: LCSH: Functional analysis. | Hyperbolic spaces. | Kreĭn, S. G. (Selim Grigor′evich), 1917- | Festschriften. | AMS: History and biography – History of mathematics and mathematicians – History of mathematics and mathematicians. msc | Algebraic geometry – Affine geometry – Affine geometry. msc | Several complex variables and analytic spaces – Holomorphic functions of several complex variables – Holomorphic functions of several complex variables. msc | Dynamical systems and ergodic theory – Topological dynamics – Topological dynamics. msc | Functional analysis – Topological linear spaces and related structures – Topological linear spaces and related structures. msc | Functional analysis – Normed linear spaces and Banach spaces; Banach lattices – Normed linear spaces and Banach spaces; Banach lattices. msc | Functional analysis – Linear function spaces and their duals – Linear function spaces and their duals. msc | Operator theory – Special classes of linear operators – Special classes of linear operators. msc | Convex and discrete geometry – General convexity – General convexity. msc | Differential geometry – Global differential geometry – Integral geometry. msc
Classification: LCC QA320 .F788 2019 | DDC 515/.7–dc23
LC record available at https://lccn.loc.gov/2019006068

Contemporary Mathematics ISSN: 0271-4132 (print); ISSN: 1098-3627 (online)

DOI: https://doi.org/10.1090/conm/733

10 9 8 7 6 5 4 3 2 1 24 23 22 21 20 19

Dedicated to the centennial of Professor Selim Grigorievich Krein

Selim Grigorievich Krein

Photograph provided by Dr. T. Voronina (nee Krein).

Contents

Preface xi

Introduction
PETER KUCHMENT AND EVGENY SEMENOV 1

My father Selim Krein
TATIANA VORONINA (NEE KREIN) 17

Kyiv, Fall of 1943 through 1946. The rebirth of mathematics
YU. M. BEREZANSKY 23

Selim Grigorievich Krein in Stony Brook
DAVID G. EBIN 29

On algebraically integrable bodies
MARK AGRANOVSKY 33

Rearrangement invariant spaces satisfying Dunford-Pettis criterion of weak compactness
SERGEY V. ASTASHKIN 45

A new method of extension of local maps of Banach spaces. Applications and examples
GENRICH BELITSKII AND VICTORIA RAYSKIN 61

Two consequences of the associativity condition for a hypercomplex system with locally compact basis
YU. M. BEREZANSKY AND A. A. KALYUZHNYI 73

Inversion formulas of integral geometry in real hyperbolic space
WILLIAM O. BRAY AND BORIS RUBIN 81

Total positivity, Grassmannian and modified Bessel functions
V. M. BUCHSTABER AND A. A. GLUTSYUK 97

A remark on the intersection of plane curves
C. CILIBERTO, F. FLAMINI, AND M. ZAIDENBERG 109

Topological billiards, conservation laws and classification of trajectories
A. T. FOMENKO AND V. V. VEDYUSHKINA 129

Hasse–Schmidt Derivations and Cayley–Hamilton theorem for exterior algebras
LETTERIO GATTO AND INNA SCHERBAK 149

Complex analysis on the real sphere, or variations on a Maxwell's theme
SIMON GINDIKIN 167

The weighted Laplace transform
GISELE RUIZ GOLDSTEIN, JEROME A. GOLDSTEIN, GIORGIO METAFUNE, AND LUIGI NEGRO 175

Surfaces with big automorphism groups
SHULIM KALIMAN 185

Some binomial formulas for non-commuting operators
PETER KUCHMENT AND SERGEY LVIN 197

Similarity of holomorphic matrices on 1-dimensional Stein spaces
JÜRGEN LEITERER 209

Weighted geometric means of convex bodies
VITALI MILMAN AND LIRAN ROTEM 233

Sobolev, Besov and Paley-Wiener vectors in Banach and Hilbert spaces
ISAAC Z. PESENSON 251

Toeplitz operators in polyanalytic Bergman type spaces
GRIGORI ROZENBLUM AND NIKOLAI VASILEVSKI 273

Complete metric space of Riemann integrable functions and differential calculus in it
SERGUEI SAMBORSKI 291

Preface

This is the first of the two volumes dedicated to the centennial of distinguished mathematician Professor Selim Grigorievich Krein. Krein has made numerous major contributions to functional analysis, operator theory, partial differential equations, fluid dynamics, and other areas. He had written several influential monographs in these areas. He had also been a prolific teacher, graduating 83 PhD students, including quite a few well known mathematicians currently scattered around the globe. Another his major activity was creation and running the annual Voronezh Winter Mathematical Schools, which has influenced significantly the mathematical life in the former Soviet Union since 1967.

The articles in this volume are written by former students and colleagues of Selim Krein, as well as lecturers and participants of Voronezh Winter Schools. They are devoted to a variety of contemporary problems of functional analysis, operator theory, several complex variables, topological dynamics, and algebraic, convex, and integral geometry. The papers present new or survey recent results in these areas.

The editors express their sincere gratitude to all the contributors to the volume and to the AMS publishing staff, in particular to Ms. Christine Thivierge and production editor Mike Saitas, for great help provided. Special thanks go to Prof. R. S. Adamova for providing important information.

The next planned volume will address problems of differential equations, fluid dynamics, and applications.

Peter Kuchment, Mathematics Department, Texas A&M University, College Station, TX 77843-3368, USA

Evgeny Semenov, Mathematics Department, Voronezh State University, Voronezh, Russia

November 24th, 2018

Contemporary Mathematics
Volume **733**, 2019
https://doi.org/10.1090/conm/733/14727

Introduction

Peter Kuchment and Evgeny Semenov

Dedicated to the centennial of Professor Selim G. Krein

1. Selim Grigorievich Krein and his mathematics

We consider it a privilege and an honor to edit the two volumes dedicated to the centennial of our teacher Professor Selim Krein, prominent mathematician and educator, who has played a major role in our lives, as well as in the lives of hundreds of other mathematicians.

Selim Krein has had a long and distinguished career as a researcher and teacher. Let us start by describing him as a teacher. Krein has always been surrounded by young mathematicians whom he taught and whose lives he influenced. He was a great lecturer. His lectures and talks were always well prepared, clearly structured, and targeted exactly the level that was appropriate for the audience. It had always been a distinct pleasure to hear him lecturing. However, the way he taught was only a small part of an explanation of his great success as a mentor of young researchers. The whole atmosphere in his school was filled with excitement and conducive to creative research.

The number of his PhD students has reached at least 83 (see the list in section 3). Quite a few of his former students have become prominent mathematicians and hold professorial positions in the USA, Bulgaria, Australia, France, Germany, Israel, Vietnam, as well as in Russia, Ukraine, and various other republics of the fSU. His first two Ph.D. students, Professors Yu. Berezanskii[1] and Yu. Daletskii, have become members of the Ukrainian Academy of Sciences.

Although Krein's major achievements were in functional analysis, operator theory, partial differential equations, differential equations in Banach spaces, fluid dynamics, and numerical analysis, he was interested in mathematics as a whole, rather than just in a specific area. In order to emphasize freedom from rigid restrictions of topics, he called one of his seminars at the Voronezh Winter Math School (see [**183**] and [**KuLin, KuLin2**]) a "seminar on higher mathematics." In his seminars (even when they were designated to some specific areas) one could deliver talks on any subject one found interesting. Though in most cases he did follow the standard procedure of assigning a problem to a PhD student, he let the strongest of them

2010 *Mathematics Subject Classification.* Primary 01A70.

[1]The editors regret to report that as this volume was being prepared for the printer, Professor Yu. M. Berezanskii, prominent mathematician and the first PhD advisee of S. Krein, passed away at the age of 94.

"off the leash" to choose their own problems of interest. As the result, topics his students worked on, besides already listed functional analysis, operator theory and PDEs, included several complex variables, non-commutative integration, K-theory, commutative algebra, ..., you name it. The amazing thing was that when a student would start to work on a problem seemingly very remote from Krein's current interests, he was able to get into a new field and help the student throughout the research and dissertation. It was also very common for his students to organize their own seminars in order to study new areas.

Another distinguishing feature of Krein as an advisor was that he helped his students not only to resolve their research problems, but also problems of their everyday life (which often were much harder). Helping former students in getting decent jobs and apartments and in overcoming unwritten restrictions imposed by the authoritarian regime were very common for him. He kept helping his former students long after their graduation. It was amazing to see that Krein remembers the names of thousands of his former (even undergraduate) students.

Krein's joyful character, readiness to joke, and ability to charm everyone added to the admiration we all felt toward him.

Selim Krein has always been eager to go an extra mile to strengthen the mathematics in Voronezh. He was the initiator of the Mathematics Research Institute, a series of lectures on contemporary topics in mathematics for professors and graduate students of numerous Voronezh institutions of higher education, and most importantly, the highly influential Voronezh Winter Mathematical School (see [**School, KuLin, KuLin2**], and [**183**]).

Let us now turn to a description of some of Krein's life and work in mathematics. One can find additional information in papers and books listed in the section 2. It is impossible to address (even without details) all the directions of Krein's research, so we will concentrate only on the major ones.

Selim Krein was born July 15, 1917 in Kiev. He got his undergraduate degree in mathematics at Kiev State University in 1935–1940 and after his graduate study in 1940–1941 defended in 1942 a PhD dissertation under advising of N. N. Bogolyubov. His first love in mathematics was functional analysis, which continued to be one of his major areas of research since then. His first works in this area were written jointly with his teacher Bogolyubov and his older brother, famous mathematician Mark Krein (1907–1989). One of the results of his dissertation was a theorem now known as the Kreins-Kakutani theorem.

When World War II started, Krein was not drafted, due to a childhood illness that damaged one of his legs and left him with a limp for the rest of his life. Being a research fellow at the Mathematics Institute of the Ukrainian Academy of Sciences (relocated to a region free of German occupation), he started to work on important war-related projects (e.g., theory of cumulative ordinances) in a group led by M. A. Lavrent'ev. This is also when his major research in fluid dynamics started. He was among the first to introduce methods of functional analysis into fluid dynamics. He studied, in particular, the motion of a vessel partially filled with fluid (one can imagine, for instance, a rocket with liquid fuel). Oscillations of a fluid in an open vessel was another topic to which he had devoted a lot of attention. All these considerations arose from important applied problems. Among the culminations of his research in this direction were his classified Doctor of Sciences dissertation, which he defended in 1950 in the Moscow Academy of Artillery Sciences, and the books [**18**, **20**, **21**].

In 1950–1951 Krein was the Head of the Numerical Mathematics Department of the Mathematics Institute of the Ukrainian Academy of Sciences. Here he developed with M. A. Krasnoselskii a minimal residual method for solving systems of linear equations. Numerical analysis has also become one of the areas to which Krein returned many times during his career. Several of his students have become numerical analysts.

It is remarkable (and little known) that at that time S. Krein was, together with S. A. Avramenko, actively involved into design of and programming for the first Soviet electronic computer MESM (Small Electronic Computing Machine), the predecessor of the much better known BESM (Big Electronic Computing Machine). In particular, they insisted on significant increase of the hardware precision of calculations. They have also been mentioned (see [Odin, pp. 53-54]) among the first Soviet computer programmers, writing MESM codes for numerical solutions of ODEs.

The end of 1940s and beginning of 1950s were the periods of a high wave of governmental anti-semitism, due to which both M. Krasnoselskii and S. Krein had to leave the Institute of Mathematics and Kiev. One can read about this in particular in [**Berez**].

After a short time spent in Krivoi Rog, Krein came to Voronezh in 1954, where he had worked since and together with M. Krasnoselski and V. Sobolev created the well known Voronezh school of functional analysis[2]. S. G. worked first as a Professor and Chairman of the Mathematics and Theoretical Mechanics Department at the Voronezh Forestry Institute (currently, Forestry Academy). He became Chair of the Department of Partial Differential Equations of Voronezh State University in 1964, and came back to his previous position at the Forestry Institute in 1971. Under his guidance, the Department of Partial Differential Equations of the Voronezh State University and the Mathematics and Theoretical Mechanics Department of the Forestry Institute had become places of very active mathematical research.

Among his first studies in functional analysis and function theory was the introduction with Yu. Daletskii the important double operator integral notion [**34**, **38**], an area further developed by M. Birman and M. Solomyak and still very active currently. Together with Yu. Berezanskii, they introduced and studied hypercomplex systems with continual basis [**35**–**37**, **59**, **67**]. This work preceded much later studies of hypergroups, when various their results were re-discovered. One should also mention works on singular integrals [**27**, **28**, **30**].

Since then, his activity in functional analysis had mostly been concentrated on the theory of interpolation of linear operators, an area that has many applications, the most important ones in PDEs. Along with Calderon, Lions, Peetre and Gagliardo, Krein was one of the creators of the contemporary theory of interpolation of linear operators. He created the so called method of scales of Banach spaces (sometimes called "the Russian method" of operator interpolation), which has become one of the major methods in this theory. Several of his students have

[2]One can find a lot of historical information about mathematics at Voronezh State University in [**Adamova**].

become leading experts in interpolation theory. The milestones of this area of Krein's research were the surveys and books [**13, 16, 91, 95, 158**].

Research on fluid dynamics and PDEs had led S. Krein to another area of functional analysis: studying Fredholm and semi-Fredholm operators and continuous and analytic families of such operators, the behavior of spectra and eigenfunctions of such families being the main properties of interest. The techniques developed by Krein and his school in this area play an important role in many pure and applied studies (wave guides and photonic crystals among the most interesting ones). The survey books and papers [**7, 15, 136**] reflect the results of this research.

Krein also initiated, edited, and co-authored a popular reference book on functional analysis [**2, 4, 5, 8, 9**].

Problems of fluid dynamics have lead Krein to studying differential equations in Banach spaces and semigroups of operators. Here he has also become one of the world leaders. Books [**3, 6**] and survey [**151**] summarize results on the correctness of the Cauchy problem, analyticity of solutions, boundary value problems, asymptotic methods, and others, many of which were obtained by S. Krein and his students. In his last two decades, Krein also worked on singularly perturbed differential equations in Banach spaces.

A large part of Krein's scientific activity was devoted to various problems of partial differential equations. Among his best are, e.g. the results on homeomorphisms created by elliptic boundary value problems [**74, 76**], boundary value problems for overdetermined systems of PDEs, and boundary value problems in variable domains. Some of these studies are described in the books [**10, 11**].

For a quarter of century, his attention was attracted to differential equations on Lie groups and manifolds. For instance, papers [**111, 131, 132, 139**] treat relations between infinite dimensional representations of Lie groups and Cauchy problems for corresponding differential equations, differential equations of second order on Lie groups, representation of algebras of germs of differential operators with analytic coefficients by integro-differential operators, construction of infinitesimal operators for operators of generalized shift, and other topics. The book [**14**] develops analogs of Floquet theory for equations on manifolds, while [**19**] is devoted to function spaces related to representations of Lie groups and Lie algebras. The books [**10, 11, 14, 19**], regretfully, have never been translated to English.

Selim Krein was always eager to share his knowledge with others. He achieved this not only through his brilliant lectures, but also through the textbooks he wrote. The elementary calculus book [**1**] has been translated into several languages. A very nice analysis textbook [**12**] (also deserving translation) treats many important topics that are usually not discussed in standard textbooks. He has also written an introductory textbook [**17**] on linear programming for economics majors.

One of the S. Krein's great influences on the mathematical life in the fSU was through the remarkable Voronezh Winter Mathematical Schools (see [**School, KuLin, KuLin2**] and [**183**]).

The current volume and the forthcoming one [**Krein2**] contain mathematical works and memories of S. Krein's students, colleagues, and participants and lecturers of the Voronezh Winter Schools.

The editors express their sincere gratitude to all the contributors to the volume and to the AMS publishing staff, in particular to Ms. Christine Thivierge for great help provided. Special thanks go to Prof. R. S. Adamova for collecting and sharing with us information about the history of the department.

2. Some references

[Adamova] R. S. Adamova, *Derpt-Yuriev-Voronezh University. Mathematics and mathematicians*, Voronezh State University Publishing, Voronezh 2018. (In Russian)

[Berez] Yu. M. Berezanskii, Kiev, Fall of 1943 through 1946. The rebirth of Mathematics, in P. Kuchment and E. Semenov, *Functional Analysis and Geometry. Selim Grigorievich Krein Centennial*, AMS, Providence, R.I. 2019.

[60] Yu. M. Berezanskii, N. N. Bogolyubov, Yu. L. Daletskii, M. A. Lavrent'ev, E. M. Semenov, Selim Grigor'evich Krein (on the occasion of his sixtieth birthday), Uspehi Mat. Nauk **33** (1978), no. 2, 217–224, MR 57 #12102)

[70] Yu. M. Berezanskii, N. N. Bogolyubov, P. A. Kuchment, B. Ya. Levin, V. P. Maslov, S. P. Novikov, E. M. Semenov, Selim Grigor'evich Krein (on his seventieth birthday), Russian Math. Surveys **42**:5 (1987), 181–183, translation from the Russian original in Uspekhi Mat. Nauk **42**:5 (1987), 223–224.

[Obit] Yu. M. Berezanskii, S. G. Gindikin, S. S. Kutateladze, P. A. Kuchment, S. P. Novikov, Yu. G. Reshetnyak, E. M. Semenov, S. A. Sklyadnev, V. M. Tikhomirov, Selim Grigor'evich Krein (obituary), Russian Math. Surveys **55**:2 (2000), 327–328, translation of the Russian originalin Uspekhi Mat. Nauk **55**:2 (2000), 125–126. DOI: 10.1070/RM2000v055n02ABEH000270

[100] V.M. Buchstaber, P.A. Kuchment, S.P. Novikov, and E.M. Semenov, To the memory of Selim Grigor'evich Krein (1917–1998), Uspekhi Mat. Nauk **73**:1 (2018), 19–193 (in Rusian). English translation in Russian Math. Surveys **73**:1 (2018), 187–190. DOI: https://doi.org/10.1070/RM9816

[School] V. A. Kostin (Editor), *Proceedings of the international conference "S. G. Krein Voronezh Winter Mathematical School - 2018"*, Nauchnaya Kniga, Voronezh 2018. (In Russian)

[2008] V. A. Kostin, B. N. Sadovskii, and E. M. Semenov (Editors), *S. G. Krein. Recollections*, Voronezh State University 2008.

[KuLin] P. Kuchment and V. Lin (Ed.), *Voronezh Winter Mathematical Schools. Dedicated to Selim Krein*, AMS Translations, Series 2, v. 184, 1998.

[KuLin2] P. Kuchment and V. Lin, *Selim Krein and Voronezh Winter Mathematical Schools*, Contemporary Mathematics, vol. 734, American Mathematical Society, 2019

[Krein2] P. Kuchment and E. Semenov (Editors), *Differential Equations, Mathematical Physics, and Applications: Selim Grigorievich Krein Centennial*, Contemporary Mathematics, vol. 734, American Mathematical Society, 2019.

[Comp] V. P. Odinets, *Sketches of History of Computer Sciences*, Komi State Pedagogical Institute, Syktyvkar 2013.

3. List of 83 S. Krein's PhD students

Artemov, Georgii; Askerov, Nazim; Atlasov, Igor; Belyaeva, Elena; Berezansky, Yurii; Bryskin, Iliya; Chan, Thu; Chernyshov, Kornelii; Chuburin, Yurii; Daletsky, Yurii; Dement'eva, Olga; Denisov, Igor; Dmitriev, Vyacheslav; Duhovnyi, Mikhail; Fam, Ki; Fomin, Vasilii; Frolov, Nikolai; Furmenko, Aleksandr; Gasanov, Nasir; Glushko, Vladimir; Gohman, Aleksei; Gorohov, Evgenii; Grabovskaya, Revekka;

Gudovich, Nikolai; Ievleva, Oksana; Ivanov, Leonid; Kan, Ngo Zuy; Kolupanova, Galina; Kopachevskii, Nikolai; Kopaneva, Vera; Kostarchuk, Victor; Kostin, Vladimir; Kotko, Lyudmila; Kozlov, Ovsei; Ksendzenko, Lyudmila; Kuchment, Peter; Kulikov, Ivan; Kurina, Galina; Kvedaras, Bronyus; Laptev, Gennadii; Litvinkov, Stepan; Livchak, Aleksei; Nguen, Shi; Nikolov, Krasemir; Nikolova, Lyudmila; Osipov, Vasilii; Ovchinnikov, Vladimir; Pankov, Aleksandr; Pavlov, Evgenii; Pesenson, Isaac; Petunin, Yurii; Plushev, Yurii; Polichka, Nina; Prozorovskaya, Olga; Russman, Isaak; Rutitskaya, Alla; Salekhov, Dmitrii; Sapronov, Ivan; Savchenko, Galina; Savchenko, Yulia; Sedaev, Aleksandr; Semenov, Evgenii; Shablitzskaya, Liliya; Shihvatov, Aleksandr; Shmulev, Igor; Shneiberg, Iosif; Simonov, Aleksandr; Sklyadnev, Sergei Sobolevskii, Pavel; Solomatina, Lyuba; Susoev, Yurii; Tetievskaya, Irina; Tovbis, Aleksandr; Trofimov, Valerii; Venevitina, Svetlana; Yakut, Lidia; Yaroshenko, Nikolai; Yatzkin, Nikolai; Zaidenberg, Mikhail; Zarubin, Anatolii; Zobin, Nahum; Zubova, Svetlana; Zyukin, Pavel.

4. List of S. Krein's publications: books and papers

References

[1] Krein S. G. and Ushakova V. N., *Mathematical Analysis of Elementary Functions*, Fizmatgiz, Moscow 1963. (Translated to German in 1966)

[2] N. Ja. Vilenkin, E. A. Gorin, A. G. Kostjučenko, M. A. Krasnosel′skiĭ, S. G. Kreĭn, V. P. Maslov, B. S. Mitjagin, Ju. I. Petunin, Ja. B. Rutickiĭ, V. I. Sobolev, V. Ja. Stecenko, L. D. Faddeev, and È. S. Citlanadze, *Funktsionalyanyĭ analiz* (Russian), Edited by S. G. Kreĭn, Izdat. "Nauka", Moscow, 1964. MR0184056

[3] S. G. Kreĭn, *Lineĭ khye differentsial′nye uravneniya v Banakhovom prostranstve* (Russian), Izdat. "Nauka", Moscow, 1967. MR0247239

[4] S. G. Krein (ed.), *Analiza funkcjonalna* (Polish), A collection of articles edited by S. G. Kreĭn. Translated from the Russian by Ryszard Bittner, Państwowe Wydawnictwo Naukowe, Warsaw, 1967. MR0220034

[5] S. G. Krein (ed.), *Functional analysis*, Edited machine translation from the Russian. Foreign Technology Division MT-65-573, U.S. Department of Commerce, National Bureau of Standards, Washington, D.C., 1968. MR0234242

[6] S. G. Kreĭn, *Linear differential equations in Banach space*, American Mathematical Society, Providence, R.I., 1971. Translated from the Russian by J. M. Danskin; Translations of Mathematical Monographs, Vol. 29. MR0342804

[7] S. G. Kreĭn, *Lineĭ nye uravneniya v banakhovom prostranstve* (Russian), Izdat. "Nauka", Moscow, 1971. MR0374949

[8] M. Sh. Birman, N. Ya. Vilenkin, E. A. Gorin, P. P. Zabreĭko, I. S. Iokhvidov, M. Ĭ. Kadets′, A. G. Kostyuchenko, M. A. Krasnosel′skiĭ, S. G. Kreĭn, B. S. Mityagin, Yu. I. Petunin, Ya. B. Rutitskiĭ, E. M. Semenov, V. I. Sobolev, V. Ya. Stetsenko, L. D. Faddeev, and È. S. Tsitlanadze, *Funktsional′nyĭ analiz* (Russian), Izdat. "Nauka", Moscow, 1972. Edited by S. G. Kreĭn; Second edition, revised and augmented; Mathematical Reference Library. MR0352920

[9] N. Ya. Vilenkin, E. A. Gorin, A. G. Kostyuchenko, S. G. Krasnosel′skiĭ, S. G. Kreĭn, V. P. Maslov, B. S. Mityagin, Yu. I. Petunin, Ya. B. Rutitskii, V. I. Sobolev, V. Ya. Stetsenko, L. D. Faddeev, and E. S. Tsitlanadze, *Functional analysis*, Wolters-Noordhoff Publishing, Groningen, 1972. Translated from the Russian by Richard E. Flaherty; English edition edited by George F. Votruba with the collaboration of Leo F. Boron. MR0390693

[10] I. S. Gudovič and S. G. Kreĭn, *Boundary value problems for overdetermined systems of partial differential equations* (Russian, with Lithuanian and English summaries), Differencial′nye Uravnenija i Primenen.—Trudy Sem. Processy **Vyp. 9** (1974), 1–145. MR0481612

[11] L. Ivanov, L. Kotko, and S. Kreĭn, *Boundary value problems in variable domains* (Russian, with Lithuanian and English summaries), Differencial′nye Uravnenija i Primenen.—Trudy Sem. Processy **Vyp. 19** (1977), 161. MR0499710

[12] Krein S. G. and Zobin N. M., *Mathematical Analysis of Smooth Functions*, Voronezh State University, Voronezh 1978

[13] S. G. Kreĭn, Ju. I. Petunin, and E. M. Semënov, *Interpolyatsiya lineĭ nykh operatorov* (Russian), "Nauka", Moscow, 1978. MR506343

[14] S. G. Kreĭn and N. I. Yatskin, *Lineĭ nye differentsial′nye uravneniya na mnogoobraziyakh* (Russian), Voronezh. Gos. Univ., Voronezh, 1980. MR640266

[15] S. G. Kreĭn, *Linear equations in Banach spaces*, Birkhäuser, Boston, Mass., 1982. Translated from the Russian by A. Iacob; With an introduction by I. Gohberg. MR684836

[16] S. G. Kreĭn, Yu. Ī. Petunīn, and E. M. Semënov, *Interpolation of linear operators*, Translations of Mathematical Monographs, vol. 54, American Mathematical Society, Providence, R.I., 1982. Translated from the Russian by J. Szűcs. MR649411

[17] S. G. Kreĭn, *Matematicheskoe programmirovanie* (Russian), Voronezhskiĭ Gosudarstvennyĭ Universitet, Voronezh, 1983. MR768927

[18] N. D. Kopachevskiĭ, S. G. Kreĭn, and Ngo Huy Can, *Operatornye metody v lineĭ noĭ gidrodinamike* (Russian), "Nauka", Moscow, 1989. Èvolyutsionnye i spektral′nye zadachi. [Evolution and spectral problems]; With an English summary. MR1037258

[19] S. G. Kreĭn and I. Z. Pesenson, *Prostranstva gladkikh èlementov, porozhdennykh predstavleniem gruppy Li* (Russian), Voronezhskiĭ Gosudarstvennyĭ Universitet, Voronezh, 1990. Interpolyatsiya i priblizhenie. [Interpolation and approximation]; With an English summary. MR1071381

[20] N. D. Kopachevsky and S. G. Krein, *Operator approach to linear problems of hydrodynamics. Vol. 1*, Operator Theory: Advances and Applications, vol. 128, Birkhäuser Verlag, Basel, 2001. Self-adjoint problems for an ideal fluid. MR1860016

[21] N. D. Kopachevsky and S. G. Krein, *Operator approach to linear problems of hydrodynamics. Vol. 2*, Operator Theory: Advances and Applications, vol. 146, Birkhäuser Verlag, Basel, 2003. Nonself-adjoint problems for viscous fluids. MR2002951

Articles

1940

[22] (with Krein M. G.), An internal characterization of the space of all continuous functions defined on Hausdorff bicompact set, Dokl. Akad. Nauk SSSR 27(1940), no.5, 427-431. (In Russian)

1941

[23] (with Vershikov I. Kh. and Tovbin A. V.), on semi-ordered rings, Dokl. Akad. Nauk SSSR 30(1941), 778-780. (In Russian)

1943

[24] M. Krein and S. Krein, *Sur l'espace des fonctions continues définies sur un bicompact de Hausdorff et ses sousespaces semiordonnés* (French, with Russian summary), Rec. Math. [Mat. Sbornik] N.S. **13(55)** (1943), 1–38. MR0012209

1946

[25] (with Bogolyubov N. N.), On positive absolutely continuous operators, Trans. Inst. Matem. Ukarin. Akad. Nauk. 9(1946), 130-139. (In Ukrainian)

1947

[26] M. Krasnosel′skiĭ and S. Kreĭn, *On the center of a general dynamical system* (Russian), Doklady Akad. Nauk SSSR (N. S.) **58** (1947), 9–11. MR0022658

1948

[27] S. G. Kreĭn and B. Ya. Levin, *On the convergence of singular integrals* (Russian), Doklady Akad. Nauk SSSR (N.S.) **60** (1948), 13–16. MR0025606

[28] S. G. Kreĭn and B. Ya. Levin, *On the strong representation of functions by singular integrals* (Russian), Doklady Akad. Nauk SSSR (N.S.) **60** (1948), 195–198. MR0025607

[29] (with Levin B. Ya.), On one problem by I. P. Natanson, Uspehi Mat. Nauk 3(1948), no.3, 183-186. (In Russian)

[30] B. I. Korenblyum, S. G. Kreĭn, and B. Y. Levin, *On certain nonlinear questions of the theory of singular integrals* (Russian), Doklady Akad. Nauk SSSR (N.S.) **62** (1948), 17–20. MR0027439

[31] (with Kac G. I.), On the limit center of a dynamical system, Trans. Inst. Matem. Ukarin. Akad. Nauk. 1948, 121-134. (In Russian)

1949

[32] M. A. Krasnosel′skiĭ and S. G. Kreĭn, *On a proof of the theorem on category of a projective space* (Russian), Ukrain. Mat. Žurnal **1** (1949), no. 2, 99–102. MR0048809

1950

[33] Yu. L. Daleckiĭ and S. G. Kreĭn, *On differential equations in Hilbert space* (Russian), Ukrain. Mat. Žurnal **2** (1950), no. 4, 71–91. MR0047925

[34] (with Daleckii Yu. L.), Some properties of operators depending upon a parameter, Dokl. Akad. Nauk. Ukrain. SSR 6(1950), 433-436. (In Ukrainian)

[35] Yu. M. Berezanskiĭ and S. G. Kreĭn, *Some classes of continuous algebras* (Russian), Doklady Akad. Nauk SSSR (N.S.) **72** (1950), 237–240. MR0036946

[36] Yu. M. Berezanskiĭ and S. G. Kreĭn, *Continuous algebras* (Russian), Dokaldy Akad. Nauk SSSR (N.S.) **72** (1950), 5–8. MR0036945

1951

[37] Yu. M. Berezanskiĭ and S. G. Kreĭn, *Hypercomplex systems with a compact basis* (Russian), Ukrain. Mat. Žurnal **3** (1951), 184–204. MR0054585

[38] (with Daleckii Yu. L.), Formulas of differentiation with respect to a parameter of functions of Hermitian operators, Doklady Akad. Nauk SSSR (N.S.) 76(1951). 13–16. MR 12,617f.

1952

[39] M. A. Krasnosel′skiĭ and S. G. Kreĭn, *An iteration process with minimal residuals* (Russian), Mat. Sbornik N.S. **31(73)** (1952), 315–334. MR0052885

[40] M. A. Krasnosel′skiĭ and S. G. Kreĭn, *Remark on the distribution of errors in the solution of a system of linear equations by means of an iterative process* (Russian), Uspehi Matem. Nauk (N.S.) **7** (1952), no. 4(50), 157–161. MR0051582

1953

[41] S. G. Kreĭn, *On functional properties of operators of vector analysis and hydrodynamics* (Russian), Doklady Akad. Nauk SSSR (N.S.) **93** (1953), 969–972. MR0061954

[42] S. G. Kreĭn, *Uniform topology in the space of transformations* (Russian), Mat. Sbornik N.S. **33(75)** (1953), 627–638. MR0059542

[43] On fixed points of a conformal mapping, Uspehi Matem. Nauk (N.S.) 8(1953). no. 1(53), 155–159. MR 14,742f.

1954

[44] On an indeterminate equation in Hilbert space and its application in potential theory, Uspehi Matem. Nauk (N.S.) 9(1954). no. 3(61), 149–153. MR 16,262g.

1955

[45] M. A. Krasnosel′skiĭ and S. G. Kreĭn, *On the principle of averaging in nonlinear mechanics* (Russian), Uspehi Mat. Nauk (N.S.) **10** (1955), no. 3(65), 147–152. MR0071596

[46] M. A. Krasnosel′skiĭ and S. G. Kreĭn, *Nonlocal existence theorems and uniqueness theorems for systems of ordinary differential equations* (Russian), Dokl. Akad. Nauk SSSR (N.S.) **102** (1955), 13–16. MR0071588

1956

[47] Mathematical problems of the theory motion of a vessel filled with a fluid, Trans. of III All-union Math. Congress, v.1, p. 205, Akad. Nauk SSSR, Moscow 1956.

[48] (with Krasnoselskii M. A.), On differential equations in Banach spaces, Trans. of III All-union Math. Congress, v.2, p. 11, Akad. Nauk SSSR, Moscow 1956.

[49] M. A. Krasnosel′skiĭ, S. G. Kreĭn, and P. E. Sobolevskiĭ, *On differential equations with unbounded operators in Banach spaces* (Russian), Dokl. Akad. Nauk SSSR (N.S.) **111** (1956), 19–22. MR0088622

[50] M. A. Krasnosel′skiĭ and S. G. Kreĭn, *On the theory of ordinary differential equations in Banach spaces* (Russian), Voronoež. Gos. Univ. Trudy Sem. Funkcional. Anal. **1956** (1956), no. 2, 3–23. MR0086191

[51] Yu. L. Daleckiĭ and S. G. Kreĭn, *Integration and differentiation of functions of Hermitian operators and applications to the theory of perturbations* (Russian), Voronež. Gos. Univ. Trudy Sem. Funkcional. Anal. **1956** (1956), no. 1, 81–105. MR0084745

[52] M. A. Krasnosel′skiĭ and S. G. Kreĭn, *On a class of uniqueness theorems for the equation* $y' = f(x, y)$ (Russian), Uspehi Mat. Nauk (N.S.) **11** (1956), no. 1(67), 209–213. MR0079152

1957

[53] S. G. Krein, *On correctness classes for certain boundary problems* (Russian), Dokl. Akad. Nauk SSSR (N.S.) **114** (1957), 1162–1165. MR0089977

[54] S. G. Kreĭn, *Differential equations in a Banach space and their application in hydromechanics* (Russian), Uspehi Mat. Nauk (N.S.) **12** (1957), no. 1(73), 208–211. MR0085418

[55] M. A. Krasnosel′skiĭ and S. G. Kreĭn, *Continuity conditions for a linear operator in terms of properties of its square* (Russian), Voronež. Gos. Univ. Trudy Sem. Funkcional. Anal. **1957** (1957), no. 5, 98–101. MR0098323

[56] S. G. Kreĭn and O. I. Prozorovskaya, *An analogue of Seidel's method for operator equations* (Russian), Voronež. Gos. Univ. Trudy Sem. Funkcional. Anal. **5** (1957), 35–38. MR0095579

[57] M. A. Krasnoselsky, S. G. Krein, and P. E. Sobolevsky, *On differential equations with unbounded operators in Hilbert space* (Russian), Dokl. Akad. Nauk SSSR (N.S.) **112** (1957), 990–993. MR0089974

[58] S. G. Kreĭn and N. N. Moiseev, *On oscillations of a vessel containing a liquid with a free surface* (Russian), Prikl. Mat. Meh. **21** (1957), 169–174. MR0089572

[59] Yu. M. Berezanskiĭ and S. G. Kreĭn, *Hypercomplex systems with continual basis* (Russian), Uspehi Mat. Nauk (N.S.) **12** (1957), no. 1(73), 147–152. MR0086272

[60] (with Krasnosel'skii M. A. and Myshkis A. D.), An extended session in march 1957 of the Voronezh Seminar on Functional Analysis, Uspehi Mat. nauk 12(1957), no.4, 241-250.

1958

[61] V. P. Gluško and S. G. Kreĭn, *Fractional powers of differential operators and imbedding theorems.* (Russian), Dokl. Akad. Nauk SSSR **122** (1958), 963–966. MR0100144

[62] S. G. Kreĭn and P. E. Sobolevskiĭ, *A differential equation with an abstract elliptical operator in Hilbert space* (Russian), Dokl. Akad. Nauk SSSR (N.S.) **118** (1958), 233–236. MR0099604

[63] (with Krasnosel'skii M. A.), On differential equations in a Banach space,, Trans. of III All-union Math. Congress, v.3, Moscow (1956), p. 73-80, Akad. Nauk SSSR, Moscow 1958.

1960

[64] S. G. Kreĭn, *On an interpolation theorem in operator theory*, Soviet Math. Dokl. **1** (1960), 61–64. MR0119094

[65] S. G. Kreĭn, *On the concept of a normal scale of spaces*, Soviet Math. Dokl. **1** (1960), 586–589. MR0121627

[66] S. G. Kreĭn, *Differential equations in Banach space and their application in hydrodynamics*, Amer. Math. Soc. Transl. (2) **16** (1960), 423–426, DOI 10.1090/trans2/016/26. MR0117452

[67] Yu. M. Berezanski and S. G. Kreĭn, *Hypercomplex systems with continuous basis*, Amer. Math. Soc. Transl. (2) **16** (1960), 358–364. MR0117591

[68] S. G. Kreĭn and O. I. Prozorovskaja, *Analytic semi-groups and incorrect problems for evolutionary equations*, Soviet Math. Dokl. **1** (1960), 841–844. MR0151862

[69] V. P. Gluško and S. G. Kreĭn, *Inequalities for norms of derivatives in weighted L_p spaces* (Russian), Sibirsk. Mat. Ž. **1** (1960), 343–382. MR0133681

1961

[70] Ill-posed problems and estimates of solutions of parabolic equations, in "Ill-posed problems of mathematics and mechanics", 84-86, Siberian Branch of Acad. of Sci., Novosibirsk 1961

[71] S. G. Kreĭn and Ju. I. Petunin, *A relationship criterion for two Banach spaces* (Russian), Dokl. Akad. Nauk SSSR **139** (1961), 1295–1298. MR0141975

[72] S. G. Kreĭn and E. M. Semenov, *A scale of spaces* (Russian), Dokl. Akad. Nauk SSSR **138** (1961), 763–766. MR0140939

1962

[73] S. G. Kreĭn and G. I. Laptev, *Boundary-value problems for an equation in Hilbert space* (Russian), Dokl. Akad. Nauk SSSR **146** (1962), 535–538. MR0156068

1963

[74] Ju. M. Berezans′kiĭ, S. G. Kreĭn, and Ja. A. Roĭtberg, *A theorem on homeomorphisms and local increase of smoothness up to the boundary for solutions of elliptic equations*, Outlines Joint Sympos. Partial Differential Equations (Novosibirsk, 1963), Acad. Sci. USSR Siberian Branch, Moscow, 1963, pp. 33–38. MR0209665

[75] S. G. Kreĭn and O. I. Prozorovskaja, *Approximate methods of solving ill-posed problems* (Russian), Ž. Vyčisl. Mat. i Mat. Fiz. **3** (1963), 120–130. MR0153125

[76] Yu. M. Berezanski, S. G. Kreĭn, and Ja. A. Roĭtberg, *A theorem on homeomorphisms and local increase of smoothness up to the boundary for solutions of elliptic equations* (Russian), Dokl. Akad. Nauk SSSR **148** (1963), 745–748. MR0146508

1964

[77] S. G. Kreĭn, *Oscillations of a viscous fluid in a container* (Russian), Dokl. Akad. Nauk SSSR **159** (1964), 262–265. MR0182238

[78] S. G. Kreĭn, *Interpolation theorems in operator theory, and embedding theorems* (Russian), Proc. Fourth All-Union Math. Congr. (Leningrad, 1961), Izdat. "Nauka", Leningrad, 1964, pp. 504–510. MR0220077

[79] (with Krasnosel'skii M. A.), Operator equations in function spaces, in Proc. Fourth All-Union Math. Congr. (Leningrad, 1961), Vol. II, 292–299, "Nauka", Leningrad 1964. (In Russian) MR 36#2926.

[80] S. G. Kreĭn and Ju. I. Petunin, *On the concept of minimal scale of spaces* (Russian), Dokl. Akad. Nauk SSSR **154** (1964), 30–33. MR0161125

[81] N. G. Askerov, S. G. Kreĭn, and G. I. Laptev, *On a class of non-selfadjoint boundary-value problems* (Russian), Dokl. Akad. Nauk SSSR **155** (1964), 499–502. MR0160133

[82] (with Laptev G. I.), Boundary value problems with parameter in the boundary condition, Proc. 3rd All-Union Symposium on Diffraction and Waves, 39-41, Tbilisi 1964

[83] (with Krasnosel'skii M. A., Rutitskii Ya. B., and Sobolev V. I.), On mathematical life in Voronezh, Uspehi Mat. nauk 19(1964), no.3, 225-245. (In Russian)

1965

[84] (with Simonov A. S.), Theorem on homomorphisms and quai-linear equations, in "Abstracts of All-union Conf. on Appli. of Funct. Anal. and Non-linear problems", 62-63, Akad. Nauk Azerb. SSR, Baku 1965.

1966

[85] S. G. Kreĭn, *Correctness of the Cauchy problem and the analyticity of solutions of the evolution equation* (Russian), Dokl. Akad. Nauk SSSR **171** (1966), 1033–1036. MR0208115

[86] S. G. Kreĭn, Ju. I. Petunin, and E. M. Semenov, *Hyperscales of Banach lattices* (Russian), Dokl. Akad. Nauk SSSR **170** (1966), 265–267. MR0203448

[87] S. G. Kreĭn and G. I. Laptev, *Boundary value problems for second order differential equations in a Banach space. I* (Russian), Differencial′nye Uravnenija **2** (1966), 382–390. MR0199518

[88] S. G. Kreĭn and G. I. Laptev, *Correctness of boundary value problems for a differential equation of the second order in a Banach space. II* (Russian), Differencial′nye Uravnenija **2** (1966), 919–926. MR0203196

[89] S. G. Kreĭn and L. N. Šablickaja, *Stability of difference schemes for the Cauchy problem* (Russian), Ž. Vyčisl. Mat. i Mat. Fiz. **6** (1966), 648–664. MR0199978

[90] S. G. Kreĭn and A. S. Simonov, *A theorem of homeomorphisms and quasilinear equations* (Russian), Dokl. Akad. Nauk SSSR **167** (1966), 1226–1229. MR0197979

[91] S. G. Kreĭn and Ju. I. Petunin, *Scales of Banach spaces* (Russian), Uspehi Mat. Nauk **21** (1966), no. 2 (128), 89–168. MR0193499

[92] (with Glushko V. P. and Mukhin V. E.), On global and local optimal plans in problems of linear programming with two-sided bounds, in "Optimal programming in industrial problems", no.2, 89-90, Voronezh State Univ. 1966. (In Russian).

[93] (with Ievleva O. B.), On oscillations of a viscous fluid in a vessel, Abstracts of the Internat. Math. Congr., Sect. 12, p.37, Moscow 1966.

[94] (with Petunin Yu. I.), New results in the theory of scales of Banach spaces, Abstracts of the Internat. Math. Congr., Sect. 5, p.56, Moscow 1966.

1967

[95] S. G. Kreĭn, Ju. I. Petunin, and E. M. Semenov, *Scales of Banach lattices of measurable functions* (Russian), Trudy Moskov. Mat. Obšč. **17** (1967), 293–322. MR0223878

[96] Sixth joint conference in physics and mathematics of the Far East, Uspehi Mat. Nauk 22(1967), no.1, 197-198. (In Russian).

[97] First Voronezh Winter Math. School, Uspehi Mat. Nauk 22(1967), no.4, 189-190. (In Russian).

1968

[98] S. G. Kreĭn and G. I. Laptev, *On the problem of the motion of a viscous fluid in an open vessel* (Russian), Funkcional. Anal. i Priložen. **2** (1968), no. 1, 40–50. MR0248462

[99] S. G. Kreĭn, *The behavior of solutions of elliptic problems under variation of the domain* (Russian), Studia Math. **31** (1968), 411–424. MR0235276

[100] The behavior of solutions of elliptic problems under variation of the domain, Proc. of 7th Math. and Phys. Conf. of Far east, p. 18, Khabarovsk 1968. (In Russian)

[101] S. G. Kreĭn, *Lineĭ nye uravneniya v banakhovom prostranstve* (Russian), Voronež. Gosudarstv. Univ., Voronezh, 1968. MR0374950

[102] S. G. Kreĭn and G. I. Laptev, *On the problem of the motion of a viscous fluid in an open vessel* (Russian), Funkcional. Anal. i Priložen. **2** (1968), no. 1, 40–50. MR0248462

[103] N. K. Askerov, S. G. Kreĭn, and G. I. Laptev, *The problem of the oscillations of a viscous liquid and the operator equations connected with it* (Russian), Funkcional. Anal. i Priložen. **2** (1968), no. 2, 21–31. MR0232233

[104] V. P. Gluško and S. G. Kreĭn, *Degenerate linear differential equations in a Banach space* (Russian), Dokl. Akad. Nauk SSSR **181** (1968), 784–787. MR0232067

1969

[105] S. G. Kreĭn and I. M. Kulikov, *The Maxwell-Leontovič operator* (Russian), Differencial′nye Uravnenija **5** (1969), 1275–1282. MR0265991

[106] S. G. Kreĭn and V. P. Trofimov, *Holomorphic operator-valued functions of several complex variables.* (Russian), Funkcional. Anal. i Priložen. **3** (1969), no. 4, 85–86. MR0262861

[107] S. G. Kreĭn and Ngo Zui Kan, *The problem of small motions of a body with a cavity partially filled with a viscous fluid*, J. Appl. Math. Mech. **33** (1969), 110–117, DOI 10.1016/0021-8928(69)90118-X. MR0260269

[108] S. G. Kreĭn and Ngo Zui Kan, *Asymptotic method in the problem of oscillations of a strongly viscous fluid*, J. Appl. Math. Mech. **33** (1969), 442–450, DOI 10.1016/0021-8928(69)90059-8. MR0259257

[109] S. G. Kreĭn and G. I. Laptev, *An abstract scheme for the examination of parabolic problems in noncylindrical regions* (Russian), Differencial′nye Uravnenija **5** (1969), 1458–1469. MR0255989

[110] III Voronezh Winter Math. School, January 26 - February 6 1969, Uspehi Mat. Nauk 24(1969), no.4, 230-231. (In Russian)

1970

[111] S. G. Kreĭn and A. M. Šihvatov, *Linear differential equations on a Lie group* (Russian), Funkcional. Anal. i Priložen. **4** (1970), no. 1, 52–61. MR0439987

[112] S. G. Kreĭn and V. P. Trofimov, *Noetherian operators that depend holomorphically on parameters* (Russian), A collection of articles on function spaces and operator equations (Proc. Sem. Functional Anal., Math. Fac., Voronezh State Univ., Voronezh, 1970)(Russian), Voronež. Gos. Univ., Voronezh, 1970, pp. 63–85. MR0430834

[113] S. G. Kreĭn and V. P. Trofimov, *The multiplicity of a characteristic point of the holomorphic operator-function* (Russian), Mat. Issled. **5** (1970), no. 4(18), 105–114. MR0312305

[114] S. G. Kreĭn and Ju. B. Savčenko, *Exponential dichotomy for partial differential equations* (Russian), Differencial′nye Uravnenija **8** (1972), 835–844. MR0304809

[115] S. G. Kreĭn and V. B. Osipov, *Ljapunov functions and Cauchy problems for certain systems of partial differential equations* (Russian), Differencial′nye Uravnenija **6** (1970), 2053–2061. MR0299930

[116] S. G. Kreĭn, Ju. Ī. Petunīn, and E. M. Semenov, *Imbedding theorems and interpolation of linear operators* (Russian), Imbedding theorems and their applications (Proc. Sympos., Baku, 1966), Izdat. "Nauka", Moscow, 1970, pp. 127–131, 245. MR0313788

[117] S. G. Kreĭn, G. I. Laptev, and G. A. Cvetkova, *The Hadamard correctness of the Cauchy problem for an evolution equation* (Russian), Dokl. Akad. Nauk SSSR **192** (1970), 980–983. MR0265728

[118] Voronezh Winter Math. School, January 26 - February 7 1970, Uspehi Mat. Nauk 25(1970), no.5, 265-266. (In Russian)

1971

[119] S. G. Kreĭn and P. A. Kučment, *A certain approach to the interpolation problem for linear operators* (Russian), Voronež. Gos. Univ. Trudy Naučn.-Issled. Inst. Mat. VGU **3** (1971), 54–60. MR0355632

[120] S. G. Kreĭn, *Interpolation of linear operators, and properties of the solutions of elliptic equations* (Russian, with English summary), Elliptische Differentialgleichungen, Band II, Akademie-Verlag, Berlin, 1971, pp. 155–166. Schriftenreihe Inst. Math. Deutsch. Akad. Wissensch. Berlin, Reihe A, Heft 8. MR0344646

[121] I. S. Gudovič and S. G. Kreĭn, *Certain boundary value problems that are elliptic in a subspace* (Russian), Mat. Sb. (N.S.) **84 (126)** (1971), 595–606. MR0282050

[122] Fifth Voronezh Winter Math. School, Uspehi Mat. Nauk 26(1971), no.5, 270-272. (In Russian)

1972

[123] S. G. Kreĭn and N. I. Jackin, *Differential form equations* (Russian), Voronež. Gos. Univ. Trudy Naučn.-Issled. Inst. Mat. VGU **Vyp. 5 Sb. Stateĭ Teor. Operator. i Differencial. Uravneniĭ** (1972), 75–79. MR0448384

[124] I. S. Gudovič and S. G. Kreĭn, *Boundary value problems for operators of exterior differentiation* (Russian), Voronež. Gos. Univ. Trudy Naučn.-Issled. Inst. Mat. VGU **Vyp. 5 Sb. Stateĭ Teor. Operator. i Differencial. Uravneniĭ** (1972), 35–45. MR0448383

[125] S. G. Kreĭn and E. M. Semenov, *A certain property of equimeasurable functions* (Russian), Voronez. Gos. Univ. Trudy Naučn.-Issled. Inst. Mat. VGU **Vyp. 5 Sb. Stateĭ Teor. Operator. i Differencial. Uravneniĭ** (1972), 70–74. MR0447512

[126] I. S. Gudovič and S. G. Kreĭn, *Elliptic boundary value problems for the system* $\binom{d_0^*}{d_1}u = f$ (Russian), Funkcional. Anal. i Priložen. **6** (1972), no. 4, 75–76. MR0318666

[127] I. S. Gudovič, S. G. Kreĭn, and I. M. Kulikov, *Boundary value problems for the Maxwell equations* (Russian), Dokl. Akad. Nauk SSSR **207** (1972), 321–324. MR0316892

[128] (with Borisovich Yu. G.), VI Voronezh Winter Math. School, Uspehi Mat. Nauk 27(1972), no.5, 273-275. (In Russian)

[129] S. G. Kreĭn and Ju. B. Savčenko, *Exponential dichotomy for partial differential equations* (Russian), Differencial′nye Uravnenija **8** (1972), 835–844. MR0304809

1973

[130] S. G. Kreĭn and S. Ja. L′vin, *A general initial problem for a differential equation in a Banach space* (Russian), Dokl. Akad. Nauk SSSR **211** (1973), 530–533. MR0342805

[131] R. Ja. Grabovskaja and S. G. Kreĭn, *A certain representation of the algebra of differential operators, and the differential equations connected with it* (Russian), Dokl. Akad. Nauk SSSR **212** (1973), 280–283. MR0342889

[132] R. Ja. Grabovskaja and S. G. Kreĭn, *The formula for the permutation of functions of operators that represent a Lie algebra* (Russian), Funkcional. Anal. i Priložen. **7** (1973), no. 3, 81. MR0336405

[133] S. G. Kreĭn and L. N. Šablickaja, *Necessary conditions for the stability of difference schemes, and the eigenvalues of difference operators* (Russian), Ž. Vyčisl. Mat. i Mat. Fiz. **13** (1973), 647–657, 812. MR0331805

[134] S. G. Kreĭn and E. M. Semenov, *Interpolation of operators of weakened type* (Russian), Funkcional. Anal. i Priložen. **7** (1973), no. 2, 89–90. MR0315429

1974

[135] N. V. Efimov, L. V. Kantorovič, S. G. Kreĭn, I. S. Iohvidov, M. A. Krasnosel′s′kiĭ, and L. A. Ljusternik, *Vladimir Ivanovič Sobolev (on the occasion of his sixtieth birthday)* (Russian), Uspehi Mat. Nauk **29** (1974), no. 1(175), 247–250. (1 plate). MR0386949

1975

[136] M. G. Zaĭdenberg, S. G. Kreĭn, P. A. Kučment, and A. A. Pankov, *Banach bundles and linear operators* (Russian), Uspehi Mat. Nauk **30** (1975), no. 5(185), 101–157. MR0415661

[137] L. A. Kotko and S. G. Kreĭn, *The completeness of a system of eigen- and associated functions of boundary value problems with a parameter in the boundary conditions* (Russian), Collection of articles on applications of functional analysis (Russian), Voronež. Tehnolog. Inst., Voronezh, 1975, pp. 71–89. MR0481622

1976

[138] S. G. Kreĭn and E. M. Semenov, *Some interpolation theorems of the theory of linear operators and their applications* (Russian), Imbedding theorems and their applications (Proc. All-Union Sympos., Alma-Ata, 1973), Izdat. "Nauka" Kazah. SSR, Alma-Ata, 1976, pp. 64–68, 188. MR0482126

[139] R. Ya. Grabovskaya and S. G. Kreĭn, *Second order differential equations with operators generating a Lie algebra representation*, Math. Nachr. **75** (1976), 9–29, DOI 10.1002/mana.19760750103. MR0481673

[140] L. A. Kotko and S. G. Kreĭn, *The completeness of the system of eigen- and associated functions of boundary value problems with a parameter in the boundary conditions* (Russian), Dokl. Akad. Nauk SSSR **227** (1976), no. 2, 288–290. MR0402296

1977

[141] Interpolation of operators in spaces of smooth functions, in "Operator theory in functional spaces", 188-205, Nauka, Novosibirsk 1977.

[142] V. I. Dmitriev, S. G. Kreĭn, and V. I. Ovčinnikov, *Fundamentals of the theory of interpolation of linear operators* (Russian), Geometry of linear spaces and operator theory (Russian), Jaroslav. Gos. Univ., Yaroslavl, 1977, pp. 31–74. MR0634076

1978

[143] S. G. Kreĭn and A. I. Furmenko, *Dynamic equivalence of linear differential operators with nilpotent operators* (Russian), Approximate methods for investigating differential equations and their applications, No. 4 (Russian), Kuĭbyshev. Gos. Univ., Kuybyshev, 1978, pp. 47–56, 134. MR578334

[144] V. I. Dmitriev and S. G. Kreĭn, *Interpolation of operators of weak type* (English, with Russian summary), Anal. Math. **4** (1978), no. 2, 83–99, DOI 10.1007/BF02116975. MR505532

1980

[145] S. G. Kreĭn and N. P. Polička, *Behavior of the solutions of boundary value problems for the Lavrent′ev-Bicadze equation under variation of the domain* (Russian), Partial differential equations (Proc. Conf., Novosibirsk, 1978), "Nauka" Sibirsk. Otdel., Novosibirsk, 1980, pp. 39–41, 247. MR601401

[146] S. G. Kreĭn and L. Ĭ. Nikolova, *Holomorphic functions in a family of Banach spaces, interpolation* (Russian), Dokl. Akad. Nauk SSSR **250** (1980), no. 3, 547–550. MR557785

1981

[147] S. G. Kreĭn and K. I. Chernyshov, *Analogue of Tikhonov's theorem for the equation* $(A+\varepsilon B)\dot{x} = C(t)x$ (Russian), Approximate methods for investigating differential equations and their applications, Kuĭbyshev. Gos. Univ., Kuybyshev, 1981, pp. 103–115. MR708060

[148] S. G. Kreĭn and K. I. Chernyshov, *Behavior of solutions of general linear systems depending meromorphically on a small parameter* (Russian), Dokl. Akad. Nauk SSSR **260** (1981), no. 3, 530–535. MR631925

[149] S. G. Kreĭn and G. A. Kurina, *Singular perturbations in problems of optimal control* (Russian), Stability of motion. Analytical mechanics. Control of motion, "Nauka", Moscow, 1981, pp. 170–178, 303. MR663404

1982

[150] S. G. Kreĭn and L. Ĭ. Nikolova, *A complex interpolation method for a family of Banach spaces* (Russian), Ukrain. Mat. Zh. **34** (1982), no. 1, 31–42, 132. MR647928

1983

[151] S. G. Kreĭn and M. I. Khazan, *Differential equations in a Banach space* (Russian), Mathematical analysis, Vol. 21, Itogi Nauki i Tekhniki, Akad. Nauk SSSR, Vsesoyuz. Inst. Nauchn. i Tekhn. Inform., Moscow, 1983, pp. 130–264. MR736523

1984

[152] S. G. Kreĭn and L. Y. Nikolova, *On the method of complex interpolation*, Complex analysis and applications '81 (Varna, 1981), Publ. House Bulgar. Acad. Sci., Sofia, 1984, pp. 298–300. MR883250

[153] S. G. Kreĭn and N. P. Polichka, *A priori estimates for solutions of a problem with displacement for the Lavrent′ev-Bitsadze equation* (Russian), Differentsial′nye Uravneniya **20** (1984), no. 12, 2112–2120. MR772808

[154] S. G. Kreĭn, V. I. Levin, and A. S. Simonov, *L. M. Likhtarnikov (on the occasion of his sixtieth birthday)* (Russian), Mat. v Shkole **3** (1984), 77. MR755110

[155] S. G. Kreĭn and K. I. Chernyshov, *Singularly perturbed differential equations in a Banach space* (Russian), Ninth international conference on nonlinear oscillations, Vol. 1 (Kiev, 1981), "Naukova Dumka", Kiev, 1984, pp. 193–197, 443. MR800427

[156] (with Chernyshov K.I., Kuchment P.A., and Lvin S. L.) On the reconstruction of functions from empirical data with a priori information, preprint, VINITI, no. 1587–84, 53 pp., 1984. (in Russian).

1985

[157] S. G. Kreĭn, *Singularly perturbed linear differential equations in a Banach space* (Russian), Differentsial′nye Uravneniya **21** (1985), no. 10, 1814–1817, 1839. MR814583

1986

[158] Yu. A. Brudnyĭ, S. G. Kreĭn, and E. M. Semënov, *Interpolation of linear operators* (Russian), Mathematical analysis, Vol. 24 (Russian), Itogi Nauki i Tekhniki, Akad. Nauk SSSR, Vsesoyuz. Inst. Nauchn. i Tekhn. Inform., Moscow, 1986, pp. 3–163, 272. Translated in J. Soviet Math **42** (1988), no. 6, 2009–2112. MR887950

[159] S. G. Kreĭn, *Asymptotic decomposition of operator equations* (Russian), Lyapunov functions and their applications (Russian), "Nauka" Sibirsk. Otdel., Novosibirsk, 1986, pp. 206–214, 248. MR892764

1987

[160] S. G. Kreĭn and S. Ya. L′vin, *Overdetermined and underdetermined equations in Hilbert spaces* (Russian), Izv. Vyssh. Uchebn. Zaved. Mat. **9** (1987), 59–66, 83. MR923845

[161] Yu. M. Berezanskiĭ, I. M. Gel′fand, M. G. Kreĭn, S. G. Kreĭn, Yu. A. Mitropol′skiĭ, and A. V. Skorokhod, *Yuriĭ L′vovich Daletskiĭ (on the occasion of his sixtieth birthday)* (Russian), Uspekhi Mat. Nauk **42** (1987), no. 4(256), 213–214. MR912076

[162] S. G. Kreĭn and S. Ya. L′vin, *Overdetermined and underdetermined elliptic problems* (Russian), Functional analysis and mathematical physics (Russian), Akad. Nauk SSSR Sibirsk. Otdel., Inst. Mat., Novosibirsk, 1985, pp. 106–116, 135. MR895832

1988

[163] S. G. Kreĭn and N. P. Polichka, *Stability of solutions of a problem with shift for the Lavrent′ev-Bitsadze equation with variation of the domain* (Russian), Applied problems in statistical analysis (Russian), Akad. Nauk SSSR Sibirsk. Otdel., Dal′nevostochn. Filial, Vladivostok, 1988, pp. 16–29, 162. MR1082384

[164] S. G. Kreĭn and S. Ya. L′vin, *Partially overdetermined and underdetermined elliptic problems* (Russian), Izv. Vyssh. Uchebn. Zaved. Mat. **10** (1988), 15–23; English transl., Soviet Math. (Iz. VUZ) **32** (1988), no. 10, 19–30. MR982602

[165] S. G. Kreĭn and S. Ya. L′vin, *Underdetermined boundary value problems* (Russian), Function spaces and equations of mathematical physics (Russian), Voronezh. Gos. Univ., Voronezh, 1988, pp. 17–24, 85. MR958951

[166] S. G. Kreĭn and S. Ya. L′vin, *An approximation approach to overdetermined and underdetermined boundary value problems* (Russian), Functional and numerical methods in mathematical physics (Russian), "Naukova Dumka", Kiev, 1988, pp. 113–117, 268. MR1038560

1989

[167] N. D. Kopachevskiĭ and S. G. Kreĭn, *A problem of the flow of a viscous fluid* (Russian, with English and German summaries), Z. Anal. Anwendungen **8** (1989), no. 6, 557–561, DOI 10.4171/ZAA/374. MR1050094

[168] S. G. Kreĭn, *An abstract scheme for the consideration of boundary value problems* (Russian), Application of new methods of analysis to differential equations (Russian), Kach. Metody Kraev. Zadach, Voronezh. Gos. Univ., Voronezh, 1989, pp. 46–51, 108. MR1038813

[169] S. G. Kreĭn and S. Ya. L′vin, *Solution of overdetermined and underdetermined elliptic problems in the case of nonsmooth data* (Russian), Ukrain. Mat. Zh. **41** (1989), no. 9, 1222–1225, 1294, DOI 10.1007/BF01056278; English transl., Ukrainian Math. J. **41** (1989), no. 9, 1053–1056 (1990). MR1026421

[170] S. G. Kreĭn and Chan Tkhu Kha, *The problem of the flow of a nonuniformly heated viscous fluid* (Russian), Zh. Vychisl. Mat. i Mat. Fiz. **29** (1989), no. 8, 1153–1158, 1260, DOI 10.1016/0041-5553(89)90127-4; English transl., U.S.S.R. Comput. Math. and Math. Phys. **29** (1989), no. 4, 127–131 (1991). MR1032745

1990

[171] S. G. Kreĭn and V. I. Fomin, *Small perturbations of singular differential equations with unbounded operator coefficients* (Russian), Dokl. Akad. Nauk SSSR **314** (1990), no. 1, 77–79; English transl., Soviet Math. Dokl. **42** (1991), no. 2, 313–315. MR1118482

[172] S. G. Kreĭn and A. I. Tovbis, *Linear singular differential equations in finite-dimensional and Banach spaces* (Russian), Algebra i Analiz **2** (1990), no. 5, 1–62; English transl., Leningrad Math. J. **2** (1991), no. 5, 931–985. MR1086444

[173] S. G. Kreĭn and E. O. Utochkina, *An implicit canonical equation in a Hilbert space* (Russian, with Ukrainian summary), Ukrain. Mat. Zh. **42** (1990), no. 3, 388–390, DOI 10.1007/BF01057021; English transl., Ukrainian Math. J. **42** (1990), no. 3, 345–347. MR1054886

1991

[174] M. Sh. Birman, S. G. Kreĭn, O. A. Ladyzhenskaya, G. V. Rozenblyum, and Yu. G. Safarov, *Mikhail Zakharovich Solomyak (on the occasion of his sixtieth birthday)* (Russian), Uspekhi Mat. Nauk **46** (1991), no. 4(280), 183–184, DOI 10.1070/RM1991v046n04ABEH002828; English transl., Russian Math. Surveys **46** (1991), no. 4, 217–219. MR1138976

[175] V. V. Vasil′ev, S. G. Kreĭn, and S. I. Piskarëv, *Operator semigroups, cosine operator functions, and linear differential equations* (Russian), Mathematical analysis, Vol. 28 (Russian), Itogi Nauki i Tekhniki, Akad. Nauk SSSR, Vsesoyuz. Inst. Nauchn. i Tekhn. Inform., Moscow, 1990, pp. 87–202, 204. Translated in J. Soviet Math. **54** (1991), no. 4, 1042–1129. MR1060534

1994

[176] E. V. Belyaeva and S. G. Krein, *Homogeneous Volterra operator equations with a regular singularity*, Russian J. Math. Phys. **2** (1994), no. 1, 3–12. MR1297937

1995

[177] S. S. Venevitina and S. G. Kreĭn, *Small motions of an elastic medium in an open immobile container* (Russian, with Russian summary), Zh. Vychisl. Mat. i Mat. Fiz. **35** (1995), no. 7, 1095–1107; English transl., Comput. Math. Math. Phys. **35** (1995), no. 7, 875–884. MR1349101

1997

[178] S. G. Kreĭn and S. S. Venevitina, *Small motions of an elastic medium in a cavity of a rotating body* (Russian), Papers from the Conference on Functional Analysis and Equations of Mathematical Physics dedicated to the eightieth birthday of Selim Grigor′evich Kreĭn (Russian), Voronezh. Gos. Univ., Voronezh, 1997, pp. 49–54. MR1724191

[179] S. G. Kreĭn and I. V. Sapronov, *On the completeness of a system of solutions of a Volterra integral equation with a singularity* (Russian), Dokl. Akad. Nauk **355** (1997), no. 4, 450–452. MR1492018

[180] Yu. M. Berezans′kiĭ, V. S. Korolyuk, S. G. Kreĭn, Yu. O. Mitropol′s′kiĭ, A. M. Samoĭlenko, A. V. Skorokhod, and Ī. V. Skripnik, *Yuriĭ L′vovich Dalets′kiĭ (on the occasion of his seventieth birthday)* (Ukrainian), Ukraïn. Mat. Zh. **49** (1997), no. 3, 323–325, DOI 10.1007/BF02487238; English transl., Ukrainian Math. J. **49** (1997), no. 3, 357–359 (1998). MR1472222

1999

[181] S. Krein, P. Kuchment, V. Ovchinnikov, and E. Semenov, *Iosif Shneiberg July 26, 1950–July 21, 1992*, Function spaces, interpolation spaces, and related topics (Haifa, 1995), Israel Math. Conf. Proc., vol. 13, Bar-Ilan Univ., Ramat Gan, 1999, pp. 1–10. MR1707354

2004

[182] N. D. Kopachevskiĭ and S. G. Kreĭn, *An abstract Green formula for a triple of Hilbert spaces, and abstract boundary value and spectral problems* (Russian, with Russian summary), Ukr. Mat. Visn. **1** (2004), no. 1, 69–97, 146; English transl., Ukr. Math. Bull. **1** (2004), no. 1, 77–105. MR2180707

2008

[183] S. Krein, Voronezh Winter Mathematical Schools, in V. A. Kostin, B. N. Sadovskii, and E. M. Semenov (Editors), *S. G. Krein. Recollections*, Voronezh State University 2008. (In Russian)

Peter Kuchment, Mathematics Department, Texas A&M University, College Station, TX 77843-3368

Evgeny Semenov, Mathematics Department, Voronezh State University, Voronezh, Russia

Contemporary Mathematics
Volume **733**, 2019
https://doi.org/10.1090/conm/733/14728

My father Selim Krein

T. Voronina (nee Krein)

My father Selim Grigor'evich Krein was born on 15th of July in 1917. It was a fatal, tragic year for Russia, the year of the Bolshevik Revolution, also known in Soviet literature as the Great October Socialist Revolution.

His name Selim means "peace," or "mir" in Russian, so the family called him, the youngest and thus most loved among seven children, Mira.

At the age of five, playing soccer with other kids, he had an accident that damaged his knee joint. As a result of this trauma, he had been confined to bed for five years, and then used crutches till his university sophomore year. He had not stopped loving and playing soccer, though. He admired all kinds of soccer moves and kicks. However, limited by his injury, he played as a goalkeeper or referee. Throughout his life, he liked playing gorodki[1], volleyball, and table tennis.

My mother remembered him as a "crazy boy" with crutches who was gliding (with friends' help, since he could not skate on these slippery floors by himself) along the polished floors of the university's hallways. Later on, his legs would get better, so he could abandon crutches and walk just with a limp.

In his youth, Mira was enchanted and inspired by the idea of creating new, just society that would be good for everyone. He was ready to fight for justice, and he wanted be a part in the life of this society. He believed that now was the right time for realization of such great projects and ideas. Alas, it turned out that his fate was to live through one of the hardest times in Russian history. His student years fell on the thirtieths, the period of people's starvation, shortages, witch hunts, arrests, executions, widespread fear, and all sorts of other horrors. That was the time of an orchestrated "public outcry" against so called "enemies of the people." There was the common dread of the nighttime knocks at your door. Fearfully, people destroyed their family archives and disavowed arrested relatives and friends.

My father has told me about the hunger and shortages of the 1930s, which was one of the hardest times. Hungry ukrainian peasants, robbed of all their harvest, would come to Kiev. Their dead bodies would litter the streets. People were happy if their son was drafted to the Red Army, since it was a chance to escape the death of hunger.

Then there was the World War II and evacuation of academia to Ufa. I know only some disjointed bits and pieces about that time of my father's life. He has told me how he had to get to the Academy hanging outside on the back of a tram; how

[1]A Russian analog of "skittles," a precursor of bowling.

there was a monthly butter ration for the whole family, which he collected and hid under a bench, while at a romantic rendezvous, and the butter was stolen; how they did not like chemist V.I.Goldansky (a future member of Academy), because he was a good dancer and had stolen away the Izia Gragerov's, father's childhood friend, girlfriend; how father's PhD supervisor, member of Academy N.N.Bogolubov, in order to be able to purchase medications for his sick wife, was selling female underwear at the flea market, wearing a jacket and a necktie over his naked chest; how he was tutoring Misha, the son of M.A. Lavrent'ev, and a future member of Academy himself.

After the war, new persecutions occurred, and more were expected. There was "The Doctor's Plot" and antisemitic campaign against "rootless cosmopolitans". My father understood well the government's intentions. They were getting ready for the "final solution" of the "Jewish Problem," following in Hitler's footsteps. He knew about concentration camps that Stalin had started building in Kazakhstan planning to deport the Jews.

S.G., as he was called by his pupils, had an innate sense of justice. The people would often choose him for getting an advice. This very same sense of justice, however, had contributed to his problems, repeatedly bringing him to the edge of a disaster. For instance, he would express his opinions against some candidates of the block of communists and non-party members, which almost led to his expulsion from the university. He was suspended from classes and expelled from Komsomol. Luckily, he was eventually somehow "forgotten."

When his research work was nominated for a major prize and he did not find his name in the list of the authors, he did not remain silent and had a fallout with administration. He consequently left the Academy of Science and the nomination was canceled. Later he had said that he regretted this decision, since the work was very strong and could and should have won the award.

Here is another story: Once my mother Eugenia Petrovna Kostrukova took part in preparing the new edition of a previously published book on chemistry of semiconductors[2]. She wrote several chapters of the book and edited the whole text. She would be giving then lectures for a special topic course and run recitations with students, using the book as a textbook. Alas, her name was not included in the list of co-authors. She was not too upset, saying that she enjoyed the work and was grateful for the opportunity to do this. Instead, it was my father who was shocked. He said that none of the people around him, people with whom he had to deal on daily basis would accept this. At times, S.G. avoided putting his own name as a coauthor, when he wanted to help junior authors who were just starting their life in science. But to appropriate the results of other people's work was an impossibility.

My father has never been faint-hearted and never tried to avoid making important decisions. He was not a nonconformist, unlike his older brother Mark Krein[3]. He was flexible enough to collaborate with the authorities, was a member of the Soviet youth organization Komsomol and then of the Communist Party. He even was chosen to be a member of the Party Committee at the University.

He did not like gossip, arguing with people, and avoided squabbles. He was able to defuse almost any situation simply by cracking a joke or by proposing a

[2]S.G. helped her with explanations of some difficult parts and although my mother was skeptical, she eventually declared, "Your father is really a very clever man indeed!"

[3]A famous mathematician. - Ed.

My parents

compromise. In some cases, though, it was necessary to confront his adversaries. He fought for the things that he consider to be right. For instance, when cybernetics was openly chastised and ridiculed, he blocked the decision of the rector and the Party's Committee that labeled cybernetics a false science and a "venal maid of imperialism."

In 1969, during the period called "bolts tightening," a new campaign against intellectuals, he was expressing his protests against all attempts to dissolve the department of mathematics of the university. At the meeting, where the rector was accused of governing the university as an extravagant, careless girl, he came out with a speech.

He also withstood pressure from a KGB general, who yelled and banged his fists on the table, keeping S. G. standing and demanding a passing grade for a student (who happened to be a KGB agent). S. G. said that he would need a written order (a known impossibility) to pass the student, who did not know the subject.

S. G. was always trying to help people who were getting a raw deal from the Soviet power-holders, e.g., who were prevented from finding a job, or from being admitted to the University. If results were positive, he would say that in some cases even the Soviet power could be helpful.

S. G. preferred to face his problems in solitude. When he read the newly published Vasily Grossman's novel, "Life and Fate," which depicted anti-semitism infecting the whole Soviet society, we were unable to speak with him for two days; he would not leave his office. The past, with its sinister injustices caught up with him then. For instance, when the Academy of Science was returning from Ufa back to Moscow after the war, some researchers who successfully and honestly worked side by side with others were not allowed to leave Ufa. It was the curse of their "fifth item" ("p'yatyj punkt")[4].

S. G. felt lack of freedom as a physical pain. It concerned the freedom of speech, opinion, access to information, and free movement. The denial of the ability to express oneself was for him as if inability to breath. He did not like even usual physical restrictions, such as neckties, and never wore them. He liked to wear his shirts over his slacks, the top button opened, which was often criticized by the university administrators. My mother would always put the collar of the shirt on top of his jacket "apache," as an open-necked collar.

S. G. had always been respectful, civil, and fair to people, even those whose opinions were opposite to his own. He was tolerant to scientists' bona fide mistakes and people's weaknesses, and openly admitted his own errors and misunderstandings. However, he could not stand people who did not take their work seriously.

I do not know how he did this, but my father could tell you off or admonish you in such a way that you would not feel humiliated, insulted or upset with him. On the contrary, it almost felt as a release from everything that was bad and gave you a ticket to go with a light heart. Once, as a kid, I was punished by being told not play with other children, but stay at home for three hours. I was sulking, sitting at home, while my friends were enjoying themselves outside, right under our windows. My father whispered into my ear conspiratorially: "There are only 40 minutes left."

Our family was living on a schedule. The biggest offence was to be late for a dinner, since it would force my mother to do extra work. Everything else was open for discussion: you could oppose, or try to get things done your way. But blackmail, crying, and tantrums were counter-productive; only convincing, logical arguments would work.

It seemed to me that my father knew everything. In the evening, after supper, when the minds need some rest, my parents would work on crossword puzzles from the popular magazine "Ogonek". My mother would get somewhat upset that S. G. was doing the puzzle by himself. He did not even read the questions aloud for us. Only those puzzles that needed consulting with a map or encyclopedia were ours. Sometimes, when mother was able to guess a word before others, my father was delighted with her: "Oh Mishkin! (his affectionate name for her), do not play games with her!" Besides his attraction to sports, he took an active interest in literature and cinematography. It was impossible to understand how he was able to find time for all these activities.

The way I remember him, he was always busy with mathematics. When he was at home, rather than at the University, we were able to see him only during our evening tea time. Even when we took a family ship cruse along the Yenisei river in Siberia, or the five rivers in the European parts of USSR (Volga, Kama, etc.), or took a trip along the Voronezh river on our boat, went to Caucasus, Carpathian mountains or any other place, anywhere, on a ship, train, bench in some backwater, or a resort, S. G. was always working: writing, thinking or correcting.

In Moscow, while in the Hospital of the Cardiology Institute, he tried to perch himself at a table in the small hospital cafeteria. Then Moscow mathematicians

[4]The fifth item of the Soviet internal passport listed the ethnic nationality of a person.

had threatened the management of the hospital that they would drag a desk into his hospital room and put it directly next to S. G.'s bed. After that, he was happy to obtain a key from an at that time empty doctor's office. There was a big table there, and it was possible to spread out his papers.

He could not exist without the mathematics and his associates. In his last days, he said that without mathematics life is empty, it equals zero. The life had meaning and colors only if there were his mathematics and his eager to learn students. The greatest of his wishes, I think, was to know that nobody was amiss for mathematics. He would be happy and proud knowing that his pupils are working to enrich mathematics, and he would not deny himself of that, not for great money, business, or fame.

I believe that mathematical texts are like songs, like poems; they can bring esthetic enjoyment and leave the imprint on your personality. You need only look at these words: Graceful resolution. Witty evidence. Laconic formulation. Perfect definition. It is not accessible for us, for those who are not enlightened. We cannot feel or understand these combinations of words. We can only trust that, indeed, these are true.

Teaching was a great part of my father's life and it gave him many positive emotions. He would say that sometimes, when he would feel sick and doubtful whether he could make it to the University, being at his lecturing pulpit, looking into the faces of his students, he would forget about infirmities and feel a quiet sense of elation, inspired to start his lecture. At exams, he was always trying to "convey telepathically" the right answers to a student's mind. He was glad and not shy to brag about it, if his students got the right answers, whereas others' did not. Sometimes S.G. would permit his students to retake exams in easy to learn and memorize increments. Sometimes he would allow up to five attempts.

His heart ached for university entrants. Once, they could not grant admittance to the mathematical department to an entrant, a young boy from provincial part of the country. The student had the highest scores in mathematics and physics, but made 23 spelling errors in his essay. S. G. then conveyed to me that there needs to be a more effective procedure for the university enrollment. He suggested to have no entrance exams and to teach all the applicants, and then, in six months to pick those who proved to be the best.

S. G. had a great sense of humor. He loved jokes and was a big master of cracking them. My mother always understood his pranks, and sometimes laughed to the point of tears in her eyes. He jokingly called us, youths, a "young degeneration" instead of generation.

S. G. did not like cheesy stories and "enlightening" conversations, and did not have time for such nonsense. It seemed to me that he was always able to catch and express a meaning momentarily. Sometimes it was enough to hear only one sentence from him, and then it was impossible to comprehend how you were unable to see this yourself before.

Once I got a job offer to become an Academic Secretary at one of the departments of the Council for Mutual Economic Assistance (COMECON). The head of this department was simultaneously the director of my institute. It was very tempting, and I was flattered. For me it seemed to have such a value, as an encouragement and evidence of respect. My father only asked: "Are you sure that this work is good?" My excitement immediately abated. My trips with inspections of the underground oil and gas pipelines in Shatlyk (a Central Asia desert), Nadym (Extreme North Siberia), Urengoy (Kazakhstan) were in his opinion more important than preparations of and then presence at COMECON conventions in Poland, Germany, and other places.

S. G. for sure could have had a more comfortable and plush home, but he did not like to be distracted from important things, and he plainly did not notice the inconveniences, preferring paying no attention to minor stuff. The best relaxation for him was to lie down on the sofa in his office, among his books. Someone once suggested that his family undervalued S. G., but he always knew that this was not true. His wife was an extraordinary, kind, and very interesting women, matching him in personality.

S. G. lived fully and enjoyed life. He loved mathematics passionately; loved his family, his home and certainly his pupils. He valued his pupils as highly as his family. They were not only pupils, but also like-minded people, associates. He had always been pleased and excited by and proud of their successes, probably even more than of his own. He would always share this with great feeling. He was convinced that interactive communication and exchange enrich people, provoking thoughts and ideas.

My father Selim Grigor'evich Krein lived in challenging times, but he lived in the real world and was happy, because he was able to be involved in a great quest and he was doing his job among and for good people. I regret that at the time I was too young and foolish, did not pay sufficient attention to, and was not interested enough in his life. My husband also regretted that he had met S. G. too late in life.

I am grateful to everyone who shared his joys and sorrows when he was alive and who remembers him today. I would like to offer my boundless gratitude to the initiators, authors, and editors of this collection and everybody who remembers S. G. and offers their support to me.

Thank you!

Email address: tatianaselimovna@gmail.com

Contemporary Mathematics
Volume **733**, 2019
https://doi.org/10.1090/conm/733/14729

Kyiv, Fall of 1943 through 1946. The rebirth of mathematics

Yu. M. Berezansky

To the dear memory of Selim Krein who helped me to become a mathematician.

ABSTRACT. In these brief notes, I describe the rebirth of Ukrainian mathematics in Kyiv in the second half of 1940s, in which Selim Krein played an important role.

In these brief recollections, I would like to go back to the Fall of 1943, when Germans were forced out of Kyiv, and in Kyiv a small group of people returned to their mathematics. I'd like to recall the mathematicians, who at that horrific time managed to involve into mathematics some young people, interested in math and by fate brought to Kyiv at that time. I do not pretend to give a full picture, writing only about those mathematicians thanks to whom I have also become one. Other people had different teachers, and their recollections probably will be different from mine.

I would like to start telling about Simon Izraelevich Zuhovitski, a scholarship in whose name has been established In Israel. This scholarship will perpetuate the memory of S.I. (as I will call him) as a well known mathematician and, in my opinion, above all as an outstanding teacher, whose brilliant lectures and even his way of life attracted to mathematics many young people. I was one of them and believe that SI played a major role in my life.

This all happened in Kyiv in Winter and Spring or 1943 – 1944. When Germans were kicked out from Kyiv, Soviet forces came instead. Locals who stayed on the occupied territory were afraid of them no less than of Germans. Before they came, there had been persistent rumors that all Ukrainians would be deported to Siberia, as Tatars were at that time from Crimea. Then there was a rumor that there was no deportation because there weren't enough trains to deport that many people, but that the corresponding deportation order had been signed by Stalin.

The mood was bleak: destroyed city, dejected population, hunger, fear of tomorrow, and talks about possible repressions. In spite of this environment, a few professors and docents[1] who stayed in Kyiv, decided to re-open Kyiv University. Enrollment was announced, no entrance exams were required. Among others, a Physics and Mathematics Department was created (physicists and mathematicians

2010 *Mathematics Subject Classification.* Primary 01.

Translated from Russian by Dr. M. Mogilevsky, with permission, from "S. G. Krein, Recollections", Voronezh State University 2008.

[1]Docent is an approximate equivalent of the Associate Professor rank. - M.M.

were pulled together due to the small number of available instructors, as well as prospective students). Around 30 to 40 students were enrolled – girls, boys maimed at war, boys who were not in the army because of an ill health (I was among them). The huge central "red" building of the university was burned down during the war, and so all science departments worked in the chemistry building, which was intact.

Students started attending lectures, while the war was still going on. Exhausted instructors taught physics, mathematics, etc. I was somewhat prepared, since before the war I had finished eight out of the ten school grades. I still had to read up. I liked mathematics, but liked physics more, maybe due to a boyish interest in radio.

And so, in the first days S.I. started lecturing in mathematical analysis. He was young, with a handsome southern face and eyes intently looking at students. A voice with a burr – something foreign was in those lectures. We knew that he was a Jew, by stroke of luck surviving during German occupation, due to the bravery of a professor who hid him. Later on we learned that this professor was the well known researcher in mathematics and mechanics Yu. D. Sokolov. I think the manners and appearance of S.I. played some role for us, but most important was how he lectured. Let me underscore again – destroyed city, hunger, and fear were around, mostly fear. And here S.I. tells us about Dedekind's theory of real numbers, as if there were nothing awful around us, just beautiful mathematical constructions, absolutely constant and eternal. And you start believing that bad times will pass, which gave strength to those boys and girls who tried to study despite all the mess and horror.

Lectures of S.I. have determined my interests: I gave up physics and decided to become a mathematician. Being still a young boy, I understood that there was science, which defied cruelty of power and regime. So, doing mathematics, one can be free, despite everything.

I was not the only one who decided to become a mathematician under the influence of S.I. But not all were lucky: some lives were ruined by NKVD[2], some could not move forward due to the stigma of "being in occupied territory". S.I. himself also could not go far in his career: soon the persecution of Jews as "cosmopolitans" would come, and he would be forced to leave the university for a pedagogical institute. Then he worked in other cities – Lutsk, and later in Moscow. But anywhere he went, he would charm students with mathematics and help (as much as he could) young mathematicians to withstand the pressure of the totalitarian USSR system, which tried to suppress all life. Later, when I got close to S.I., I learned that he was a deeply religious person, and that his dream was to live and to die in the land of his ancestors. And I am glad that his dream was fulfilled, and that the last years of his life he spent and then forever stayed in Israel.

Certainly, S.I. was not the only mathematician who influenced me during the first two years of the University. Analytic geometry was read by Boris Jakovlevich Bukreev, an elderly (85 years old) professor. Listening to him, we were impressed by the clarity of his mind and originality of his exposition – often, deeply in thought, he drew graphs on a black table standing in front of the blackboard, and we could not see them. Some academism in his exposition of analytic geometry was rectified during the recitations. Those in geometry and analysis were lead by a nice and friendly woman, Docent Vera Petrovna Belousova, who had just returned from

[2]NKVD is the older name of the Soviet KGB, currently FSB. - M. M.

evacuation. She was probably the first person to notice that I could understand something. Next year she was not conducting recitations, but was lecturing in analytic and differential geometry.

Algebra was taught by Professor Nickolaj Petrovich Sokolov, an elderly man, whom my family knew when we all were under German occupation. He was nice to me and for some reason he thought, even before I entered the university, that I ought to be a mathematician (in my mind, I had not given any particular reason for that). These were all my mathematics teachers during the first two[3] years of the university studies.

There were other lectures – on mandatory Marxism-Leninism, foreign languages, physics, and astronomy. I went to all of them and passed all exams with an "A", but all of these were outside of my interests. I remember only Astronomy, where I got the first "B" in my life. I took a make-up exam, but Prof. Vsehsvyatskij was implacable, and this B stayed in my transcript.

Three semesters passed, and the war was over. I've got to love mathematics and read random math books that I could find. But math still was something external for me. Neither me, nor my friends understood at the time how to pick a problem to solve, a theorem to prove. Maybe the war and barely alive teachers were the reason, but all this mathematics stayed "outside" me.

The war ended, and Kyiv mathematicians who were in evacuation or at the frontline, started coming back. The number of faculty at Mech-Math[4] (not Phys-Math anymore) increased, and the life got livelier. We already studied mostly at the main, "red" building of the University (rebuilt by German POW).

And so, a young person, limping and with gray hair and radiant eyes appeared at the Mech-Math in the beginning of the third year, in the Fall of 1945. This was Selim Grigor'evich Krein (S. G.). He read us functions of complex variable. In his lectures, there were no fatigue, usual for S.I. Zukhovitsky. He taught in such a way, that each of us felt that we can be involved in discovering something new in mathematics. All the girls (and not only from Mech-Math) were in love with him. Despite his limp, he superbly played volleyball, and all the girls went to those games to watch S.G. My future wife was among them, as well. He organized a seminar for young mathematicians and students (M.A. Krasnosel'sky, B.I. Korenblyum, S.A. Avramenko, N.I. Pol'sky, G.I Katz, Yu. L. Daletsky). We, third year students, did not even know at that time what a seminar was. I was one of the first speakers at the seminar, and not a success as such: I did not quite understand, I think, the Hopf's book on dynamical systems. Nevertheless, S.G. somehow managed to correct me quite gently. At that seminar I first learned the concepts of a Hilbert space, self–adjoint operator, spectral decomposition, etc.

At his time, S.G. suggested to me the problem of developing a theory of a generalized shift and the corresponding convolution in such a way that it would become a natural generalization of constructions of finite-dimensional hypercomplex systems. He brought this problem from Moscow, where B.M Levitan was developing a theory of operators of generalized shift. This theory started with I.M. Gelfand. S.G. brought me four pages inscribed in pencil personally by I.M. Gelfand, in his small handwriting. The story goes that I.M. made a bet with someone that he

[3]Or rather one and a half years, since the first year consisted of only one semester – the university started working in the beginning of 1944.

[4]Department of Mechanics and Mathematics. - M. M.

would sit and write a research paper on the spot, and those pages were the result of this bet.

In any case, S.G and I started actively discussing those questions. This resulted in the theory of commutative hypercomplex systems with continual basis, published in DAN USSR (Soviet Doklady) in 1950. These were my first publications. Many our constructions were rediscovered later, in 20-30 years, in the West, when the theory of hypergroups was being created.

In 1945-1946, many other well known professors appeared at Mech-Mat. For instance, lectures in our 3rd and 4th years were given (besides of those mentioned earlier) by Yu.D. Sokolov, A.S. Smorgorzhevsky, N.N. Bogolyubov, B.V. Gnedenko, I.Z. Shtokalo. But the impact of S.G. on me was still huge. With him I felt that I could discover something new, and this was quite exciting.

If thanks to S.I. Zukhovitsky I became a mathematician, then thanks to S.G. Krein I started actively working in mathematics. My gratitude to these two mathematicians is immense. S.G. not only involved me into active work in mathematics, he also helped me with various practical arrangements of my future life. If not for him, I would not had been able to get into any graduate school. Who was I by the summer of 1946? A graduating with honors 23 year old young man, kind of successfully doing mathematics, but having no publications. I had been on occupied territory and was not a member of Komsomol[5]. The only my "virtue" (significant in the USSR in 1948) was that I was not Jewish.

I had a conflict with Komsomol: Lida Kostyanaya (later L.M. Kisilevskaya), who probably was the head of the party group at the department, asked me, as a straight A student, to join the Komsomol. I responded that I did not want to, since I did not believe in that thing. She was absolutely taken aback, asked what I did under Germans, and left. Nevertheless, she apparently happened to be a decent person and did not report me. As the attitude of the administration to me was at least indifferent, they did give me a recommendation for the graduate school. This was insufficient, though. S.G. Krein helped me significantly, enlisting the support of N.N. Bogolyubov and convincing the director of Institute of Mathematics of Academy of Science of Ukraine M.A. Lavrent'ev to help me. I thus got into the graduate school of the Institute, and my advisors became brothers M.G. and S.G. Kreins.

The influence on me by M.G. Krein is a separate story. I will only tell here that I first met him, probably, during the same third year of the University. He still was working (while living in Odessa) at Kyiv Institute of Mathematics of Academy of Sciences of Ukraine, often coming to Kyiv and lecturing to us, graduate students. He was a live incarnation of my old perception that doing mathematics you remain free, despite all the abominations of the state system. He did not accept the system, and it rejected him. Nevertheless, he had became one of the most prominent mathematicians, creators of functional analysis.

S.G. was, with respect to communism and power, very different from M.G. He piously believed in the communist ideas, and even once was the secretary of a party group. It seems to me, he continued to believe even when he was expelled from the Institute of Mathematics of AN of Ukraine in 1951 and had to work at KrivojRog Mining Institute(1952-1954). He then moved to Voronezh, where he,

[5]The young communist branch of the communist party with practically mandatory membership between ages of 14 and 27. Not being its member was a serious handicap. - M. M.

jointly with M.A. Krasnoselskii created a well known school of functional analysis and its applications. I am trying to remember what was my attitude toward those political views of S.G. Probably, being young, I just did not pay attention to them. S.G. himself, I think, was confident that to do good, one may be a member of the Party. I regarded the Party and the regime negatively, probably since my childhood. Here was the influence of my parents (in their youth they served in the Army of Simon Petliura and all their life had been hiding that fact and were afraid of being discovered). But probably most of all I was influenced by the famine of 1933 in Ukraine. As a boy, I saw corpses of peasants on streets of Kyiv. And at that time I understood that there was something wrong with the communism, that it was a system leading to destruction. $\langle \dots \rangle$

Whatever was the reason, we did not discuss those topics with S.G., intuitively avoiding possible confrontation (with M.G. we did talk about them). Now I think that maybe S.G. was correct: to do good at those times, one needed to be a member of the Communist Party. Not being a member, M.G. certainly would not be able to help me to get into any graduate program.

My goal was to describe in these notes the formation of young mathematicians in Kyiv in the period from the Fall of 1943 till 1946, when Jewish mathematicians in Kyiv felt quite confident. But after that period, the horrors began for S.G, for M.A. Krasnoselskii, B.I. Korenblyum, and B.A. Trakhtenberg, working at the Institute of Mathematics, as for all Jews at that time in the USSR.

The accusations against S.G. and M.A. Krasnoselskii were absurd. Besides doing pure mathematics, they were working on applied defense problems, supervised by the special part of administration dealing with secret tasks like that. There were no computers then, but lots of computation were needed. And so, S.G. and M. A. would give math majors an opportunity to make some money, asking them to do simple computations, from which it was impossible to learn any state secrets. But formally, there was an element of "a disclosure of secrets", and the administration used this. In 1951, a campaign was organized against S.G. and M.A. with invitations of employees to attend (including myself, I still remember that conversation). It was clear as a day that they wanted to get rid of good, but undesirable mathematicians.

Anti-semitic ideas "seized the masses" at that time. Why were these ideas very strong in Ukraine then (stronger than in Russia)? I have my own explanation. One reason is the obedience of Ukrainians, for whom, as a rule, the opinion of a boss is the law (that's why in USSR there were many Ukrainians in the Army). But more seriously, it is the famine of 1933, when about 10 million peasants died. I want to underscore – peasants, while in cities people were not dying. When relatives and descendants of those peasants came to cities after the war, their attitude towards city dwellers and intelligentsia in general was actively negative. Nowadays, there is no antisemitism in Ukraine, neither among people, nor on the state level. I claim this with full confidence. Descendants of those peasants of 1933 became city dwellers and often members of intelligentsia themselves, they forgot about famine and became "competitive". But in 1951, it was different, and "one cannot erase words from a fairy tale".

Certainly, not only the mathematicians I have mentioned above played important roles in my life. For instance, G.E. Shilov worked in Kyiv in 1950-1954, thanks to whom and to his student A.G. Kostyuchenko I started working on generalized functions and lecturing at Kyiv University. Young probabilists V.S.Korolyuk and

A.B. Skorokhod came, contacts with whom were very fruitful. But all this is a different story, which I am not writing now. My goal in these notes was to describe Kyiv mathematics scene from the Fall 1943 till 1946.

Institute of Mathematics National Academy of Sciences of Ukraine, Tereschenkivs'ka 3, Kyiv, 01601 Ukraine

Email address: mfat@imath.kiev.ua

Email address: akalyuz@gmail.com

Contemporary Mathematics
Volume **733**, 2019
https://doi.org/10.1090/conm/733/14730

Selim Gregorievich Krein in Stony Brook

David G. Ebin

Find yourself a teacher and acquire for yourself a friend.
Ethics of the Fathers, 1:6

An account of Krein's stay in Stony Brook requires a prologue, which we now present: Krein was living in Voronezh in central Russia and had had three heart attacks. Seeking medical assistance, he consulted specialists in Moscow who advised that he needed cardiac surgery. However, the necessary procedures were not yet well-developed in the Soviet Union. In particular, the Soviet doctors had never done heart surgery on a 73 year-old man. In response, Krein's friends and associates, particularly I. Ts. Gohberg and S. I. Korenblum searched for an appropriate foreign medical facility. Inquiries in Europe revealed that the costs of the surgery would be prohibitive despite the eagerness of a number of mathematicians to contribute.

Then, Gohberg, who had visited Stony Brook a number of times, asked me if we could somehow arrange for the surgery at our University Hospital which had just recently started a cardiac surgery service. Coincidentally with Gohberg's mission to find treatment for Krein, I had been collecting contributions to start a Jewish school in our area. Gohberg, who had stayed with us through some of the period of the school's development, had provided a generous contribution to the project despite his being a recent immigrant from the USSR to Israel and his having many fewer resources than most Americans. So naturally, I was determined to do whatever I could to accommodate Gohberg's request.

Also at about the same time, our colleague and friend, Irwin Kra, had recently undergone a cardiac procedure. He referred me to Dr. Peter Cohn, Stony Brook's Chief of Cardiology. Dr. Cohn proceeded to discuss the treatment of Krein with the hospital administration. Together they decided that this was an opportunity to save the life of a prominent scientist and also to promote the new cardiac surgery service at Stony Brook. They agreed to provide the surgery with surgeon Dr. Constantine Anagnostopoulos at no charge, if the patient would agree to have an article about it published in the local newspaper. Krein readily agreed. (See [**1**].)

On May 21, 1990 Krein arrived at Kennedy Airport. He came to our home where he was able to relax before the surgery. He would remain with us after his surgery until he was ready to travel home. He became a part of our family, an uncle to our children. We learned about his wife, daughter and stepson. He was not fluent in English, but he and I managed in Russian. Our children were delighted

---

2010 *Mathematics Subject Classification.* Primary 01A70.

when he joined in our singing at the Sabbath table sharing the only Jewish song he remembered from his childhood, a song from the Passover Haggadah.

Krein's Stony Brook family quickly grew to include our local Russian speaking mathematicians, Misha and Lilya Lyubich, Borya and Rita Solomyak, and Irina Neymotin who was accompanied by her mother Ida Adolfovna Vulis, also a recent cardiac surgery patient. They assisted in the hospital stay, aiding in translating and in the aftercare in our home. The home group was augmented by visitors including B. I. Korenbloom, P. A. Kuchment, S. Ya. Lvin, and I. Z. Pesenson.

In searching for memories of this time, my wife and I recall the steady stream of helpers who brought good cheer and the needed support in the recovery process. Borya Solomyak recalled that Krein was released from the hospital on June 8th, ten days after the surgery. His reminiscences included that he, Rita, and Misha Lyubich served as interpreters for the detailed explanations that were given for the surgical procedures. They noted that this is very different than what one would have expected in the USSR, where the patient is usually told little of what will happen.

The cardiac by-pass surgery reportedly lasted seven hours. Sergey Lvin recalled that he arrived at the hospital after the surgery finding Krein very weak, but his normal self. That is, Krein inquired about Lvin and his family, calling him Seriozha, but using the respectful "vi" instead of the more familiar "ti" form of "you." Borya Solomyak also recalled that in the post surgical period in the hospital Krein was already working at math and flirting with the nurses. Lilya Lyubich recalled that a nurse asked her how to say, "How are you?" in Russian. When the nurse repeated Lilya's words, Krein responded by giving her a kiss, showing how appreciation can be expressed without language. After Krein was released from the hospital, Dusa McDuff and Jack Milnor held a reception for him at their home. It was a friendly affair through which Krein was welcomed by the Stony Brook mathematical community.

The generosity to Krein was contagious. While recovering, he asked if he could use his stay in America as an opportunity to obtain new glasses. We took him to an optometrist, who upon learning Krein's story and circumstances, provided the glasses at no charge. Krein's stay thus engendered a pro bono atmosphere in which medical care, hospitality, personal attention, and even a pair of glasses were freely given.

The pro bono actions continued outward. At the time of Krein's surgery there was an increasing number of AIDS cases in the Soviet Union and increasing fear of AIDS transmission through multi-use hypodermic needles. Single-use needles were not readily available in the Soviet Union. Krein had heard that Soviets traveling abroad were encouraged to bring disposable needles back with them. To this end, we took Krein to the Stony Brook Apothecary. Our local druggist was happy to provide Krein with packages of needles when he heard of the need.

In releasing Krein from the hospital, Dr. Cohn pronounced Krein's prognosis to be excellent and said that he should live to be one hundred. Although he did not live that long, the present volume attests that he is with us at one hundred in the minds of his students and his collaborators. Krein's students came from diverse places, showing their respect and devotion to their teacher. Krein's students and friends worked together with the medical experts to make it all happen. The students clearly had found themselves a teacher and Krein had acquired many friends.

References

[1] J. Quittner, For Soviet, Surgery on LI Adds Up to Big Success. Prominent mathematician gets heart operation, Newsday, Thursday, June 7th, 1990.

Department of Mathematics, Stony Brook University, Stony Brook, New York 11794-3651

Email address: ebin@math.sunysb.edu

Contemporary Mathematics
Volume **733**, 2019
https://doi.org/10.1090/conm/733/14731

On algebraically integrable bodies

Mark Agranovsky

Dedicated to 100th anniversary of S. G. Krein

ABSTRACT. Let K be a bounded body in $\mathbb{R}^n, n$ is odd, with infinitely smooth boundary. We prove that if the volume cut off from the body by a hyperplane is a free of real singularities algebraic function of the parameters of the hyperplane then the body is an ellipsoid. This partially answers a question of V.I. Arnold: whether odd-dimensional ellipsoids are the only algebraically integrable bodies?

1. Introduction

1.1. Formulation of the problem. Let K be a bounded domain in $\mathbb{R}^n$, with infinitely smooth boundary. For any affine hyperplane $\{x \in \mathbb{R}^n : \langle \omega, x \rangle = t\}$, defined by the parameters $\omega \in \mathbb{R}^n \setminus \{0\}, t \in \mathbb{R}$, denote $V_K^{\pm}(\omega, t)$ the volumes of the portions of the domain K on each side from the hyperplane. Here $\langle , \rangle$ is the inner product in $\mathbb{R}^n$.

DEFINITION 1.1. A bounded domain K in $\mathbb{R}^n$ is called **algebraically integrable** if the two-valued function $V_K^{\pm}$ is algebraic. This means that the function $V_K^{\pm}$ can be obtained as a solution of an algebraic equation, i.e., there exists a nonzero polynomial $Q(\omega_1, ..., \omega_n, t, w)$ of $n+2$ variables such that

$$Q(\omega_1, ..., \omega_n, t, V_K^{\pm}(\omega_1, ..., \omega_n, t)) = 0, \tag{1.1}$$

for all $(\omega, t) \in (\mathbb{R}^n \setminus \{0\}) \times \mathbb{R}$ for which $V_K^{\pm}(\omega, t)$ are naturally defined, i.e., such that $K \cap \{x \in \mathbb{R}^n : \langle \omega, x \rangle = t\} \neq \emptyset$.

In the sequel, we will be considering the volume function V_K defined as

$$V_K(\omega, t) = V_K^{+}(\omega, t) = vol_n(\{x \in K : \langle \omega, x \rangle \leq t\}) = \int_{\{x \in K : \langle \omega, x \rangle \leq t\}} dx. \tag{1.2}$$

The problem of describing algebraically integrable domains goes back to Newton **[9]**,**[15]**. In connection with Kepler's law in celestial mechanics, Newton established that there are no algebraically integrable convex bodies (ovals) in $\mathbb{R}^2$ (for the fact to be true one has to assume the infinite smoothness of the boundaries). On

---

2010 *Mathematics Subject Classification.* Primary 44A12; Secondary 51M99.

The author thanks Mikhail Zaidenberg for drawing the author's attention to the subject and stimulating discussions, and Victor Vassiliev for helpful references. The author thanks the referee for useful remarks.

the other hand, balls and, more generally, ellipsoids in odd-dimensional spaces are algebraically integrable.

V.I. Arnold raised the problem of generalization of Newton's lemma for higher dimensions:

PROBLEM 1.2. ([**2**],1990-27, 1987-14) Do there exist algebraically integrable smooth ovaloids different from ellipsoids in $\mathbb{R}^n$ with odd n?

In [**2**], ovaloid means a closed hypersurface bounding a convex body. V. Vassiliev [**13**],[**3**],[**4**],[**12**] proved that there are no algebraically integrable bounded domains (no convexity is required) with C^∞ boundaries in $\mathbb{R}^n$ for even n. In odd dimensions, the question whether ellipsoids are the only algebraically integrable domains, remains unanswered. In this article we partially answer this question in affirmative, under the condition (satisfied for ellipsoids) that the algebraic extension of the function V_K has no real singular points. No convexity of the domain is a priori assumed.

1.2. Main notions. We will be writing the equation of an affine hyperplane in the normal form $\langle x, \xi\rangle = t$, where $|\xi| = 1$ and $t \in \mathbb{R}$. Correspondingly, the volume function $V_K(\xi, t)$ becomes a function defined on the cylinder $S^{n-1} \times \mathbb{R}$.

Furthermore, in our considerations, the algebraicity of V_K with respect to t, rather than with respect to the direction variable ξ, will be playing role. More precisely, the equation $Q(\xi, t, w) = 0$ is assumed algebraic only with respect to t. In other words, we consider elements Q of the polynomial ring $C(S^{n-1})[z, w]$ over the coefficient algebra $C(S^{n-1})$. This ring consists of all polynomials of two (complex) variables z, w with coefficients - continuous functions on the unit sphere S^{n-1}.

Let us give the exact definition.

DEFINITION 1.3. Let K be a bounded body in $\mathbb{R}^n$. We will call K **algebraically** t**-integrable** if the volume function $V_K(\xi, t)$ satisfies the equation

$$Q(\xi, t, V_K(\xi, t)) = 0, \tag{1.3}$$

where

$$Q(\xi, t, w) = \sum_{i=0}^{N} \sum_{k=0}^{k_i} q_{i,k}(\xi) t^k w^i$$

and the coefficients $q_{i,k} \in C(S^{n-1})$.

Thus, our condition for the body K is even weaker than the algebraic integrability, because we assume the algebraic dependence of the volume function $V_K(\xi, t)$ only with respect to t.

Denote

$$Disc_Q(\xi, t) = q_N(\xi, t)^{2N-1} \Pi_{i<j} (w_i(\xi, t) - w_j(\xi, t))^2,$$

the discriminant of the polynomial $Q(\xi, t, w)$ with respect to the variable t, where $w = w_i(\xi, t)$ are the roots of the algebraic equation $Q(\xi, t, w) = q_0(\xi, t) + ... + q_N(\xi, t) w^N = 0$. Note, that our definition slightly differs from the standard one, where the factor in front of the product is q_N^{2N-2}.

Since the discriminant is a polynomial of the coefficients of Q, which are, in turn, polynomials of t, we have:

$$Disc_Q(\xi, t) = \sum_{j=0}^{M} d_j(\xi) t^j,$$

where $d_j(\xi) \in C(S^{n-1})$.

DEFINITION 1.4. We will call a algebraically t-integrable domain $K \subset \mathbb{R}^n$ **free of real singularities** if its volume function satisfies the equation $Q(\xi, t, V_K(\xi, t)) = 0$, where the function $Q(\xi, t, w)$ is continuous in $\xi \in S^{n-1}$ and a polynomial in t, w, such that

(1) $Disc_Q(\xi, t) \neq 0$ for all $(\xi, t) \in S^{n-1} \times \mathbb{R}$.
(2) The leading coefficient of $Disc_Q(\xi, t)$ satisfies $d_M(\xi) \neq 0, \xi \in S^{n-1}$.

Note that if n is even, then there is no domain K in $\mathbb{R}^n$ with C^∞ boundary for which the volume function $V_K(\xi, t)$ extends smoothly through the tangent planes to ∂K. Indeed, at the elliptic (having nonzero principal curvatures of the same sign) points $x_0 \in \partial K$, the volume $V_K(\nu_{x_0}, t)$ behaves as $c(t - t_0)^{\frac{n+1}{2}}$ [**5**], where ν_{x_0} is the normal vector and t_0 is the distance of the tangent plane $T_{x_0}(\partial K)$ from the origin. If n is even then $\frac{n+1}{2}$ is non-integer and hence the function $V_K(\nu_{x_0}, t)$ is not in the class $C^k, k > \frac{n+1}{2}$, at $t = t_0$.

1.3. Formulation of the main result. The main result of this article is the following

THEOREM 1.5. *Let $n \geq 3$ be odd and $K \subset \mathbb{R}^n$ be a bounded domain with infinitely smooth boundary. Then K is an algebraically t-integrable domain, free of real singularities, if and only if K is an ellipsoid.*

Let us illustrate Theorem 1.5 for the partial case of rationally integrable domains.

COROLLARY 1.6. *Let K be a bounded domain in $\mathbb{R}^n, n$ is odd, with infinitely smooth boundary. Suppose that the volume function $V_K(\xi, t)$ is a rational function in t without real poles:*

$$V_K(\xi, t) = \frac{A(\xi, t)}{B(\xi, t)},$$

where $A(\xi, t), B(\xi, t) \in C(S^{n-1})[t],\ B(\xi, t) \neq 0, b_k(\xi) \neq 0, \xi \in S^{n-1}, t \in \mathbb{R},\ b_k(\xi)$ is the leading coefficient of $B(\xi, t)$. Then K is an ellipsoid.

In this case the algebraic equation for V_K is

$$Q(\xi, t, V_K(\xi, t)) = 0,$$

where

$$Q(\xi, t, w) = B(\xi, t)w - A(\xi, t).$$

The polynomial Q has no multiple roots w and the discriminant is

$$Disc_Q(\xi, t) = B(\xi, t).$$

All the conditions of Theorem 1.5 are fulfilled due to the conditions for the denominator $B(\xi, t)$. The partial case $B(\xi, t) = 1$ corresponds to polynomially integrable domains.

REMARK 1.7. In the Newton's proof, a contradiction with algebraicity is obtained for family of sections of the planar domain by straight lines rotating around a fixed point (see the article [**11**], where an interesting discussion of the Newton's proof in historical aspect is presented). In our terms, the Newton'a arguments lead to a contradiction with the assumption of algebraic dependence on ξ. On the other hand, in the article [**13**], families of parallel cross-sections are exploited in the proof, i.e., the assumption of the algebraicity with respect to t plays a leading role. Our condition of the t-integrability of a domain is just of the second kind.

1.4. Examples.

(1) Take $K = B^n$ be the unit ball in $\mathbb{R}^n$. Then

$$V_{B^n}(\xi, t) = V_{B^n}(t) = c\int_{-1}^{t}(1-s^2)^{\frac{n-1}{2}}\,ds.$$

This Abelian integral represents a polynomial $P(\xi, t)$ in t if n is odd, and is a transcendental function if n is even. The algebraic equation for $w = V_K(\xi, t)$ is

$$Q(\xi, t, w) = w - P(\xi, t) = 0.$$

In this case $N = 1, q_N(\xi, t) = 1, Disc_Q(\xi, t) = 1$ and therefore B^n is algebraically t-integrable and free of real singularities.

(2) More generally, consider the case of ellipsoid in the odd-dimensional space, i.e. an image $E = \mathcal{A}(B_n)$ of the unit ball under a non-degenerate affine transformation. Then

$$V_E(\xi, t) = det\mathcal{A} \cdot V_{B^n}\Big(\frac{t}{|\mathcal{A}^{-1}(\xi)|}\Big).$$

Correspondingly, the algebraic equation for the volume function $w = V_E$ is

$$Q(\xi, t, w) = w - P\Big(\frac{t}{|\mathcal{A}^{-1}(\xi)|}\Big) = 0.$$

Here Q is a polynomial in t, w with the coefficients that are continuous functions of $\xi \in S^{n-1}$. Thus, E is algebraically t-integrable. The conditions 1,2 in Definition 1.4 are fulfilled, like in the case of the ball.

Thus, the conditions in Theorem 1.5 for the domain K to be ellipsoid are necessary.

2. Preliminaries

From now on, we fix a domain K in $\mathbb{R}^n, n \geq 3$ is odd, satisfying all the conditions of Theorem 1.5. Without loss of generality, we can assume that $0 \in K$.

The natural domain of definition of the volume function $V_K(\xi, t)$, defined in (1.2) is

$$h_-(\xi) \leq t \leq h_+(\xi),$$

where

$$h_-(\xi) = \inf_{x \in K}\langle \xi, x\rangle, h_+(\xi) = \sup_{x \in K}\langle \xi, x\rangle,$$

are the support functions of the domain K.

Since $V_K(-\xi,-t) = vol(K) - V_K(\xi,t)$, we have

$$\frac{d^m V_K}{dt^m}(-\xi,-t) = (-1)^{m-1}\frac{d^m V_K}{dt^m}(\xi,t), m \geq 1. \tag{2.1}$$

According to our assumption, $Q(\xi,t,V_K(\xi,t)) = 0$ is satisfied in the domain of definition of V_K. Since $0 \in K$, this domain includes the set $S^{n-1} \times (-\varepsilon,\varepsilon)$, where $\varepsilon > 0$ is sufficiently small.

The algebraic with respect to z, w equation $Q(\xi,z,w) = 0$ defines $w = w_\xi(z) = w(\xi,z)$ as a multi-valued algebraic function of $z \in \mathbb{C}$. Indeed, the equation $Q(\xi,z,w) = 0$ has $N = N(\xi,z)$ roots $w = w_j(\xi,z), j = 1,...,N$. The condition for the leading coefficient of the resultant yields that the degree N does not depend on ξ and z. The function $V_K(\xi,t)$ coincides in its domain of definition with one of the branches $w_j(\xi,t)$.

The branch points of the algebraic function $w(\xi,z)$ are the values (ξ,z) at which two or more roots coincide. The set of branch points is the zero set of the discriminant $Disc_Q(\xi,z) = 0$.

Another type of singular points are poles, i.e., the points (ξ_0,z_0) such that $\lim_{(\xi,z)\to(\xi_0,z_0)} w(\xi,z) = \infty$. The leading coefficient of the polynomial Q vanishes at the poles: $q_N(\xi_0,z_0) = 0$ and hence $Disc_Q(\xi,z_0) = 0$.

Thus, the condition for $Disc_Q(\xi,t)$ of non-having real zeroes provides that the algebraic function $w(\xi,z)$ has neither poles, nor branch points for z on the real axis, $Im z = 0$. The branches $w_j(\xi,z)$ are locally holomorphic functions of z away from the singular points.

Introduce the *parallel section function*

$$A_K(\xi,t) = vol_{n-1}(K \cap \{\langle\xi,x\rangle = t\}) = \int\limits_{K\cap\{\langle\xi,x\rangle=t\}} dx.$$

On one hand, $A_K(\xi,t)$ is the t-derivative of $V_K(\xi,t)$:

$$A_K(\xi,t) = \frac{dV_K}{dt}(\xi,t), h_-(\xi) \leq t \leq h_+(\xi).$$

On the other hand, $A_K(\xi,t)$ is the Radon transform of the characteristic function χ_K of the domain K :

$$A_K(\xi,t) = R\chi_K(\xi,t) = \int\limits_{\langle\xi,x\rangle=t} \chi_K(x)dx.$$

The inversion formula for Radon transform yields

$$1 = \chi_K(x) = c\int\limits_{|\xi|=1} \frac{d^{n-1}}{dt^{n-1}}A_K(\xi,\langle\xi,x\rangle)dS(\xi), x \in K,$$

where $dS(\xi)$ is the normalized Lebesgue surface measure on S^{n-1}.

Let us apply Laplace operator to the both sides of the equality. Then we have $\Delta\chi_K(x) = 0, x \in K$, in the left hand side. Applying Laplace operator to the right hand side results in twice differentiating in t and increasing by two the order of the t-derivative of the function A_K under the sign of the integral. Therefore we obtain:

$$\int\limits_{S^{n-1}} \frac{d^{n+1}}{dt^{n+1}}A_K(\xi,\langle\xi,x\rangle)dS(\xi) = \int\limits_{S^{n-1}} \frac{d^{n+2}V_K}{dt^{n+2}}(\xi,\langle\xi,x\rangle)dS(\xi) = 0, x \in K. \tag{2.2}$$

3. Proof of Theorem 1.5

3.1. Outline of the proof. The proof is based on a reduction to the case of polynomially integrable domains, which has been solved recently ([**10**],[**1**]).

Assuming that $0 \in K$, we derive from the inversion formula (2.2) for the Radon transform that the Fourier coefficients of the expansion of $V_K(\xi, z)$, where z is near 0, into spherical harmonics on the unit sphere $\xi \in S^{n-1}$ are polynomials in z.

Using the conditions imposed on the discriminant we prove that the volume function $V_K(\xi, t)$ has an algebraic continuous extension along two closed paths $\Gamma^{\pm}$ surrounding the singular set in the upper and in lower half-planes, respectively. Then, using the above established polynomiality of the Fourier coefficients we prove that those continuous branches holomorphically extend, in z, inside the paths. This removes singularities (branching and poles) and shows that the germ $V_K(\xi, z)$, for z near 0, belongs to a branch which is an entire function of $z \in \mathbb{C}$.

The last step is applying Great Picard Theorem which implies that entire algebraic functions are polynomials. Therefore the germ $V_K(\xi, t)$ extends to the real line as a polynomial in t and we conclude that K is a polynomially integrable domain. According to ([**10**], [**1**]), K is an ellipsoid.

3.2. Local polynomiality of Fourier coefficients of the volume function $V_K(\xi, t)$. Applying translation, we can assume that $0 \in K$. For sufficiently small t, the hyperplanes $\langle \xi, x\rangle = t$ intersect K for all $\xi \in S^{n-1}$ and therefore the function $V_K(\xi, t)$ is well defined for $|t| < \varepsilon$ if ε is sufficiently small.

Expand the function $V_K(\xi, t)$ into Fourier series with respect to the variable $\xi \in S^{n-1}$:

$$V_K(\xi, t) = \sum_{k=0}^{\infty} \sum_{\alpha=1}^{d(k)} v_{k,\alpha}(t) Y_k^{\alpha}(\xi), \tag{3.1}$$

where $\{Y_k^{\alpha}\}_{\alpha=1}^{d(k)}$ is an orthonormal basic in the space H_k of all spherical harmonics of degree k.

The corollary of the inversion formula is the following lemma which plays a key role in the proof of the main result.

LEMMA 3.1. *The Fourier coefficients $v_{k,\alpha}(t)$ in (3.1) are polynomials in the interval $(-\varepsilon, \varepsilon)$.*

Proof We will be using the notation

$$V_K^{(m)}(\xi, t) = \frac{d^m V_K}{dt}(\xi, t).$$

The algebraic function $w(\xi, z) = V_K(\xi, z)$ and its z-derivatives are holomorphic in a small disc $|z| < \varepsilon$. By Cauchy formula

$$V_K^{(n+2)}(\xi, t) = \frac{1}{2\pi i} \int_{|\zeta|=\varepsilon_1} \frac{V_K^{(n+2)}(\xi, \zeta) d\zeta}{\zeta - t}, \tag{3.2}$$

where $0 < \varepsilon_1 < \varepsilon, |t| < \varepsilon_1$.

Substituting this representation into (2.2) yields:

$$\frac{1}{2\pi i} \int_{|\xi|=1} \int_{|\zeta|=\varepsilon_1} \frac{V_K^{(n+2)}(\xi, \zeta)}{\zeta - \langle x, \xi\rangle} d\zeta dS(\xi) = 0,$$

where $x \in K$ is in a neighborhood of $0 \in K$.

Let $0 < \varepsilon_2 < \varepsilon_1$. If $|x| < \varepsilon_2$ then $|\langle x, \xi\rangle| < \varepsilon_2$ and the expansion of the Cauchy kernel into the power series converges uniformly with respect to ξ. Therefore integration the series with respect to ξ is possible and one obtains:

$$\sum_{j=0}^{\infty} \int_{|\xi|=1} b_j(\xi)\langle x, \xi\rangle^j dS(\xi) = 0,$$

where

$$b_j(\xi) = \frac{1}{2\pi i} \int_{|\zeta|=\varepsilon_1} \frac{V_K^{(n+2)}(\xi, \zeta)d\zeta}{\zeta^{j+1}}.$$

Since n is odd, it follows from the relation (2.1) with $m = n+2$ that

$$b_j(-\xi) = (-1)^j b_j(\xi). \tag{3.3}$$

By homogeneity, each term in the series is zero:

$$\int_{|\xi|=1} b_j(\xi)\langle x, \xi\rangle^j dS(\xi) = 0, j = 0, 1, ...$$

When x runs over an open ε- neighborhood of $x = 0$, the homogeneous polynomials $\psi_{x,j}(\xi) = \langle x, \xi\rangle^j$ span the space $\mathcal{P}_j$ of all homogeneous polynomials in $\mathbb{R}^n$ of degree j. The restriction of this space to the unit sphere $|\xi| = 1$ decomposes into the orthogonal sum

$$span\{\psi_{x,j}|_{S^{n-1}}, |x| < \varepsilon\} = \mathcal{P}_j|_{S^{n-1}} = \oplus_{s=0}^{[\frac{j}{2}]} H_{j-2s},$$

where H_k is the space of all spherical harmonics of degree k. Thus, we obtain

$$b_j \perp \oplus_{s=0}^{[\frac{j}{2}]} H_{j-2s}.$$

On the other hand, the symmetry relation (3.3) implies that b_j decomposes on S^{n-1} into the spherical harmonics of the same parity as j. Therefore, b_j is orthogonal to *all* spherical harmonics of degree $k \leq j$:

$$b_j \perp \oplus_{k=0}^{j} H_k. \tag{3.4}$$

Denote $p_{k,\alpha}(t)$ the Fourier coefficients, with respect to ξ, of $V_K^{(n+2)}(\xi, t)$:

$$p_{k,\alpha}(t) = \int_{|\xi|=1} V_K^{(n+2)}(\xi, t) Y_k^{\alpha}(\xi) dS(\xi).$$

Substitute the integral representation (3.2) of $V_K^{(n+2)}$:

$$p_{k,\alpha}(t) = \frac{1}{2\pi i} \int_{|\xi|=1} \int_{|\zeta|=\varepsilon_1} \frac{V_K^{(n+2)}(\xi, \zeta)}{\zeta - t} Y_k^{\alpha}(\xi) d\zeta dS(\xi). \tag{3.5}$$

Expanding once again the Cauchy kernel $\frac{1}{\zeta - t}$ into the power series for $|t| < \varepsilon_2$ yields

$$p_{k,\alpha}(t) = \sum_{j=0}^{\infty} \Big(\int_{|\xi|=1} b_j(\xi) Y_k^{\alpha}(\xi) dS(\xi) \Big) t^j.$$

The orthogonality relation (3.4) yields that all the terms with $k \leq j$ vanish and hence $p_{k,\alpha}(t)$ contains only terms with $j < k-1$. Hence $p_{k,\alpha}$ is a polynomial of degree $< k-1$. Since

$$p_{k,\alpha}(t) = \frac{d^{n+2}v_{k,\alpha}}{dt^{n+2}}(t),$$

the repeated integration of the equality with respect to t implies that the functions $v_{k,\alpha}(t)$ are polynomials as well.

3.3. Removing complex singularities. Denote $Reg(Q) = \{z \in \mathbb{C} : Disc_Q(\xi, z) \neq 0, \forall \xi \in S^{n-1}\}$ and $Sing(Q) = \mathbb{C} \setminus Reg(Q)$.

LEMMA 3.2. *$Sing(Q)$ is a bounded set.*

Proof By the condition of Theorem 1.5, the leading coefficient $d_M(\xi)$ in the expansion

$$Disc_Q(\xi, t) = \sum_{j=0}^{M} d_j(\xi)t^j$$

does not vanish for all $\xi \in S^{n-1}$. Since $d_M \in C(S^{n-1})$, there exists $c_M > 0$ such that $|d_M(\xi)| \geq c_M, \xi \in S^{n-1}$. Then for any $z \in \mathbb{C}$ holds

$$|Disc_Q(\xi, z)| \geq c_M|z|^M - c_{M-1}|z|^{M-1} - \dots - c_0,$$

where $c_j = \|d_j\|_{C(S^{n-1})}$. Therefore if $|z| > R$ and $R > 0$ is sufficiently large then $Disc_Q(\xi, z) \neq 0$ for all $\xi \in S^{n-1}$ and hence $Sing(Q) \subset \{|z| \leq R\}$. Lemma is proved.

Note that the condition of Theorem 1.5 implies that the real line consists of regular points: $\{Imz = 0\} \subset Reg(Q)$.

LEMMA 3.3. *Let $\Gamma \subset Reg(Q)$ be a Jordan curve , containing a real segment $I_\varepsilon = \{-\varepsilon \leq Rez \leq \varepsilon, Imz = 0\}$ and enclosing the singular set $Sing(Q)$. Fix a point $z_0 \in \Gamma \setminus I_\varepsilon$. Then the volume function $V_K(\xi, t)$ extends from $(\xi, t) \in S^{n-1} \times [-\varepsilon, \varepsilon]$ to $S^{n-1} \times (\Gamma \setminus \{z_0\})$ as a continuous function $W_\Gamma(\xi, z)$ satisfying $Q(\xi, t, W_\Gamma(\xi, z)) = 0$ for $(\xi, z) \in S^{n-1} \times (\Gamma \setminus \{z_0\})$.*

Proof Let $z = \varphi(s), s \in [0, 1], \varphi(0) = \varphi(1) = z_0$, be a parametrization of the closed curve Γ. Then the algebraic equation $Q(\xi, z, w) = 0$ takes on Γ the form

$$Q(\xi, \varphi(s), w) = \sum_{j=0}^{N} q_j(\xi, \varphi(s))w^j = 0.$$

The leading coefficient $q_N(\xi, \varphi(s)) \neq 0$ for $(\xi, s) \in S^{n-1} \times [0, 1]$ because $Disc_Q(\xi, z) = q_N(\xi, z)Disc(Q)(\xi, z)$ and $Disc_Q(\xi, z) \neq 0$ when $z \in \Gamma$.

Therefore the equation can be written in the monic form by dividing by the leading coefficient:

$$f_0(\xi, s) + f_1(\xi, s)w + \dots + f_{N-1}(\xi, s)w^{N-1} + w^N = 0, \tag{3.6}$$

where

$$f_j(\xi, s) = \frac{q_j(\xi, \varphi(s))}{q_N(\xi, \varphi(s))}.$$

Thus, we deal with an algebraic *monic* equation for w with the coefficients-continuous functions on the cylinder $S^{n-1} \times [0, 1]$. Since $Disc_Q(\xi, z) \neq 0$ on Γ, the equation has no multiple roots.

The monodromy theorem (see Thm.16.2 in [**8**], or [**6**]) implies that the algebraic equation (3.6) is completely solvable. This means that there is no monodromy on the cylinder $S^{n-1}\times[0,1]$ and there exist N continuous functions $W_1(\xi,s),...,W_N(\xi,s)$ on $S^{n-1}\times[0,1]$ satisfying the equation (3.6).

Indeed, consider the mapping $p:E\to B$ of the space

$$E=\{(\lambda_1,...,\lambda_N)\in\mathbb{C}^N:\lambda_i\neq\lambda_j,i,j=1,...,N.\}$$

to the space

$$B=\{(a_0,...,a_{N-1})\in\mathbb{C}^N:Disc_P\neq 0,P(w)=w^N+a_{N-1}w^{N-1}+\cdots+a_0\},$$

defined as follows: $p(\lambda)$ is the vector of the coefficients of the monic polynomial of degree N with the roots $\lambda_1,...,\lambda_N$. By Implicit Function Theorem, p is a $N!$ - covering map. It is regular due to $a_N=1$. Since $n\geq 3$, the fundamental group $\pi(S^{n-1}\times[0,1])=0$ and by the monodromy theorem the continuous map $f:S^{n-1}\times[0,1]$, $f(\xi,s)=(f_0(\xi,s),\cdots,f_{N-1}(\xi,s))$ can be lifted to a continuous mapping $W:S^{n-1}\times[0,1]\to E$ such that $pW=f$. Then

$$W(\xi,s)=(W_1(\xi,s),\cdots,W_N(\xi,s)),$$

where $W_i(\xi,s)$ are the continuous roots of (3.6).

Let us regard the functions $W_i(\xi,s)$ as the continuous functions $W_i(\xi,z),z=\varphi(s)$, of $(\xi,z)\in S^{n-1}\times(\Gamma\setminus\{z_0\})$. A monodromy can occur at the initial-end point z_0 after moving along the closed curves Γ, so that one can assign to $W_i(\xi,z_0)$ two values , corresponding to the values $s=0$ and $s=1$ of the parameter on the curves.

Since the curve Γ belongs to the regular set $Reg(Q)$, the functions $W_i(\xi,z)$ are locally holomorphic in a neighborhood $U_{i,\xi,z}$ of $z\neq z_0$. Moreover, due to compactness of S^{n-1}, the neighborhood can be chosen the same for all $\xi\in S^{n-1}$.

The germ $w=V_K(\xi,t),|t|<\varepsilon$, of the algebraic function $Q(\xi,t,w)=0$, coincides with one of the functions $W_i,i=1,...,N$. This function $W_\Gamma(\xi,t)$ is just what we need. Lemma is proved.

LEMMA 3.4. *For every $\xi\in S^{n-1}$, the function $V_K(\xi,z)$ extends from $|z|<\varepsilon$ to the whole complex plane as an entire function $F(\xi,z)$ of $z\in\mathbb{C}$.*

Proof Choose in Lemma 3.3 the Jordan curve $\Gamma=\Gamma^+\subset Reg(Q)$ in the upper half-plane $Im z\geq 0$. For instance, take

$$\Gamma^+=[-R,R]\cup C_R^+,$$

where C_R^+ is the open upper half-circle of radius R with the center 0. The condition of Theorem 1.5 and Lemma 3.2 imply that the curve Γ^+ encloses the singular set $Sing(Q)$ if $R>0$ is sufficiently large. The marked point z_0 is taken in the half-circle C_R^+ and the parametrization $\varphi:[0,1]\to\Gamma^+$ satisfies $\varphi(0)=\varphi(1)=z_0$. Let W_{Γ^+} be the branch of $w(\xi,z)$ along $\Gamma\setminus\{z_0\}$ from Lemma 3.3.

Now fix a basis $Y_k^\alpha,\alpha=1,\cdots,d(k)$, in the space of spherical harmonics of degree k and consider the functions

$$I_{k,\alpha}(z)=\int\limits_{\xi\in S^{n-1}}Y_k^\alpha(\xi)W_{\Gamma^+}(\xi,z)dA(\xi).$$

They are holomorphic in a neighborhood of $\Gamma^+\setminus\{z_0\}$. When z is real, $z=t$, and $|z|=|t|<\varepsilon$, then $W_{\Gamma^+}(\xi,z)=V_K(\xi,t)$ and therefore

$$I_{k,\alpha}(t)=v_{k,\alpha}(t)$$

is the Fourier coefficient of the volume function $V_K(\xi, t)$.

By Lemma 3.1, the Fourier coefficient $v_{k,\alpha}(t)$ is a polynomial in t near $t = 0$ (of degree depending on k). Therefore, $I_{k,\alpha}(t)$ are polynomials in t in a real neighborhood of $t = 0$ and by analyticity, $I_{k,\alpha}(z)$ coincides with a complex polynomial $P_{k,\alpha}(z)$ on $\Gamma \setminus \{z_0\}$. By continuity it happens at the point z_0 as well.

In particular, the one-sided limits

$$I_{k,\alpha}(z_0 - 0) = \lim_{s \to 1-0} I_{k,\alpha}(\varphi(s)), I_{k,\alpha}(z_0 + 0) = \lim_{s \to 0+0} I_{k,\alpha}(\varphi(s))$$

at the point $z_0 \in \Gamma^+$ along the curve Γ^+ coincide:

$$I_{k,\alpha}(z_0 - 0) = I_{k,\alpha}(z_0 + 0) = P_{k,\alpha}(z_0).$$

Going back to the definition of $I_{k,\alpha}(z)$ we have:

$$\int_{S^{n-1}} Y_k^\alpha(\xi) W_{\Gamma^+}(\xi, z_0 - 0) dS(\xi) = \int_{S^{n-1}} Y_k^\alpha(\xi) W_\Gamma(\xi, z_0 + 0) dS(\xi).$$

Since it is true for all basic spherical harmonics, we conclude that

$$W_{\Gamma^+}(\xi, z_0 - 0) = W_\Gamma(\xi, z_0 + 0),$$

for all $\xi \in S^{n-1}$ and thus $W_{\Gamma^+}(\xi, z)$ is continuous on the entire curve Γ^+, including the point z_0. .

Furthermore, for any $m = 0, 1, \ldots$ one can write

$$\int_{\Gamma^+} I_{k,\alpha}(z) z^m dz = \int_{\Gamma^+} P(z) z^m dz = 0.$$

Substituting the expression for $I_{k,\alpha}(z)$ and rewriting the left hand side by Fubini theorem one obtains

$$\int_{S^{n-1}} Y_k^\alpha(\xi) \Big(\int_{\Gamma^+} W_{\Gamma^+}(\xi, z) z^m dz \Big) dS\xi) = 0.$$

Due to the arbitrariness of the basic spherical harmonic Y_k^α we conclude

$$\int_\Gamma W_{\Gamma^+}(\xi, z) z^m dz = 0,$$

for any $\xi \in S^{n-1}$. Vanishing of all complex moments on Γ means that for any $\xi \in S^{n-1}$ the function $W_{\Gamma^+}(\xi, z)$ is the boundary value of a function $F_\xi^+(z)$, holomorphic in the domain Ω^+ in the upper half-plane, bounded by Γ^+. Thus, $W_{\Gamma^+}(\xi, z)$ extends from a neighborhood of the curve Γ^+ inside the domain Ω^+ as a single-valued holomorphic function $F_\xi^+(z)$.

Repeating the same argument for the lower half-plane $Imz \leq 0$ and for a corresponding closed curve Γ^-, we conclude that $V_K(\xi, z)$ continuously extends from the small disc $|z| < \varepsilon$ as a function $F_\xi^-(z)$ which is a single-valued holomorphic function of z in the domain Ω^- in the lower half-plane, bounded by Γ^-.

Since $F_\xi^+(t) = F_\xi^-(t) = V_K(\xi, t)$ for $|t| < \varepsilon$, the two functions define a function $F(\xi, z)$, holomorphic with respect to z in $\Omega = \Omega^+ \cup \Omega^-$.

The function $F(\xi, z)$ solves the equation $Q(\xi, t, F(\xi, t)) = 0$ for $|t| < \varepsilon$ and, by analyticity, for all $z \in \Omega$. There is no singularities of the branch $F(\xi, z)$ in the domain Ω. However, by the construction, the domain Ω contains *all* singular points of the algebraic function $w = w(\xi, z)$. Therefore, the branch $F(\xi, z)$ (as an algebraic function of z) has no poles and branch points in $\mathbb{C}$ and hence is an entire function of $z \in \mathbb{C}$. Lemma is proved.

3.4. End of the proof of Theorem 1.5. We provide for completeness the proof of the following well known statement:

LEMMA 3.5. *Let $F(z)$ be an entire algebraic function in the complex plane. Then F is a polynomial.*

PROOF. If F is not a polynomial then ∞ is an essential singularity of G. According to Great Picard Theorem, $F(z)$ takes in a neighborhood of ∞ any value, with possibly one exception, infinitely many times. Therefore, if F is not a polynomial and satisfies a nontrivial algebraic equation $Q(z, F(z)) = 0, Q$ is a polynomial, then for all $a \in \mathbb{C}$, except for at most one, we have $Q(z_\nu, a) = Q(z_\nu, F(z_\nu)) = 0$ for an infinite sequence of $z_\nu \in \mathbb{C}$. Since Q is a polynomial , $Q(z, a) = 0$ for all $z \in \mathbb{C}$ and for all but one $a \in \mathbb{C}$. Therefore Q is identically zero which is not the case. Lemma is proved.

DEFINITION 3.6. A body K is called polynomially integrable if its volume function is a polynomial with respect to t :

$$V_K(\xi, t) = \sum_{j=1}^{N} a_j(\xi) t^j. \tag{3.7}$$

Our discussion above implies the following statement:

COROLLARY 3.7. *The body K in the formulation of Theorem* 1.5 *is polynomially integrable.*

PROOF. We have proven in Lemma 3.4 that the germ $V_K(\xi, z), |z| < \varepsilon$, extends to a single-valued branch $F(\xi, z), z \in \mathbb{C}$, of the algebraic function $w = w(\xi, z)$ defined by the equation $Q(\xi, z, w) = 0$. This single-valued branch is an entire function in z. By Lemma 3.5, for every $\xi \in S^{n-1}$ the function $F(\xi, z)$ is a polynomial of z. The degree $N(\xi)$ is equal to the number of zeroes of $F(\xi, z)$ and hence does not exceed the maximal degree of the polynomials $Q(\xi, z, 0)$ with respect to z, where Q is the defining polynomial for the algebraic function V_K. Therefore, the degrees $N(\xi), \xi \in S^{n-1}$, are bounded and hence $V(\xi, t) = F(\xi, t)$ has the form (3.7). Lemma is proved.

To finish the proof, we refer to the two recent results:

THEOREM 3.8. [**1**] *Let n be odd. If K is a bounded polynomially integrable body in $\mathbb{R}^n$ with infinitely smooth boundary then K is convex.*

THEOREM 3.9. [**10**] *Let n be odd. Let K be a bounded convex polynomially integrable body in $\mathbb{R}^n$ with infinitely smooth boundary. Then K is an ellipsoid.*

Corollary 3.7 states that if a domain K satisfies the conditions of Theorem 1.5 then K is polynomially integrable. By Theorem 3.8, K is convex, and then by Theorem 3.9 the body K is an ellipsoid. Theorem 1.5 is proved.

It is worth noting that condition 1 for the discriminant in Definition 1.4 does not allow both leaves $w = V_K^{\pm}(\xi, t)$ to satisfy the same equation $Q(\xi, t, w) = 0$ because in this case we would have multiple roots whenever the hyperplane $\langle x, \xi \rangle = t$ equally divides the volume of the body K. A consistent with Definition 1.1 version of Theorem 1.5, admitting two valued solution $V_K^{\pm}(\xi, t)$, can be obtained by replacing condition 1 by (even weaker) direct condition of real analytic extendibility $V_K(\xi, t)$ to all real t. The proof of such a version of Theorem 1.5 is essentially the same.

References

[1] M. Agranovsky, *On polynomially integrable domains in Euclidean spaces*, Complex analysis and dynamical systems, Trends Math., Birkhäuser/Springer, Cham, 2018, pp. 1–21, DOI 10.1007/978-3-319-70154-7_1. MR3784162

[2] V. I. Arnold, *Arnold's problems*, Springer-Verlag, Berlin; PHASIS, Moscow, 2004. Translated and revised edition of the 2000 Russian original; With a preface by V. Philippov, A. Yakivchik and M. Peters. MR2078115

[3] V. I. Arnol′d and V. A. Vasil′ev, *Newton's Principia read 300 years later*, Notices Amer. Math. Soc. **36** (1989), no. 9, 1148–1154. MR1024727

[4] V.I. Arnold, V.A. Vassiliev, *Addendum to [3]*, Notices AMS, 37:1, 144.

[5] I. M. Gel'fand, M. I. Graev, N. Ya. Vilenkin, *Generalized Functions, Volume 5: Integral Geometry and Representation Theory*,Volume: 381; 1966; 449 pp.

[6] E. A. Gorin and V. Ja. Lin, *Algebraic equations with continuous coefficients, and certain questions of the algebraic theory of braids* (Russian), Mat. Sb. (N.S.) **78 (120)** (1969), 579–610. MR0251712

[7] S. Helgason, *Groups and geometric analysis: Integral geometry, invariant differential operators, and spherical functions*, Pure and Applied Mathematics, vol. 113, Academic Press, Inc., Orlando, FL, 1984. MR754767

[8] S.-t. Hu, *Homotopy theory*, Pure and Applied Mathematics, Vol. VIII, Academic Press, New York-London, 1959. MR0106454

[9] I. S. Newton, *Philosophiae naturalis principia mathematica* (Latin), William Dawson & Sons, Ltd., London, 1687. MR0053865

[10] A. Koldobsky, A. S. Merkurjev, and V. Yaskin, *On polynomially integrable convex bodies*, Adv. Math. **320** (2017), 876–886, DOI 10.1016/j.aim.2017.09.028. MR3709123

[11] B. Pourciau, *The integrability of ovals: Newton's Lemma 28 and its counterexamples*, Arch. Hist. Exact Sci. **55** (2001), no. 5, 479–499, DOI 10.1007/s004070000034. MR1827869

[12] V. A. Vassiliev, *Applied Picard-Lefschetz theory*, Mathematical Surveys and Monographs, vol. 97, American Mathematical Society, Providence, RI, 2002. MR1930577

[13] V. A. Vassiliev, *Newton's lemma XXVIII on integrable ovals in higher dimensions and reflection groups*, Bull. Lond. Math. Soc. **47** (2015), no. 2, 290–300, DOI 10.1112/blms/bdv002. MR3335123

[14] C. T. C. Wall, *Singular points of plane curves*, London Mathematical Society Student Texts, vol. 63, Cambridge University Press, Cambridge, 2004. MR2107253

[15] Wikipedia, *Newton's theorem about ovals*.

Bar-Ilan University, Holon Institute of Technology
Email address: agranovs@math.biu.ac.il

Contemporary Mathematics
Volume **733**, 2019
https://doi.org/10.1090/conm/733/14732

Rearrangement invariant spaces satisfying Dunford-Pettis criterion of weak compactness

Sergey V. Astashkin

Dedicated to the memory of Selim G. Krein

ABSTRACT. We survey old and recent results related to rearrangement invariant spaces X, where an analogue of the classical Dunford-Pettis characterization of relatively weakly compact subsets in L_1 holds. This is precisely the class of spaces such that for every weakly null sequence $\{x_n\}_{n=1}^{\infty} \subset X$ of pairwise disjoint functions we have $\|x_n\|_X \to 0$. In these spaces Cesaro means of any weakly null martingale difference sequence are norm null. Moreover, all reflexive subspaces of such a space X are strongly embedded in X. We simplify the proofs of some known results and prove a few new ones.

1. Introduction

Recall that a bounded set $K \subset L_1(\mu) = L_1(\Omega, \Sigma, \mu)$, where (Ω, Σ, μ) is a probability space, is called *equi-integrable* (or *uniformly integrable*) if

$$\lim_{\delta \to 0} \sup_{m(E)<\delta} \sup_{x(\omega)\in K} \int_E |x(\omega)|\, d\mu(\omega) = 0.$$

The following famous description of relatively weakly compact subsets in $L_1(\mu)$ as equi-integrable sets is due to N. Dunford and B. J. Pettis (see [**9**], [**8**, Ch. VII], or [**2**, Theorem 5.2.9]).

THEOREM 1.1. *Let K be a bounded set in $L_1(\mu)$. The following conditions are equivalent:*
(i) K is relatively weakly compact;
(ii) K is equi-integrable.

This theorem was published in 1940 and at once it has become a powerful tool in the study of L_1-spaces and their (especially, reflexive) subspaces. This result has connections with the theory of countably additive measures on some σ-algebra of sets, the Radon-Nikodym property, the Kadets-Pełczyński alternative and with many other results from Banach space theory.

2010 *Mathematics Subject Classification.* Primary 46E30, 46B50; Secondary 46B20, 46B42.
Key words and phrases. weakly compact set, equi-integrable set, equi-absolutely continuous norms, rearrangement invariant space, Orlicz space, Marcinkiewicz space, Lorentz space, positive Schur property, martingale difference, strongly embedded subspace, reflexive subspace.
*The work was supported by the Ministry of Education and Science of the Russian Federation, project 1.470.2016/1.4 and by the RFBR grant 17-01-00138.

The concept of equi-integrability easily can be generalized to Banach lattices X of measurable functions on (Ω, Σ, μ). We shall say that a set $K \subset X$ has *equi-absolutely continuous norms* in X if

$$\lim_{\delta\to 0} \sup_{\mu(E)<\delta} \sup_{x\in K} \|x\chi_E\|_X = 0.$$

In this paper we restrict ourselves to the case of rearrangement invariant (r.i.) spaces X on $[0,1]$ with the usual Lebesgue measure m (see the definition below). We will say that an r.i. space X satisfies *Dunford-Pettis criterion of weak compactness (shortly $X \in (WDP)$)* if every relatively weakly compact subset of X has equi-absolutely continuous norms in X. Observe that the converse assertion holds in every r.i. space.

It is worth to note that the (WDP)-property really has been considered previously under a different approach in the more general setting of Banach lattices. As an excellent source of relevant information we indicate the papers [**29**]–[**31**] and references herein (see also a discussion related to the notion of positive Schur property in Section 3). Nevertheless, the study of r.i. spaces satisfying Dunford-Pettis criterion of weak compactness has specific features and is of particular interest.

By Theorem 1.1, L_1 satisfies Dunford-Pettis criterion of weak compactness. At the same time, it is easy to see that $L_p \notin (WDP)$ if $p > 1$. Indeed, let $f_n := m(E_n)^{-1/p}\chi_{E_n}$, where E_n are pairwise disjoint subsets of $[0,1]$, $n = 1, 2, \ldots$ Then $f_n \stackrel{w}{\to} 0$ in L_p but, obviously, the set $\{f_n\}_{n=1}^\infty$ fails to have equi-absolutely continuous norms in L_p.

From the historical point of view, it is interesting to note that even a little bit earlier than Theorem 1.1 was published, in 1936, W. Orlicz proved that each Orlicz space $L_F = L_F[0,1]$ such that

$$\lim_{t\to\infty} \frac{G(2t)}{G(t)} = \infty,$$

where G is the complementary function to F, satisfies Dunford-Pettis criterion of weak compactness [**23**, assertion 1.5]. It was a first step to introduction of the following special class of non-reflexive Orlicz spaces on $[0,1]$ denoted now usually by (∇_3): an Orlicz space $L_F \in (\nabla_3)$ if and only if

$$\lim_{t\to\infty} \frac{G(Ct)}{G(t)} = \infty$$

for some $C > 1$. This class was defined explicitly much later by D. Leung [**18**] in connection with the study of the weak Banach-Saks property. In 1994, J. Alexopoulos [**3**] proved that Dunford-Pettis criterion of weak compactness holds in every space $L_F \in (\nabla_3)$, provided that the function F satisfies the Δ_2-condition at infinity (it is likely that the author of [**3**] was not aware of the above Orlicz's result).

On the other hand, in [**5**] S. Astashkin, N. Kalton, and F. Sukochev showed that the class (∇_3) coincides exactly with the intersection of the family of all Orlicz spaces with the class of r.i. spaces such that the conditions $\{x_n\}_{n=1}^\infty \subset X$, $x_n \stackrel{w}{\to} 0$, and $x_n \stackrel{\mu}{\to} 0$ imply $\|x_n\|_X \to 0$ (following [**16**], the latter property is referred to in [**5**] as the (Wm)-property). Moreover, as is proved in [**5**], an r.i. space X has the (Wm)-property (shortly $X \in (Wm)$) if and only if $X \in (WDP)$, which gives a characterization of r.i. spaces satisfying Dunford-Pettis criterion of weak compactness. Also, it is shown there that an Orlicz space $L_F \in (Wm)$ if and only if either $L_F = L_1$ or $L_F \in (\nabla_3)$.

The main objective of this paper is to survey above and also more recent results related to some properties of r.i. spaces satisfying Dunford-Pettis criterion of weak compactness. Let us describe them shortly. Recall that an r.i. space X has the martingale difference Cesaro mean property (MDCMP) if for every weakly null martingale difference sequence $\{d_n\}_{n=1}^{\infty}$ in X we have

$$\lim_{n\to\infty} \frac{1}{n} \left\| \sum_{k=1}^{n} d_k \right\|_X = 0.$$

As far as we know, the first result related to the latter property was obtained by F. J. Freniche [**11**]. He has shown that Cesaro means of any weakly null martingale difference sequence in $L_p[0,1], 1 \le p < \infty$, are norm null. Later, in [**1**], the analogous result was proved for Orlicz spaces from the class (∇_3). An extension of this is contained in the paper [**5**], where by using results of [**11**] it is shown that each r.i. space $X \in (WDP)$ has the (MDCMP)-property.

We say that a closed subspace H of the r.i. space X on $[0,1]$ is *strongly embedded into* X if, in H, the convergence in X-norm is equivalent to the convergence in measure. It is worth emphasizing that originally this concept was introduced in the case $X = L_p$ (with a different name: $\Lambda(p)$-space) by W. Rudin [**26**], in the special setting of Fourier analysis on the circle group $[0, 2\pi)$.

In particular, according to the Khintchine inequality [**2**, Theorem 6.2.3], the closed linear span $[r_k]$ of Rademacher functions in L_1 is strongly embedded into L_p for every $1 \le p < \infty$ (recall that $r_k(t) = \text{sign}(\sin 2^k \pi t)$, $k \in \mathbb{N}$, $t \in [0,1]$).

In 2008, E. Lavergne proved that all reflexive subspaces of an Orlicz space $L_F \in (\nabla_3)$ are strongly embedded into L_F [**17**]. In [**6**] this result is extended to all r.i. spaces with the (Wm)-property (or, equivalently, satisfying Dunford-Pettis criterion of weak compactness). At the same time, there is a wide class of r.i. spaces X, which fail to have the (Wm)-property, but however all reflexive subspaces of X are strongly embedded into this space.

The author would like to thank the referee for his (her) helpful remarks and comments.

2. Preliminaries

In this section, we shall briefly list the definitions and notions used throughout this paper. For more detailed information, we refer the reader to the monographs [**15, 20**].

A Banach space $(X, \|\cdot\|_X)$ of real-valued Lebesgue measurable functions (with identification m-a.e.) on the interval $[0,1]$ will be called *rearrangement invariant (r.i.)* if

(i). X is an ideal lattice, that is, if $y \in X$ and x is any measurable function on $[0,1]$ with $|x| \le |y|$, then $x \in X$ and $\|x\|_X \le \|y\|_X$;

(ii). X is rearrangement invariant in the sense that if $y \in X$, and if x is any measurable function on $[0,1]$ with $x^* = y^*$, then $x \in X$ and $\|x\|_X = \|y\|_X$.

Here, x^* denotes the non-increasing, right-continuous rearrangement of x given by

$$x^*(t) = \inf\{\ s \ge 0 : m\{u \in [0,1] : |x(u)| > s\} \le t\ \}, \quad t > 0,$$

where m denotes the usual Lebesgue measure.

For any r.i. space X on $[0,1]$ we have $L_\infty[0,1] \subseteq X \subseteq L_1[0,1]$. The fundamental function ϕ_X of an r.i. space X is defined by $\phi_X(t) := \|\chi_{[0,t]}\|_X$. Here and in what follows χ_E denotes the characteristic function of a set E.

The *Köthe dual* (or the *associated*) space $X^\times$ of an r.i. space X consists of all measurable functions y, for which

$$\|y\|_{X^\times} := \sup\Big\{\int_0^1 |x(t)y(t)|dt : \ x \in X, \ \|x\|_X \le 1\Big\} < \infty.$$

If X^* denotes the Banach dual of X, then $X^\times \subset X^*$ and $X^\times = X^*$ if and only if the norm $\|\cdot\|_X$ is order-continuous, i.e., from $\{x_n\} \subseteq X, x_n \downarrow_n 0$, it follows that $\|x_n\|_X \to 0$. Note that the norm $\|\cdot\|_X$ of the r.i. space X is order-continuous if and only if X is separable. We denote by X_0 the closure of L_∞ in X (the *separable part* of X). The space X_0 is r.i., and it is separable if $X \ne L_\infty$.

Let us recall some classical examples of r.i. spaces on $[0,1]$. Each increasing concave function φ on $[0,1]$, $\varphi(0)=0$, generates the Lorentz space $\Lambda(\varphi)$ (resp. the Marcinkiewicz space $M(\varphi)$) endowed with the norm

$$\|x\|_{\Lambda(\varphi)} = \int_0^1 x^*(t)d\varphi(t)$$

(resp.

$$\|x\|_{M(\varphi)} = \sup_{0<\tau\le 1} \frac{1}{\varphi(\tau)} \int_0^\tau x^*(t)dt).$$

The space $M(\varphi)$ is not separable provided that $\lim_{\tau\to 0}\varphi(\tau) = 0$ (or, equivalently, $M(\varphi) \ne L_1$). At the same time, its subspace $M_0(\varphi)$, consisting of all $x \in M(\varphi)$ such that

$$\lim_{\tau\to 0} \frac{1}{\varphi(\tau)} \int_0^\tau x^*(t)dt = 0,$$

is a separable r.i. space which, in fact, coincides with the separable part $(M(\varphi))_0$.

Let $F(u)$ be an increasing convex function on $[0,\infty)$ such that $F(0)=0$. Denote by L_F the Orlicz space on $[0,1]$ (see e.g. [**14**], [**24**]) endowed with the Luxemburg norm

$$\|x\|_{L_F} = \inf\{\lambda > 0 : \int_0^1 F(|x(t)|/\lambda)dt \le 1\}.$$

In particular, if $F(u) = u^p$, $1 \le p < \infty$, we obtain L_p. An Orlicz space L_F is separable if and only if the function F satisfies the Δ_2-condition at infinity, i.e., there are $C > 0$ and $u_0 > 0$ such that $F(2u) \le CF(u)$ for all $u \ge u_0$. In this case $L_F^* = L_G$, where G is the complementary function to F defined by

$$G(u) = \sup\{uv - F(v) : v \ge 0\}, \ \ u \ge 0.$$

Convergence in measure (resp. in weak topology) of a sequence of measurable functions $\{x_n\}_{n=1}^\infty$ (resp. from an r.i. space X) to a measurable function x (resp. from X) is denoted by $x_n \xrightarrow{\mu} x$ (resp. $x_n \xrightarrow{w} x$). Moreover, throughout the paper $\|f\|_p := \|f\|_{L_p[0,1]}$, $1 \le p \le \infty$.

3. Some characterizations of the class of r.i. spaces satisfying Dunford-Pettis criterion of weak compactness

Here, we identify the class of r.i. spaces satisfying Dunford-Pettis criterion of weak compactness in terms of weak convergence and convergence in measure. Moreover, we show that this class can be characterized also via suitable properties of normalized disjoint sequences (i.e., sequences of pairwise disjoint functions).

Recall that a Banach lattice E has the *positive Schur property* if every weakly null sequence $\{x_n\}$ of positive vectors in E is norm convergent, see [**29**]–[**31**]. It follows from, e.g., [**22**, Corollary 2.3.5] that it suffices to verify the latter condition only for disjoint sequences.

Further, we shall make use of the following simple result.

PROPOSITION 3.1. *If an r.i. space X has the positive Schur property, then X is separable.*

PROOF. If X is non-separable, then there exists a sequence of pairwise disjoint normalized functions $\{x_n\}_{n=1}^{\infty}$, which is equivalent in X to the canonical unit vector basic sequence in l_∞ [**19**, 1.a.7]. Observe that

$$l_\infty^* = \pi(l_1) \oplus \pi(l_1)^\perp,$$

where $\pi(l_1)$ is the subspace of all order continuous functionals on l_∞ and $\pi(l_1)^\perp$ is the subspace of singular functionals (and so $x^*(x_n) = 0$ for every $x^* \in \pi(l_1)^\perp$ and all $n \in \mathbb{N}$) [**13**, Theorem 10.3.6]. Therefore, $x_n \overset{w}{\to} 0$ in X and so X fails to have the positive Schur property, □

REMARK 3.2. Note that an r.i. space X has the positive Schur property if and only if $X \in (Wm)$ (see the Introduction). Clearly, it suffices to check the implication: if X has the positive Schur property, then $X \in (Wm)$. Indeed, let $\{x_n\} \subset X$, $x_n \overset{w}{\to} 0$ in X and $x_n \overset{\mu}{\to} 0$. Assuming the contrary, we have $\|x_{n_k}\|_X \geq \varepsilon_0$ for some subsequence $\{x_{n_k}\}$ and $\varepsilon_0 > 0$. By the assumption and Proposition 3.1, X is separable. Then, according to the classical Kadec-Pełczyński alternative (see e.g. [**12**], [**20**, Proposition 1.c.8] and [**2**, Lemma 5.2.1]), there is a further almost disjoint subsequence $\{x_{n_{k_j}}\}$, i.e., such that for some disjoint sequence $\{u_j\}$ in X we have

$$\|x_{n_{k_j}} - u_j\|_X \to 0.$$

Clearly, $\{u_j\}$ is a seminormalized sequence in X (i.e., $C^{-1} \leq \|u_j\|_X \leq C$ for some $C > 0$ and all $j \in \mathbb{N}$) and $u_j \overset{w}{\to} 0$ in X, which contradicts the fact that X has the positive Schur property.

THEOREM 3.3. *An r.i. space X has the positive Schur property if and only if X satisfies Dunford-Pettis criterion of weak compactness, i.e., for each relatively weakly compact set $K \subset X$ we have*

$$\lim_{\delta\to 0} \sup_{m(E)<\delta} \sup_{x\in K} \|x\chi_E\|_X = 0. \tag{3.1}$$

PROOF. First, assume that condition (3.1) holds for each relatively weakly compact set $K \subset X$. Let $\{x_n\}_{n=1}^{\infty} \subset X$ be a sequence of pairwise disjoint functions such that $x_n \overset{w}{\to} 0$. Since the set $\{x_n\}_{n=1}^{\infty}$ is relatively weakly compact and $m(supp\, x_n) \to 0$, it follows from the assumption that

$$\|x_n\|_X = \|x_n \chi_{supp\, x_n}\|_X \to 0.$$

To prove the converse assertion, assume that X has the positive Schur property and that there exists a relatively weakly compact set $K \subset X$, for which (3.1) does not hold. Due to Proposition 3.1 the space X is separable, and so, using standard arguments, we can find $\eta > 0$, $\{\tilde{x}_n\}_{n=1}^\infty \subset K$ and a sequence of pairwise disjoint sets $\{E_n\}_{n=1}^\infty \subset [0,1]$, $n = 1, 2, \dots$, such that

$$\|\tilde{x}_n \chi_{E_n}\|_X \geq \eta, \quad n = 1, 2, \dots . \tag{3.2}$$

Clearly, the functions $y_n := \tilde{x}_n \chi_{E_n}$ are pairwise disjoint. Moreover, from the Eberlein-Šmulian Theorem (e.g. [**2**, Theorem 1.6.3]) and [**4**, Theorem 4.34] it follows that the sequence $\{y_n\}_{n=1}^\infty$ is weakly null in X. Therefore, by the assumption, $\|y_n\|_X \to 0$. Since this contradicts (3.2), the proof is complete. □

A close result to the next theorem was proved in [**10**] (see Proposition 4.9).

THEOREM 3.4. *An r.i. space X on $[0,1]$ satisfies Dunford-Pettis criterion of weak compactness if and only if each normalized sequence of pairwise disjoint functions from X contains a subsequence equivalent to the unit vector basis in l_1.*

PROOF. Assume that $X \notin (WDP)$. Then, from Theorem 3.3 it follows that there exists a normalized sequence of pairwise disjoint functions $\{x_n\}$ in X such that $x_n \overset{w}{\to} 0$ in X. It is obvious that $\{x_n\}$ has no subsequences equivalent to the unit vector basis of l_1.

Conversely, let X satisfy the Dunford-Pettis criterion of weak compactness. Our argument will be based on applying Rosenthal's l_1 Theorem (see e.g. [**25**] or [**2**, Theorem 10.2.1]). Take a normalized disjoint sequence $\{x_n\}$ in X and suppose that it contains a weakly Cauchy subsequence $\{x_{n_k}\}$. Since X has the positive Schur property (see Theorem 3.3), X does not contain any subspace isomorphic to c_0 and, therefore, X is weakly sequentially complete (see e.g. [**20**, Theorem 1.c.4]). Hence, $\{x_{n_k}\}$ converges weakly and so, by [**2**, Lemma 1.6.1], it is weakly null. As a result we get that $\{x_{n_k}\}$ is norm null, which is a contradiction. This shows that $\{x_n\}$ has no weakly Cauchy subsequences and, therefore, by Rosenthal's l_1 Theorem, $\{x_n\}$ has a subsequence equivalent to the unit vector basis of l_1. The proof is completed. □

As a consequence we obtain the following identification of Lorentz spaces and separable parts of Marcinkiewicz spaces satisfying Dunford-Pettis criterion of weak compactness (for Lorentz spaces this result, for the first time, was proved in [**27**]).

COROLLARY 3.5. *Let φ be an increasing concave function on $[0,1]$.*

(a) The Lorentz space $\Lambda(\varphi) \in (WDP)$ if and only if $\lim_{s\to 0}\varphi(s) = 0$ (or equivalently $\Lambda(\varphi) \neq L_\infty$);

(b) The only space $M_0(\varphi)$ satisfying Dunford-Pettis criterion of weak compactness is L_1, i.e., when $\lim_{s\to 0}\varphi(s) > 0$.

PROOF. (*a*) It is sufficient to combine Theorem 3.4 with the fact that, if $\lim_{s\to 0}\varphi(s) = 0$, each sequence of pairwise disjoint normalized functions from $\Lambda(\varphi)$ contains a subsequence equivalent to the unit vector basis of l_1 [**28**, Proposition 1].

(*b*) Similarly, since every sequence of pairwise disjoint normalized functions from $M(\varphi)$ contains a subsequence equivalent to the unit vector basis of c_0 provided that $\lim_{s\to 0}\varphi(s) = 0$ [**28**, Proposition 1]. On the other hand, if $\lim_{s\to 0}\varphi(s) > 0$, we have $M_0(\varphi) = L_1$ and we can apply Theorem 1.1. □

4. Orlicz spaces satisfying Dunford-Pettis criterion of weak compactness

The following result characterizes Orlicz spaces satisfying the Dunford-Pettis criterion of weak compactness.

THEOREM 4.1. [**5**] *Let F be an increasing convex function on $[0,\infty)$ such that $F(0)=0$. The Orlicz space $L_F \in (WDP)$ if and only if either*

(i) $L_F = L_1$, *or*

(ii) *the complementary function G to F satisfies*

$$\lim_{t\to+\infty} \frac{G(Ct)}{G(t)} = \infty, \tag{4.1}$$

for some $C>0$.

PROOF. At first, we show that from (ii) it follows that $L_F \in (WDP)$ (by Theorem 1.1, this holds for L_1). To this end, according to Theorem 3.3 it suffices to prove that each weakly null sequence $\{x_n\} \subset L_F$ of pairwise disjoint nonnegative functions is norm null in L_F. To the contrary, passing to a subsequence, we will have $\|x_n\|_{L_F} \geq \varepsilon_0$ for some $\varepsilon_0 > 0$.

Next, let $E_n := supp\, x_n$, $n=1,2,\dots$ Then, since $(L_F)^\times = L_G$ [**14**, Ch. 2, § 14] (see also [**14**, Ch. 2, § 9, inequality (9.20)]), for each $n=1,2,\dots$ we can find a function $y_n \in L_G$, $y_n \geq 0$, such that $supp\, y_n \subset E_n$,

$$\int_0^1 G(y_n)\, ds = 1,$$

and

$$\int_0^1 x_n y_n\, ds \geq \varepsilon_0, \quad n=1,2,\dots \tag{4.2}$$

Let us show that

$$\lim_{n\to\infty} \int_0^1 G(y_n/C)\, ds = 0. \tag{4.3}$$

Fixing $\varepsilon>0$, by (4.1), we can choose $t_0>0$ such that

$$\frac{G(t/C)}{G(t)} \leq \frac{\varepsilon}{2}$$

whenever $t \geq t_0$. Since $m(E_n) \to 0$ as $n\to\infty$, there is a positive integer N such that for all $n \geq N$ we have

$$m(E_n) \leq \frac{\varepsilon}{2G(t_0/C)}.$$

Thus, for every $n \geq N$ we get

$$\begin{aligned}\int_0^1 G(y_n/C)\, ds &= \int_{\{y_n<t_0\}} G(y_n/C)\, ds + \int_{\{y_n\geq t_0\}} G(y_n/C)\, ds \\ &\leq G(t_0/C)m(E_n) + \frac{\varepsilon}{2}\int_0^1 G(y_n)\, ds \leq \varepsilon,\end{aligned}$$

and (4.3) is proved.

Thanks to (4.3), we can choose a subsequence $\{y_{n_k}\}$ so that

$$\sum_{k=1}^{\infty}\int_0^1 G(y_{n_k}/C)\,ds<\infty.$$

Then the function $y:=\sum_{k=1}^{\infty}y_{n_k}$ is well-defined and $y\in L_G$. Since $\{x_n\}$ is weakly null, we must have

$$\int_0^1 x_{n_k}y\,ds=\int_0^1 x_{n_k}y_{n_k}\,ds\to 0\ \text{ as }\ k\to\infty,$$

which contradicts (4.2).

For the converse implication, it suffices to show that the conditions $L_F\in(WDP)$ and $L_F\neq L_1$ imply (4.1). Firstly, by Theorem 3.3, L_F has the positive Schur property. Moreover, since $L_F\neq L_1$, we may assume that F is an N-function, i.e.,

$$\lim_{t\to 0+}\frac{F(t)}{t}=\lim_{t\to+\infty}\frac{t}{F(t)}=0.$$

This and separability of L_F (see Proposition 3.1) allows us to employ the Ando's criterion of relative weak compactness in Orlicz spaces [**24**, p. 144]. By that criterion, the set $\{x_n\}_{n=1}^{\infty}\subset L_F$ is relatively weakly compact if and only if

$$\lim_{u\to 0}\sup_{n\in\mathbb{N}}\frac{1}{u}\int_0^1 F(u|x_n(t)|)\,dt=0. \tag{4.4}$$

The plan of the proof is the following. Assuming that (4.1) does not hold, we shall find a sequence of normalized pairwise disjoint functions $\{x_n\}_{n=1}^{\infty}\subset L_F$, satisfying condition (4.4). The latter combined together with [**2**, Lemma 1.6.1] would immediately imply that $\{x_n\}$ is weakly null and hence L_F fails to have the positive Schur property, which contradicts our assumption.

If (4.1) does not hold, then for any $C>0$ there exists $M=M(C)>0$ such that

$$G(Ct_k)\le MG(t_k),\quad k=1,2,\dots, \tag{4.5}$$

for some $t_k\to\infty$. Passing to the inverse function, we obtain

$$CG^{-1}(\tau_k)\le G^{-1}(M\tau_k),\quad k=1,2,\dots, \tag{4.6}$$

where $\tau_k=G(t_k)\to\infty$. Recall that the fundamental function ϕ_{L_F} of the Orlicz space L_F is given by

$$\phi_{L_F}(t)=\|\chi_{(0,t)}\|_{L_F}=\frac{1}{F^{-1}(1/t)},\quad t>0$$

[**14**, Ch2, § 9, p. 72]. Therefore, since $(L_F)^{\times}=L_G$ [**14**, Ch. 2, § 14] and

$$\phi_X(t)\cdot\phi_{X^{\times}}(t)=t,\ \ 0\le t\le 1$$

[**15**, Ch. 2,§ 4, p. 144], we obtain

$$G^{-1}(\tau)\cdot F^{-1}(\tau)=\tau$$

for all $\tau\ge 1$. Combining this together with (4.6), we infer that for any $C>0$ there exists $M=M(C)>0$ such that

$$F(v_k)\ge MF\Big(\frac{Cv_k}{M}\Big),\quad k=1,2,\dots,$$

for some $v_k \uparrow \infty$ as $k \to \infty$. The latter guarantees that for every $n \geq 1$ there exists $M_n > 0$ and $v_k^n \uparrow \infty$ as $k \to \infty$ such that

$$F(v_k^n) \geq M_n F\Big(\frac{nv_k^n}{M_n}\Big), \quad n, k = 1, 2, \dots . \tag{4.7}$$

Clearly, we can assume that $M_n \uparrow \infty$ as $n \to \infty$ (see (4.5)). Moreover, v_k^n may be chosen in such a way that $v_k^n \uparrow \infty$ as $k \to \infty$ and that

$$\alpha_k^n := \frac{1}{F(v_k^n)}$$

satisfy the following condition:

$$\sum_{k=1}^{\infty} \alpha_k^n \leq 2^{-n}, \quad n = 1, 2 \dots . \tag{4.8}$$

Since the function $F(t)/t$ is increasing, from (4.7) it follows that for all $u \in (0, n/M_n)$

$$\frac{\alpha_k^n}{u} F(uF^{-1}(1/\alpha_k^n)) \leq \frac{\alpha_k^n}{n} M_n F(\frac{n}{M_n} v_k^n) \leq \frac{1}{n}, \quad n, k = 1, 2, \dots .$$

Thus, putting $\beta_k := \alpha_k^k$, $k = 1, 2, \dots$, we get

$$\frac{\beta_k}{u} F(uF^{-1}(1/\beta_k)) \leq \frac{1}{n}, \quad 0 < u < \frac{n}{M_n}, \quad k \geq n. \tag{4.9}$$

By (4.8), we have $\sum_{n=1}^{\infty} \beta_n \leq 1$, and so there exist pairwise disjoint sets $E_n \subset [0, 1]$, such that $m(E_n) = \beta_n$, $n \in \mathbb{N}$.

Now, setting $x_n := F^{-1}(1/\beta_n)\chi_{E_n}$, $n = 1, 2, \dots$, we see that x_n are pairwise disjoint and $\|x_n\|_{L_F} = 1$. Moreover, let us observe that condition (4.4) may be rewritten for the chosen x_n, $n = 1, 2, \dots$, as follows

$$\lim_{u \to 0} \sup_{n=1,2,\dots} \frac{\beta_n}{u} F(uF^{-1}(\beta_n^{-1})) = 0. \tag{4.10}$$

For any $\varepsilon > 0$, let $n \in \mathbb{N}$ be such that $n > 1/\varepsilon$. Since $F(t)/t \to 0$ as $t \to 0$, there exists $\delta > 0$ such that for any $u \in (0, \delta)$ we have

$$\frac{\beta_k}{u} F(uF^{-1}(\beta_k^{-1})) < \varepsilon, \quad k = 1, 2, \dots, n-1.$$

Due to (4.9) and also to the choice n, we see that the latter inequality also holds for $k \geq n$, whenever $0 < u < n/M_n$. Thus, the sequence $\{x_n\}_{n=1}^{\infty}$ satisfies (4.10) and hence condition (4.4) of Ando's criterion of relative weak compactness. This completes the proof. □

REMARK 4.2. If an Orlicz space L_F coincides with some Lorentz space Λ_ψ, then, by Corollary 3.5, L_F satisfies the Dunford-Pettis criterion of weak compactness. Note that the converse implication fails. In other words, as the following example shows, there exist an Orlicz space L_F, whose norm is not equivalent to the norm of any Lorentz space, but the complementary function G to the function F satisfies (4.1) and so $L_F \in (WDP)$.

EXAMPLE 4.3. Let $G(t) = t^{\ln \ln \ln t}$ for sufficiently large $t > 0$. An immediate computation shows that $G'(t) \geq 0$ and $G''(t) \geq 0$, if t is large enough. Consequently, we may assume that $G(t)$ is equivalent to an increasing convex function on $[0, \infty)$ such that $G(0) = 0$. It is easy to check that this function satisfies condition (4.1) and therefore, by Theorem 4.1, the Orlicz space L_F, where F is the complementary function to G, satisfies the Dunford-Pettis criterion of weak compactness.

Let us show that L_F does not coincide with any Lorentz space. To this end, by the well-known Lorentz result [**21**, Theorem 2], it is sufficient to check that

$$\int_{G^{-1}(1)}^{\infty} \frac{G(ct)}{G(t)^2}\, dG(t) = \infty \tag{4.11}$$

for each $c > 0$.

Since for sufficiently large $t > 0$

$$G'(t) = G(t)\frac{1}{t}\Big(\ln\ln\ln t + \frac{1}{\ln\ln t}\Big),$$

we have

$$\frac{G'(t)}{G(t)^2} = \frac{1}{t \cdot t^{\ln\ln\ln t}}\Big(\ln\ln\ln t + \frac{1}{\ln\ln t}\Big),$$

and, therefore, condition (4.11) is equivalent to the following:

$$\int_{G^{-1}(1)}^{\infty} \frac{1}{t} \cdot \frac{t^{\ln\ln\ln(ct)}}{t^{\ln\ln\ln t}} c^{\ln\ln\ln(ct)}\Big(\ln\ln\ln t + \frac{1}{\ln\ln t}\Big)\, dt = \infty.$$

Thus, it suffices to establish that

$$\int_{G^{-1}(1)}^{\infty} \frac{1}{t} \cdot \Big(\frac{1}{t}\Big)^{\ln\ln\ln t - \ln\ln\ln(ct)} c^{\ln\ln\ln(ct)}\, dt = \infty \tag{4.12}$$

for each $c > 0$.

We may (and will) assume that $c < 1$. Then,

$$c^{\ln\ln\ln(ct)} \geq c^{\ln\ln\ln t} = e^{\ln\ln\ln t \cdot \ln c} = (\ln\ln t)^{\ln c}. \tag{4.13}$$

Next, we have

$$\begin{aligned}
\ln\ln\ln t - \ln\ln\ln(ct) &= \ln\Big(\frac{\ln\ln t}{\ln\ln(ct)}\Big) = \ln\Big(\frac{\ln\ln t}{\ln(\ln t + \ln c)}\Big) \\
&= \ln\Bigg(\frac{\ln\ln t}{\ln\ln t + \ln(1 + \frac{\ln c}{\ln t})}\Bigg) \\
&= -\ln\Bigg(1 + \frac{\ln(1 + \frac{\ln c}{\ln t})}{\ln\ln t}\Bigg).
\end{aligned}$$

Since $\ln(1-u) \geq -2u$ if $0 \leq u \leq 1/2$, then for all sufficiently large $t > 0$ we have

$$\ln\Bigg(1 + \frac{\ln(1 + \frac{\ln c}{\ln t})}{\ln\ln t}\Bigg) \geq \ln\Big(1 + \frac{2\ln c}{\ln t \ln\ln t}\Big) \geq \frac{4\ln c}{\ln t \ln\ln t},$$

and hence

$$\Big(\frac{1}{t}\Big)^{\ln\ln\ln t - \ln\ln\ln(ct)} \geq \Big(\frac{1}{t}\Big)^{-\frac{4\ln c}{\ln t \ln\ln t}} = e^{\frac{4\ln c}{\ln\ln t}} \geq \frac{1}{2}$$

if $t > 0$ is large enough. This estimate combined with inequality (4.13) yields (4.12). Thus, (4.11) holds for every $c > 0$ and so L_F does not coincide with any Lorentz space.

5. The martingale difference Cesaro mean property in r.i. spaces

Let $\{\Sigma_n\}$ be a non-decreasing sequence of sub-σ-algebras of the σ-algebra of Lebesgue measurable sets in $[0,1]$. If, for each n, d_n is a Σ_n-measurable function such that the conditional expectation $\mathbb{E}(d_n|\Sigma_{n-1})$ with respect to Σ_{n-1} equals zero, the sequence $\{d_n\}_{n=1}^\infty$ is said to be a *martingale difference sequence (m.d.s.) adapted to* $\{\Sigma_n\}$. If every d_n belongs to the r.i. space X on $[0,1]$, we say that $\{d_n\}$ is a m.d.s. in X. Recall (see the Introduction) that an r.i. space X has the (MDCMP)-property if for every weakly null m.d.s. $\{d_n\}_{n=1}^\infty$ in X we have

$$\lim_{n\to\infty}\frac{1}{n}\|\sum_{k=1}^n d_k\|_X = 0.$$

In the paper [**11**] it was proved that $L_p = L_p[0,1]$ has the (MDCMP)-property for every $1 \le p < \infty$ (however, in the case $p > 1$ the proof of this fact is essentially contained in the classical Banach and Saks' paper [**7**]). Next, we restrict ourselves to the case $p = 1$.

THEOREM 5.1. [**11**]. *Let $\{f_n\}$ be an equi-integrable m.d.s. in L_1, $\|f_n\|_1 \le 1$. Then we have*

$$\lim_{n\to\infty}\frac{1}{n}\Big\|\sum_{k=1}^n f_k\Big\|_1 = 0.$$

Therefore, L_1 has the (MDCMP)-property.

PROOF. Suppose that $\{f_n\}$ is adapted to the sequence $\{\Sigma_n\}$. Let $\varepsilon > 0$ be arbitrary and let M be a constant, which will be fixed later. We set $g_n := f_n\chi_{\{|f_n|\le M\}}$ $(n = 1,2,\dots)$, $h_1 := g_1$, and $h_n := \mathbb{E}(g_n|\Sigma_{n-1})$ for $n > 1$. Then h_n is a Σ_{n-1}-measurable function if $n > 1$, $\{g_n - h_n\}$ is a m.d.s. adapted to $\{\Sigma_n\}$ and $\|g_n - h_n\|_\infty \le 2M$.

Denoting

$$S_n := \sum_{k=1}^n f_k,\ S_n' := \sum_{k=1}^n (f_k - g_k),\ S_n'' := \sum_{k=1}^n (g_k - h_k),\ S_n''' := \sum_{k=1}^n h_k,$$

we have

$$\begin{aligned}\|S_n\|_1 &\le \|S_n'\|_1 + \|S_n''\|_1 + \|S_n'''\|_1\\ &\le \sum_{k=1}^n \|f_k\chi_{\{|f_k|>M\}}\|_1 + \|S_n''\|_2 + \sum_{k=1}^n \|h_k\|_1.\end{aligned} \tag{5.1}$$

Let us estimate each of summands from the right-hand side of this inequality separately.

First, since the m.d.s. $\{g_n - h_n\}$ is an orthogonal sequence in L_2, we have

$$\|S_n''\|_2^2 = \sum_{k=1}^n \|g_k - h_k\|_2^2 \le 4nM^2.$$

Next, as $\{f_n\}$ is equi-integrable, we can find $\delta > 0$ such that for any measurable set $A \subset [0,1]$ with $m(A) \le \delta$ we have

$$\int_A |f_n(x)|\,dx \le \frac{\varepsilon}{3}$$

for each n. Then, fixing M such that $M > \delta^{-1}$, we obtain

$$m\{|f_n| > M\} \le M^{-1}\|f_n\|_1 \le \delta,$$

and therefore

$$\|f_n\chi_{\{|f_n|>M\}}\|_1 \le \varepsilon/3, \quad n \in \mathbb{N}. \tag{5.2}$$

Further, let u be a Σ_{n-1}-measurable function from L_∞. Since $\{f_n\}$ is a m.d.s. adapted to $\{\Sigma_n\}$, we have

$$\int_0^1 f_n(x)u(x)\,dx = \int_0^1 \mathbb{E}(f_n u|\Sigma_{n-1})(x)\,dx = \int_0^1 u(x)\mathbb{E}(f_n|\Sigma_{n-1})(x)\,dx = 0.$$

This and the definition of conditional expectation yield

$$\begin{aligned}\int_0^1 h_n(x)u(x)\,dx &= \int_0^1 \mathbb{E}(g_n u|\Sigma_{n-1})(x)\,dx \\ &= \int_0^1 f_n(x)u(x)\chi_{\{|f_n|\le M\}}(x)\,dx \\ &= -\int_0^1 f_n(x)u(x)\chi_{\{|f_n|>M\}}(x)\,dx.\end{aligned}$$

Thus,

$$\Big|\int_0^1 h_n(x)u(x)\,dx\Big| \le \|u\|_\infty\|f_n\chi_{\{|f_n|>M\}}\|_1,$$

and combining this with (5.2) and with the fact that h_n is Σ_{n-1}-measurable, we obtain $\|h_n\|_1 \le \varepsilon/3$.

Putting all the parts together, we see that from (5.1) it follows

$$\|S_n\|_1 \le n\varepsilon/3 + 2M\sqrt{n} + n\varepsilon/3,$$

and the first assertion of the theorem is proved. The second assertion is an immediate consequence of the first one and Theorem 1.1.

□

In 2000, P. Abraham, J. Alexopoulos, and S. J. Dilworth proved that each Orlicz space from the class (∇_3) has the (MDCMP)-property [**1**]. Applying Theorem 5.1, we can easily extend the latter result to all r.i. spaces $X \in (WDP)$.

COROLLARY 5.2. [**5**] *Each r.i. space X satisfying Dunford-Pettis criterion of weak compactness has the (MDCMP)-property.*

PROOF. If $\{d_n\}_{n=1}^\infty$ is a weakly null m.d.s. in X, then it is weakly null also in L_1 and so is equi-integrable. Therefore, by Theorem 5.1, we have

$$\lim_{n\to\infty} \frac{1}{n}\Big\|\sum_{k=1}^n d_k\Big\|_1 = 0.$$

This implies that $\frac{1}{n}\sum_{k=1}^n d_k \to 0$ in measure. Since the sequence $\{\frac{1}{n}\sum_{k=1}^n d_k\}_{n=1}^\infty$ is weakly null in X and X has the (Wm)-property (see Theorem 3.3 and Remark 3.2), we obtain

$$\lim_{n\to\infty} \frac{1}{n}\Big\|\sum_{k=1}^n d_k\Big\|_X = 0.$$

□

6. Reflexive subspaces of r.i. spaces satisfying Dunford-Pettis criterion of weak compactness

In 2008, E. Lavergne proved that every Orlicz space L_F from the class (∇_3) has the following property: each reflexive subspace of L_F is embedded strongly into L_F (i.e., the L_F-convergence in this subspace coincides with the convergence in measure) [**17**]. In [**6**], this result was extended (with a different proof) to all r.i. spaces $X \in (WDP)$.

THEOREM 6.1. [**6**] *If an r.i. space X satisfies Dunford-Pettis criterion of weak compactness, then each reflexive subspace of X is embedded strongly into X.*

The proof of this theorem is based on using the following result, which was obtained also in [**6**].

PROPOSITION 6.2. *Let X be an r.i. space on $[0,1]$. The following conditions are equivalent:*

(a) every reflexive subspace of the r.i. space X is embedded strongly into X;

(b) for each almost disjoint sequence $\{u_n\}_{n=1}^{\infty} \subset X$, $\|u_n\|_X = 1$, $n = 1, 2, \ldots$, the closed linear span $[u_n]$ is a non-reflexive subspace of the space X.

PROOF OF THEOREM 6.1. Assuming the contrary, by Proposition 6.2, we can find an almost disjoint sequence $\{u_n\}_{n=1}^{\infty} \subset X$ such that $\|u_n\|_X = 1$, $n = 1, 2, \ldots$, and $[u_n]$ is a reflexive space. So, we have $\|u_n - v_n\|_X \to 0$ as $n \to \infty$ for some disjoint sequence $\{v_n\}$. Further, in view of the principle of small perturbations [**2**, Theorem 1.3.9] the sequences $\{u_n\}$ and $\{v_n\}$ contain subsequences (we denote them still by $\{u_n\}$ and $\{v_n\}$) equivalent in X. Therefore, the subspace $[v_n]$ is reflexive together with $[u_n]$. Moreover, for some $C > 0$ we have

$$C^{-1} \le \|v_n\|_X \le C, \quad n = 1, 2, \ldots \tag{6.1}$$

Hence, there exists a subsequence $\{v_{n_k}\} \subset \{v_n\}$ such that $v_{n_k} \to v$ weakly for some function $v \in X$. From [**2**, Lemma 1.6.1] it follows that $v = 0$, i.e., $v_{n_k} \overset{w}{\to} 0$. Thus, by the assumption and Theorem 3.3, we infer that $\|v_{n_k}\|_X \to 0$. Since this contradicts (6.1), the proof is completed. □

In conclusion, note that the converse assertion to Theorem 6.1 fails for a wide class of separable r.i. spaces. More precisely, in [**6**] the following result is proved.

THEOREM 6.3. *Let $\psi(t)$ be a concave increasing function on $[0,1]$ such that $\psi(t) \ge ct \ln^{1/2}(e/t)$ for some $c > 0$ and all $0 < t \le 1$.*

Then, each reflexive subspace of the space $M_0(\psi)$ is embedded strongly into this space but $M_0(\psi)$ does not satisfy Dunford-Pettis criterion of weak compactness.

References

[1] P. Abraham, J. Alexopoulos, and S. J. Dilworth, *On the convergence in mean of martingale difference sequences*, Quaest. Math. **23** (2000), no. 2, 193–202, DOI 10.2989/16073600009485968. MR1795736

[2] F. Albiac and N. J. Kalton, *Topics in Banach space theory*, Graduate Texts in Mathematics, vol. 233, Springer, New York, 2006. MR2192298

[3] J. Alexopoulos, *de la Vallée Poussin's theorem and weakly compact sets in Orlicz spaces*, Quaestiones Math. **17** (1994), no. 2, 231–248. MR1281594

[4] C. D. Aliprantis, O. Burkinshaw, *Positive operators,* Springer, Dodrecht, 2006.

[5] S. V. Astashkin, N. Kalton, and F. A. Sukochev, *Cesaro mean convergence of martingale differences in rearrangement invariant spaces*, Positivity **12** (2008), no. 3, 387–406, DOI 10.1007/s11117-007-2146-y. MR2421142

[6] S. V. Astashkin and S. I. Strakhov, *On symmetric spaces with convergence in measure on reflexive subspaces*, Izvesiya vuzov. Matematika 2018, no. 8, 3–11 (In Russian); English transl. in Russian Mathematics **62** (2018), no. 8, 1–8.

[7] S. Banach and S. Saks, *Sur la convergence forte dans les champs L^p*, Studia Math. **2** (1930), 51–57.

[8] J. Diestel, *Sequences and series in Banach spaces*, Graduate Texts in Mathematics, vol. 92, Springer-Verlag, New York, 1984. MR737004

[9] N. Dunford and B. J. Pettis, *Linear operations on summable functions*, Trans. Amer. Math. Soc. **47** (1940), 323–392, DOI 10.2307/1989960. MR0002020

[10] J. Flores, F. L. Hernández, E. Spinu, P. Tradacete, and V. G. Troitsky, *Disjointly homogeneous Banach lattices: duality and complementation*, J. Funct. Anal. **266** (2014), no. 9, 5858–5885, DOI 10.1016/j.jfa.2013.12.024. MR3182963

[11] F. J. Freniche, *Cesàro convergence of martingale difference sequences and the Banach-Saks and Szlenk theorems*, Proc. Amer. Math. Soc. **103** (1988), no. 1, 234–236, DOI 10.2307/2047557. MR938674

[12] M. I. Kadec and A. Pełczyński, *Bases, lacunary sequences and complemented subspaces in the spaces L_p*, Studia Math. **21** (1961/1962), 161–176, DOI 10.4064/sm-21-2-161-176. MR0152879

[13] L. V. Kantorovich and G. P. Akilov, *Functional analysis*, 2nd ed., Pergamon Press, Oxford-Elmsford, N.Y., 1982. Translated from the Russian by Howard L. Silcock. MR664597

[14] M. A. Krasnosel′skiĭ and Ja. B. Rutickiĭ, *Convex functions and Orlicz spaces*, Translated from the first Russian edition by Leo F. Boron, P. Noordhoff Ltd., Groningen, 1961. MR0126722

[15] S. G. Kreĭn, Yu. Ī. Petunīn, and E. M. Semënov, *Interpolation of linear operators*, Translations of Mathematical Monographs, vol. 54, American Mathematical Society, Providence, R.I., 1982. Translated from the Russian by J. Szűcs. MR649411

[16] A. V. Krygin, E. M. Sheremet'ev, and F. A. Sukochev, *Conjugation of weak and measure convergences in noncommutative symmetric spaces*, Dokl. AN UzSSSR, 1993, no. 2, 8–9 (In Russian).

[17] E. Lavergne, *Reflexive subspaces of some Orlicz spaces*, Colloq. Math. **113** (2008), no. 2, 333–340, DOI 10.4064/cm113-2-13. MR2425092

[18] D. H. Leung, *On the weak Dunford-Pettis property*, Arch. Math. (Basel) **52** (1989), no. 4, 363–364, DOI 10.1007/BF01194411. MR998412

[19] J. Lindenstrauss and L. Tzafriri, *Classical Banach spaces. I: Sequence spaces*, Ergebnisse der Mathematik und ihrer Grenzgebiete, vol. 92, Springer-Verlag, Berlin-New York, 1977. MR0500056

[20] J. Lindenstrauss and L. Tzafriri, *Classical Banach spaces. II, Function Spaces,* Springer-Verlag, Berlin, 1979.

[21] G. G. Lorentz, *Relations between function spaces*, Proc. Amer. Math. Soc. **12** (1961), 127–132, DOI 10.2307/2034138. MR0123182

[22] P. Meyer-Nieberg, *Banach lattices*, Universitext, Springer-Verlag, Berlin, 1991. MR1128093

[23] W. Orlicz, *Über Raüme (L^M)*, Bull. Acad. Polon. Sci. Ser. A (1936), 93–107.

[24] M. M. Rao and Z. D. Ren, *Theory of Orlicz spaces*, Monographs and Textbooks in Pure and Applied Mathematics, vol. 146, Marcel Dekker, Inc., New York, 1991. MR1113700

[25] H. P. Rosenthal, *A characterization of Banach spaces containing l^1*, Proc. Nat. Acad. Sci. U.S.A. **71** (1974), 2411–2413. MR0358307

[26] W. Rudin, *Trigonometric series with gaps*, J. Math. Mech. **9** (1960), 203–227. MR0116177

[27] A. A. Sedaev, F. A. Sukochev, and V. I. Chilin, *Weak compactness in Lorentz spaces*, Uzb. Math. J., 1993, no. 1, 84–93 (In Russian).

[28] E. V. Tokarev, *Subspaces of certain symmetric spaces* (Russian), Teor. Funkciĭ Funkcional. Anal. i Priložen. **Vyp. 24** (1975), 156–161. MR0626854

[29] W. Wnuk, *A note on the positive Schur property*, Glasgow Math. J. **31** (1989), no. 2, 169–172, DOI 10.1017/S0017089500007692. MR997812

[30] W. Wnuk, *Some characterizations of Banach lattices with the Schur property*, Rev. Mat. Univ. Complut. Madrid **2** (1989), no. suppl., 217–224. Congress on Functional Analysis (Madrid, 1988). MR1057221

[31] W. Wnuk, *Banach lattices with properties of the Schur type—a survey*, Confer. Sem. Mat. Univ. Bari **249** (1993), 25 pp. MR1230964

DEPARTMENT OF MATHEMATICS, SAMARA NATIONAL RESEARCH UNIVERSITY, MOSKOVSKOYE SHOSSE 34, 443086, SAMARA, RUSSIA

Email address: astash56@mail.ru

Contemporary Mathematics
Volume **733**, 2019
https://doi.org/10.1090/conm/733/14733

A new method of extension of local maps of Banach spaces. Applications and examples

Genrich Belitskii and Victoria Rayskin

To the memory of Selim Krein-prominent scientist, distinguished teacher, outstanding organizer, and remarkable person

ABSTRACT. A known classical method of extension of smooth local maps of Banach spaces uses smooth bump functions. However, such functions are absent in the majority of infinite-dimensional Banach spaces. This is an obstacle in the development of local analysis, in particular in the questions of extending local maps onto the whole space. We suggest an approach that substitutes bump functions with special maps, which we call blid maps. It allows us to extend smooth local maps from non-smooth spaces, such as $C^q[0,1], q = 0, 1, \ldots$. As an example of applications, we show how to reconstruct a map from its derivatives at a point, for spaces possessing blid maps. We also show how blid maps can assist in finding global solutions to cohomological equations having linear transformation of argument.

1. Introduction

With the advancement of dynamical systems and analysis, the complexity of global analysis became evident. This stimulated the development of techniques for the study of local properties of a global problem. One of the methods of localization is based on the functions with bounded support. The history of applications of functions vanishing outside of a bounded set goes back to the works of S. Sobolev [**S**] on generalized functions.[1] Later, functions with bounded support were used by K.O. Friedrichs in his paper of 1944 [**F**]. His colleague, D.A. Flanders, suggested the name mollifiers. K.O. Friedrichs himself acknowledged S. Sobolev's work on mollifiers stating that: "These mollifiers were introduced by Sobolev and the author". A special type of mollifier, which is equal to 1 in the area of interest and smoothly vanishes outside of a bigger set, we call a bump function.

Bump functions are ubiquitous in all areas of analysis. We are particularly interested in their applications to studying local properties of dynamical systems in R^n (cf. e.g. Z. Nitecki [**N**], S. Sternberg [**St**]). J. Palis in his work [**P**] considers bump functions in Banach spaces. He proves the existence of Lipschitz-continuous

2010 *Mathematics Subject Classification.* Primary 26E15; Secondary 46B07, 58Bxx.
The authors thank the referees for helpful suggestions.

[1]S. Sobolev was a student of N.M. Guenter, and this work was probably influenced by N.M. Guenter. Professor N. Guenter was accused in the development of "abstract" science at the time when the USSR was desperate for theories applicable to manufacturing atomic weapons. N.M. Guenter was forced to resign from his job.

extensions of local maps with the help of Lipschitz-continuous bump functions. However, Z. Nitecki points out that generally speaking, the smoothness of these extensions may not be higher than Lipschitz.

This is an obstacle in the local analysis of dynamical systems in infinite-dimensional spaces. The majority of infinite-dimensional Banach spaces do not have smooth bump functions. In the works of the authors [**B**], [**B-R**], [**R**], the conditions of a local C^∞-conjugation between C^∞ diffeomorphisms on some Banach spaces are discussed. To construct the conjugation, we use bounded smooth locally identical maps. We call them blid maps. Blid maps are the maps that substitute bump functions and allow localization on Banach spaces.

The main objectives of this paper are to present blid maps (Section 2) and to introduce the questions of existence of smooth blid maps on various infinite dimensional Banach spaces and their subsets (Sections 3, 5). One of the important questions that arises in this topic is the following: Which infinite dimensional spaces possess smooth blid maps?

As an example of an application, for spaces possessing blid maps, we prove an infinite-dimensional version of the Borel lemma on a reconstruction of C^∞ map from the derivatives at a point (Section 4.1). We also show how blid maps assist in the proof of decomposition lemmas, frequently used in local analysis (Section 4.2). Finally, in Section 4.3 we apply blid maps to the investigation of cohomological equations with a linear transformation of an argument, which frequently arise in the normal forms theory.

We discuss a possibility of extension of a map $f : U \to Y$, where U is a neighborhood of a point z in a space X. More precisely, does there exist a mapping defined on the entire X, which coincides with f in some (smaller) neighborhood? Because we do not specify this neighborhood, there arises the following notion of a germ at a point.

Let X be a real Banach space, Y be either a real or complex one, and $z \in X$ be a point. Two maps f_1 and f_2 from neighborhoods U_1 and U_2 of the point z into Y are called *equivalent* if there is a neighborhood $V \subset U_1 \cap U_2$ of z such that both of the maps coincide on V. A *germ at* z is an equivalence class. Therefore, every local map f from a neighborhood of z into Y defines a germ at z. Sometimes in the literature it is denoted by $[f]$, although in general we will use the same notation, f, as for the map.

We consider Fréchet C^q-maps with $q = 0, 1, 2..., \infty$. All notions and notations of differential calculus in Banach spaces we borrow from H. Cartan [**C**].

For a given C^q-germ f at z, we pose the following questions.

QUESTION 1.1. Does its global C^q-representative (i.e., a C^q-map defined on the whole X) exist?

QUESTION 1.2. Assume that f has local representatives with bounded derivatives. Does there exist a global one with the same property?

It was shown in the works of S. D'Alessandro and P. Hajek [**DH**] and P. Hajek and M. Johanis [**HJ**] that there exist separable Banach spaces that do not allow C^2-smooth extension of a local C^2-representative[2].

Below, without loss of generality we assume that $z = 0$.

[2]We thank reviewers of this paper for the reference to this result.

Usually for extension of local maps described in Question 1.1 and Question 1.2 bump functions are used. The classical definition of a bump function [**S**] is a non-zero bounded C^q-function from X to $\mathbb{R}$ having a bounded support. We use a similar modified definition which is more suitable for our aims. Namely, a bump function at 0 is a C^q-map $\delta_U : X \to \mathbb{R}$ which is equal to 1 in a neighborhood of 0 and vanishing outside of a lager neighborhood U. If f is a local representative of a C^q-germ defined in a neighborhood V ($\overline{U} \subset V$) then

$$F(x) = \begin{cases} \delta_U(x)f(x), & x \in U \\ 0, & x \notin U \end{cases} \tag{1.1}$$

is a global C^q-representative of the germ f, and it solves at least the Question 1.1. If, in addition, all derivatives of δ_U are bounded on the entire X, then (1.1) solves both of the Questions. If these functions do exist for any U, then every germ has a global representative.

A continuous bump function exists in any Banach space. It suffices to set $\delta(x) = \tau(||x||)$, where τ is a continuous bump function at zero on the real line. Let $p = 2n$ be an even integer. Then

$$\delta(x) = \tau(||x||^p)$$

is a C^∞-bump function at zero on l_p. Here τ is a C^∞-bump function on the real line.

However, it is proved in the work [**M**] of V.Z. Meshkov that if p is not an even integer, then l_p space does not have C^q-smooth $(q > p)$ bump functions. The Banach-Mazur theorem states that any real separable Banach space is isometrically isomorphic to a closed subspace of $C[0,1]$. Consequently, the space of $C[0,1]$ does not have smooth bump functions at all (see J. Kurzweil [**K**] or [**M**]). Following [**M**], we will say that a space is C^q-smooth, if it possesses a C^q-bump function.[3]

EXAMPLE 1.3. The real function

$$f(x) = \int_0^1 \frac{dt}{1 - x(t)}, \quad x \in C[0,1]$$

defines a C^∞ (which is even analytic) germ at zero. In spite of the absence of smooth bump functions, the germ has a global C^∞ representative. To show this, let h be a real C^∞-function on the real line such that

$$h(s) = \begin{cases} s, & |s| < 1/3 \\ 0, & |s| > 1/2. \end{cases} \tag{1.2}$$

Then the C^∞-function

$$F(x) = \int_0^1 \frac{dt}{1 - h(x(t))}$$

coincides with f in the ball $||x|| < 1/3$ and is a global representative of the germ with bounded derivatives of all orders.

[3]Usually, a space is called smooth if it satisfies a similar property related to the smoothness of a norm (see, for example, R. Fry, S. McManus [**F-M**]). However, we will adopt the definition of [**M**].

2. Blid maps

DEFINITION 2.1. A *C^q-blid map* for a Banach space X is a global **B**ounded **L**ocal **Id**entity at zero C^q-map $H : X \to X$.

In other words, H is a global representative of the germ at zero of identity map such that

$$\sup_x ||H(x)|| < \infty.$$

The existence of blid maps allows locally defined mappings to be extended to the whole space.

THEOREM 2.2. *Let a space X possesses a C^q-blid map H. Then for every Banach space Y and any C^q-germ f at zero from X to Y there exists a global C^q-representative. Moreover, if all derivatives of H are bounded, and f contains a local representative bounded together with all its derivatives, then it has a global one with the same property.*

PROOF. Let $||H(x)|| < N$ for all $x \in X$, and $H(x) = x$ for $||x|| < n$. Further, let f be a representative of a germ defined on a neighborhood U, and let a closed ball $B_\epsilon = \{x : ||x|| \leq \epsilon\} \subset U$. The map

$$H_1(x) = \frac{\epsilon}{N} H\left(\frac{N}{\epsilon} x\right) \tag{2.1}$$

is a C^q-blid map also, and its image is contained in U. Therefore the map

$$F(x) = f(H_1(x)) \tag{2.2}$$

is well-defined on the whole space X, and it coincides with the map f in the neighborhood $||x|| < \frac{\epsilon n}{N}$. The map F is a global C^q-representative of the germ f. If both of the maps H and f are bounded together with all of their derivatives, then F possesses the same property. This completes the proof. □

3. Examples

Let us present spaces having blid maps.

1. Let X be C^q-smooth, and let $\delta(x)$ be a C^q-bump function at zero. Then

   $$H(x) = \delta(x)x$$

   is a C^q-blid map. If the bump function is bounded together with all its derivatives, then H has the same property.
2. Let $X = C(M)$ be the space (a Banach algebra) of all continuous functions on a compact Hausdorff space M with

   $$||x|| = \max_t |x(t)|, \quad t \in M,$$

   and let h be a C^∞-bump function on the real line. Then the map

   $$H(x)(t) = h(x(t))x(t), \quad x \in X \tag{3.1}$$

   is C^∞-blid map with bounded derivatives of all orders.

   Indeed, since h is locally equal to 1, H is a local identity. Let ϵ be a positive real number such that $h(\tau) \equiv 0$ for all $|\tau| > \epsilon$. We can always find such ϵ, because bump functions have bounded support. Then,

   $$||H(x)(t)|| = ||h(x(t))x(t)|| \leq \epsilon.$$

Also,

$$||H'(x)(t)|| = ||h'(x(t))x(t) + h(x(t))|| \leq \epsilon \sup |h'| + 1.$$

Similarly, one can show boundedness of all higher order derivatives.

3. More generally, let $X \subset C(M)$ be a subspace such that

$$x(t) \in X \implies h(x(t))x(t) \in X \tag{3.2}$$

for any C^∞ bump function h on the real line.

Then (3.1) defines a C^∞-blid map with bounded derivatives. For example, any ideal X of the algebra satisfies (3.2).

4. Let $X = C^n[0,1]$ be the space (which is also a Banach algebra) of all C^n-functions on $[0,1]$. Then a simple generalization of (3.1) gives a C^∞-blid map with bounded derivatives. The same holds for any closed subspace $X_1 \subset C^n[0,1]$ satisfying (3.2). As above, X_1 may be an ideal of the algebra.
5. Let X possess a C^q-blid map H, and a subspace[4] X_1 of X be H-invariant. Then the restriction $H_1 = H|X_1$ is a C^q-blid map on X_1.
6. Let $\pi : X \to X$ be a bounded projector and X possess C^q-blid map H. Then the restriction $\pi(H)|Im(\pi)$ is a C^q- blid map on $Im(\pi)$, while the restriction $(H-\pi(H))|Ker(\pi)$ is a C^q-blid map on $Ker(\pi)$. Consequently, if $X_1 \subset X$ is a subspace, such that there exists another subspace of X, so that these two form a complementary pair, then X_1 possesses a blid map.

COROLLARY 3.1. *Let a space X be as in items* 1-6. *Then for any Banach space Y and any C^q-germ at zero there is a global C^q-representative. If the germ contains a local representative with bounded derivatives, then there is a global one with the same property.*

4. Applications

4.1. The Borel lemma for Banach spaces. Let X be a linear space over a field $\Bbbk$ ($char\ \Bbbk = 0$) and Y be a linear space over a field $\Bbbk$ ($\mathbb{K} \subset \mathbb{K}$). A map $P_j : X \to Y$ is called a polynomial homogeneous map of degree j if there is a j-linear map

$$g : \underbrace{X \times X \times ... \times X}_{j} \to Y$$

such that $P_j(x) = g(x, x, ..., x)$. For the given $P_j(x)$, the map g is not unique, but there is a unique symmetric one. We will assume that g is symmetric. Then, the first derivative of P_j at a point $z \in X$, is a linear map $X \to Y$, and can be calculated by the formula

$$P_j'(z)(x) = jg(\underbrace{z, ..., z}_{j-1}, x)$$

In general, the derivative of order $n \leq j$, is a homogeneous polynomial map of degree n and equals

$$P_j^{(n)}(z)(x)^n = j(j-1)...(j-n+1)g(\underbrace{z, ..., z}_{j-n}, \underbrace{x, ...x}_{n})$$

[4]By definition, a subspace of a Banach space is always closed.

In particular,

$$P_j^{(j)}(z)(x)^j = j!g(x, ..., x) = j!P_j(x)$$

does not depend on z. And lastly, for $n > j$, $P_j^{(n)}(z) = 0$.

It follows that all derivatives of P_j at zero are zero, except for the order j.

The latter equals $P_j^{(j)}(0)(x)^j = j!P_j(x)$.

Now, let X and Y be Banach spaces. Recall that X must be real, while Y can be real or complex. Let $f : X \to Y$ be a local C^∞ map. Then

$$P_j(x) = f^{(j)}(0)(x)^j \tag{4.1}$$

for any $j = 0, 1, ...$ is a polynomial map of degree j, and it is continuous and even C^∞. Therefore, for some $c_j > 0$ we have the estimate

$$||P_j(x)|| \le c_j||x||^j, \quad x \in X. \tag{4.2}$$

QUESTION 4.1. Given a sequence $\{P_j\}_{j=0}^\infty$ of continuous polynomial maps from X to Y of degree j, does there exist a C^∞-germ $f : X \to Y$, which satisfies (4.1) for all $j = 0, 1, ...$?

The classical Borel lemma states, that given a sequence of real numbers $\{a_n\}$, there is a C^∞ function f on the real line such that $f^{(n)}(0) = a_n$. This means that the answer to the Question 4.1 is affirmative for $X = Y = \mathbb{R}$. The same is true for finite-dimensional X and Y.

THEOREM 4.2 (The Borel lemma). *Let a Banach space X possess a C^∞-blid map with bounded derivatives of all orders. Then for any Banach space Y and any sequence $\{P_j\}_{j=0}^\infty$ of continuous homogeneous polynomial maps from X to Y there is a C^∞-map $f : X \to Y$ with bounded derivatives of all orders such that* (4.1) *is satisfied for all $j = 0, 1, ...$*

PROOF. Let H be a C^∞-blid map at zero with bounded derivatives on X. Set $H_j(x) = \epsilon_j H(x/\epsilon_j)$. For a given ϵ_j the map $H_j(x)$ is a C^∞-blid map also.

Then the map $P_j(H_j(x))$ belongs to $C^\infty(X, Y)$, and all its derivatives at 0 are zero, except for the order j. The latter equals to $P_j^{(j)}(0)(x)^j = j!P_j(x)$. In addition, all derivatives of the map are bounded, and the derivative of order n allows the following estimate

$$||(P_j(H_j(x))^{(n)}|| \le \epsilon_j^{j-n} c_{j,n}$$

with constants $c_{j,n}$ depending only on the maps P_j, H, and not depending on a choice of ϵ_j. Therefore, under an appropriate choice of ϵ_j the series

$$f(x) = \sum_0^\infty \frac{1}{j!} P_j\left(H_j(x)\right)$$

converges in C^∞ topology to a map from X to Y. It is clear that

$$f^{(n)}(0)(x)^n = P_n(x).$$

This equality proves the statement. □

COROLLARY 4.3. *Let a space X be as in items* 1-6 *of Section* 3. *Then for any Banach space Y and any sequence $\{P_j\}_{j=0}^\infty$ of continuous homogeneous polynomial maps from X to Y there is a C^∞-map $f : X \to Y$ with bounded derivatives of all orders such that* (4.1) *is satisfied for all $j = 0, 1, ...$*

4.2. Decomposition lemmas. In this section we will state lemmas that are useful in local and global analysis. We will use them in Section 4.3, where we present solution of the cohomological equation.

Let X possesses a C^∞-blid map H, and can be decomposed into a sum of two subspaces,

$$X = X_+ + X_-.$$

Then there is a bounded projector $\pi_+ : X \to X$ with $Im(\pi_+) = X_+$, and the bounded projector $\pi_- = id - \pi_+$ onto $Ker(\pi_+) = X_-$. One can write

$$x = (x_+, x_-),\ x \in X,\ x_\pm \in X_\pm.$$

Denote by $H_+(x_+)$ and $H_-(x_-)$ the corresponding C^∞-blid maps on X_+ and X_-.

Let now Y be a Banach space. Recall that a C^∞-map $f : X \to Y$ is called flat on a subset S if it vanishes on S together with all its derivatives.

LEMMA 4.4. *Let all derivatives of H be bounded on X. Let a map $f_0 : X \to Y$ be bounded with all derivatives on every bounded subset, and be flat at zero. Then there is a decomposition $f_0 = f_+ + f_-$ and a neighborhood U of zero such that f_+ (f_-) has the same boundedness property and is flat on the intersection $X_+ \cap U$ ($X_- \cap U$).*

PROOF. Let $\epsilon_j < 1$ and

$$P_j(x) = \left(\frac{\partial^j}{\partial x_-^j} f_0\left(H(x)\right)_{|x_-=0} \right) \left(\epsilon_j H_-(x_-/\epsilon_j)\right)^j.$$

Then P_j are flat on X_+ in a neighborhood

$$U = \{x : H(x) = x\},$$

and

$$\left(\frac{\partial^{p+q}}{\partial x_+^p \partial x_-^q} \left(P_j(x)\right) \right) |_{x_-=0} = 0, \quad q \neq j, \tag{4.3}$$

while

$$\left(\frac{\partial^{p+q}}{\partial x_+^p \partial x_-^q} \left(P_j(x)\right) \right) |_{x_-=0} = j! \left(\frac{\partial^{p+q}}{\partial x_+^p \partial x_-^q} f_0(H(x)) \right) |_{x_-=0}, \quad q = j. \tag{4.4}$$

Additionally, the maps satisfy an estimate

$$||P_j^{(n)}(x)|| \le c_{n,j}\epsilon_j^{j-n}.$$

Therefore, the series

$$f_+(x) = \sum_0^\infty \frac{1}{j!} P_j(x)$$

converges in the C^∞-topology under an appropriate choice of ϵ_j. Its sum f_+ is a C^∞-map from $X \to Y$ flat on X_+. Equations (4.3) and (4.4) imply

$$f_+^{(n)}(x)|_{x_-=0} = \left((f_0(H(x)))^{(n)} \right) |_{x_-=0}$$

As a result, the map

$$f_-(x) = f_0(x) - f_+(x)$$

is flat on X_- when $H(x) = x$. □

LEMMA 4.5. *Let a C^∞-map $v : X \to Y$ vanish in a neighborhood of zero. Then there is a decomposition*

$$v = v_+ + v_-$$

into a sum of maps vanishing on strips $S_+(\epsilon) = \{x : ||x_+|| < \epsilon\}$ and $S_-(\epsilon) = \{x : ||x_-|| < \epsilon\}$ respectively.

PROOF. Let $v(x) = 0$ as $||x|| < \delta$. One can choose $\epsilon < \delta$ and a C^∞-blid map H to be such that

$$||H(x)|| < \delta, \quad x \in X, \quad H(x) = x \text{ as } ||x|| < \epsilon.$$

Then the C^∞-map

$$v_+(x) = v(x_+, H_-(x_-))$$

vanishes as $||x_+|| < \epsilon$, while the map

$$v_-(x) = v(x) - v_+(x)$$

vanishes as $||x_-|| < \epsilon$. □

Note that in the Lemma 4.5 we do not assume boundedness of derivatives of map H.

4.3. Cohomological equations in Banach spaces. Given a map $F : X \to X$, the equation

$$g(Fx) - g(x) = f(x) \tag{4.5}$$

will be called cohomological equation with respect to the unknown C^∞ function $g : X \to \mathbb{C}$.

Various versions of this equation are well-known and have been studied in multiple articles. Yu.I. Lyubich in the article [**L**] presents a very broad overview of this area and considers (4.5) in a non-smooth category. For a discussion of smooth cohomological equations we recommend the book [**B-T**] of G. Belitskii, V. Tkachenko.

In the present work, we consider a linear space X over a field K, and a linear map $F = A : X \to X$. These cohomological equations are often studied in the theory of normal forms.

Consider a homogeneous polynomial map $f(x) = P_n(x)$ of degree $n > 0$ and linear map $F = A$. Let us also look for a solution in a polynomial form, $g(x) = Q_n(x)$. Then we arrive at the equality

$$(L_n - id)Q_n(x) = P_n(x), \tag{4.6}$$

where $L_nQ_n(x) = Q_n(Ax)$. If the operator $L_n - id$ is invertible, then (4.6) has a solution $Q_n(x)$ for every $P_n(x)$. Otherwise, additional restrictions on $P_n(x)$ arise. If $X = K^m$ is finite-dimensional, then the invertibility of $L_n - id$ is provided by the absence of the resonance relations

$$1 = a_1^{p_1} a_2^{p_2} \dots a_m^{p_m}, \tag{4.7}$$

$a_k \in specA$, $\sum_1^m p_i = n$.

Therefore, the absence of resonance relations for all $n > 0$ ensures the solvability of (4.6) for any $n > 0$ and any P_n. If $K = R$, then the mentioned condition implies also that A is hyperbolic, i.e., its spectrum does not intersect the unit circle in $\mathbb{C}$. Consider now (4.5) with real finite-dimensional space X, and $f \in C^\infty$. If all equations (4.6) are solvable for $P_n(x) = f^{(n)}(0)(x)^n$, then (4.5) is called formally solvable at zero. It is known that if A is hyperbolic, invertible automorphism, then formal solvability implies C^∞ solvability. Our aim is to prove a similar assertion in the infinite-dimensional case.

So, let $A : X \to X$ be a continuous invertible hyperbolic linear operator. Then, there is a direct decomposition $X = X_+ + X_-$ in a sum of A-invariant subspaces such that spec $A_+ = specA|X_+$ lies inside of the circle, while spec $A_- = specA|X_-$ lies outside. Moreover, there is an equivalent norm in X such that

$$\begin{aligned} &||A_+x|| < q||x_+||, \\ &||A_-^{-1}x|| < q||x_-|| \end{aligned} \tag{4.8}$$

with some $q < 1$. If (4.5) has a solution, then $f(0) = 0$, and we assume this condition to be fulfilled.

PROPOSITION 4.6. *Let one of the subspaces X_+, X_- be trivial, then* (4.5) *has a global C^∞-solution for any C^∞-function f, $f(0) = 0$ with bounded derivatives on every bounded subset S.*

PROOF. Assume $X_- = 0$. Then the series

$$g(x) = -\sum_0^\infty f(A^k x)$$

converges in the space of C^∞-functions, since

$$||(f(A^k x))^{(n)}|| \le q^{kn} c_n(S)$$

for every bounded subset $S \subset X = X_+$.

Similarly, if $X_+ = 0$, then the series

$$g(x) = \sum_1^\infty f(A^{-k} x)$$

leads to a solution we need. □

If both of the spaces X_+, X_- are non-trivial, then the existence of solutions requires additional assumptions and constructions. As above, $P_n(x) = f^{(n)}(0)(x)^n$ is a continuous homogeneous polynomial of degree n. Let g be a solution of (4.5). Differentiating both of its sides, we arrive at equation

$$A_n Q_n(x) - Q_n(x) = P_n(x), \quad n = 1, 2, 3...., \tag{4.9n}$$

where $Q_n(x) = g^{(n)}(0)(x)^n$, and A_n is a linear map in the Banach space of continuous homogeneous polynomials. It arises after n-multiple differentiation of the function $g(Ax)$. If all equations (4.9n) have continuous solutions, then we say that (4.5) is formally solvable at zero. The latter a-priory takes place if identity does not belong to $specA_n$. The opposite condition can be considered as an infinite-dimensional version of resonance relations.

THEOREM 4.7. *Let A be a hyperbolic linear automorphism, and X possesses a C^∞-blid map with bounded derivatives on X. If all derivatives of f are bounded on every bounded subset, and* (4.5) *is formally solvable at zero (i.e. all equations* (4.9n) *have continuous solutions), then there exists a global C^∞-solution $g(x)$.*

PROOF. First, we will show that a formal solvability implies a local one, i.e., implies existence of (a global) C^∞-function g such that (4.5) holds in a neighborhood of zero.

Let $Q_n(x)$ be continuous solutions of (4.9n). We will build a C^∞-function $g_0(x)$, following the Borel lemma. The substitution $g = w + g_0$ reduces (4.5) to the equation

$$w(Ax) - w(x) = f_0(x), \tag{4.10}$$

where $f_0(x) = f(x) - g_0(Ax) + g_0(x)$ is flat at zero. Lemma 4.4 provides the decomposition

$$f_0(x) = f_+(x) + f_-(x)$$

together with a neighborhood U. Let H be a blid map. One can assume that $H(X) \subset U$. Consider the pair of equations:

$$\begin{aligned} w_+(Ax) - w_+(x) &= f_+(H(x)), \\ w_-(Ax) - w_-(x) &= f_-(H(x)). \end{aligned} \tag{4.11}$$

Then, the estimates (4.8) imply that the series

$$w_+(x) = -\sum_0^\infty f_+(H(A^k x)),\ w_-(x) = \sum_1^\infty f_-(H(A^{-k}x))$$

converge since $f_\pm$ is flat on $X_\pm$. The series present solutions to the first and second equations (4.11) correspondingly.

The (global) function $w_1(x) = w_-(x) + w_+(x)$ satisfies (4.10) in a neighborhood of the origin, i.e. it is local solution for this equation. In turn, the function $g = w_1 + g_0$ is a local solution of the initial equation (4.5).

Now we will prove that local solvability implies a global one. Let γ_0 be a local solution. The substitution $g = h + \gamma_0$ reduces (4.5) to the equation

$$h(Ax) - h(x) = v(x) \tag{4.12}$$

where $v(x) = f(x) - \gamma(Ax) + \gamma(x)$ vanishes in a neighborhood of zero. Let $v = v_+ + v_-$ be a decomposition described by Lemma 4.5. Consider the series

$$h_+(x) = -\sum_0^\infty v_+(A^k x). \tag{4.13}$$

Since $||A_+|| < 1$, for any bounded set D there is a number $k_0(D)$ such that $A^k(D) \subset S_+(\epsilon)$ for $k > k_0(D)$. Hence, if $k > k_0(D)$, then $v_+(A^k x) = 0$ for all $x \in D$, and (4.13) represents a global smooth function. By the same arguments the series

$$h_-(x) = \sum_1^\infty v_-(A^{-k}x)$$

is a global smooth function. The sum $h = h_+ + h_-$ is a smooth global solution of (4.12). This construction completes the proof. □

5. More examples and open questions

One of the important questions of extensions of local maps on Banach spaces is the following. Do Banach spaces without smooth blid maps exist? Recently, affirmative answer was presented in [**DH**] (also see [**HJ**]). The authors proved that there exist Banach spaces that do not allow C^2-extension (and hence the C^2-blid map).

QUESTION 5.1. For which spaces do smooth blid maps exist? Do they exist on l_p, with non-even p?

Some other open questions related to the generalization of Theorem 2.2 we discuss below.

How can we extend germs of maps defined at a closed subset $S \subset X$? For this construction we need to define smooth blid maps at S. More precisely, generalizing the definition of germs at a point, we will say that maps f_1 and f_2 from neighborhoods U_1 and U_2 of S into Y are *equivalent*, if they coincide in a (smaller) neighborhood of S. Every equivalence class is called a *germ at* S. We pose the same question. Given a C^q-germ at S, does there exist a global representative? Assume there exist a C^q-map $H : X \to X$ whose image $H(X)$ is contained in a neighborhood U of S and which is equal to the identity map in a smaller neighborhood. Such maps we call *smooth blid maps at* S. Then every local map f defined in U can be extended on the whole X. It suffices to set $F(x) = f(H(x))$.

In the next example, we construct the map H for a segment (in particular, for a ball).

EXAMPLE 5.2. Let $S(A)$ be a set of all functions $x \in C[0,1]$ whose graphs $(t, x(t))$ are contained in a closed $A \subset \mathbb{R}^2$, where A is chosen in such a way that $S(A) \neq \emptyset$. Let $h(t,x)$ be a C^∞-function, which is equals to 1 in a neighborhood of A and vanishes outside of a bigger set. Then, for an arbitrary $y \in C[0,1]$

$$H_y(x)(t) = y(t) + h(t, x(t))(x(t) - y(t))$$

is a C^∞-blid map for $S(A)$.

If $A = \{\{t,x\} : \min(\psi(t), \phi(t)) \leq x \leq \max(\psi(t), \phi(t))\}$ for some $\phi, \psi \in C[0,1]$, then $S(A)$ can be thought of as a segment $[\phi, \psi] \subset C[0,1]$.

In particular, given $z \in C[0,1]$ and a constant $r > 0$, setting $\phi = z - r$ and $\psi = z + r$, we obtain the ball $B_r(z) = \{x : ||x - z|| \leq r\} \subset C[1,0]$.

Every C^q-germ at $[\phi, \psi] \subset C[0,1]$ contains a global representative.

Note, this example has an obvious generalization to segments and balls in $C^k[0,1]$.

The Question 5.1 and Example 5.2 bring us to the next question.

QUESTION 5.3. For which pairs (S, X) do similar constructions exist? In particular, can a smooth blid map be constructed for any bounded subset S of a space X possessing a smooth blid map? For example, we do not know whether a smooth blid map can be constructed for a sphere $S = \{x \in C[0,1] : ||x|| = r\}$.

QUESTION 5.4. The Borel lemma for finite-dimensional spaces is a particular case of the well-known Whitney extension theorem from a closed set $S \subset \mathbb{R}^n$. What is an infinite-dimensional version of the Whitney theorem?

References

[B] G. R. Belitskiĭ, *The Sternberg theorem for a Banach space* (Russian), Funktsional. Anal. i Prilozhen. **18** (1984), no. 3, 71–72, DOI 10.1007/BF01086163. MR757253

[B-T] G. Belitskii and V. Tkachenko, *One-dimensional functional equations*, Operator Theory: Advances and Applications, vol. 144, Birkhäuser Verlag, Basel, 2003. MR1994638

[B-R] G. Belitskii, V. Rayskin, *Equivalence of families of diffeomorphisms on Banach spaces*, Math. preprint archive, UT Austin, 07-71. `https://www.ma.utexas.edu/mp_arc-bin/mpa?yn=07-71`

[C] H. Cartan, *Calcul différentiel* (French), Hermann, Paris, 1967. MR0223194

[DH] S. D'Alessandro and P. Hájek, *Polynomial algebras and smooth functions in Banach spaces*, J. Funct. Anal. **266** (2014), no. 3, 1627–1646, DOI 10.1016/j.jfa.2013.11.017. MR3146828

[F] K. O. Friedrichs, *The identity of weak and strong extensions of differential operators*, Trans. Amer. Math. Soc. **55** (1944), 132–151, DOI 10.2307/1990143. MR0009701

[F-M] R. Fry and S. McManus, *Smooth bump functions and the geometry of Banach spaces: a brief survey*, Expo. Math. **20** (2002), no. 2, 143–183, DOI 10.1016/S0723-0869(02)80017-2. MR1904712

[HJ] P. Hájek and M. Johanis, *Smooth analysis in Banach spaces*, De Gruyter Series in Nonlinear Analysis and Applications, vol. 19, De Gruyter, Berlin, 2014. MR3244144

[M] V. Z. Meshkov, *Smoothness properties in Banach spaces*, Studia Math. **63** (1978), no. 2, 111–123, DOI 10.4064/sm-63-2-111-123. MR511298

[K] J. Kurzweil, *On approximation in real Banach spaces*, Studia Math. **14** (1954), 214–231 (1955), DOI 10.4064/sm-14-2-214-231. MR0068732

[L] Yu. I. Lyubich, *The cohomological equations in nonsmooth categories*, Études opératorielles, Banach Center Publ., vol. 112, Polish Acad. Sci. Inst. Math., Warsaw, 2017, pp. 221–272, DOI 10.4064/bc112-0-13. MR3754081

[N] Z. Nitecki, *Differentiable dynamics. An introduction to the orbit structure of diffeomorphisms*, The M.I.T. Press, Cambridge, Mass.-London, 1971. MR0649788

[R] V. Rayskin, *Theorem of Sternberg-Chen modulo the central manifold for Banach spaces*, Ergodic Theory Dynam. Systems **29** (2009), no. 6, 1965–1978, DOI 10.1017/S0143385708000989. MR2563100

[P] J. Palis, *On the local structure of hyperbolic points in Banach spaces*, An. Acad. Brasil. Ci. **40** (1968), 263–266. MR0246331

[S] S. Soboleff *Sur un théorème d'analyse fonctionnelle*, Rec. Math. [Mat. Sbornik] N.S., **4(46):3** (1938), 471–497.

[St] S. Sternberg, *On the structure of local homeomorphisms of Euclidean n-space. II.*, Amer. J. Math. **80** (1958), 623–631, DOI 10.2307/2372774. MR0096854

Department of Mathematics and Computer Science, Ben Gurion University of the Negev, P.O.B. 653, Beer Sheva, 84105, Israel
Email address: genrich@cs.bgu.ac.il

Department of Mathematics, Tufts University, Medford, Massachusetts 02155-5597
Email address: victoria.rayskin@tufts.edu

Contemporary Mathematics
Volume **733**, 2019
https://doi.org/10.1090/conm/733/14734

Two consequences of the associativity condition for a hypercomplex system with locally compact basis

Yu. M. Berezansky and A. A. Kalyuzhnyi

To the dear memory of Selim Krein who helped us to become mathematicians.

ABSTRACT. Under an associativity condition for a hypercomplex system with a locally compact basis, we prove existence of a multiplicative measure. We then provide an invariant measure construction on a locally compact group.

1. Introduction

Starting in 1938 it was observed by J. Delsarte [**8**] and then B. M. Levitan [**11**] that several results in the classical harmonic analysis allow generalizations where the exponential functions $e^{i\lambda q}$, $\lambda, q \in \mathbb{R}$, are replaced with functions $\chi(q, \lambda)$ of a more general type. Namely, these are functions that satisfy the relation

$$(T_q\chi(\cdot, \lambda))(p) = \chi(p, \lambda)\chi(q, \lambda),$$

where the points p, q belong to a topological space Q, and $\{T_q\}_{q\in Q}$ is a family of "generalized shift" operators. These operators act on the space of complex-valued functions on Q and have the property that the mapping Δ:

$$(\Delta f)(p, q) := (T_q f)(p)$$

is a *coassociative comultiplication* with a right *counit*[1] ε, $\varepsilon(f); = f(e)$, where e is a uniquely fixed point in Q. In the case where Q is a locally compact group G, we have

$$(T_q f)(p) = f(pq).$$

In 1950, in a research initiated by I. M. Gelfand, it was observed that in the Delsarte-Levitan theory a basic role is played by a convolution type operation related to the family of generalized shift operators T_p, $p \in Q$. It generalizes the classical group convolution. This has lead to development of a theory of hypercomplex systems with a locally compact basis Q, where the structure constants $c_{j,k,l}$ of an ordinary hypercomplex system with finite basis were replaced with a function $c(p, q, r)$ in the variables $p, q, r \in Q$.

2010 *Mathematics Subject Classification.* Primary 43A62.

Key words and phrases. Hypercomplex system, multiplicative measure, Haar measure.

The authors express their gratitude to the reviewer and to editor for useful remarks and references.

[1]These notions will be briefly explained later on in the article.

The first article on this topic by Yu. M. Berezansky and S. G. Krein [**4**] was published in 1950. It was then extended by these authors, in a series of Russian language articles published between 1950 and 1953, to a general theory of (mainly commutative) hypercomplex systems with compact and discrete bases Q. A short survey of these results was published in 1957 [**5**].

At that time, the articles were not available in English, and the area in general seemed somewhat esoteric and did not attracted much attention. Its authors S. G. Krein and Yu. M. Berezansky have also moved to other mathematical problems. The situation started changing in of 1970s, when Ch. Dunkl [**9**], R. Spector [**13**], and R. Jewett [**10**] introduced the concept of a *hypergroup*, which was almost the same as that of a hypercomplex system with locally compact basis, and essentially rediscovered independently the results of [**4, 5**]. This has triggered a renewed interest in hypercomplex systems and hypergroups in Kyiv and some universities in Russia. A number of young (then) Kyiv mathematicians began to work in this direction in 1980s, among them A. A. Kalyuzhnyi, L. I. Vainerman, I. D. Olshanetskii, G. B. Podkolsin, and Yu. A. Chapovsky.

One can find an account of the theory of hypercomplex systems with locally compact basis in [**1, 2**], as well as in §2.2 of Section 4 of the book [**3**], where generalized functions with the basis Q were considered. For an account of the theory of hypergroups, see [**6**]. The existence of (left) Haar measures for commutative hypergroups is proven in [**14**], and for general hypergroups under some conditions in [**7**].

Let us now describe briefly the content of this paper. Existence of a multiplicative measure (see its definition in Section 2) is of fundamental importance for the theory of hypercomplex systems. A general proof of existence of such a measure is available only in the commutative case [**2**]. Theorem 3.1 of this article provides some sufficient conditions for existence of such a measure.

The second result Theorem 3.2) addresses the problem of existence of a left invariant measure on a locally compact group G. It turns out that, if one starts with an arbitrary positive measure μ on G and constructs the corresponding convolution of functions on G, then associativity of this convolution implies left invariance of the measure μ. This observation might provide a new approach to proving existence of a left invariant measure on some locally compact groups.

2. Definition of a hypercomplex system with a locally compact basis

Let Q be a locally compact space (whose points will be denoted by $p, q, r, \dots$), $\mathcal{B}(Q)$ – a σ-algebra of Borel sets, and $\mathcal{B}_0(Q)$ – a subring of $\mathcal{B}(Q)$ consisting of sets with compact closures. We shall consider *Borel measures* on $\mathcal{B}(Q)$, i.e. positive regular measures on $\mathcal{B}(Q)$, finite on compact sets. We will also denote by $C(Q)$ the space of continuous functions on Q and subspaces $C_b(Q)$, $C_c(Q)$ consisting, respectively, of bounded and compactly supported functions in $C(Q)$.

A *hypercomplex system with a basis* Q is defined via its structure measure $c(A, B, r)$, where $A, B \in \mathcal{B}(Q)$, $r \in Q$.

Definition 2.1. A *structure measure* $c(A, B, r)$ is a nonnegative Borel measure with respect to A (respectively B) if any fixed B, r (respectively A, r) that satisfies the following properties:

(H1) $\forall A, B \in \mathcal{B}_0(Q)$, the function $c(A, B, r)$ of r belongs to $C_c(Q)$;

(H2) $\forall A, B, C \in \mathcal{B}_0(Q)$ and $s \in Q$, the following *associativity relation* holds[2]:

$$\int_Q c(A, B, r) d_r c(E_r, C, s) = \int_Q c(B, C, r) d_r c(A, E_r, s) \tag{2.1}$$

DEFINITION 2.2. A measure m is said to be a *multiplicative*, if it is positive on any open set and satisfies the relation

$$\int_Q c(A, B, r)\, dm(r) = m(A)m(B) \text{ for all } A, B \in \mathcal{B}_0(Q). \tag{2.2}$$

A multiplicative measure always exists in commutative case, i.e. when $c(A, B, r) = c(B, A, r)$, $A, B \in \mathcal{B}_0(Q)$, $r \in Q$.

For a multiplicative measure m, we denote by $L_1(Q, m)$ the space of functions absolutely integrable with respect to m and endowed with the standard L_1-norm.

One can show [**2, 4, 5**] that for any $f, g \in L_1(Q, m)$ the convolution

$$(f * g)(r) = \int_Q f(p) d_p \left(\int_Q g(q) d_q c(E_p, E_q, r) \right) \tag{2.3}$$

is well defined and condition $(H2)$ implies its associativity. The space $L_1(Q, m)$ with this convolution is a Banach algebra.

DEFINITION 2.3. This algebra is called the *hypercomplex system with the basis* Q.

It is obvious that in this case

$$c(A, B, r) = (\kappa_A * \kappa_B)(r), \tag{2.4}$$

where $\kappa_A(r)$ denotes the indicator function of the set A.

We want to emphasize two features of the hypercomplex system:

(1) nonnegativity of the structure measure;
 and
(2) continuity and compactness of the support of the convolution $(\kappa_A * \kappa_B)(\cdot)$, where A, B are Borel sets with compact closures.

It follows from (2.3) and (2.4) that the associativity condition 2.1) in $(H2)$ implies that

$$(\kappa_A * \kappa_B) * \kappa_C = \kappa_A * (\kappa_B * \kappa_C)$$

for all $A, B, C \in \mathcal{B}_0(Q)$.

A more detailed account of the theory of hypercomplex systems with a locally compact basis and its various important examples can be found in [**2**].

Let us now consider a hypercomplex system that is the group algebra of a locally compact group G. Points of G will be denoted by $p, q, r, \ldots$. The group operation is $G \ni p, q \mapsto pq \in G$, and the corresponding left invariant measure is m:

$$\mathcal{B}(G) \ni A \mapsto m(A) \geq 0, \ m(pA) = m(A)$$

for every $p \in G$ and $A \in \mathcal{B}(G)$ (see, e.g. [**12**]). The space $L_1(G, m)$ with the convolution

$$(f * g)(r) = \int_G f(q) g(q^{-1}r) dm(q), \quad f, g \in L_1(G, m), r \in G, \tag{2.5}$$

[2] Here and later the expression $\int_Q f(r) d_r C(E_r, B, s)$ for $f \in C_c(Q)$ means $\int_Q f(r) c(dr, B, s)$.

i.e. the *group algebra* of the group G, is an example of a hypercomplex system with the locally compact basis $Q = G$, and the left invariant measure m is multiplicative.

The relation (2.5) implies easily that in this case the structure measure has the form

$$c(A,B,r) = (\kappa_A * \kappa_B)(r) = m(A \cap rB^{-1}), \tag{2.6}$$
$$A, B \in \mathcal{B}_0(G), \qquad r \in G, \qquad B^{-1} = \{q^{-1} | q \in B\},$$

where we used that $\kappa_B(q^{-1}r) = \kappa_{rB^{-1}}(q)$ for $q \in G$.

The condition $(H1)$ is satisfied. Indeed, for fixed compact sets A, B one can find a "sufficiently large" compact set C, such that for $r \in G \setminus C$ one has $A \cap rB^{-1} = \emptyset$. Also, of $c(A,B,r)$ is continuous with respect to r, since the convolution of bounded functions from $L_1(G,m)$ is continuous. Then the associativity condition $(H2)$ follows from associativity of the convolution (2.5). From (2.6), we get (2.2), and thus measure m is multiplicative.

Clearly, one can deal similarly with right- instead of left-invariant measures.

DEFINITION 2.4.

- A hypercomplex system is said to be *normal*, if there exists a multiplicative measure m and an involutive homeomorphism $Q \ni p \mapsto p^* \in Q$, such that $m(A) = m(A^*)$ and

$$c(A,B,C) = c(C,B^*,A), \;\; c(A,B,C) = c(A^*,C,B) \;\; (A,B,C \in \mathcal{B}_0(Q)),$$

  where $c(A,B,C) = \int_C c(A,B,r)\,dm(r)$.
- A point $e \in Q$ is said to be a *basis unity* of a normal hypercomplex system, if

$$c(A,B,e) = m(A^* \cap B) \text{ for any } A, B \in \mathcal{B}_0(Q). \tag{2.7}$$

For a normal hypercomplex system with a basis unity the Banach space $L^1(Q,m)$ is a $*$-Banach algebra with an approximate identity.

In the group algebra case, the neutral element of the group is a basis unity of the hypercomplex system.

For a normal hypercomplex system with a basis unity e, there is a family of right and left generalized shift operators. In the modern terms, one introduces

DEFINITION 2.5.

- a *coassociative comultiplication*

$$\Delta : C_b(Q) \to C_b(Q \times Q),$$

  defined by relation

$$((\Delta f)(q,\cdot), g)_{L^2(Q,m)} = (f * \check{g})(q), \quad f, g \in C_0(Q),$$

  where $\check{g}(q) = g(q^*)$;
  and
- a *counit* $\varepsilon : C(Q) \to \mathbb{C}$, where $\varepsilon(f)(p) = f(e)$.

Note that the comultiplication Δ is a positive mapping, but not a homomorphism of the algebra of functions under pointwise multiplication.

Note that (see, e.g. Theorem 2.1 and (2.6) in Chapter 1 of [**2**]) the multiplicative measure m of a normal hypercomplex system with a basis unity is left invariant with respect to the comultiplication, i.e.,

$$\int_Q (\Delta f)(p,q)dm(q) = \int_Q f(q)dm(q), \quad f \in C_0(Q). \tag{2.8}$$

3. The results

The first theorem addresses existence of a multiplicative measure m for a general hypercomplex system with a locally compact basis Q.

THEOREM 3.1. *Let the structure measure of a hypercomplex system satisfy the following conditions:*

(i) *For any $A \in \mathcal{B}_0(Q)$, the function $c(A,Q,r)$ takes only finite values and does not depend on $r \in Q$;*
(ii) *$c(O,Q,r) > 0$ for any open $O \in \mathcal{B}(Q)$.*

Define a measure m as $\mathcal{B}(Q) \ni A \mapsto m(A) = c(A,Q,r)$. Then m is a multiplicative measure of the hypercomplex system.

PROOF. Using the associativity relation, condition $(H2)$, and condition (i) of the theorem, we have for any $A, B \in \mathcal{B}_0(Q)$ and $r, s \in Q$, that

$$\begin{aligned}
\int_Q c(A,B,r)dm(r) &= \int_Q c(A,B,r)d_r c(E_r,Q,s) \\
&= \int_Q c(B,Q,r)d_r c(A,E_r,s) \\
&= c(B,Q,r)\int_Q d_r c(A,E_r,s) \\
&= c(B,Q,r)c(A,Q,s) = m(B)m(A).
\end{aligned}$$

The measure m is positive on open sets due to condition (ii). Therefore m is multiplicative. □

The condition (i) holds for any hypercomplex system with the basis $Q = G$, where G is a locally compact group. Indeed, from the equality (2.6) we get that $c(A,Q,r) = m(A \cap rQ^{-1}) = m(A)$, $A \in \mathcal{B}_0(G)$, where m is a left invariant measure. Condition (ii) also holds.

Conditions (i) and (ii) are also satisfied for the wide class of normal hypercomplex systems with basis unity. Indeed, using left invariance of the multiplicative measure (2.8) we have for all $A \in \mathcal{B}_0(Q)$, $r \in Q$

$$\begin{aligned}
c(A,Q,r) &= (\kappa_A * \kappa_Q)(r) = \int_Q (\Delta\kappa_A)(q,r)\kappa_Q(q^*)dm(q) \\
&= \int_Q (\Delta\kappa_A)(q,r)dm(q) = \int_Q \kappa_A(q)dm(q) = m(A).
\end{aligned}$$

Let G be locally compact group, $\mathcal{B}(G)$ be, as before the σ-algebra of Borel sets on G, and $\mathcal{B}_0(G)$ be its subring of sets with compact closure. Consider an *arbitrary* positive regular measure μ that is positive on open subsets of G and satisfies the following property:

$$\forall A \in \mathcal{B}_0(G) \text{ a function } G \ni r \mapsto \mu(rA) \text{ is continuous.} \tag{3.1}$$

And introduce a convolution $*_\mu$ of type (2.5) but using the measure μ instead of a left invariant measure m,

$$(f *_\mu g)(r) = \int_G f(q) g(q^{-1}r) d\mu(q). \tag{3.2}$$

This convolution exists for bounded measurable functions f and g with compact supports. Consider the corresponding structure measure,

$$\begin{aligned} c_\mu(A,B,r) &= (\kappa_A *_\mu \kappa_B)(r) \\ &= \int_G \kappa_A(q)\kappa_B(q^{-1}r)d\mu(q) = \int_G \kappa_A(q)\kappa_{rB^{-1}}(q)d\mu(q) \\ &= \mu(A \cap rB^{-1}), \ A,B \in \mathcal{B}_0(G), r \in G. \end{aligned} \tag{3.3}$$

The function $c_\mu(A,B,r)$ in (3.3), for fixed B and r (A and r), is a measure on $\mathcal{B}(G)$. It has compact support with respect to $r \in G$ with A and B being fixed, since $A \cap rB^{-1} = \emptyset$, if r is outside from a big compact.

THEOREM 3.2. *Assume that the structure measure $c_\mu(A,B,r)$, constructed from measure μ according to (3.3), satisfies the associativity condition (H2). Then μ is left invariant.*

PROOF. For fixed $A,B,C \in \mathcal{B}_0(G)$ and $s \in G$, using expression (3.3) we obtain for the left hand side integral in (2.1):

$$\begin{aligned} \int_G c_\mu(A,B,r)d_r c_\mu(E_r,C,s) &= \int_G (\kappa_A *_\mu \kappa_B)(r) d\mu(E_r \cap sC^{-1}) \\ &= \int_{sC^{-1}} (\kappa_A *_\mu \kappa_B)(r) d\mu(r) \\ &= \int_G (\kappa_A *_\mu \kappa_B)(r)\kappa_{sC^{-1}}(r) d\mu(r) \\ &= \int_G (\kappa_A *_\mu \kappa_B)(r)\kappa_C(r^{-1}s) d\mu(r) \\ &= ((\kappa_A *_\mu \kappa_B) *_\mu \kappa_C)(s) \end{aligned}$$

For the right hand side integral, we have

$$\begin{aligned} \int_G c_\mu(B,C,r)d_r c_\mu(A,E_r,s) &= \int_G (\kappa_B *_\mu \kappa_C)(r) d\mu(A \cap sE_r^{-1}) \\ &= \int_G (\kappa_B *_\mu \kappa_C)(r^{-1}s) d\mu(A \cap r) \\ &= \int_A (\kappa_B *_\mu \kappa_C)(r^{-1}s) d\mu(r) \\ &= \int_G \kappa_A(r)(\kappa_B *_\mu \kappa_C)(r^{-1}s) d\mu(r) \\ &= (\kappa_A *_\mu (\kappa_B *_\mu \kappa_C))(s). \end{aligned}$$

Thus, the associativity relation (H2) implies that the convolution $*_\mu$ is associative. Then,

$$\begin{aligned} ((\kappa_A *_\mu \kappa_B) *_\mu \kappa_C)(s) &= \int_G (\kappa_A *_\mu \kappa_B)(r_1)\kappa_C(r_1^{-1}s) d\mu(r_1) \\ &= \int_G \int_G \kappa_A(r_2)\kappa_B(r_2^{-1}r_1)\kappa_C(r_1^{-1}s) d\mu(r_1) d\mu(r_2) \end{aligned}$$

and

$$\begin{aligned}(\kappa_A *_\mu (\kappa_B *_\mu \kappa_C))(s) &= \int_G \kappa_A(r_2)(\kappa_B *_\mu \kappa_C)(r_2^{-1}s)d\mu(r_2)\\ &= \int_G\int_G \kappa_A(r_2)\kappa_B(r_1)\kappa_C(r_1^{-1}r_2^{-1}s)d\mu(r_1)d\mu(r_2).\end{aligned}$$

Since the expressions in the left-hand sides of these two relations are equal, for $A, B, C \in \mathcal{B}_0(G)$ and $s \in G$ we get the equality

$$\begin{aligned}(3.4)\quad &\int_G\int_G \kappa_A(r_2)\kappa_B(r_2^{-1}r_1)\kappa_C(r_1^{-1}s)d\mu(r_1)d\mu(r_2)\\ &\qquad = \int_G\int_G \kappa_A(r_2)\kappa_B(r_1)\kappa_C(r_1^{-1}r_2^{-1}s)d\mu(r_1)d\mu(r_2).\end{aligned}$$

The left hand side of (3.4) contains the integral with respect to r_2 of the function

$$\begin{aligned}f_1(r_2) &= \mu(r_2B\cap sC^{-1}) = \int_G \kappa_{r_2B}(r_1)\kappa_{sC^{-1}}(r_1)d\mu(r_1)\\ &= \int_G \kappa_B(r_2^{-1}r_1)\kappa_C(r_1^{-1}s)d\mu(r_1).\qquad (3.5)\end{aligned}$$

Similarly, in the right hand side of (3.4) we see the integral with respect to r_2 of the function

$$\begin{aligned}f_2(r_2) &= \mu(B\cap r_2^{-1}sC^{-1}) = \int_G \kappa_B(r_1)\kappa_{r_2^{-1}sC^{-1}}(r_1)d\mu(r_1)\\ &= \int_G \kappa_B(r_1)\kappa_C(r_1^{-1}r_2^{-1}s)d\mu(r_1).\qquad (3.6)\end{aligned}$$

For fixed $B, C \in \mathcal{B}_0(G)$ and $s \in G$, the expression (3.5) is a continuous function of $r_2 \in G$, due to the condition (3.1). Analogously, the expression (3.6) is a continuous function of $r_2^{-1} \in G$, i.e. of $r_2 \in G$.

Thus, we have two continuous functions $f_1(r_2)$ and $f_2(r_2)$, defined by (3.6), for which the equality (3.4) implies

$$(3.7)\qquad \int_A f_1(r_2)d\mu(r_2) = \int_A f_2(r_2)d\mu(r_2),\ \text{for all } A \in \mathcal{B}_0(G).$$

The measure μ is positive on open sets from G. Therefore, (3.7) implies that $f_1(r_2) = f_2(r_2)$ for all $r_2 \in G$. From (3.5) and (3.6), we conclude that for $B, C \in \mathcal{B}_0(G)$ and $r_2, s \in G$ we have the equality

$$\mu(r_2B\cap sC^{-1}) = \mu(B\cap r_2^{-1}sC^{-1}).$$

Let us now fix $B \in \mathcal{B}_0(G)$ and $r_2, s \in G$ and choose $C \in \mathcal{B}_0(G)$ so "large" that $r_2B \subset sC^{-1}$ and $B \subset r_2^{-1}sC^{-1}$. Then the previous equality yields

$$\mu(r_2B) = \mu(B),\quad B \in \mathcal{B}_0(G),\ r_2 \in G,$$

and thus μ is left invariant. □

References

[1] Yu. M. Berezansky and A. A. Kalyuzhnyĭ, *Hypercomplex systems and hypergroups: connections and distinctions*, Applications of hypergroups and related measure algebras (Seattle, WA, 1993), Contemp. Math., vol. 183, Amer. Math. Soc., Providence, RI, 1995, pp. 21–44, DOI 10.1090/conm/183/02052. MR1334770

[2] Yu. M. Berezansky and A. A. Kalyuzhnyi, *Harmonic analysis in hypercomplex systems*, Mathematics and its Applications, vol. 434, Kluwer Academic Publishers, Dordrecht, 1998. Translated from the 1992 Russian original by P. V. Malyshev and revised by the authors. MR1627482

[3] Yu. M. Berezanskiĭ and Yu. G. Kondrat′ev, *Spektral′nye metody v beskonechnomernom analize* (Russian), "Naukova Dumka", Kiev, 1988. MR978630

[4] Yu. M. Berezanskiĭ and S. G. Kreĭn, *Continuous algebras* (Russian), Dokaldy Akad. Nauk SSSR (N.S.) **72** (1950), 5–8. MR0036945

[5] Yu. M. Berezanskiĭ and S. G. Kreĭn, *Hypercomplex systems with continual basis* (Russian), Uspehi Mat. Nauk (N.S.) **12** (1957), no. 1(73), 147–152. MR0086272

[6] W. R. Bloom and H. Heyer, *Harmonic analysis of probability measures on hypergroups*, De Gruyter Studies in Mathematics, vol. 20, Walter de Gruyter & Co., Berlin, 1995. MR1312826

[7] Yu. Chapovsky. Existence of an invariant measure on a hypergroup, ArXiv: 1212:6571.

[8] J. Delsarte. Sur une extension de la formule de Taylor, J. Math. Pure et Appl., **17** (1938), No. 3, 213–231.

[9] C. F. Dunkl, *Structure hypergroups for measure algebras*, Pacific J. Math. **47** (1973), 413–425. MR0336225

[10] R. I. Jewett, *Spaces with an abstract convolution of measures*, Advances in Math. **18** (1975), no. 1, 1–101, DOI 10.1016/0001-8708(75)90002-X. MR0394034

[11] B. M. Levitan. Theory of Generalized Translation Operators, Moskow: Nauka, 1973 (in Russian).

[12] M. A. Naĭmark, *Normed rings*, Translated from the first Russian edition by Leo F. Boron, P. Noordhoff N. V., Groningen, 1959. MR0110956

[13] M. Spector. *Aperçu de la theorie des Hypergroupes*, Lect. Notes in Math. **497** (1975), 643–673, Springer, Berlin. MR0447974

[14] R. Spector, *Mesures invariantes sur les hypergroupes* (French, with English summary), Trans. Amer. Math. Soc. **239** (1978), 147–165, DOI 10.2307/1997851. MR0463806

Institute of Mathematics National Academy of Sciences of Ukraine, Tereschenkivs'ka 3, Kyiv, 01601 Ukraine

Email address: mfat@imath.kiev.ua

Institute of Mathematics National Academy of Sciences of Ukraine, Tereschenkivs'ka 3, Kyiv, 01601 Ukraine

Email address: akalyuz@gmail.com

Contemporary Mathematics
Volume **733**, 2019
https://doi.org/10.1090/conm/733/14735

Inversion formulas of integral geometry in real hyperbolic space

William O. Bray and Boris Rubin

To the memory of Selim Grigorievich Krein on the occasion of his centennial celebration

ABSTRACT. This expository article is a brief survey of authors' results related to inversion of Radon transforms in the n-dimensional real hyperbolic space. The exposition is focused on horospherical and totally geodesic transforms over the corresponding submanifolds of arbitrary fixed dimension d, $1 \le d \le n-1$. Our main objective is explicit inversion formulas for these transforms on L^p functions and smooth functions with suitable behavior at infinity.

1. Introduction

The subject of this article is an outgrowth of the pioneering papers by Funk [**17**], Radon [**45**] and John [**32**]. In the Euclidean setting, the problem is to reconstruct a function on $\mathbb{R}^n$ from its integrals over d-dimensional planes in $\mathbb{R}^n$, where d is a fixed integer, $1 \le d \le n-1$. The case $d = 1$ corresponds to lines in $\mathbb{R}^n$. This problem paves the way to numerous investigations, modifications, generalizations, and applications to diverse areas in mathematics; see, e.g., the books by Gardner [**18**], Gelfand, Graev, and Vilenkin [**20**], Gelfand, Gindikin, and Graev [**19**], Helgason [**26–28**], Kuchment [**35**], Natterer [**44**], where many other references can be found.

In the present article we give a brief survey of our recent results related to similar problems in the n-dimensional real hyperbolic space $\mathbb{H}^n$. Integral geometry in hyperbolic space exhibits a crucial difference in contrast with Euclidean space: there are two different analogues of the Euclidean lines in the real hyperbolic geometry - geodesics and horocycles. In higher dimensions we correspondingly have two substitutes for the Euclidean planes, namely, the totally geodesic submanifolds and horospheres.

The plan of the paper is the following. Section 2 contains basic notation and preliminaries. Section 3 deals with the horospherical transforms over d-dimensional horospheres in $\mathbb{H}^n$, $1 \le d \le n-1$. The case $d = n-1$ agrees with the well-known Gelfand-Graev transform, when the horosphere is a cross-section of the hyperboloid

---

2010 *Mathematics Subject Classification.* Primary 44A12; Secondary 44A15.
Key words and phrases. Real hyperbolic space, Radon transforms, inversion formulas.

The second-named author had a privilege to participate in the famous Voronezh Winter Mathematical School in the former USSR. Professor Krein was the founder and also the heart and the soul of that wonderful school which was so inspiring for many generations of mathematicians, now working all over the world.

$\mathbb{H}^n$ by the hyperplane whose normal lies in the asymptotic cone. This transform was studied by Gelfand and his collaborators on smooth rapidly decreasing functions in the framework of the theory of distributions; see [**20**], [**59**, p. 162]. Our approach not only covers horospheres of arbitrary dimension but also provides new explicit inversion formulas which are applicable to both smooth and L^p functions. In Section 4 we consider similar problems for totally geodesic transforms, when the integration is performed over d-dimensional totally geodesic submanifolds of $\mathbb{H}^n$. In all cases, our main objective is explicit inversion formulas.

For detailed proofs and related results, the reader is referred to our original papers and other publications mentioned in the text.

2. Preliminaries

The pseudo-Euclidean space $\mathbb{E}^{n,1}$, $n \ge 2$, is the $(n+1)$-dimensional real vector space of points in $\mathbb{R}^{n+1}$ with the inner product

$$[\mathbf{x}, \mathbf{y}] = -x_1y_1 - \ldots - x_ny_n + x_{n+1}y_{n+1}. \tag{2.1}$$

We denote by $e_1, \ldots, e_{n+1}$ the coordinate unit vectors in $\mathbb{E}^{n,1}$; S^{n-1} is the unit sphere in the coordinate plane $\mathbb{R}^n = \{x \in \mathbb{E}^{n,1} : x_{n+1} = 0\}$; $\sigma_{n-1} = 2\pi^{n/2}/\Gamma(n/2)$ is the surface area of S^{n-1}. For $\theta \in S^{n-1}$, $d\theta$ denotes the surface element on S^{n-1}.

The n-dimensional real hyperbolic space $\mathbb{H}^n$ is realized as the upper sheet of the two-sheeted hyperboloid in $\mathbb{E}^{n,1}$, that is,

$$\mathbb{H}^n = \{\mathbf{x} \in \mathbb{E}^{n,1} : [\mathbf{x}, \mathbf{x}] = 1, \ x_{n+1} > 0\}.$$

In the following, the points of $\mathbb{H}^n$ will be denoted by the non-boldfaced letters, unlike the generic points in $\mathbb{E}^{n,1}$. The geodesic distance between points x and y in $\mathbb{H}^n$ is defined by $d(x, y) = \cosh^{-1}[x, y]$. The point $x_0 = (0, \ldots, 0, 1) \sim e_{n+1}$ serves as the origin of $\mathbb{H}^n$;

$$\Gamma = \{\mathbf{x} \in \mathbb{E}^{n,1} : \ [\mathbf{x}, \mathbf{x}] = 0, \, x_{n+1} > 0\}$$

is the asymptotic cone for $\mathbb{H}^n$. The notation

$$G = SO_0(n, 1)$$

is used for the identity component of the special pseudo-orthogonal group $SO(n, 1)$ preserving the bilinear form (2.1); I_m is the identity $m \times m$ matrix.

We will be using different coordinate systems on $\mathbb{H}^n$. Every point $x \in \mathbb{H}^n$ is represented in the *hyperbolic coordinates* $(\theta, r) \in S^{n-1} \times [0, \infty)$ as

$$x = \theta \sinh r + e_{n+1} \cosh r. \tag{2.2}$$

In the *horospherical coordinates* $(v, t) \in \mathbb{R}^{n-1} \times \mathbb{R}$, we have

$$\begin{aligned} x &= n_v a_t x_0 = a_t n_{e^{-t}v} x_0 && (2.3)\\ &= (e^{-t}v, \, \sinh t + \frac{|v|^2}{2} e^{-t}, \, \cosh t + \frac{|v|^2}{2} e^{-t}), && (2.4)\end{aligned}$$

where $v = (v_1, \ldots, v_{n-1})^T \in \mathbb{R}^{n-1}$ is a column vector,

$$n_v = \begin{bmatrix} I_{n-1} & -v & v \\ v^T & 1-|v|^2/2 & |v|^2/2 \\ v^T & -|v|^2/2 & 1+|v|^2/2 \end{bmatrix}; \tag{2.5}$$

$$a_t = \begin{bmatrix} I_{n-1} & 0 & 0 \\ 0 & \cosh t & \sinh t \\ 0 & \sinh t & \cosh t \end{bmatrix}, \qquad t \in \mathbb{R}, \tag{2.6}$$

is the hyperbolic rotation in the coordinate plane (x_n, x_{n+1}); cf. [**59**, p. 13]. The subgroups of matrices a_t and n_v will be denoted by A and N, respectively. Let also

$$N_d = \{n_v \in N : v_1 = \ldots = v_{n-1-d} = 0\} \tag{2.7}$$

and

$$N_{n-1-d} = \{n_v \in N : v_d = \ldots = v_{n-1} = 0\} \tag{2.8}$$

be the subgroups of N generated by vectors $v \in \mathbb{R}^{n-1}$ with the corresponding zero coordinates. If $v = u + w \in \mathbb{R}^{n-1}$ with $u \in \mathbb{R}^{n-1-d}$ and $w \in \mathbb{R}^d$, then a straightforward matrix multiplication yields

$$n_v = n_u n_w. \tag{2.9}$$

In the following,

$$K = \left\{ \begin{bmatrix} k & 0 \\ 0 & 1 \end{bmatrix} : k \in SO(n) \right\}. \tag{2.10}$$

Abusing notation, we identify $k \in SO(n)$ with the corresponding matrix

$$\begin{bmatrix} k & 0 \\ 0 & 1 \end{bmatrix} \in G.$$

We fix a G-invariant measure dx on $\mathbb{H}^n$, which has the following form in the coordinates (2.2):

$$dx = \sinh^{n-1} r \, dr d\theta. \tag{2.11}$$

The Haar measure dg on G will be normalized in a consistent way by the formula

$$\int_G f(gx_0)\, dg = \int_{\mathbb{H}^n} f(x)\, dx. \tag{2.12}$$

As usual, $C(\mathbb{H}^n)$ denotes the space of continuous functions on $\mathbb{H}^n$;

$$C_\mu(\mathbb{H}^n) = \{f \in C(\mathbb{H}^n) : f(x) = O(x_{n+1}^{-\mu})\}; \tag{2.13}$$

$C_0(\mathbb{H}^n)$ is the subspace of $C(\mathbb{H}^n)$ which consists of functions vanishing at infinity; $C_c^\infty(\mathbb{H}^n)$ denotes the space of infinitely differentiable compactly supported functions on $\mathbb{H}^n$. If $\Omega = \{\mathbf{x} \in E^{n,1} : [\mathbf{x}, \mathbf{x}] > 0,\ x_{n+1} > 0\}$ is the interior of the cone Γ, then $C_c^\infty(\mathbb{H}^n)$ is formed by the restrictions of functions belonging to $C_c^\infty(\Omega)$ onto $\mathbb{H}^n$.

We say that an integral under consideration exists in the Lebesgue sense if it is finite when the corresponding integrand is replaced by its absolute value.

3. Horospherical radon transforms

This section follows our papers [**11, 52**], where the reader can find further details.

3.1. Horospheres. An $(n-1)$-dimensional horosphere in $\mathbb{H}^n$ is defined as the cross-section of the hyperboloid $\mathbb{H}^n$ by the hyperplane $[\mathbf{x}, \mathbf{b}] = 1$, where $\mathbf{b}$ is a point of the cone Γ. The correspondence between the set Ξ_{n-1} of all $(n-1)$-horospheres and the set of all points in Γ is one-to-one. One can equivalently define Ξ_{n-1} as the set of all G-orbits

$$\Xi_{n-1} = \{g\xi^0_{n-1} : g \in G\} \tag{3.1}$$

of the "basic" horosphere ξ^0_{n-1} corresponding to the point

$$\mathbf{b}_0 = (0, \dots, 0, 1, 1) \in \Gamma.$$

To define lower-dimensional horospheres corresponding to $d < n-1$, we set

$$\mathbb{R}^{n-d-1} = \mathbb{R}e_1 \oplus \cdots \oplus \mathbb{R}e_{n-d-1} \quad (\text{if } d = n-1 \text{ this set is empty}), \tag{3.2}$$

$$\mathbb{R}^d = \mathbb{R}e_{n-d} \oplus \cdots \oplus \mathbb{R}e_{n-1}, \quad \mathbb{R}^{d+1} = \mathbb{R}^d \oplus \mathbb{R}e_n, \tag{3.3}$$

$$\mathbb{E}^{d+1,1} \simeq \mathbb{R}^{d+2} = \mathbb{R}^{d+1} \oplus \mathbb{R}e_{n+1}, \qquad \mathbb{H}^{d+1} = \mathbb{H}^n \cap \mathbb{R}^{d+2}; \tag{3.4}$$

$$\xi_0 = \mathbb{H}^{d+1} \cap \{x \in \mathbb{H}^n : [x, \mathbf{b}_0] = 1\}. \tag{3.5}$$

The last formula defines a d-dimensional horosphere in $\mathbb{H}^{d+1}$. We call it the basic one. The set Ξ_d of all d-dimensional horospheres in $\mathbb{H}^n$ (d-horospheres, for short) is defined as the collection of all G-orbits

$$\Xi_d = \{g\xi_0 : g \in G\}. \tag{3.6}$$

3.2. Horospherical transforms.

DEFINITION 3.1. Let $1 \le d \le n-1$. Given $\xi = g\xi_0 \in \Xi_d$, $g \in G$, the d-horospherical transform of a sufficiently good function f on $\mathbb{H}^n$ is defined by

$$\hat{f}(\xi) = \int_{\mathbb{R}^d} f(gn_w x_0)\, dw. \tag{3.7}$$

One can show that this definition is independent of the choice of $g : \xi_0 \to \xi$. The case when g is the identity map, corresponds to the integral of f over the basic horosphere ξ_0. If, by the Iwasawa decomposition, $g = ka_t n_v$, then (2.9) yields

$$\hat{f}(\xi) \equiv \hat{f}(g\xi_0) = \hat{f}(ka_t n_u \xi_0), \tag{3.8}$$

where $k \in K$, a_t is the matrix (2.6), and $n_u \in N_{n-1-d}$.

The definition (3.7) can be put in group-theoretic terms as follows. Since $\mathbb{H}^n$ is identified as the homogeneous space G/K, a function f on $\mathbb{H}^n$ becomes the right K-invariant function $g \to f(gK)$ on G. Let $M_d \subset G$ be the subgroup of matrices of the form

$$m_{\alpha,\beta} = \begin{bmatrix} \tilde{m}_{\alpha,\beta} & 0 \\ 0 & I_2 \end{bmatrix}$$

where

$$\tilde{m}_{\alpha,\beta} = \begin{bmatrix} \alpha & 0 \\ 0 & \beta \end{bmatrix} \in S(O(n-d-1) \times O(d)),$$

and let N_d be the subgroup of G defined by (2.7). Denoting group elements in N_d by n_d and the Haar measure on N_d by dn_d, we can write (3.7) as

$$\hat{f}(gM_dN_d) = \int_{N_d} f(gn_dK)\, dn_d. \tag{3.9}$$

LEMMA 3.2. *Let $1 \le d \le n-1$. If $f \in L^p(\mathbb{H}^n)$, $1 \le p < 2(n-1)/d$, then the integral $\hat{f}(\xi) \equiv \hat{f}(ka_t n_u \xi_0)$ is finite for almost all $k \in K$, almost all $t \in \mathbb{R}$ and all $u \in \mathbb{R}^{n-1-d}$. If $f \in C_\mu(\mathbb{H}^n)$, $\mu > d/2$, then $\hat{f}(\xi)$ is finite for every $\xi \in \Xi_d$. The conditions $1 \le p < 2(n-1)/d$ and $\mu > d/2$ are sharp.*

This lemma was proved in [**51**, Section 6.3.2]) for $d = n-1$ and in [**11**, Section 3] for all $1 \le d \le n-1$.

3.3. Hyperbolic convolutions and spherical means. All details related to this subsection can be found in [**51**, Section 6.1.2] and [**52**].

Given a measurable function k on $[1,\infty)$, the corresponding hyperbolic convolution on $\mathbb{H}^n$ is defined by

$$(Kf)(x) = \int_{\mathbb{H}^n} k([x,y]) f(y)\, dy, \qquad x \in \mathbb{H}^n. \tag{3.10}$$

If this integral exists in the Lebesgue sense, then, by Fubini's theorem,

$$\int_{\mathbb{H}^n} k([x,y]) f(y)\, dy = \sigma_{n-1} \int_0^\infty k(\cosh r)\ (M_x f)(\cosh r)\ \sinh^{n-1} r\, dr, \tag{3.11}$$

where $M_x f$ is the spherical mean

$$(M_x f)(s) = \frac{(s^2-1)^{(1-n)/2}}{\sigma_{n-1}} \int_{\{y \in \mathbb{H}^n :\, [x,y]=s\}} f(y)\, d\sigma(y), \qquad s > 1, \tag{3.12}$$

$d\sigma(y)$ being the relevant induced measure. We can also write (3.12) in the "more geometric" form as

$$(M_x f)(\cosh t) = \int_K f(r_x k a_t x_0)\, dk, \tag{3.13}$$

where $r_x \in G$ takes x_0 to x and a_t is the matrix (2.6).

LEMMA 3.3. ([**51**, p. 370], [**42**, pp. 131-133]). *Let $f \in L^p(\mathbb{H}^n)$, $1 \le p \le \infty$. Then*

$$\sup_{s>1} \|(M_{(\cdot)} f)(s)\|_p \le \|f\|_p. \tag{3.14}$$

If $1 \le p < \infty$, then $(M_x f)(s)$ is a continuous L^p-valued function of $s \in [1,\infty)$ and

$$\lim_{s \to 1} \|(M_{(\cdot)} f)(s) - f\|_p = 0. \tag{3.15}$$

If f is a continuous function vanishing at infinity, then $(M_x f)(s)$ is a continuous function of $(x,s) \in \mathbb{H}^n \times (1,\infty)$ and $(M_x f)(s) \to f(x)$ as $s \to 1$, uniformly on $\mathbb{H}^n$.

An important example of convolutions (3.10) arising in the study of the horospherical transforms is the analytic family of the potential type operators

$$(Q^\alpha f)(x)=\zeta_{n,\alpha}\int\limits_{\mathbb{H}^n} f(y)\,\frac{([x,y]-1)^{(\alpha-n)/2}}{([x,y]+1)^{n/2-1}}\,dy, \tag{3.16}$$

$$\zeta_{n,\alpha}=\frac{\Gamma((n-\alpha)/2)}{2^{\alpha/2+1}\,\pi^{n/2}\Gamma(\alpha/2)}, \qquad Re\,\alpha>0, \quad \alpha-n\neq 0,2,4,\ldots .$$

LEMMA 3.4. [**51**, p. 385] *If*

$$f\in L^p(\mathbb{H}^n), \qquad 1\le p<\infty, \qquad 0<\alpha<2(n-1)/p,$$

then $(Q^\alpha f)(x)$ exists as an absolutely convergent integral for almost all $x\in\mathbb{H}^n$.

LEMMA 3.5. [**51**, p. 386] *Let $f\in C_c^\infty(\mathbb{H}^n)$,*

$$D_\alpha=-\Delta_H-\alpha(2n-2-\alpha)/4, \qquad \alpha\ge 2,$$

where Δ_H is the Beltrami-Laplace operator on $\mathbb{H}^n$. If $\alpha-n\neq 0,2,4,\ldots$, then

$$D_\alpha Q^\alpha f=Q^{\alpha-2}f \qquad (Q^0 f=f). \tag{3.17}$$

In particular, if $\alpha=2\ell$ is even and $2\ell-n\neq 0,2,4,\ldots$, then

$$\mathcal{P}_\ell(\Delta_H)Q^{2\ell}f=f, \qquad \mathcal{P}_\ell(\Delta_H)=D_2D_4\ldots D_{2\ell}. \tag{3.18}$$

3.4. Inversion formulas.

3.4.1. *Gelfand-Graev-Vilenkin's formulas.* Using the Dirac delta function, in the case $d=n-1$, for a compactly supported smooth function f one can write

$$\hat f(\xi)=\int\limits_{\mathbb{E}^{n,1}} f(x)\,\delta([x,\xi]-1)\,dx, \qquad \xi\in\Gamma. \tag{3.19}$$

The following inversion formulas can be found in Gelfand, Graev, and Vilenkin [**20**] (see also Vilenkin and Klimyk [**59**, p. 162]):

$$f(x)=\begin{cases}\dfrac{(-1)^m}{2(2\pi)^{2m}}\displaystyle\int\limits_\Gamma \delta^{(2m)}([x,\xi]-1)\,\hat f(\xi)\,d\xi & \text{if } n=2m+1,\\[2ex] \dfrac{(-1)^m\Gamma(2m)}{(2\pi)^{2m}}\displaystyle\int\limits_\Gamma ([x,\xi]-1)^{-2m}\,\hat f(\xi)\,d\xi & \text{if } n=2m.\end{cases}$$

The divergent integrals in these formulas are given precise meaning in the framework of the theory of distributions. In this subsection we suggest alternative inversion formulas which do not contain divergent integrals and are applicable not only to smooth functions, but also to $f\in L^p(\mathbb{H}^n)$.

3.4.2. *The method of mean value operators.* An idea of this inversion method is to average $\hat f(\xi)$ over all d-horospheres ξ at a fixed positive distance from a given point $x\in\mathbb{H}^n$. Inverting a simple Abel type fractional integral, we obtain the spherical mean (3.12) that gives $f(x)$ after passing to the limit according to Lemma 3.3.

Let

$$(I_-^\alpha\psi)(r)=\frac{1}{\Gamma(\alpha)}\int\limits_r^\infty \frac{\psi(s)\,ds}{(s-r)^{1-\alpha}} \qquad \alpha>0, \tag{3.20}$$

be the Riemann-Liouville fractional integral. Numerous properties of such integrals, including inversion formulas in the context of their application to integral geometric problems are discussed in [**51**, Chapter 2] (see also [**53**]). We introduce the mean value operator

$$\check{\varphi}_x(t) = \int\limits_K \varphi(r_x k a_t \xi_0)\, dk, \qquad x \in \mathbb{H}^n, \quad t \in \mathbb{R}, \tag{3.21}$$

where $r_x \in G$ is an arbitrary transformation satisfying $r_x x_0 = x$.

LEMMA 3.6. *If $\varphi = \hat{f}$, then*

$$\check{\varphi}_x(t) = c\, e^{-td/2}\, (I_-^{d/2} M_x f)(\cosh t), \qquad c = 2^{d/2-1}\, \sigma_{d-1}\Gamma(d/2), \tag{3.22}$$

where $M_x f$ is the spherical mean (3.12). It is assumed that the integral on the right-hand side of (3.22) exists in the Lebesgue sense.

THEOREM 3.7. (cf. Theorem 6.79 in [**51**] and Theorem 4.10 in [**11**]) *Let $1 \le d \le n-1$, $f \in L^p(\mathbb{H}^n)$, $1 \le p < 2(n-1)/d$. Suppose that $\varphi = \hat{f}$ and set*

$$\psi_x(r) = c^{-1} e^{td/2} \check{\varphi}_x(t)\big|_{t=\cosh^{-1} r}, \qquad c = 2^{d/2-1}\, \sigma_{d-1}\Gamma(d/2).$$

Then

$$f(x) = \lim_{s \to 1} (\mathcal{D}_-^{d/2} \psi_x)(s), \tag{3.23}$$

where the derivative $\mathcal{D}_-^{d/2}\psi_x$ is defined by following formulas.
(i) *If d is even, $d = 2m$, then*

$$(\mathcal{D}_-^{d/2}\psi_x)(s) = (-1)^m \left(\frac{d}{ds}\right)^m \psi_x(s). \tag{3.24}$$

(ii) *If d is odd, $d = 2m-1$, then*

$$(\mathcal{D}_-^{d/2}\psi_x)(s) = (-1)^m s^{1/2} \left(\frac{d}{ds}\right)^m \left[s^{m-1/2}\, I_-^{1/2}\, s^{-m}\, \psi_x\right], \tag{3.25}$$

$$= (-1)^m s^{1/2} \frac{d}{ds} \left[s^{1/2}\, I_-^{1/2}\, s^{-1}\, \psi_x^{(m-1)}\right]. \tag{3.26}$$

The limit in (3.23) is understood in the L^p-norm. Moreover, if $f \in C_\mu(\mathbb{H}^n)$, then (3.23) holds uniformly.

REMARK 3.8. It would be interesting to relate the Abel-type inversion formula (3.23) to the one obtained by Andersen, Olafsson and Schlichtrull [**1**], which uses Heckman-Opdam hypergeometric shift operators.

3.4.3. *Inversion by polynomials of the Beltrami-Laplace operator.* We consider the mean value operator (3.21) with $t = 0$ and denote

$$\check{\varphi}(x) = \int\limits_K \varphi(r_x k \xi_0)\, dk. \tag{3.27}$$

This is the *dual horospherical transform* that integrates a function φ on Ξ_d over all d-horospheres passing through the given point $x \in \mathbb{H}^n$. The next lemma is a modification of Lemma 3.6 corresponding to $t = 0$ and can be regarded as a horospherical analogue of the celebrated Fuglede result for d-plane Radon-John transforms [**16**]. It establishes important connection between the d-horospherical transform, the dual transform (3.27) and the analytic family of potentials (3.16).

LEMMA 3.9. *The following equality holds provided that either side of it exists in the Lebesgue sense:*

$$(\hat{f})^{\vee}(x) = c\,(Q^d f)(x), \qquad c = \frac{2^d\,\pi^{d/2}\,\Gamma(n/2)}{\Gamma((n-d)/2)}. \tag{3.28}$$

Lemmas 3.9 and 3.5 imply the following inversion result.

THEOREM 3.10. *Let $\varphi = \hat{f}$, $f \in C_c^\infty(\mathbb{H}^n)$, $1 \le d \le n-1$. If d is even, then*

$$f = c\,\mathcal{P}(\Delta_H)\,\check{\varphi}, \tag{3.29}$$

where

$$\mathcal{P}(\Delta_H) = \prod_{i=1}^{d/2} \big[\Delta_H + i(n-1-i)\big], \quad c = \frac{(-1)^{d/2}\,\Gamma((n-d)/2)}{2^d\,\pi^{d/2}\,\Gamma(n/2)}.$$

If d is odd, then a local inversion formula, like (3.29), is unavailable in principle. Both even and odd cases can be treated in the framework of a certain analytic family of operators generalizing (3.27).

Let

$$h_\alpha(t) = e^{td/2}\,(\cosh t - 1)^{\alpha/2-1}\,|\sinh t|, \qquad \alpha > 0,$$

$$(\overset{*}{\mathfrak{H}}{}^\alpha \varphi)(x) = c_\alpha \int\limits_{K\times\mathbb{R}} \varphi(r_x k a_t \xi_0)\,h_\alpha(t)\,dkdt, \tag{3.30}$$

$$c_\alpha = \frac{\Gamma((n-\alpha-d)/2)}{2^{\alpha/2+d+1}\,\pi^{d/2}\,\Gamma(\alpha/2)\,\Gamma(n/2)}, \qquad \alpha > 0, \quad \alpha + d - n \ne 0, 2, 4, , \dots .$$

PROPOSITION 3.11. (see [**11**, Proposition 4.14]) *If φ is a compactly supported continuous function on Ξ_d, then*

$$\lim_{\alpha\to 0} \overset{*}{\mathfrak{H}}{}^\alpha \varphi = c^{-1}\,\check{\varphi} \tag{3.31}$$

where c is a constant from (3.28).

In the case $\alpha = n-d$, which was excluded in (3.30), we set

$$(\overset{*}{\mathfrak{H}}{}^{n-d} \varphi)(x) = c_{n,d} \int\limits_{K\times\mathbb{R}} \varphi(r_x k a_t \xi_0)\,\tilde{h}_{n-d}(t)\,dkdt, \tag{3.32}$$

where

$$c_{n,d} = -\frac{1}{2^{(n+d)/2+1}\,\pi^{d/2}\,\Gamma(n/2)\,\Gamma((n-d)/2)},$$

$$\tilde{h}_{n-d}(t) = e^{td/2}\,|\sinh t|\,(\cosh t - 1)^{(n-d)/2-1}\,\log(\cosh t - 1).$$

THEOREM 3.12. *Let $\varphi = \hat{f}$, $f \in C_c^\infty(\mathbb{H}^n)$,*

$$\mathcal{P}_\ell(\Delta_H) = (-1)^\ell \prod_{i=1}^{\ell} \big[\Delta_H + i(n-1-i)\big], \qquad \ell \in \mathbb{N}. \tag{3.33}$$

(i) *If n is odd, then*

$$f = \mathcal{P}_\ell(\Delta_H)\overset{*}{\mathfrak{H}}{}^{2\ell-d}\varphi \qquad \forall \ell \ge d/2. \tag{3.34}$$

(ii) *If $n = 2$, then*

$$f = -\Delta_H \overset{*}{\mathfrak{H}}{}^1 \varphi + \frac{1}{4\pi} \int\limits_{\mathbb{R}} \varphi(a_t \xi_0)\, dt. \tag{3.35}$$

(iii) *If $n = 4, 6, \ldots$, then*

$$f = \mathcal{P}_{n/2}(\Delta_H) \overset{*}{\mathfrak{H}}{}^{n-d} \varphi. \tag{3.36}$$

4. The totally geodesic Radon transforms

The Radon transforms in this section are defined by integration over totally geodesic submanifolds of $\mathbb{H}^n$. These transforms are called *totally geodesic Radon transforms*. The exposition in this section mainly relies on our works [**8**, **47**, **48**]. Many related results can be found in the bibliography at the end of the paper.

4.1. Preliminaries. For the sake of convenience, we will abuse notation and denote by Ξ_d the set of all d-dimensional totally geodesic submanifolds $\xi \subset \mathbb{H}^n$, $1 \le d \le n-1$. This should cause no confusion on the readers part as section 2 and this section are independent. Each ξ represents a cross-section of the hyperboloid $\mathbb{H}^n$ by a $(d+1)$-dimensional plane through the origin of $\mathbb{E}^{n,1}$.

We write $\mathbb{E}^{n,1} = \mathbb{R}^{n-d} \times \mathbb{E}^{d,1}$, where

$$\mathbb{R}^{n-d} = \mathbb{R}e_1 \times \ldots \times \mathbb{R}e_{n-d}, \qquad \mathbb{E}^{d,1} \sim \mathbb{R}^{d+1} = \mathbb{R}e_{n-d+1} \times \ldots \times \mathbb{R}e_{n+1},$$

$e_1, \ldots, e_{n+1}$ being coordinate unit vectors. Everywhere in the following, $x_0 = (0, \ldots, 0, 1)$ and $\xi_0 = \mathbb{H}^n \cap \mathbb{E}^{d,1} = \mathbb{H}^d$ denote the origins in $\mathbb{H}^n$ and Ξ_d, respectively. The subgroup $H = SO(n-d) \times SO_0(d,1)$ is the isotropy subgroup of ξ_0, so that $\Xi_d = G/H$.

Once the invariant measures dg on G and dh on H have been determined in a canonical way, we can define an invariant measure $d\xi = d(gH)$ on $\Xi_d = G/H$ normalized by

$$\int\limits_G \Phi(g)\, dg = \int\limits_{G/H} d(gH) \int\limits_H \Phi(gh)\, dh \quad \text{(see [26, p. 91])}. \tag{4.1}$$

The following statement gives this measure precise meaning.

LEMMA 4.1. [**8**, p. 43] *If $\varphi \in L^1(\Xi_d)$, then*

$$\int\limits_{\Xi_d} \varphi(\xi)\, d\xi = \sigma_{n-d-1} \int\limits_0^\infty \Big[\int\limits_K \varphi(k g_\theta^{-1} \xi_0) dk \Big]\, d\nu(\theta), \tag{4.2}$$

where $d\nu(\theta) = (\sinh\theta)^{n-d-1} (\cosh\theta)^d\, d\theta$,

$$g_\theta = \begin{bmatrix} \cosh\theta & 0 & \sinh\theta \\ 0 & I_{n-1} & 0 \\ \sinh\theta & 0 & \cosh\theta \end{bmatrix}. \tag{4.3}$$

Given $x \in \mathbb{H}^n$ and $\xi \in \Xi_d$, let r_x and r_ξ $(\in G)$ be pseudo-rotations satisfying $r_x x_0 = x$, $r_\xi \xi_0 = \xi$. We set

$$f_\xi(x) = f(r_\xi x), \qquad \varphi_x(\xi) = \varphi(r_x \xi).$$

In this notation, the totally geodesic Radon transform and its dual are defined by the formulas

$$(Rf)(\xi) = \int_{SO_0(d,1)} f_\xi(\gamma x_0)\, d\gamma, \qquad (R^*\varphi)(x) = \int_K \varphi_x(k\xi_0)\, dk. \tag{4.4}$$

One can show [**8**, p. 48] that

$$\int_{\Xi_d} (Rf)(\xi) \frac{d\xi}{(\cosh d(e_{n+1}, \xi))^n} = \int_{\mathbb{H}^n} f(x) \frac{dx}{x_{n+1}^{n-d}}. \tag{4.5}$$

THEOREM 4.2. ([**8**, p. 48], see also [**57**]) *Let*

$$f \in L^p(\mathbb{H}^n), \qquad 1 \le p < (n-1)/(d-1), \qquad 1 \le d \le n-1.$$

Then $(Rf)(\xi)$ *is finite for almost all* $\xi \in \Xi_d$. *If* $f \in C_\mu(\mathbb{H}^n)$, $\mu > d-1$, *then* $(Rf)(\xi)$ *is finite for all* $\xi \in \Xi_d$.

The conditions $p < (n-1)/(d-1)$ and $\mu > d-1$ in this theorem are sharp. For example, if $p \ge (n-1)/(d-1)$, the function

$$f_0(x) = (x_{n+1}^2 - 1)^{(2-n)/2p} (1 + x_{n+1})^{-1/p} (\log(1 + x_{n+1}))^{-1}$$

belongs to $L^p(\mathbb{H}^n)$, but $(Rf_0)(\xi) \equiv \infty$.

4.2. Inversion formulas. There exist a variety of different inversion formulas for the totally geodesic Radon transform Rf. Some of these formulas work well for any $f \in L^p(\mathbb{H}^n)$, $1 \le p < (n-1)/(d-1)$, while others are applicable only to sufficiently smooth f having "good" behavior at infinity. All known methods actually rely on certain averaging procedure combined with differentiation, the order of which can be half-integer, depending on parity of d. The basic idea behind all statements in this subsection is to reconstruct the hyperbolic spherical mean (3.12) and then pass to the limit according to Lemma 3.3.

The standard averaging operator is the *shifted dual Radon transform*, which averages a function φ on Ξ_d over all $\xi \in \Xi_d$ at a fixed geodesic distance θ from x. This operator has the form

$$(R_x^*\varphi)(r) = \int_K \varphi_x(\gamma g_\theta^{-1}\xi_0)\, d\gamma, \qquad \sinh\theta = r. \tag{4.6}$$

The case $r = 0$ corresponds to the usual dual Radon transform as in (4.4). The first use of this operator in the hyperbolic setting is due to Helgason [**25**]. The term *shifted* is due to Rouvière [**46**] who studied such operators in the more general group-theoretic setting.

To formulate the inversion result, we invoke the Erdélyi-Kober type fractional integrals

$$(I_{-,2}^\alpha u)(t) = \frac{2}{\Gamma(\alpha)} \int_t^\infty \frac{u(s)\, s\, ds}{(s^2 - t^2)^{1-\alpha}}, \qquad \alpha > 0,$$

which are modifications of the Riemann-Liouville integrals (3.20). The corresponding fractional derivative satisfying $D_{-,2}^\alpha I_{-,2}^\alpha u = u$ can be represented in different forms depending on the class of functions u and the value of α. More information on fractional differentiation can be found in [**51**, **53**]. For our purposes, when $\alpha = d/2$, we set

$$D^{d/2}_{-,2} u = (-D)^{d/2} u, \qquad D = \frac{1}{2t}\frac{d}{dt},$$

if d is even, and

$$D^{d/2}_{-,2} u = t^{2-d+2m} (-D)^{m+1} t^d v, \quad v = I^{1-d/2+m}_{-,2} t^{-2m-2} u, \quad m = [d/2],$$

for any $1 \le d \le n-1$.

THEOREM 4.3. [**50**, p. 192] *A function $f \in L^p(\mathbb{H}^n)$, where $1 \le p < (n-1)/(d-1)$, can be recovered from $\varphi = Rf$ by the formula*

$$f(x) = \lim_{t\to 0} \pi^{-d/2} (D^{d/2}_{-,2} R^*_x \varphi)(t), \tag{4.7}$$

where the limit is understood in the L^p-norm. If $f \in C_\mu(\mathbb{H}^n)$, $\mu > d-1$, then (4.7) holds uniformly.

Alternative inversion formulas for totally geodesic Radon transform of L^p functions in terms of wavelet transforms or hypersingular integrals of the Marchaud type were obtained in [**7**, **8**, **47**, **49**]. In these formulas, due to the properties of the corresponding maximal functions, the existence of the limit is guaranteed not only in the L^p-norm, as in Theorem 4.3, but also in the almost everywhere sense. For example, in the case $d = 1$, the hyperbolic X-ray transform can be inverted by the following theorem.

THEOREM 4.4. [**47**, p. 3018] *Let $\varphi = Rf$, $f \in L^p(\mathbb{H}^n)$, $1 \le p < \infty$. Then*

$$f(x) = \frac{1}{\pi} \int_0^\infty \frac{(R^*\varphi)(x) - (R^*_x \varphi)(r)}{r^2}\, dr, \qquad \int_0^\infty = \lim_{\varepsilon \to 0} \int_\varepsilon^\infty, \tag{4.8}$$

where the limit is understood in the L^p-norm and in the a.e. sense. If $f \in C_\mu(\mathbb{H}^n)$ for some $\mu > 0$, then the limit in (4.8) holds uniformly.

4.2.1. *Inversion on smooth functions.* The approach presented below relies on application of polynomials of the Beltrami-Laplace operator Δ_H to the corresponding hyperbolic potentials. An idea of this approach belongs to Helgason [**28**, p. 125] who considered the case of d even. To cover all $0 < d < n$, we proceed as in [**48**] and introduce the following fractional analogues of the Radon transforms (4.4):

$$(R^\alpha f)(\xi) = \int_{\mathbb{H}^n} f(x)\, r_\alpha(x,\xi)\, dx, \quad (\overset{*}{R}{}^\alpha \varphi)(x) = \int_{\Xi_d} \varphi(\xi)\, r_\alpha(x,\xi)\, d\xi.$$

Here $Re\,\alpha > 0$ and the kernel $r_\alpha(x,\xi)$ is defined as follows. If $\alpha + d - n \ne 0, 2, 4, \ldots$, then

$$r_\alpha(x,\xi) = \gamma_{n,d}(\alpha) \big(\sinh d(x,\xi)\big)^{\alpha+d-n},$$

$$\gamma_{n,d}(\alpha) = \frac{2^{-\alpha-d}\,\Gamma((n-\alpha-d)/2)\,\Gamma((n-d)/2)}{\pi^{n/2}\Gamma(n/2)\,\Gamma(\alpha/2)}.$$

If $\alpha + d - n = 0, 2, 4, \ldots$, then

$$r_\alpha(x,\xi) = \gamma'_{n,d}(\alpha) \big(\sinh d(x,\xi)\big)^{\alpha+d-n} \log\big(\sinh d(x,\xi)\big),$$

$$\gamma'_{n,d}(\alpha) = \frac{2^{1-\alpha-d}(-1)^{(\alpha+d-n)/2+1}\,\Gamma((n-d)/2)}{\pi^{n/2}\,\Gamma(n/2)\,\Gamma(\alpha/2)\,((\alpha+d-n)/2)!}.$$

Analytic families $\{R^\alpha\}$ and $\{\overset{*}{R}{}^\alpha\}$ include the Radon transform and its dual in the following sense.

LEMMA 4.5. [**48**, p. 217] *If $f(x)$ and $\varphi(\xi)$ are compactly supported continuous functions, then*

$$\lim_{\alpha\to 0} R^\alpha f = cRf, \qquad \lim_{\alpha\to 0} \overset{*}{R}{}^\alpha \varphi = cR^*\varphi, \qquad c = \frac{2^{-d}\Gamma\big((n-d)/2\big)}{\pi^{d/2}\Gamma(n/2)}.$$

We also need the following integral operators of the potential type which play the same role in our consideration as the operators Q^α (see (3.16)) in the study of the horospherical transforms. For $Re\,\alpha > 0$, $\alpha - n \neq 0, 2, 4, \dots$, let

$$(K^\alpha f)(x) = c_{n,\alpha} \int_{\mathbb{H}^n} f(y)\big(\sinh d(x,y)\big)^{\alpha-n} dy, \tag{4.9}$$

$$c_{n,\alpha} = \frac{2^{-\alpha}\Gamma\left((n-\alpha)/2\right)}{\pi^{n/2}\,\Gamma(\alpha/2)}.$$

In the case $\alpha = n, n+2, n+4, \dots$, we set

$$(K^\alpha f)(x) = c'_{n,\alpha} \int_{\mathbb{H}^n} f(y)\big(\sinh d(x,y)\big)^{\alpha-n} \log\big(\sinh d(x,y)\big) dy, \tag{4.10}$$

$$c'_{n,\alpha} = \frac{2^{1-\alpha}(-1)^{(\alpha-n)/2+1}}{((\alpha-n)/2)!\;\pi^{n/2}\Gamma(\alpha/2)}.$$

The expression (4.10) can be obtained from (4.9) by continuity $\lim\limits_{\alpha\to n+2m} K^\alpha f = K^{n+2m} f$ if we assume that

$$\int_{\mathbb{H}^n} f(x) x^j\, dx = 0 \quad \forall\, |j| \le m.$$

LEMMA 4.6. [**48**, p. 218] *For $\alpha > 0$, $\alpha + d - n \neq 0, 2, 4, \dots$,*

$$\overset{*}{R}{}^\alpha R f = K^{\alpha+d} f, \qquad R^* R^\alpha f = K^{\alpha+d} f \tag{4.11}$$

provided that either side of the corresponding equality is finite when f is replaced by $|f|$. In the case $a = 0$, we have

$$R^* R f = K^d f. \tag{4.12}$$

For $\alpha = n - d$,

$$\overset{*}{R}{}^{n-d} R f = R^* R^{n-d} f = K^n f + c_f, \tag{4.13}$$

$$c_f = \frac{\psi(n/2) - \psi((n-d)/2)}{2^n \pi^{n/2}\,\Gamma(n/2)} \int_{\mathbb{H}^n} f(x)\, dx, \qquad \psi(z) = \frac{\Gamma'(z)}{\Gamma(z)}.$$

THEOREM 4.7. [**48**, p. 219] *Let $f \in C_c^\infty(\mathbb{H}^n)$,*

$$\Delta_\alpha = -\Delta_H + (\alpha - n)(\alpha - 1), \qquad \alpha \ge 2.$$

If $\alpha - n \neq 0, 2, 4, \dots$, then

$$\Delta_\alpha K^\alpha f = K^{\alpha-2} f \qquad (K^0 f = f). \tag{4.14}$$

For $\alpha = n + 2m$, $m = 0, 1, 2, \ldots$, this equality holds provided

$$\int_{\mathbb{H}^n} f(y)\, y^j\, dy = 0 \qquad \forall\, |j| = 0, 2, \ldots, 2m.$$

For $\alpha = n$,

$$-\Delta_H K^n f = a_f + \begin{cases} K^{n-2} f, & \text{if } n > 2, \\ f, & \text{if } n = 2, \end{cases}$$

$$a_f = \frac{2^{1-n}(n-1)}{\pi^{n/2}\,\Gamma(n/2)} \int_{\mathbb{H}^n} f(x)\, dx.$$

Let

$$P_m(\Delta_H) = \prod_{i=1}^{m} \big[-\Delta_H + (2i-n)(2i-1) \big].$$

If d is even, then (4.12) and (4.14) give the following nice formula which is due to Helgason (cf. [**28**, p. 125]):

$$cf = P_{d/2}(\Delta_H) R^* R f, \qquad c = \frac{(-4\pi)^{d/2}\,\Gamma(n/2)}{\Gamma\big((n-d)/2\big)}. \tag{4.15}$$

In the general case, Theorem 4.7 yields the following inversion result.

THEOREM 4.8. [**48**, p. 208] *Let*

$$\varphi = Rf, \qquad f \in C_c^\infty(\mathbb{H}^n), \qquad 1 \le d \le n-1.$$

(i) *If n is odd, then*

$$f = P_m(\Delta)\, \overset{*}{R}{}^{2m-d} \varphi \quad \forall\, m \ge d/2. \tag{4.16}$$

(ii) *If n is even, then (4.16) is applicable only for $d/2 \le m \le n/2 - 1$, and another inversion formula holds:*

$$\begin{aligned} f(x) &= -P_{n/2}(\Delta_H)\Big[\frac{2^{1-n}}{\pi^{n/2}\Gamma(n/2)} \int_{\Xi_d} \varphi(\xi) \log\big(\sinh d(x,\xi)\big)\, d\xi\Big] \\ &+ \frac{(-1)^{n/2}\Gamma\big((n+1)/2\big)}{\pi^{(n+1)/2}} \int_{\Xi_d} \varphi(\xi)\, d\xi. \end{aligned} \tag{4.17}$$

EXAMPLE 4.9. In the case $n = 2$, the Radon transform $\varphi = Rf$ can be inverted by the formula

$$f(x) = \Delta_H \Big[\frac{1}{2\pi} \int_{\Xi_d} \varphi(\xi) \log\big(\sinh d(x,\xi)\big) d\xi\Big] + \frac{1}{2\pi} \int_{\Xi_d} \varphi(\xi) d\xi. \tag{4.18}$$

Acknowledgment

The authors are thankful to the referee for valuable comments and suggestions.

References

[1] N. B. Andersen, G. Ólafsson, and H. Schlichtkrull, *On the inversion of the Laplace and Abel transforms for causal symmetric spaces*, Forum Math. **15** (2003), no. 5, 701–725, DOI 10.1515/form.2003.038. MR2010031

[2] C. Berenstein, *Radon transforms, wavelets, and applications*, Integral geometry, Radon transforms and complex analysis (Venice, 1996), Lecture Notes in Math., vol. 1684, Springer, Berlin, 1998, pp. 1–33, DOI 10.1007/BFb0096089. MR1635610

[3] C. A. Berenstein and E. Casadio Tarabusi, *Inversion formulas for the k-dimensional Radon transform in real hyperbolic spaces*, Duke Math. J. **62** (1991), no. 3, 613–631, DOI 10.1215/S0012-7094-91-06227-7. MR1104811

[4] C. A. Berenstein and E. Casadio Tarabusi, *On the Radon and Riesz transforms in real hyperbolic spaces*, Geometric analysis (Philadelphia, PA, 1991), Contemp. Math., vol. 140, Amer. Math. Soc., Providence, RI, 1992, pp. 1–21, DOI 10.1090/conm/140/1197583. MR1197583

[5] C. A. Berenstein and E. C. Tarabusi, *Range of the k-dimensional Radon transform in real hyperbolic spaces*, Forum Math. **5** (1993), no. 6, 603–616, DOI 10.1515/form.1993.5.603. MR1242891

[6] C. A. Berenstein and E. Casadio Tarabusi, *An inversion formula for the horocyclic Radon transform on the real hyperbolic space*, Tomography, impedance imaging, and integral geometry (South Hadley, MA, 1993), Lectures in Appl. Math., vol. 30, Amer. Math. Soc., Providence, RI, 1994, pp. 1–6. MR1297561

[7] C. A. Berenstein and B. Rubin, *Radon transform of L^p-functions on the Lobachevsky space and hyperbolic wavelet transforms*, Forum Math. **11** (1999), no. 5, 567–590, DOI 10.1515/form.1999.014. MR1705903

[8] C. A. Berenstein and B. Rubin, *Totally geodesic Radon transform of L^p-functions on real hyperbolic space*, Fourier analysis and convexity, Appl. Numer. Harmon. Anal., Birkhäuser Boston, Boston, MA, 2004, pp. 37–58. MR2087237

[9] W. O. Bray, *Aspects of harmonic analysis on real hyperbolic space*, Fourier analysis (Orono, ME, 1992), Lecture Notes in Pure and Appl. Math., vol. 157, Dekker, New York, 1994, pp. 77–102. MR1277819

[10] W. O. Bray and B. Rubin, *Inversion of the horocycle transform on real hyperbolic spaces via a wavelet-like transform*, Analysis of divergence (Orono, ME, 1997), Appl. Numer. Harmon. Anal., Birkhäuser Boston, Boston, MA, 1999, pp. 87–105. MR1731261

[11] W. O. Bray and B. Rubin, *Radon transforms over lower-dimensional horospheres in real hyperbolic space*, Transactions of the Amer. Math. Soc., vol. 372, Number 2, 2019, 1091–1112.

[12] W. O. Bray and D. C. Solmon, *The horocycle transform and harmonic analysis on the Poincaré disk*, Preprint (1988).

[13] J. W. Cannon, W. J. Floyd, R. Kenyon, and W. R. Parry, *Hyperbolic geometry*, Flavors of geometry, Math. Sci. Res. Inst. Publ., vol. 31, Cambridge Univ. Press, Cambridge, 1997, pp. 59–115. MR1491098

[14] B. Fridman, P. Kuchment, K. Lancaster, D. Ma, M. Mogilevsky, S. Lissianoi, V. Papanicolaou, and I. Ponomaryov, *Numerical harmonic analysis on the hyperbolic plane*, Proceedings of the ISAAC Congress, Delaware, June 1997.

[15] B. Fridman, P. Kuchment, D. Ma, and V. G. Papanicolaou, *Solution of the linearized inverse conductivity problem in a half space via integral geometry*, Voronezh Winter Mathematical Schools, Amer. Math. Soc. Transl. Ser. 2, vol. 184, Amer. Math. Soc., Providence, RI, 1998, pp. 85–95, DOI 10.1090/trans2/184/06. MR1729927

[16] B. Fuglede, *An integral formula*, Math. Scand. **6** (1958), 207–212, DOI 10.7146/math.scand.a-10545. MR0105724

[17] P. Funk, *Über Flächen mit lauter geschlossenen geodätischen Linien* (German), Math. Ann. **74** (1913), no. 2, 278–300, DOI 10.1007/BF01456044. MR1511763

[18] R. J. Gardner, *Geometric tomography*, 2nd ed., Encyclopedia of Mathematics and its Applications, vol. 58, Cambridge University Press, New York, 2006. MR2251886

[19] I. M. Gelfand, S. G. Gindikin, and M. I. Graev, *Selected topics in integral geometry*, Translations of Mathematical Monographs, vol. 220, American Mathematical Society, Providence, RI, 2003. Translated from the 2000 Russian original by A. Shtern. MR2000133

[20] I. M. Gel′fand, M. I. Graev, and N. Ya. Vilenkin, *Generalized functions. Vol. 5: Integral geometry and representation theory*, Translated from the Russian by Eugene Saletan, Academic Press, New York-London, 1966. MR0207913

[21] S. Gindikin, *Integral geometry on real quadrics*, Lie groups and Lie algebras: E. B. Dynkin's Seminar, Amer. Math. Soc. Transl. Ser. 2, vol. 169, Amer. Math. Soc., Providence, RI, 1995, pp. 23–31, DOI 10.1090/trans2/169/03. MR1364451

[22] S. Gindikin, *Integral geometry on hyperbolic spaces*, Harmonic analysis and integral geometry (Safi, 1998), Chapman & Hall/CRC Res. Notes Math., vol. 422, Chapman & Hall/CRC, Boca Raton, FL, 2001, pp. 41–46. MR1789143

[23] S. G. Gindikin, *Horospherical transform on symmetric Riemannian manifolds of noncompact type* (Russian, with Russian summary), Funktsional. Anal. i Prilozhen. **42** (2008), no. 4, 50–59, 111–112, DOI 10.1007/s10688-008-0042-2; English transl., Funct. Anal. Appl. **42** (2008), no. 4, 290–297. MR2492426

[24] S. Gindikin, *Local inversion formulas for horospherical transforms* (English, with English and Russian summaries), Mosc. Math. J. **13** (2013), no. 2, 267–280, 363. MR3134907

[25] S. Helgason, *The totally-geodesic Radon transform on constant curvature spaces*, Integral geometry and tomography (Arcata, CA, 1989), Contemp. Math., vol. 113, Amer. Math. Soc., Providence, RI, 1990, pp. 141–149, DOI 10.1090/conm/113/1108651. MR1108651

[26] S. Helgason, *Groups and geometric analysis*: *Integral geometry, invariant differential operators, and spherical functions*, Pure and Applied Mathematics, vol. 113, Academic Press, Inc., Orlando, FL, 1984. MR754767

[27] S. Helgason, *Geometric analysis on symmetric spaces*, 2nd ed., Mathematical Surveys and Monographs, vol. 39, American Mathematical Society, Providence, RI, 2008. MR2463854

[28] S. Helgason, *Integral geometry and Radon transforms*, Springer, New York, 2011. MR2743116

[29] J. Hilgert, A. Pasquale, and E. B. Vinberg, *The dual horospherical Radon transform for polynomials* (English, with English and Russian summaries), Mosc. Math. J. **2** (2002), no. 1, 113–126, 199. MR1900587

[30] J. Hilgert, A. Pasquale, and E. B. Vinberg, *The dual horospherical Radon transform as a limit of spherical Radon transforms*, Lie groups and symmetric spaces, Amer. Math. Soc. Transl. Ser. 2, vol. 210, Amer. Math. Soc., Providence, RI, 2003, pp. 135–143, DOI 10.1090/trans2/210/10. MR2018358

[31] S. Ishikawa, *The range characterizations of the totally geodesic Radon transform on the real hyperbolic space*, Duke Math. J. **90** (1997), no. 1, 149–203, DOI 10.1215/S0012-7094-97-09006-2. MR1478547

[32] F. John, *The ultrahyperbolic differential equation with four independent variables*, Duke Math. J. **4** (1938), no. 2, 300–322, DOI 10.1215/S0012-7094-38-00423-5. MR1546052

[33] A. Katsevich, *An inversion formula for the dual horocyclic Radon transform on the hyperbolic plane*, Math. Nachr. **278** (2005), no. 4, 437–450, DOI 10.1002/mana.200310251. MR2121570

[34] E. Kelly, *Lower-dimensional horocycles and the Radon transforms on symmetric spaces*, Preprint, 1974.

[35] P. Kuchment, *The Radon transform and medical imaging*, CBMS-NSF Regional Conference Series in Applied Mathematics, vol. 85, Society for Industrial and Applied Mathematics (SIAM), Philadelphia, PA, 2014. MR3137435

[36] Á. Kurusa, *The Radon transform on hyperbolic space*, Geom. Dedicata **40** (1991), no. 3, 325–339, DOI 10.1007/BF00189917. MR1137086

[37] Á. Kurusa, *The invertibility of the Radon transform on abstract rotational manifolds of real type*, Math. Scand. **70** (1992), no. 1, 112–126, DOI 10.7146/math.scand.a-12389. MR1174206

[38] Á. Kurusa, *Support theorems for totally geodesic Radon transforms on constant curvature spaces*, Proc. Amer. Math. Soc. **122** (1994), no. 2, 429–435, DOI 10.2307/2161033. MR1198457

[39] Á. Kurusa, *The totally geodesic Radon transform on the Lorentz space of curvature* -1, Duke Math. J. **86** (1997), no. 3, 565–583, DOI 10.1215/S0012-7094-97-08618-X. MR1432309

[40] Á. Kurusa, *Orbital integrals on the Lorentz space of curvature* -1, Arch. Math. (Basel) **75** (2000), no. 2, 132–146, DOI 10.1007/PL00000433. MR1767164

[41] S. Lissianoi and I. Ponomarev, *On the inversion of the geodesic Radon transform on the hyperbolic plane*, Inverse Problems **13** (1997), no. 4, 1053–1062, DOI 10.1088/0266-5611/13/4/010. MR1463592

[42] P. I. Lizorkin, *Direct and inverse theorems in approximation theory for functions on a Lobachevskiĭ space* (Russian), Trudy Mat. Inst. Steklov. **194** (1992), no. Issled. po Teor. Differ. Funktsiĭ Mnogikh Peremen. i ee Prilozh. 14, 120–147; English transl., Proc. Steklov Inst. Math. **4(194)** (1993), 125–151. MR1289652

[43] M. Morimoto, *Sur les transformations horosphériques généralisées dans les espaces homogènes* (French), J. Fac. Sci. Univ. Tokyo Sect. I **13** (1966), 65–83 (1966). MR0209407

[44] F. Natterer, *The mathematics of computerized tomography*, Classics in Applied Mathematics, vol. 32, Society for Industrial and Applied Mathematics (SIAM), Philadelphia, PA, 2001. Reprint of the 1986 original. MR1847845

[45] J. Radon, *Über die Bestimmung von Funktionen durch ihre Integralwerte längs gewisser Mannigfaltigkeiten*, Ber. Verh. Sächs. Akad. Wiss. Leipzig, Math. - Nat. Kl., **69** (1917), 262–277 (Russian translation in the Russian edition of S. Helgason, The Radon transform, Moscow, Mir, 1983, pp. 134–148).

[46] F. Rouvière, *Inverting Radon transforms: the group-theoretic approach*, Enseign. Math. (2) **47** (2001), no. 3-4, 205–252. MR1876927

[47] B. Rubin, *Helgason-Marchaud inversion formulas for Radon transforms*, Proc. Amer. Math. Soc. **130** (2002), no. 10, 3017–3023, DOI 10.1090/S0002-9939-02-06554-1. MR1908925

[48] B. Rubin, *Radon, cosine and sine transforms on real hyperbolic space*, Adv. Math. **170** (2002), no. 2, 206–223, DOI 10.1006/aima.2002.2074. MR1932329

[49] B. Rubin, *Convolution-backprojection method for the k-plane transform, and Calderón's identity for ridgelet transforms*, Appl. Comput. Harmon. Anal. **16** (2004), no. 3, 231–242, DOI 10.1016/j.acha.2004.03.003. MR2054281

[50] B. Rubin, *On the Funk-Radon-Helgason inversion method in integral geometry*, Geometric analysis, mathematical relativity, and nonlinear partial differential equations, Contemp. Math., vol. 599, Amer. Math. Soc., Providence, RI, 2013, pp. 175–198, DOI 10.1090/conm/599/11908. MR3202479

[51] B. Rubin, *Introduction to Radon transforms: With elements of fractional calculus and harmonic analysis*, Encyclopedia of Mathematics and its Applications, vol. 160, Cambridge University Press, New York, 2015. MR3410931

[52] B. Rubin, *New inversion formulas for the horospherical transform*, J. Geom. Anal. **27** (2017), no. 1, 908–946, DOI 10.1007/s12220-016-9704-0. MR3606575

[53] S. G. Samko, A. A. Kilbas, and O. I. Marichev, *Fractional integrals and derivatives: Theory and applications*, Gordon and Breach Science Publishers, Yverdon, 1993. Edited and with a foreword by S. M. Nikol′skiĭ; Translated from the 1987 Russian original; Revised by the authors. MR1347689

[54] V. I. Semjanistyĭ, *Homogeneous functions and some problems of integral geometry in the spaces of constant curvature*, Soviet Math. Dokl. **2** (1961), 59–62. MR0133006

[55] V. I. Semjanistyĭ, *Some integral transformations and integral geometry in an elliptic space* (Russian), Trudy Sem. Vektor. Tenzor. Anal. **12** (1963), 397–441. MR0166606

[56] V. I. Semjanistyĭ, *Certain problems of integral geometry in pseudo-Euclidean and non-Euclidean spaces* (Russian), Trudy Sem. Vektor. Tenzor. Anal. **13** (1966), 244–302. MR0216290

[57] R. S. Strichartz, *L^p estimates for Radon transforms in Euclidean and non-Euclidean spaces*, Duke Math. J. **48** (1981), no. 4, 699–727, DOI 10.1215/S0012-7094-81-04839-0. MR782573

[58] N. Ja. Vilenkin, *Special functions and the theory of group representations*, Translated from the Russian by V. N. Singh. Translations of Mathematical Monographs, Vol. 22, American Mathematical Society, Providence, R. I., 1968. MR0229863

[59] N. Ja. Vilenkin and A. U. Klimyk, *Representation of Lie groups and special functions. Vol. 2: Class I representations, special functions, and integral transforms*, Mathematics and its Applications (Soviet Series), vol. 74, Kluwer Academic Publishers Group, Dordrecht, 1993. Translated from the Russian by V. A. Groza and A. A. Groza. MR1220225

Department of Mathematics, Missouri State University
Email address: wbray@missouristate.edu

Department of Mathematics, Louisiana State University, Baton Rouge, Louisiana 70803
Email address: borisr@math.lsu.edu

Contemporary Mathematics
Volume **733**, 2019
https://doi.org/10.1090/conm/733/14736

Total positivity, Grassmannian and modified Bessel functions

V. M. Buchstaber and A. A. Glutsyuk

To Selim Grigorievich Krein with gratitude for his mathematical results and Voronezh Winter Mathematical Schools

ABSTRACT. A rectangular matrix is called *totally positive,* (according to F. R. Gantmacher and M. G. Krein) if all its minors are positive. A point of a real Grassmannian manifold $G_{l,m}$ of l-dimensional subspaces in $\mathbb{R}^m$ is called *strictly totally positive* (according to A. E. Postnikov) if one can normalize its Plücker coordinates to make all of them positive. The totally positive matrices and the strictly totally positive Grassmannians, that is, the subsets of strictly totally positive points in Grassmannian manifolds arise in many areas: in classical mechanics (see the book of F. R. Gantmacher and M. G. Krein); in a wide context of analysis, differential equations and probability theory (see the book of S. Karlin); in physics, for example, in construction of solutions of the Kadomtsev-Petviashvili (KP) partial differential equation (see a paper by T. M. Malanyuk, a paper by M. Boiti, F. Pemperini, A. Pogrebkov, a paper of Y. Kodama, L. Williams). Different problems of mathematics, mechanics and physics led to constructions of totally positive matrices by many mathematicians, including F. R. Gantmacher, M. G. Krein, I. J. Schoenberg, S. Karlin, A. E. Postnikov and ourselves. One-dimensional families of totally positive matrices whose entries are modified Bessel functions of the first kind have arisen in our study (in collaboration with S. I. Tertychnyi) of model of the overdamped Josephson effect in superconductivity and double confluent Heun equations related to it.

In the present paper we give a new construction of multidimensional families of totally positive matrices different from the above-mentioned families. Their entries are again formed by values of modified Bessel functions of the first kind, but now with non-negative integer indices. Their columns are numerated by the indices of the modified Bessel functions, and their rows are numerated by their arguments. This yields new multidimensional families of strictly totally positive points in all the Grassmannian manifolds. These families represent images of explicit injective mappings of the convex open simplex $\{x = (x_1, \dots, x_l) \in \mathbb{R}^l \mid 0 < x_1 < \cdots < x_l\} \subset \mathbb{R}^l$ to the Grassmannian manifolds $G_{l,m}$, $l < m$.

2010 *Mathematics Subject Classification.* Primary 33C10; Secondary 15B48.
Supported by part by RFBR grants 14-01-00506 and 17-01-00192.
Supported by part by RFBR grants 16-01-00748, 16-01-00766.

1. Introduction

1.1. Brief survey on totally positive matrices. Main result. The following notion was introduced in the classical books [**16**, **17**] in the context of the classical mechanics.

DEFINITION 1.1. [**1**, **16**, **28**], [**17**, p. 289 of the Russian edition] A rectangular $l \times m$-matrix is called *totally positive* (nonnegative), if its minors of all sizes are positive (nonnegative).

EXAMPLE 1.2. It is known that every generalized Vandermonde matrix

$$(f(x_i, y_j))_{i=1,\dots,m;\ j=1,\dots,n},\ f(x,y) = x^y,$$
$$0 < x_1 < \dots < x_m,\ 0 \leq y_1 < y_2 < \dots < y_n$$

is totally positive [**13**, chapter XIII, section 8].

The study of $n \times n$ matrices with positive elements goes back to Perron [**27**] who had shown that for such a matrix the eigenvalue that is largest in the modulus is simple, real and positive, and the corresponding eigenvector can be normalized to have all the components positive (1907).

In 1908 this result was generalized by G. Frobenius [**13**, chapter 13, section 2] to those matrices with non-negative coefficients that are block-non-decomposable. For each of these matrices he proved that its complex eigenvalues of maximal modulus are roots of a polynomial $P(\lambda) = \lambda^h - r^h$, all of them are simple and one of them is real and positive.

In 1935–1937 F. R. Gantmacher and M. G. Krein [**14**, **15**] observed that if the matrix satisfies a stronger condition of total positivity (in fact a weaker, oscillation property is sufficient), then *all* its eigenvalues are simple, real and positive.

Earlier in 1930 I. Schoenberg [**30**] studied a more general class of matrices including totally positive ones: namely, the $m \times n$-matrices such that for every $k \leq \min\{m, n\}$ all the non-zero minors of order k have the same sign (either all positive, or all negative). He proved important results relating the latter property to variation-diminishing property of the corresponding linear transformations $\mathbb{R}^n \to \mathbb{R}^m$. (Recall that a linear transformation $\mathbb{R}^n \to \mathbb{R}^m$ is called *variation-diminishing,* if it does not increase the number of sign changes in the sequence of coordinates of any vector.) Further results in this direction were obtained by Motzkin in his 1933 dissertation [**26**] and most complete results were obtained by Gantmacher and Krein [**17**].

A two-sided sequence $(a_j)_{j\in\mathbb{Z}}$ of real numbers is called *totally nonnegative (positive),* if the infinite matrix $(a_{j-i})_{i,j\in\mathbb{Z}}$ is totally nonnegative (positive). There is a remarkable result on characterization of totally nonnegative sequences. It says that for each totally nonnegative sequence distinct from a geometric progression the corresponding generating function $\sum_j a_j z^j$ converges as a Laurent series in some annulus, extends meromorphically to all of $\mathbb{C}^*$ and is a product of $\exp(q_1 z + q_2 z^{-1})$ and an infinite product of fractions of appropriate linear functions; here $q_1, q_2 \geq 0$. In 1948 I. Schoenberg [**31**] proved sufficiency: total nonnegativity of Laurent series of the latter functions. The converse (necessity) was proved by A. Edrei in 1953 [**8**]. See also [**20**, Theorem 8.9.5] and references in this book and in [**8**, **31**].

The characterization of all totally nonnegative two-sides sequences is a basic fundamental result used in the description of the characters of representations of the infinite unitary group $U(\infty) = \varinjlim U(n)$ and the infinite symmetric group $S(\infty) =$

$\varinjlim S(n)$. See papers by E. Thoma [**33**], D. Voiculescu [**36**], joint papers by A. M. Vershik and S. V. Kerov [**34, 35**] and references therein.

Many results on characterization and properties of totally positive matrices and their relations to other areas of mathematics (e.g., combinatorics, dynamical systems, geometry and topology, probability theory, Fourier analysis, representation theory), mechanics and physics are given in [**1–4, 8–24, 28, 29, 31, 34, 35**] (see also references therein). F. R. Gantmacher and M. G. Krein [**16, 17**] considered totally positive matrices in the context of applications to mechanical problems. S. Karlin [**20**] considered them in a wide context of analysis, differential equations and probability theory. In 2008 G. Lusztig suggested an analogue of the theory of total positivity in the Lie group context [**22**].

Total positivity was used to construct solutions of the Kadomtsev-Petviashvili (KP) differential equation in a paper by T. M. Malanyuk [**25**], a paper of M. Boiti, F. Pempinelli, A. Pogrebkov [**4**, section II], and a paper of Y. Kodama, L. Williams [**21**]. S. Fomin's talk at the ICM-2010 [**11**] was devoted to deep relations between total positivity and cluster algebras. There exist several approaches to construction of totally positive matrices, see [**5, 16, 20, 29**], [**17**, p. 290 of Russian edition]. In the previous paper [**5**] we have constructed a class of explicit one-dimensional families of totally positive matrices given by a finite collection of double-sided infinite vector functions, whose components are modified Bessel functions of the first kind[1]. Matrices of such kind arise in a paper of V. M. Buchstaber and S. I. Tertychnyi [**6**] in the construction of solutions of the non-linear differential equations in a model of overdamped Josephson junction in superconductivity. It was shown in [**5**] that the nature of the modified Bessel functions as coefficients of appropriate generating function implies that the infinite vector formed by appropriate minors of the above-mentioned matrices satisfies the *differential-difference heat equation* with positive constant potential in the Hilbert space l_2.

In the present paper we provide a new construction of multidimensional family of totally positive matrices formed by a finite collection of one-sided infinite vector functions. This family is parameterized by a domain in $\mathbb{R}^l$. The rows of the matrix correspond to coordinates x_i in $\mathbb{R}^l$, and their entries are modified Bessel functions of the corresponding coordinates.

DEFINITION 1.3. (see [**20**, chapter 2, Definition 1.1, p. 46]). A function $K(x, y)$ on a product of two totally ordered sets $X \times Y$ is called a *totally positive (strictly totally positive) kernel* of order $r \in \mathbb{N}$, if for every $1 \leq m \leq r$, $x_1 < \cdots < x_m$ in X and $y_1 < \cdots < y_m$ in Y the determinant of the matrix $(K(x_i, y_j))_{i,j=1}^m$ is nonnegative (respectively, positive).

We recall that the modified Bessel functions $I_j(y)$ of the first kind are Laurent series coefficients for the family of analytic functions

$$g_y(z) := e^{\frac{y}{2}(z+\frac{1}{z})} = \sum_{j=-\infty}^{+\infty} I_j(y) z^j.$$

[1]After our paper [**5**] was published, we have found that total positivity of the matrices considered there, i.e., [**5**, Theorem 1.3], was not new, and it follows from [**20**, Theorem 10.1 (a), p.428]. At the same time, in [**5**] we have elaborated a new method of proof of their total positivity. This method is again used in the present paper.

Equivalently, they are defined by the integral formulas

$$I_j(y) = \frac{1}{\pi}\int_0^{\pi} e^{y\cos\phi}\cos(j\phi)d\phi,\ j\in\mathbb{Z}.$$

REMARK 1.4. It is known that the infinite matrix $(A_{km})_{k,m\in\mathbb{Z}}$ with $A_{km} = I_{m-k}(x)$ is totally positive for every $x>0$, see [**20**, Theorem 10.1 (a), p. 428], [**5**, Theorem 1.3] and Footnote 1.

THEOREM 1.5. *For every $r\in\mathbb{N}$ the function*

$$K(x,j) = I_j(x),\quad j\in\mathbb{Z}_{\geq 0},\ x>0$$

is a strictly totally positive kernel of order r with $X=\mathbb{R}_+$ and $Y=\mathbb{Z}_{\geq 0}$.

Let us reformulate Theorem 1.5 in a more explicit way. To do this, set

$$X_l = \{x=(x_1,\dots,x_l)\in\mathbb{R}_+^l \mid x_1<x_2<\dots<x_l\};$$
$$K_m = \{k=(k_1,\dots,k_m)\in\mathbb{Z}_{\geq 0}^m \mid k_1<k_2<\dots<k_m\}.$$

For every $x\in X_l$ and $k\in K_m$ set

$$A_{k,x} = (a_{ij})_{i=1,\dots,l;\ j=1,\dots m},\ a_{ij} = I_{k_j}(x_i). \tag{1.1}$$

In the special case, when $l=m$, set

$$f_k(x) = \det A_{k,x}. \tag{1.2}$$

THEOREM 1.6. *For every $m\in\mathbb{N}$, $k\in K_m$ and $x\in X_m$ one has $f_k(x)>0$.*

Theorem 1.6 is equivalent to Theorem 1.5 and will be proved in Section 2. The following statement is an immediate consequence of the theorem.

COROLLARY 1.7. *For every $x=(x_1,\dots,x_l)\in X_l$ the one-sided infinite matrix formed by the values $a_{ij}=I_j(x_i)$, $i=1,\dots,l$, $j=0,1,2,\dots$ is totally positive.*

REMARK 1.8. Various necessary and sufficient conditions on a kernel K to be (strictly) totally positive were stated and proved in S. Karlin's book [**20**, chapter 2]. If $K(x,y)$ is defined on the product of two intervals and is smooth enough, a sufficient condition for its strict total positivity is that a suitable matrix of partial derivatives of K (which is a matrix function of (x,y)) is strictly totally positive for all (x,y) [**20**, chapter 2, Theorem 2.6, p. 55]. (The same condition written in the form of non-strict inequality is necessary for total positivity, see [**20**, chapter 2, Theorem 2.2, p. 51].) In [**20**, chapter 3, p. 109] S. Karlin presented a construction of totally positive kernel coming from a single modified Bessel function of the first kind. Namely, set

$$\kappa_\alpha(x;\lambda) = \begin{cases} e^{-(x+\lambda)}(\frac{x}{\lambda})^{\frac{\alpha}{2}}I_\alpha(2\sqrt{x\lambda}) \text{ for } x\geq 0 \\ 0 \text{ for } x<0 \end{cases},$$

$$K_\alpha(x,y) = \kappa_\alpha(x-y;\lambda).$$

It was shown in [**20**, chapter 3, p. 109] that for every $\alpha>1$ and every $r<\alpha+2$ the function $K_\alpha(x,y)$ is a totally positive kernel of order r. S. Karlin also presented constructions of totally positive matrices coming from the Green function of a linear differential operator presented as a product of n first order linear differential operators of appropriate type. All his operators act on functions of a variable x. Karlin's constructions dealt with the fundamental solution $\phi(x,t)$ of the corresponding linear differential equation: $\phi(x,t)=0$ for all $x\leq t$, $\phi_x^{(j)}(t,t)=0$ for all

$j < n-1$, the derivative $\phi_x^{(n-1)}(x,t)$ has a discontinuity at $x = t$. Karlin showed [**20**, p. 503] that appropriate class of matrices of type $\phi(x_i, t_j)$ have positive determinants. A similar construction of totally positive matrices was associated to any given classical Sturm–Liouville differential operator [**20**, pp. 535–538].

Let us emphasize that each totally positive matrix given by our construction includes values of modified Bessel functions with different indices, which are solutions of Sturm–Liouville equations with different parameters. On the other hand, each totally positive matrix from the above-mentioned Karlin's construction is expressed via either just one given modified Bessel function, or via fundamental solutions of just one given linear equation. Therefore, our result is not covered by Karlin's constructions.

It is known that the modified Bessel functions $I_\nu(x)$ of the first kind are given by the series

$$I_\nu(x) = (\frac{1}{2}x)^\nu \sum_{k=0}^{\infty} \frac{(\frac{1}{4}x^2)^k}{k!\Gamma(\nu+k+1)},$$

and the latter series extends them to all the real values of the index ν.

Thus, the modified Bessel functions of the first kind yield examples of totally positive kernels of two following different kinds. For every $r \in \mathbb{N}$, Karlin's example, see Remark 1.8, yields a totally positive kernel $K_\alpha(x,y) = \kappa_\alpha(x-y;\lambda)$ of order r constructed from just one modified Bessel function I_α with arbitrary real index $\alpha > 1$ such that $r < \alpha + 2$. Our main result (Theorem 1.5) gives other, strictly totally positive kernel $K(y,s) = I_s(y)$ depending on $y \in \mathbb{R}_+$ and $s \in \mathbb{Z}_{\geq 0}$.

Open Question. *Is it true that the determinants $f_k(x)$ in (1.2) with $x \in X_m$ are all positive for every $m \in \mathbb{N}$ and every $k = (k_1, \dots, k_m)$ with* **(may be non-integer)** k_j and $k_1 \in \mathbb{R}_{\geq 0}$, $k_1 < \cdots < k_m$? *In other words, is it true that the kernel $K(y,s) = I_s(y)$ is strictly totally positive as a function in $(y,s) \in \mathbb{R}_+ \times \mathbb{R}_{\geq 0}$?*

1.2. A brief survey on total positivity in Grassmannian manifolds and Lie groups. A point L of Grassmannian manifold $G_{l,m}$ of l-subspaces in $\mathbb{R}^m$, $m > l$ is represented by an $l \times m$-matrix, whose rows form a basis of the subspace represented by the point L. Recall that the *Plücker coordinates* of the point L are the l-minors of this matrix. The Plücker coordinates of the point are well-defined up to multiplication by a common factor, and they are considered as homogeneous coordinates representing a point of the projective space $\mathbb{RP}^N$, $N = \binom{m}{l} - 1$. The Plücker coordinates induce the Plücker embedding of the Grassmannian manifold to $\mathbb{RP}^N$.

DEFINITION 1.9. A point $L \in G_{l,m}$ is called *strictly totally positive,* if it can be represented by a matrix with all the maximal minors positive.

A. E. Postnikov's paper [**29**] deals with the matrices $l \times m$, $m \geq l$ of rank l satisfying the condition of nonnegativity of its minors of the maximal size. One of its main results provides an explicit combinatorial cell decomposition of the corresponding subset in the Grassmannian $G_{l,m}$, called the *totally nonnegative Grassmannian.* The cells are coded by combinatorial types of appropriate planar networks. K. Talaska [**32**] developed further and generalized Postnikov's result. In particular, for a given point of the totally nonnegative Grassmannian the results of [**32**] allow to decide what is its ambient cell and what are its affine coordinates in the cell. S. Fomin and A. Zelevinsky [**12**] studied a more general notion of total positivity (nonnegativity) for elements of a semisimple complex Lie group with a given double Bruhat cell decomposition. They have proved that the totally positive parts of the double Bruhat cells are bijectively parameterized by the product of the positive quadrant $\mathbb{R}_+^m$ and the positive subgroup of the maximal torus. For other results on totally positive (nonnegative) Grassmannians see [**19**].

Theorem 1.6 of the present paper implies the following corollary.

COROLLARY 1.10. *For every $l, m \in \mathbb{N}$, $l < m$, and every $k \in K_m$ the mapping $H_k : X_l \to G_{l,m}$ sending x to the l-subspace in $\mathbb{R}^m$ generated by the vectors*

$$v_k(x_i) = (I_{k_1}(x_i) \dots I_{k_m}(x_i)), \ i = 1, \dots, l$$

is well-defined and injective. Its image is contained in the open subset of strictly totally positive points.

PROOF. The well-definedness and positivity of Plücker coordinates are obvious, since the l-minors of the matrix $A_{k,x}$ are positive, by Theorem 1.6. Let us prove injectivity. First let us fix two distinct points $x, y \in X_l$. Let us show that $H_k(x) \neq H_k(y)$. Fix a component y_i that is different from every component x_j of the vector x. Then the vectors $v_k(x_1), \dots, v_k(x_l), v_k(y_i)$ are linearly independent, since every $(l+1)$-minor of the matrix formed by them is non-zero, by Theorem 1.6. Hence, $v_k(y_i)$ is not contained in the l-subspace $H_k(x)$, which is generated by the vectors $v_k(x_1), \dots, v_k(x_l)$. Thus, $H_k(y) \neq H_k(x)$. □

EXAMPLE 1.11. Consider the infinite matrix with elements

$$A_{ms} = I_{s-m}(x), \quad m, s \in \mathbb{Z}.$$

It is known, see [**20**, Theorem 10.1 (a), p. 428], [**5**, Theorem 1.3], that this matrix is totally positive for every $x > 0$. Therefore, the subspace generated by any of its l rows is l-dimensional, and it represents a strictly totally positive point of the infinite-dimensional Grassmannian manifold of l-subspaces in the infinite-dimensional vector space. Every submatrix in A_{ms} given by its l rows and a finite number $r > l$ of columns represents a strictly totally positive point of the Grassmannian manifold $G_{l,r}$.

2. Positivity. Proof of Theorem 1.6

We prove Theorem 1.6 by induction in m. In its proof we will use the following classical properties of the modified Bessel functions I_j of the first kind [**37**, section 3.7].

$$I_j = I_{-j}; \tag{2.1}$$

$$I_j|_{\{y>0\}} > 0; \ I_j(0) = 0 \text{ for } j \neq 0; \ I_0(0) > 0; \tag{2.2}$$

$$I_0' = I_1; \ I_j' = \frac{1}{2}(I_{j-1} + I_{j+1}). \tag{2.3}$$

For $m = 1$ the statement of the theorem follows obviously from (2.2).

Let the statement of the theorem be proved for $m = m_0$. Let us prove it for $m = m_0 + 1$. To do this, consider the sequence of the determinants $f_k(x)$ for all $k \in K_m$ as an infinite-dimensional vector function

$$(f_k(x_1, w))_{k \in K_m} \tag{2.4}$$

in the new variables

$$(x_1, w), \quad w = (w_2, \dots, w_m), \ w_j = x_j - x_1; \quad w \in X_{m-1}.$$

The proof of the induction step is analogous to the arguments from [**5**, section 2]. We show (the next two propositions and corollary) that the vector function (2.4) with fixed w and variable x_1 satisfies a linear bounded autonomous ordinary differential equation on the Hilbert space l_2 with coordinates f_k, $k \in K_m$. We show that the positive quadrant $\{f_k \geq 0 \mid k \in K_m\} \subset l_2$ is invariant under the positive flow of the vector field defining this differential equation, and that the initial value $(f_k(0, w))_{k \in K_m}$ lies there. This implies that $f_k(x_1, w) \geq 0$ for all $x_1 \geq 0$, and then we easily deduce that the latter inequality is strict for $x_1 > 0$.

Let us recall the notion of the discrete Laplacian Δ_{discr} acting on functions on the Cayley graph of the additive group $\mathbb{Z}^m$. Namely, for every $j = 1, \dots, m$ let T_j denote the corresponding shift operator:

$$(T_j f)(k) = f(k_1, \dots, k_{j-1}, k_j - 1, k_{j+1}, \dots, k_l).$$

Then

$$\Delta_{discr} = \frac{1}{2} \sum_{j=1}^{m} (T_j + T_j^{-1} - 2). \tag{2.5}$$

Thus, one has

$$(\Delta_{discr} f)(p) = \frac{1}{2} \sum_{s=1}^{m} (f(p_1, \dots, p_{s-1}, p_s - 1, p_{s+1}, \dots, p_m)$$

$$+ f(p_1, \dots, p_{s-1}, p_s + 1, p_{s+1}, \dots, p_m)) - m f(p). \tag{2.6}$$

REMARK 2.1. We will deal with the class of sequences $f(k)$ with the following properties:

(i) $f(k) = 0$, whenever $k_i = k_j$ for some $i \neq j$;

(ii) $f(k)$ is an even function in each component k_i.

This class includes $f(k) = f_k(x_1, w)$: statement (i) is obvious; statement (ii) follows from equality (2.1). In this case the discrete Laplacian is well-defined by the above formulas (2.5), (2.6) on the restrictions of the latter sequences $f(k)$ to $k \in K_m$, as in [**5**, Remark 2.1]. (Each sequence $(f(k))_{k \in K_m}$ can be extended to a sequence $(f(k))_{k \in \mathbb{Z}^m}$ satisfying (ii) and antisymmetric with respect to permutation of components, hence satisfying (i).) In more detail, it suffices to check well-definedness of formula (2.6) for $p = (p_1, \dots, p_m) \in K_m$ with $p_1 = 0$. All the terms of the sum in (2.6) except for the first one are well-defined, since they are numerated by indices $k \in K_m$ and maybe some indices k with equal components $k_j = k_{j+1}$, for which $f(k) = 0$, by (i). The first term equals $f(-1, p_2, \dots, p_m) + f(1, p_2, \dots, p_m) = 2f(1, p_2, \dots, p_m)$, by (ii), and hence, is well-defined.

PROPOSITION 2.2. *(analogous to* [**5**, Proposition 2.2]*). For every $m \geq 1$ and $w \in \mathbb{R}^{m-1}$ the vector function $(f(x_1,k) = f_k(x_1,w))_{k\in K_m}$ satisfies the following linear differential-difference equation*[2]*:*

$$\frac{\partial f}{\partial x_1} = \Delta_{discr} f + mf. \tag{2.7}$$

PROOF. Each function $f_k(x_1,w)$ is the determinant of the matrix whose columns are the vector functions

$$V_{k_j}(x_1,w) = (I_{k_j}(x_1), I_{k_j}(x_1+w_2), \dots . I_{k_j}(x_1+w_m)), \quad j = 1,\dots,m.$$

The derivative $\frac{\partial f_k(x_1,w)}{\partial x_1}$ thus equals the sum over $j = 1,\dots,m$ of the same determinants, where the column V_{k_j} is replaced by its derivative $\frac{\partial V_{k_j}}{\partial x_1}$. According to (2.3), the latter derivative equals $\frac{1}{2}(V_{k_j-1}+V_{k_j+1})$. Therefore,

$$\frac{\partial f}{\partial x_1} = \frac{1}{2}\sum_{j=1}^{m}(T_j + T_j^{-1})f = (\Delta_{discr} f + mf). \tag{2.8}$$

□

REMARK 2.3. (analogous to [**5**, Remark 23]). For every $k \in K_m$ the k-th component of the right-hand side in (2.7) is a linear combination with strictly positive coefficients of the components $f(x_1,k')$ with $k' \in K_m$ obtained from $k = (k_1,\dots,k_m)$ by adding ± 1 to some k_i. This follows from (2.8).

PROPOSITION 2.4. [**5**, Proposition 2.4]. *For every constant $R > 1$ and every $j \geq R^2$ one has*

$$|I_j(z)| < \frac{R^j}{j!} \text{ for every } 0 \leq z \leq R. \tag{2.9}$$

COROLLARY 2.5. *(analogous to* [**5**, Corollary 2.6]*). For every $w \in \mathbb{R}^{m-1}$ and $x_1 \in \mathbb{R}$ one has $(f_k(x_1,w))_{k\in K_m} \in l_2$. Moreover, there exists a function $C(R) > 0$ in $R > 1$ such that*

$$\sum_{k\in K_m} |f_k(x_1,w)|^2 < C(R) \text{ whenever } |x_1| + |w| \leq R, \tag{2.10}$$

here $|w| = |w_2| + \dots + |w_m|$.

PROOF. The proof of Corollary 2.5 repeats the proof of [**5**, Corollary 2.6] with minor changes. Namely, let us fix a number $R > 1$ and set

$$M := \max_{j\in\mathbb{Z},\ 0\leq z\leq R} |I_j(z)|.$$

By (2.9) and [**5**, Remark 2.5], M is finite. Recall that $0 \leq k_1 < \dots < k_m$ for every $k = (k_1,\dots,k_m) \in K_m$. For every $(x_1,w) \in \mathbb{R}^m$ with $|x_1| + |w| \leq R$ one has

$$|f_k(x_1,w)| \leq m!M^m \text{ for every } k \in K_m, \tag{2.11}$$

$$|f_k(x_1,w)| < m!\frac{R^{k_m}}{k_m!}M^{m-1} \text{ whenever } k_m \geq R^2. \tag{2.12}$$

Indeed, the modulus of each element of the $m \times m$-matrix $A_{k,x}$ is no greater than M, whenever $|x_1| + |w| \leq R$. Therefore, the modulus $|f_{k,n}(x)|$ of its determinant

[2] In similar formula (2.7) in [**5**] there is a misprint: the right-hand side should be multiplied by $\frac{1}{2}$.

defined as the sum of $m!$ products of its elements is no greater than $m!M^m$. This proves (2.11). The last column of the matrix $A_{k,x}$ consists of the values $I_{k_m}(x_i) = I_{k_m}(x_1 + w_i)$. If $k_m \geq R^2$ and $|x_1| + |w| \leq R$, then their moduli are no greater than $\frac{R^{k_m}}{k_m!}$, by inequality (2.9). Therefore, each one of the $m!$ above-mentioned products of matrix elements in the expression of determinant has module no greater than $M^{m-1}\frac{R^{k_m}}{k_m!}$. This implies (2.12). The number of those $k \in K_m$ for which $k_m < R^2$ is less than R^{2m}. Therefore, the sum in (2.10) through $k \in K_m$ does not exceed[3]

$$C(R) := R^{2m}(m!M^m)^2 + (m!M^{m-1})^2 \sum_{k \in K_m} \frac{R^{2k_m}}{(k_m!)^2} < +\infty.$$

□

DEFINITION 2.6. [**5**, Definition 2.7]. Let Ω be the closure of an open convex subset in a Banach space. For every $x \in \partial\Omega$ consider the union of all the rays issued from x that intersect Ω in at least two distinct points (including x). The closure of the latter union of rays is a convex cone, which will be here referred to as the *generating cone* $K(x)$.

PROPOSITION 2.7. [**5**, Proposition 2.8]. *Let H be a Banach space, $\Omega \subset H$ be the closure of an open convex subset. Let v be a C^1 vector field on a neighborhood of the set Ω in H such that $v(x) \in K(x)$ for every $x \in \partial\Omega$. Then the set Ω is invariant under the flow of the field v: each positive semitrajectory starting at Ω is contained in Ω.*

Now the proof of the induction step in Theorem 1.6 is analogous to the argument in [**5**, end of section 2]. The right-hand side of differential equation (2.7) is a bounded linear vector field on the Hilbert space l_2 of sequences $(f_k)_{k \in K_m}$. We will denote the latter vector field by v. Let $\Omega \subset l_2$ denote the "positive quadrant" defined by the inequalities $f_k \geq 0$. For every point $y \in \partial\Omega$ the vector $v(y)$ lies in its generating cone $K(y)$: the components of the field v are non-negative on Ω, by Remark 2.3. The vector function $(f_k(x_1) = f_k(x_1, w))_{k \in K_m}$ in $x_1 \geq 0$ is an l_2-valued solution of the corresponding differential equation, by Corollary 2.5. One has $(f_k(0))_{k \in K_m} \in \Omega$:

$$f_k(0) = 0 \text{ whenever } k_1 > 0;$$

$$f_{(0,k_2,\dots,k_m)}(0) = I_0(0) f_{(k_2,\dots,k_m)}(w_2, \dots w_m) > 0. \tag{2.13}$$

The latter equality and inequality follow from definition, the properties (2.2) and the induction hypothesis. This together with Proposition 2.7 implies that

$$f_k(x_1, w) \geq 0 \text{ for every } k \in K_m \text{ and } x_1 \geq 0. \tag{2.14}$$

Now let us prove that the inequality is strict for all $k \in K_m$ and $x_1 > 0$. Indeed, let $f_p(x_0) = 0$ for some $p = (p_1, \dots, p_m) \in K_m$ and $x_0 > 0$. All the derivatives of the function f_p are non-negative, by (2.7), Remark 2.3 and (2.14). Therefore, $f_p \equiv 0$ on the segment $[0, x_0]$. This together with (2.7), Remark 2.3 and (2.14) implies that $f_{p'} \equiv 0$ on $[0, x_0]$ for every $p' \in K_m$ obtained from p by adding ± 1 to some component. We then get by induction that $f_{(0,k_2,\dots,k_m)}(0) = 0$, – a contradiction to (2.13). The proof of Theorem 1.6 is complete.

---

[3] In paper [**5**] there is a misprint in a similar formula for the constant $C(R)$ on p. 3863: in its right-hand side one should replace the factor $l!M^{l-1}$ and the term $\frac{R^{|k|_{n,\max}}}{(|k|_{n,\max})!}$ by their squares.

3. Acknowledgments

The authors are grateful to A. M. Vershik, F. V. Petrov, A. I. Bufetov, P. G. Grinevich, V. G. Gorbunov, A. K. Pogrebkov, V. M. Tikhomirov and V. Yu. Protasov for helpful discussions. The authors are grateful to P. A. Kuchment and the referee for helpful remarks on the paper.

References

[1] S. Abenda and P. G. Grinevich, *Rational degenerations of* M*-curves, totally positive Grassmannians and KP2-solitons*, Comm. Math. Phys. **361** (2018), no. 3, 1029–1081, DOI 10.1007/s00220-018-3123-y. MR3830261

[2] A. Berenstein, S. Fomin, and A. Zelevinsky, *Parametrizations of canonical bases and totally positive matrices*, Adv. Math. **122** (1996), no. 1, 49–149, DOI 10.1006/aima.1996.0057. MR1405449

[3] A. Berenstein and A. Zelevinsky, *Total positivity in Schubert varieties*, Comment. Math. Helv. **72** (1997), no. 1, 128–166, DOI 10.1007/PL00000363. MR1456321

[4] M. Boiti, F. Pempinelli, and A. K. Pogrebkov, *Heat operator with pure soliton potential: properties of Jost and dual Jost solutions*, J. Math. Phys. **52** (2011), no. 8, 083506, DOI 10.1063/1.3621715. MR2858062

[5] V. M. Buchstaber and A. A. Glutsyuk, *On determinants of modified Bessel functions and entire solutions of double confluent Heun equations*, Nonlinearity **29** (2016), no. 12, 3857–3870, DOI 10.1088/0951-7715/29/12/3857. MR3580333

[6] V. M. Bukhshtaber and S. I. Tertychnyĭ, *A remarkable sequence of Bessel matrices* (Russian, with Russian summary), Mat. Zametki **98** (2015), no. 5, 651–663, DOI 10.4213/mzm10801; English transl., Math. Notes **98** (2015), no. 5-6, 714–724. MR3438522

[7] Y. Choquet-Bruhat, C. DeWitt-Morette, and M. Dillard-Bleick, *Analysis, manifolds and physics*, North-Holland Publishing Co., Amsterdam-New York-Oxford, 1977. MR0467779

[8] A. Edrei, *On the generation function of a doubly infinite, totally positive sequence*, Trans. Amer. Math. Soc. **74** (1953), 367–383, DOI 10.2307/1990808. MR0053989

[9] S. M. Fallat and C. R. Johnson, *Totally nonnegative matrices*, Princeton Series in Applied Mathematics, Princeton University Press, Princeton, NJ, 2011. MR2791531

[10] S. Fallat, C. R. Johnson, and A. D. Sokal, *Total positivity of sums, Hadamard products and Hadamard powers: results and counterexamples*, Linear Algebra Appl. **520** (2017), 242–259, DOI 10.1016/j.laa.2017.01.013. MR3611466

[11] S. Fomin, *Total positivity and cluster algebras*, Proceedings of the International Congress of Mathematicians. Volume II, Hindustan Book Agency, New Delhi, 2010, pp. 125–145. MR2827788

[12] S. Fomin and A. Zelevinsky, *Double Bruhat cells and total positivity*, J. Amer. Math. Soc. **12** (1999), no. 2, 335–380, DOI 10.1090/S0894-0347-99-00295-7. MR1652878

[13] F. R. Gantmacher, *The theory of matrices. Vols. 1, 2*, Translated by K. A. Hirsch, Chelsea Publishing Co., New York, 1959. MR0107649

[14] F. Gantmacher and M. Krein, *Sur les matrices oscillatoires.* Comptes Rendus Acad. Sci. Paris **201** (1935), 577–579.

[15] F. Gantmakher and M. Krein, *Sur les matrices complètement non négatives et oscillatoires* (French), Compositio Math. **4** (1937), 445–476. MR1556987

[16] F. R. Gantmacher and M. G. Krein, *Oscillation matrices and small oscillations of mechanical systems* (Russian), Moscow-Leningrad, 1941. MR0005985

[17] F. R. Gantmaher and M. G. Kreĭn, *Oscillyacionye matricy i yadra i malye kolebaniya mehaničeskih sistem* (Russian), Gosudarstv. Isdat. Tehn.-Teor. Lit., Moscow-Leningrad, 1950. 2nd ed. MR0049462. English translation: *Oscillation matrices and kernels and small vibrations of mechanical systems*, American Mathematical Society, Providence, 2002.

[18] M. Gasca and J. M. Peña, *On factorizations of totally positive matrices*, Total positivity and its applications (Jaca, 1994), Math. Appl., vol. 359, Kluwer Acad. Publ., Dordrecht, 1996, pp. 109–130. MR1421600

[19] I. M. Gel′fand, R. M. Goresky, R. D. MacPherson, and V. V. Serganova, *Combinatorial geometries, convex polyhedra, and Schubert cells*, Adv. in Math. **63** (1987), no. 3, 301–316, DOI 10.1016/0001-8708(87)90059-4. MR877789

[20] S. Karlin, *Total positivity. Vol. I*, Stanford University Press, Stanford, Calif, 1968. MR0230102

[21] Y. Kodama and L. Williams, *KP solitons and total positivity for the Grassmannian*, Invent. Math. **198** (2014), no. 3, 637–699, DOI 10.1007/s00222-014-0506-3. MR3279534

[22] G. Lusztig, *A survey of total positivity*, Milan J. Math. **76** (2008), 125–134, DOI 10.1007/s00032-008-0083-2. MR2465988

[23] G. Lusztig, *Total positivity in reductive groups*, Lie theory and geometry, Progr. Math., vol. 123, Birkhäuser Boston, Boston, MA, 1994, pp. 531–568, DOI 10.1007/978-1-4612-0261-5_20. MR1327548

[24] G. Lusztig, *Total positivity in partial flag manifolds*, Represent. Theory **2** (1998), 70–78, DOI 10.1090/S1088-4165-98-00046-6. MR1606402

[25] T. M. Malanyuk, *A class of exact solutions of the Kadomtsev-Petviashvili equation* (Russian), Uspekhi Mat. Nauk **46** (1991), no. 3(279), 193–194; , DOI 10.1070/RM1991v046n03ABEH002792; English transl., Russian Math. Surveys **46** (1991), no. 3, 225–227. MR1134101

[26] T. S. Motzkin, *Beiträge zur Theorie der Linearen Ungleichungen.* (Dissertation, Basel, 1933), Jerusalem, 1936.

[27] O. Perron, *Zur Theorie der Matrices* (German), Math. Ann. **64** (1907), no. 2, 248–263, DOI 10.1007/BF01449896. MR1511438

[28] A. Pinkus, *Totally positive matrices*, Cambridge Tracts in Mathematics, vol. 181, Cambridge University Press, Cambridge, 2010. MR2584277

[29] Postnikov A., *Total positivity, Grassmannians, and networks.* Preprint. `https://arxiv.org/abs/math/0609764`

[30] I. Schoenberg, *Über variationsvermindernde lineare Transformationen* (German), Math. Z. **32** (1930), no. 1, 321–328, DOI 10.1007/BF01194637. MR1545169

[31] I. J. Schoenberg, *Some analytical aspects of the problem of smoothing*, Studies and Essays Presented to R. Courant on his 60th Birthday, January 8, 1948, Interscience Publishers, Inc., New York, 1948, pp. 351–370. MR0023309

[32] K. Talaska, *Combinatorial formulas for Γ-coordinates in a totally nonnegative Grassmannian*, J. Combin. Theory Ser. A **118** (2011), no. 1, 58–66, DOI 10.1016/j.jcta.2009.10.006. MR2737184

[33] E. Thoma, *Die unzerlegbaren, positiv-definiten Klassenfunktionen der abzählbar unendlichen, symmetrischen Gruppe* (German), Math. Z. **85** (1964), 40–61, DOI 10.1007/BF01114877. MR0173169

[34] A. M. Vershik and S. V. Kerov, *Characters and factor-representations of the infinite unitary group* (Russian), Dokl. Akad. Nauk SSSR **267** (1982), no. 2, 272–276. MR681202. English translation in Soviet Math. Dokl. **26** (1982), no. 3, 570–574.

[35] A. M. Vershik and S. V. Kerov, *Characters and factor representations of the infinite symmetric group* (Russian), Dokl. Akad. Nauk SSSR **257** (1981), 1037–1040. MR614033. English translation in Soviet Math. Dokl. **23** (1981), no. 2, 389–392.

[36] D. Voiculescu, *Représentations factorielles de type II1 de $U(\infty)$* (French), J. Math. Pures Appl. (9) **55** (1976), no. 1, 1–20. MR0442153

[37] G.N. Watson, *A treatise on the theory of Bessel functions (2nd. ed.).* Vol. 1, Cambridge University Press, 1966.

Steklov Mathematical Institute, 8, Gubkina street, 119991, Moscow, Russia

All-Russian Scientific Research Institute for Physical and Radio-Technical Measurements (VNIIFTRI)
Email address: `buchstab@mi.ras.ru`

CNRS, France (UMR 5669 (UMPA, ENS de Lyon) and UMI 2615 (Interdisciplinary Scientific Center J.-V.Poncelet))

National Research University Higher School of Economics (HSE), Moscow, Russia
Email address: `aglutsyu@ens-lyon.fr`

Contemporary Mathematics
Volume **733**, 2019
https://doi.org/10.1090/conm/733/14737

A remark on the intersection of plane curves

C. Ciliberto, F. Flamini, and M. Zaidenberg

To the memory of Professor Selim Grigorievich Krein

ABSTRACT. Let D be a very general curve of degree $d = 2\ell - \varepsilon$ in $\mathbb{P}^2$, with $\varepsilon \in \{0, 1\}$. Let $\Gamma \subset \mathbb{P}^2$ be an integral curve of geometric genus g and degree m, $\Gamma \neq D$, and let $\nu : C \to \Gamma$ be the normalization. Let δ be the degree of the *reduction modulo 2* of the divisor $\nu^*(D)$ of C (see §2.1). In this paper we prove the inequality $4g + \delta \geqslant m(d - 8 + 2\varepsilon) + 5$. We compare this with similar inequalities due to Geng Xu ([**88, 89**]) and Xi Chen ([**17.18**]). Besides, we provide a brief account on genera of subvarieties in projective hypersurfaces.

CONTENTS

Introduction
1. Focal loci
2. Double planes
3. The main result
4. Genera of subvarieties: a survey
References

2010 *Mathematics Subject Classification.* Primary 14J70, 14D05; Secondary 14C17, 14C20, 14H30.

Key words and phrases. Projective hypersurfaces, intersection number, foci, geometric genus, algebraic hyperbolicity.

The first author acknowledges the MIUR Excellence Department Project awarded to the Department of Mathematics, University of Rome Tor Vergata, CUP E83C18000100006. The work of the third author was partially supported by the grant 346300 for IMPAN from the Simons Foundation and the matching 2015-2019 Polish MNiSW fund. This research started during a visit of the third author at the Dept. of Mathematics of Univ. Roma "Tor Vergata" (Sept.-Dec. 2015 – supported by this Department, the INdAM "F. Severi" in Rome and the cooperation program GDRE-GRIFGA) and of the second author at the Institut Fourier (July 2016 – supported by INdAM-GNSAGA and within the context of the Laboratory Ypatia of Mathematical Sciences LIA LYSM AMU-CNRS-ECM-INdAM). Collaboration has also benefitted of funds *Mission Sustainability 2017 - Fam Curves* CUP E81-18000100005 (Tor Vergata University). The authors thank all these institutions and programs for the support and the excellent working conditions.

Introduction

Given an effective divisor $D \in |\mathcal{O}_{\mathbb{P}^n}(d)|$ and an integral (i.e., reduced and irreducible) projective curve Γ of degree m in $\mathbb{P}^n$, which is not contained in $\mathrm{supp}(D)$, let $j(D,\Gamma)$ be the order of $\Gamma \cap D$. Assume D is very general and set

$$j(n,d,m) := \min\{j(D,\Gamma) \,|\, \Gamma \subset \mathbb{P}^n \text{ as above}\} \quad \text{and} \quad j(n,d) := \min_{m \geqslant 1}\{j(n,d,m)\}.$$

Similarly, with Γ and D as before, let $i(D,\Gamma)$ stand for the *number of places* of Γ on D, that is, the number of centers of local branches of the curve Γ on D. Then, set

$$i(n,d,m) := \min\{i(D,\Gamma) \,|\, \Gamma \subset \mathbb{P}^n \text{ as above}\} \quad \text{and} \quad i(n,d) := \min_{m \geqslant 1}\{i(n,d,m)\}.$$

The problem of computing $j(n,d)$ and $i(n,d)$ has been considered in [**17**, **88**, **89**] (basically devoted to $n = 2$ case) and [**18**] (where the case $n \geqslant 2$ is considered). The relations of this with the famous Kobayashi problem on hyperbolicity of the complement of a very general hypersurface in $\mathbb{P}^n$ is well known (see section 4).

Geng Xu ([**88**, Thm. 1]) proved that

$$j(2,d) = d - 2, \quad \text{for any } d \geqslant 3,$$

where the equality is attained either by a bitangent line or by an inflectional tangent line of D, i.e. the minimum is achieved by $m = 1$. Moreover, for $d = 3$, he also proved in [**89**, Corollary] that, for any integer $m \geqslant 1$, the number of rational curves of degree m which meet set-theoretically a given (arbitrary) smooth plane cubic curve D at exactly one point is finite and positive. Therefore, for $d = 3$ the minimum $j(2,3) = 1$ is achieved by any integer $m \geqslant 1$.

Xi Chen ([**17**, Thm. 1.2]) proved that, for $d > m$, one has

$$j(2,d,m) \geqslant \min\left\{dm - \frac{m(m+3)}{2},\ 2dm - 2m^2 - 2\right\}.$$

Furthermore (cf. [**17**, Cor. 1.1]), for $d \geqslant \max\{\frac{3m}{2} - 1,\ 3\}$ one has

$$j(2,d,m) = dm - \dim(|\mathcal{O}_D(m)|) = dm - \frac{m(m+3)}{2}.$$

In addition, he conjectured (see [**17**, Conj. 1.1]) that

$$j(2,d,m) = dm - \dim(|\mathcal{O}_D(m)|) \quad \text{if} \quad d > \max\{m, 2\}.$$

In arbitrary dimension $n \geqslant 2$, Xi Chen ([**18**, Thm 1.7]) proved that, for D very general and Γ as above, one has

$$2g - 2 + i(D,\Gamma) \geqslant (d - 2n)m\,, \tag{0.1}$$

where g is the *geometric genus* of Γ, i.e., the arithmetic genus of its normalization.

In this paper we obtain a new inequality of type (0.1), although only in the case $n = 2$ (see Theorem 3.1). Indeed, let D be a very general curve of degree $d = 2\ell - \varepsilon$ in $\mathbb{P}^2$, with $\varepsilon \in \{0,1\}$. Let Γ be an integral curve in $\mathbb{P}^2$ of geometric genus g and degree m, $\Gamma \neq D$, and let $\nu : C \to \Gamma$ be the normalization. Let $\delta(D,\Gamma)$ be the degree of the *reduction modulo 2* of the divisor $\nu^*(D)$ on C (cf. § 2.1). In Theorem 3.1 we prove that

$$4g + \delta(D,\Gamma) \geqslant m(d + 2\varepsilon - 8) + 5. \tag{0.2}$$

Note that $\delta(D,\Gamma) \leqslant i(D,\Gamma)$, and the equality holds if and only if at any place p of Γ on D, the local intersection multiplicity of D and Γ at p is odd. This happens, for instance, if Γ intersects D transversely. In the latter case $\delta(D,\Gamma) = i(D,\Gamma) = md$ and both (0.1) and (0.2) are uninteresting. On the other hand, (0.1) and (0.2) become interesting when $\delta(D,\Gamma)$ and $i(D,\Gamma)$ are small. Though the difference between the two quantities is a priori unpredictable, one may expect that, generally speaking, $\delta(D,\Gamma)$ is strictly smaller than $i(D,\Gamma)$. Unfortunately, the genus g works against us in (0.2); however, for $g = 0, 1$ and d even, (0.2) is better than (0.1). Further related problems have been recently considered in [**19, 65, 66**].

As a final remark, note that (0.2) is more useful than (0.1) if one looks, as we do in this paper, at the geometric genera of curves contained in a *double plane* X_d, that is, a cyclic double cover of $\mathbb{P}^2$ branched along a very general plane curve D of even degree d. For instance, letting $g = 0$, $\delta(D,\Gamma) = 0, 2$ and d even, we are looking actually for rational curves on X_d. By (0.2) we see that such a rational curve over Γ might exist, as expected, only for $d \leqslant 6$ (for low m one has even smaller bounds on d). The case $d = 6$ corresponds to a $K3$ surface, which always contains a rational curve. In contrast, it follows from (0.2) that the double planes with very general branching curves of even degree $d \geqslant 8$ ($d \geqslant 10$, respectively) do not carry any rational curve (any rational or elliptic curve, respectively, hence are *algebraically hyperbolic*). For $d = 8$ and $d = 10$ these double planes are *Horikawa surfaces* H_8 and H_{10}, that is, their Chern numbers satisfy $c_2 = 5c_1^2 + 36$ (in other words, (c_1^2, c_2) lies on the Noether line). The algebraic hyperbolicity of H_{10} was established first by X. Roulleau and E. Rousseau ([**75**]). J. Liu ([**66**]) showed that some of the Horikawa surfaces H_{10} are even Kobayashi hyperbolic, whereas there is no hyperbolic H_8. Indeed, the Horikawa double planes H_8 carrying elliptic curves are dense in the set of all such surfaces, while the Kobayashi hyperbolicity is open in the Hausdorff topology.

The proof of Theorem 3.1 presented in §3 follows, with minor variations due to the different setting, the basic ideas exploited in [**21**] (and later in [**22**]). These are based on a smart use of the theory of *focal loci*, see e.g. [**20**]. For the reader's convenience, we recall in § 1 the basic notions and results of this theory. We apply this technique to families of double covers of $\mathbb{P}^2$ branched along a very general plane curve D or along D plus a general line, according to whether the degree of D is even (see § 2.3 and § 3.2.1) or odd (see § 2.4 and § 3.2.2).

In the last §4 we provide a short survey on genera of subvarieties in projective varieties, with accent on projective hypersurfaces.

1. Focal loci

For the reader's convenience, we recall here some basic notions from [**20**, **21**].

Let X be a smooth projective variety of dimension $n+1$. Assume we have a flat, projective family $\mathcal{D} \xrightarrow{p} \mathcal{B}$ of effective divisors on X over a smooth, irreducible, quasiprojective base $\mathcal{B}$, with irreducible general fiber. Up to shrinking $\mathcal{B}$ to a suitable Zariski dense, open subset, we may suppose that for any closed point $b \in \mathcal{B}$ the fiber D_b of $\mathcal{D} \xrightarrow{p} \mathcal{B}$ over b is irreducible.

Assume we have a commutative diagram

$$
\begin{array}{ccc}
\mathfrak{C} & \overset{i}{\hookrightarrow} & \mathcal{D} \\
& \underset{\mathfrak{q}}{\searrow} & \downarrow p \\
& & \mathcal{B}
\end{array} \tag{1.1}
$$

where $\mathfrak{q} : \mathfrak{C} \to \mathcal{B}$ is a flat projective family such that, for all $b \in \mathcal{B}$, the fiber Γ_b over b is a reduced curve of *geometric genus* g, and where i is an inclusion: so, for any $b \in \mathcal{B}$, one has $\Gamma_b \subset D_b$ via the inclusion i_b.

By a result of Tessier (see [**80**, Théorème 1]), there is a simultaneous normalization

$$
\begin{array}{ccc}
\mathcal{C} & \overset{\nu}{\longrightarrow} & \mathfrak{C} \\
& \underset{q}{\searrow} & \downarrow \mathfrak{q} \\
& & \mathcal{B}
\end{array} \tag{1.2}
$$

such that $\mathcal{C}$ is smooth and, for every $b \in \mathcal{B}$, the fiber C_b of $q : \mathcal{C} \to \mathcal{B}$ is the normalization $\nu_b : C_b \to \Gamma_b$ of Γ_b. For any $b \in \mathcal{B}$, the curve C_b is smooth of *(arithmetic) genus* g.

Composing with the inclusion $\mathcal{D} \overset{j}{\hookrightarrow} \mathcal{B} \times X$, we get the commutative diagram

(1.3)

$$
\begin{array}{ccccccc}
\mathcal{C} & \overset{\nu}{\longrightarrow} & \mathfrak{C} & \overset{i}{\hookrightarrow} & \mathcal{D} & \overset{j}{\hookrightarrow} & \mathcal{B} \times X \\
& \underset{q}{\searrow} & \downarrow \mathfrak{q} & & \downarrow p & \underset{\mathrm{pr}_1}{\swarrow} & \downarrow \mathrm{pr}_2 \\
& & \mathcal{B} & \underset{\mathrm{id}}{\longrightarrow} & \mathcal{B} & & X
\end{array}
$$

where pr_i is the projection onto the ith factor, for $i = 1, 2$.

We set

$$s := j \circ i \circ \nu \colon \mathcal{C} \to \mathcal{B} \times X\,,$$

and let $\mathcal{N} := \mathcal{N}_s$ be the normal sheaf to s, defined by the exact sequence

$$0 \longrightarrow \mathcal{T}_{\mathcal{C}} \xrightarrow{ds} s^*(\mathcal{T}_{\mathcal{B}\times X}) \longrightarrow \mathcal{N} \longrightarrow 0\,,$$

where $\mathcal{T}_Y$ stands for the tangent sheaf of a smooth variety Y.

For $b \in \mathcal{B}$ we set

$$N_b := \mathcal{N}|_{C_b} = \mathcal{N} \otimes \mathcal{O}_{C_b} \quad \text{and} \quad s_b = s|_{C_b} : C_b \to \{b\} \times X = X\,.$$

In addition, we let

$$\varphi := \mathrm{pr}_2 \circ s : \mathcal{C} \to X\,.$$

Then $\varphi_b = \varphi|_{C_b}$ coincides with s_b for any $b \in \mathcal{B}$, that is,

$$\varphi_b = s_b : C_b \xrightarrow{\nu_b} \Gamma_b \xhookrightarrow{i_b} D_b \xhookrightarrow{(\mathrm{pr}_2 \circ j)_b} X .$$

As in [**21**, § 2], we set

$$z(\mathcal{C}) := \dim\,(\varphi(\mathcal{C})), \tag{1.4}$$

so that $z(\mathcal{C}) \leqslant n+1 = \dim(X)$. If $z(\mathcal{C}) = n+1$ one says that $\mathfrak{C} \xrightarrow{q} \mathcal{B}$, or $\mathcal{C} \xrightarrow{q} \mathcal{B}$, is a *covering family*.

Proposition 1.1 (See [**20**, Prop. 1.4 and p. 98]). *In the above setting, we have:*

(a) *for any $b \in \mathcal{B}$, the sheaf N_b fits into the exact sequence*

$$0 \longrightarrow \mathcal{T}_{C_b} \xrightarrow{ds_b} s_b^*(\mathcal{T}_X) \longrightarrow N_b \longrightarrow 0$$

and $\mathcal{C} \xrightarrow{q} \mathcal{B}$ induces on C_b a characteristic map

$$\chi_b \colon T_{\mathcal{B},b} \otimes \mathcal{O}_{C_b} \longrightarrow N_b \,,$$

where $T_{\mathcal{B},b}$ denotes the tangent space to $\mathcal{B}$ at b;

(b) *if $b \in \mathcal{B}$ and $x \in C_b$ are general points, then*

$$\dim\,(N_{b,x}) = \dim\,(s_b^*(\mathcal{T}_X)_x) - \dim\,(\mathcal{T}_{C_b,x}) = n \quad \textit{and} \quad \mathrm{rk}\,(\chi_{b,x}) = z(\mathcal{C}) - 1 \,.$$

Definition 1.2 (See [**20**, Def. (1.5)]). Given $b \in \mathcal{B}$, the *focal set* of $\mathcal{C} \xrightarrow{q} \mathcal{B}$ on C_b is the closed subscheme Φ_b of C_b defined as

$$\Phi_b := \{x \in C_b |\, \mathrm{rk}(\chi_{b,x}) < z(\mathcal{C}) - 1\}.$$

If $b \in \mathcal{B}$ is general, then Φ_b is a proper subscheme of C_b. The points in Φ_b are called *focal points* of $\mathcal{C} \xrightarrow{q} \mathcal{B}$ on C_b. We denote by Φ_b^{sm} the open subset of Φ_b consisting of the points $x \in \Phi_b$ which map to smooth points of Γ_b via the normalization morphism $\nu_b \colon C_b \to \Gamma_b$.

Proposition 1.3 ([**21**, Prop. 2.3 and Prop. 2.4]). *Let $\mathcal{C} \longrightarrow \mathcal{B}$ be a covering family. Then the following hold.*

(i) *Suppose that for $x \in C_b$ the point $s(x)$ is smooth in both Γ_b and D_b. Assume also that $s(x)$ is a* fundamental point *of the family $\mathcal{D} \xrightarrow{p} \mathcal{B}$, i.e. it is a base point of the family. Then $x \in \Phi_b^{\mathrm{sm}}$.*

(ii) *For a general point $b \in \mathcal{B}$ one has*

$$\deg(\Phi_b^{\mathrm{sm}}) \leqslant 2g - 2 - K_X \cdot \Gamma_b \,. \tag{1.5}$$

2. Double planes

In this section we collect useful material for the proof of our main result. The result itself is stated and proven in §3. The contents of this section, which suffice for our applications, can be easily adapted to the higher dimensional case.

2.1. The δ–invariant. Let C be any smooth, irreducible, projective curve, and let $\Delta = \sum_i m_i p_i$ be an effective divisor on C. We set $\Delta_2 := \sum_i \overline{m}_i p_i$, where $\overline{m}_i \in \{0, 1\}$ is the residue of the integer m_i modulo 2. We also set $\delta_2(\Delta) := \deg(\Delta_2)$.

For any smooth curve $D \subset \mathbb{P}^2$ and any integral curve $\Gamma \subset \mathbb{P}^2$, $\Gamma \neq D$, with normalization $\nu\colon C \to \Gamma$, we set

$$\delta(D, \Gamma) := \delta_2(\nu^*(D)) . \tag{2.1}$$

We notice that

$$\delta(D, \Gamma) \leqslant i(D, \Gamma).$$

2.2. Basics on a certain weighted projective 3-space. For any positive integer ℓ, we denote by $\mathcal{L}_\ell$ the linear system $|\mathcal{O}_{\mathbb{P}^2}(\ell)|$ of plane curves of degree ℓ, and by $\mathcal{U}_\ell$ its open dense subset of points corresponding to smooth curves. We let $N_\ell = \dim(\mathcal{L}_\ell) = \frac{\ell(\ell+3)}{2}$. We denote by $\mathcal{D}_\ell \to \mathcal{L}_\ell$ the universal curve, and we use the same notation $\mathcal{D}_\ell \to \mathcal{U}_\ell$ for its restriction to $\mathcal{U}_\ell$.

The linear system $\mathcal{L}_\ell$ determines the ℓth *Veronese embedding* $\mathbb{P}^2 \overset{v_\ell}{\hookrightarrow} \mathbb{P}^{N_\ell}$, whose image is the ℓ–*Veronese surface* V_ℓ in $\mathbb{P}^{N_\ell}$. Let $[x] = [x_0, x_1, x_2]$ be homogeneous coordinates in $\mathbb{P}^2$, and let

$$[x^I], \text{ where } I = (i_0, i_1, i_2) \text{ is a multiindex such that } |I| = i_0 + i_1 + i_2 = \ell,$$

be homogeneous coordinates in $\mathbb{P}^{N_\ell}$. In these coordinates the Veronese map is given by

$$\mathbb{P}^2 \ni [x] \xrightarrow{v_\ell} [x^I]_{|I|=\ell} \in \mathbb{P}^{N_\ell}, \quad \text{where } x^I := x_0^{i_0} x_1^{i_1} x_2^{i_2} .$$

We equip the weighted projective 3-space $\mathbb{P}(1, 1, 1, \ell)$ with weighted homogeneous coordinates $[x, z] := [x_0, x_1, x_2, z]$, where x_0, x_1, x_2 [resp. z] have *weigth* 1 [resp. has weight ℓ]. We introduce as well coordinates $[x^I, z]_{|I|=\ell}$ in $\mathbb{P}^{N_\ell+1}$ and embed $\mathbb{P}^{N_\ell}$ in $\mathbb{P}^{N_\ell+1}$ as the hyperplane Π with equation $z = 0$. Then $\mathbb{P}(1, 1, 1, \ell)$ can be identified with the cone $W_\ell \subset \mathbb{P}^{N_\ell+1}$ over the l–Veronese surface V_ℓ with vertex $P = [0, \ldots, 0, 1]$. Blowing P up yields a minimal resolution

$$\rho : Z_\ell \to W_\ell \cong \mathbb{P}(1, 1, 1, \ell),$$

with exceptional divisor $E \cong V_\ell \cong \mathbb{P}^2$. The projection from P induces a $\mathbb{P}^1$–bundle structure

$$\pi\colon Z_\ell \to V_\ell \cong \mathbb{P}^2 .$$

Let f be the class of a fiber of π. One has

$$Z_\ell \cong \mathbb{P}(\mathcal{O}_{\mathbb{P}^2}(\ell) \oplus \mathcal{O}_{\mathbb{P}^2}) \quad \text{and} \quad \mathcal{O}_{Z_\ell}(1) = \rho^*(\mathcal{O}_{W_\ell}(1)).$$

For every integer m, we set

$$\mathcal{O}_\ell(m) := \pi^*(\mathcal{O}_{\mathbb{P}^2}(m)) \quad \text{and} \quad \mathcal{L}_\ell(m) := |\mathcal{O}_\ell(m)| . \tag{2.2}$$

Note that

$$\mathcal{O}_{Z_\ell}(1) \cong \mathcal{O}_\ell(\ell) \otimes \mathcal{O}_{Z_\ell}(E). \tag{2.3}$$

Since

$$E \cong \mathbb{P}^2, \ \mathcal{O}_{Z_\ell}(1) \otimes \mathcal{O}_E \cong \mathcal{O}_E, \ \text{and } \mathcal{O}_\ell(\ell) \otimes \mathcal{O}_E \cong \mathcal{O}_{\mathbb{P}^2}(\ell),$$

we deduce

$$\mathcal{O}_{Z_\ell}(E) \otimes \mathcal{O}_E \cong \mathcal{O}_{\mathbb{P}^2}(-\ell). \tag{2.4}$$

Finally, we denote by K_ℓ the canonical sheaf of Z_ℓ.

LEMMA 2.1. *One has*

$$K_\ell \cong \mathcal{O}_\ell(\ell-3)\otimes\mathcal{O}_{Z_\ell}(-2) \cong \mathcal{O}_\ell(-\ell-3)\otimes\mathcal{O}_{Z_\ell}(-2E)\,.$$

PROOF. The Picard group $\mathrm{Pic}(Z_\ell)$ is freely generated by the classes $\mathcal{O}_\ell(1)$ and $\mathcal{O}_{Z_\ell}(E)$, and also by $\mathcal{O}_\ell(1)$ and $\mathcal{O}_{Z_\ell}(1)$, see (2.3). Let H [resp. L] be a general member of $|\mathcal{O}_{Z_\ell}(1)|$ [resp. of $\mathcal{L}_\ell(1)$]. Write $K_\ell \sim \alpha H + \beta L$, where $\alpha, \beta \in \mathbb{Z}$. From the relations $K_\ell \cdot f = -2$, $H \cdot f = 1$, and $L \cdot f = 0$ one gets $\alpha = -2$.

By adjunction formula and (2.4) we obtain

$$\mathcal{O}_{\mathbb{P}^2}(-3) \cong K_E = (K_\ell + E)|_E \cong (-2H + \beta L + E)|_E \cong \mathcal{O}_{\mathbb{P}^2}(\beta - \ell)\,.$$

So, $\beta = \ell - 3$, as desired. □

Finally, let $\mathbb{G}_\ell$ be the group of all automorphisms of $\mathbb{P}(1,1,1,\ell)$ which stabilize the divisor with equation $z = 0$. This group is naturally isomorphic to the automorphism group of the pair (W_ℓ, V_ℓ), i.e., the group of all automorphisms of W_ℓ stabilizing V_ℓ, where V_ℓ is cut out on W_ℓ by Π. In turn, the latter group is isomorphic to the automorphism group of the pair $(Z_\ell, \rho^*(V_\ell))$. One has the exact sequence

$$1 \to \mathbb{C}^* \to \mathbb{G}_\ell \to \mathrm{PGL}(3,\mathbb{C}) \to 1. \tag{2.5}$$

2.3. The even degree case. Let D be a smooth curve in $\mathbb{P}^2$ of even degree $d = 2\ell \geqslant 2$ which, in the homogeneous coordinate system fixed in §2.2, is given by equation $f(x_0, x_1, x_2) = 0$, where f is a homogeneous polynomial of degree d. Viewed as a hypersurface of V_ℓ, D is cut out on V_ℓ by a quadric with equation $Q(x^I)_{|I|=\ell} = 0$, where Q is a homogeneous polynomial of degree 2 in the variables $\{x^I\}_{|I|=\ell}$.

The *double plane* associated with D is the double cover $\psi : \mathbb{D}^* \to \mathbb{P}^2$ branched along D. It can be embedded in $\mathbb{P}(1,1,1,\ell)$ as a hypersurface $\mathbb{D}^*_a$ defined by a (weighted homogeneous) equation of the form $az^2 = f(x_0, x_1, x_2)$, for any $a \in \mathbb{C}^*$. Under the identification of $\mathbb{P}(1,1,1,\ell)$ with W_ℓ, we see that $\mathbb{D}^*_a$ is cut out on W_ℓ by a quadric in $\mathbb{P}^{N_\ell+1}$ of the form $az^2 = Q(x^I)_{|I|=\ell}$.

Consider the sublinear system $\mathcal{Q}_\ell$ of $|\mathcal{O}_{W_\ell}(2)|$ of surfaces cut out on W_ℓ by the quadrics of $\mathbb{P}^{N_\ell+1}$ with equation of the form $az^2 = Q(x^I)_{|I|=\ell}$.

When $a \neq 0$, the quadrics in question are such that their polar hyperplane with respect to P has equation $z = 0$. When $a = 0$, such a quadric is singular at P, it represents the cone with vertex at P over the quadric in $\Pi = \{z = 0\} \cong \mathbb{P}^{N_\ell}$ with equation $Q(x^I)_{|I|=\ell} = 0$ and it cuts out on W_ℓ a cone, with vertex at P, over a quadric section of V_ℓ.

In particular, $\mathcal{Q}_\ell$ contains the codimension 1 sublinear system $\mathcal{Q}^c_\ell$ of all such cones with vertex at P over a quadric section of V_ℓ, thus $\dim(\mathcal{Q}_\ell) = N_d + 1$. Moreover $\mathcal{Q}_\ell$ is stable under the action of $\mathbb{G}_\ell$ on W_ℓ.

We set $\widetilde{\mathcal{Q}}_\ell := \rho^*(\mathcal{Q}_\ell)$, which is a sublinear system of $|\mathcal{O}_{Z_\ell}(2)|$. Note that $\widetilde{\mathcal{Q}}_\ell$ contains the sublinear system $\widetilde{\mathcal{Q}}^c_\ell = \rho^*(\mathcal{Q}^c_\ell)$ of all divisors of the form $2E$ plus a divisor in $\mathcal{L}_\ell(d)$.

We denote by $\mathcal{Q}^*_\ell$ the dense open subset of $\mathcal{Q}_\ell$ of points corresponding to smooth surfaces. Since no surface in $\mathcal{Q}^*_\ell$ passes through P, we may and will identify $\mathcal{Q}^*_\ell$ with its pull–back via ρ on Z_ℓ, which is a dense open subset of $\widetilde{\mathcal{Q}}_\ell$ sitting in the complement of $\widetilde{\mathcal{Q}}^c_\ell$. We denote by $\mathcal{D}^*_\ell \to \mathcal{Q}^*_\ell$ the universal family.

A surface $\mathbb{D}^* \in \mathcal{Q}^*_\ell$ cuts out on V_ℓ a smooth curve $D \in \mathcal{U}_\ell$ and conversely; indeed the projection from P realizes $\mathbb{D}^*$ as the double cover of $\mathbb{P}^2$ branched along D. This yields a surjective morphism

$$\mathcal{Q}^*_\ell \ni \mathbb{D}^* \stackrel{\beta}{\longrightarrow} \mathbb{D}^* \cap V_\ell := D \in \mathcal{U}_d\,,$$

which sends the double plane $\mathbb{D}^*$ to its branching divisor D. This morphism is equivariant under the actions of $\mathbb{G}_\ell$ on both $\mathcal{Q}^*_\ell$ and $\mathcal{U}_d$, where $\mathbb{G}_\ell$ acts on $\mathcal{U}_d$ via the natural action of the quotient group $\mathrm{PGL}(3,\mathbb{C})$, see (2.5). The morphism β is not injective, its fibers being isomorphic to $\mathbb{C}^*$.

As an immediate consequence of Lemma 2.1, we have:

LEMMA 2.2. *Let D be a smooth curve in $\mathbb{P}^2$ of even degree $d = 2\ell \geqslant 2$, let $\psi : \mathbb{D}^* \to \mathbb{P}^2$ be the double cover branched along D, let $\Gamma \subset \mathbb{P}^2$ be a projective curve of degree m not containing D, and let Γ^* be its pull–back via ψ considered as a curve in Z_ℓ. One has*

$$K_\ell \cdot \Gamma^* = -m(d+6)\,. \tag{2.6}$$

In the setting of Lemma 2.2, consider the diagram

$$\begin{array}{ccc} C^* & \xrightarrow{\ \nu^*\ } & \Gamma^* \\ {\scriptstyle \psi'}\downarrow & & \downarrow{\scriptstyle \psi} \\ C & \xrightarrow[\ \nu\ ]{} & \Gamma \end{array} \tag{2.7}$$

where ν and ν^* are the normalization morphisms and ψ and ψ' have degree 2 (to ease notation, here we have identified ψ with its restriction to Γ^*).

Let $\delta := \delta(D,\Gamma)$. It could be that C^* splits into two components both isomorphic to C; in this case $\delta = 0$. If $\delta = 0$ and the genus of C is zero, then C^* certainly splits. Suppose that C^* is irreducible, and let g and g^* be the geometric genera of Γ and Γ^* (i.e. the arithmetic genera of C and C^*, respectively). Since ψ' has exactly δ ramification points, the Riemann-Hurwitz formula yields

$$2(g^*-1) = 4(g-1) + \delta\,. \tag{2.8}$$

2.4. The odd degree case. Fix a line $h \in |\mathcal{O}_{\mathbb{P}^2}(1)|$, and let D be a smooth curve in $\mathbb{P}^2$ of degree $d = 2\ell - 1 \geqslant 1$, which intersects h transversely. We denote by $\mathcal{U}^h_d$ the open subset of $\mathcal{U}_d$ consisting of such curves.

For each $D \in \mathcal{U}^h_d$, we consider the reducible curve of degree $d+1 = 2\ell$

$$\Delta := D + h \in |\mathcal{O}_{\mathbb{P}^2}(d+1)|$$

and the double cover $\psi \colon \mathbb{D}^* \to \mathbb{P}^2$, branched along Δ. The difference with the even degree case is that $\mathbb{D}^*$ is no longer smooth, but it has double points at the d points in $D \cap h$. In any event, as in the even degree case, we can consider the set $\mathcal{Q}^*_{\ell;h} \subset |\mathcal{O}_{Z_\ell}(2)|$ of all such surfaces $\mathbb{D}^*$, with its universal family $\mathcal{D}^*_{\ell;h} \to \mathcal{Q}^*_{\ell;h}$ which parameterizes all double planes $\psi \colon \mathbb{D}^* \to \mathbb{P}^2$ as above. We still have the morphism

$$\beta : \mathcal{Q}^*_{\ell;h} \to \mathcal{U}^h_d$$

associating to $\mathbb{D}^*$ the branching divisor Δ of $\varphi \colon \mathbb{D}^* \to \mathbb{P}^2$ minus h.

The group acting here is no longer the full group $\mathbb{G}_\ell$ but its subgroup $\mathbb{G}_{\ell;h}$ which fits in the exact sequence

$$1 \to \mathbb{C}^* \to \mathbb{G}_{\ell;h} \to \mathrm{Aff}(2,\mathbb{C}) \to 1,$$

where $\mathrm{Aff}(2,\mathbb{C})$ is the *affine group* of all projective transformations in $\mathrm{PGL}(3,\mathbb{C})$ stabilizing h.

Keeping the setting and notation of §2.3, Lemma 2.2 still holds, as well as diagram (2.7). If Γ intersects h at m distinct points which are off D, then the double cover $\psi' : C^* \to C$ has $\delta + m \geqslant m > 0$ ramification points, where $\delta = \delta(D,\Gamma)$ as above. In particular, C^* is irreducible, and (2.8) is replaced by

$$2(g^* - 1) = 4(g-1) + \delta + m. \tag{2.9}$$

3. The main result

In this section we prove the following:

THEOREM 3.1. *Let $\delta \geqslant 0$ be an integer such that, for a very general curve D in $\mathbb{P}^2$ of degree $d = 2\ell - \varepsilon$, where $\varepsilon \in \{0,1\}$, there exists an integral curve $\Gamma \subset \mathbb{P}^2$, $\Gamma \neq D$, of geometric genus g and degree m with $\delta(D,\Gamma) = \delta$. Then*

$$4g + \delta \geqslant m(d + 2\varepsilon - 8) + 5. \tag{3.1}$$

The proof of Theorem 3.1 will be done in §3.2. First we need some more preliminaries, which we collect in the next subsection. We keep all notation and conventions introduced so far.

3.1. Constructing appropriate families. Fix integers $m \geqslant 1$ and $g \geqslant 0$. Let $\mathcal{H}$ be the locally closed subset of $\mathcal{L}_m$, whose points correspond to integral curves $\Gamma \subset \mathbb{P}^2$ of degree m and geometric genus g; $\mathcal{H}$ is a quasiprojective variety. We let $\mathcal{U} \to \mathcal{H}$ be the universal curve.

3.1.1. *The even degree case.* Fix an even positive integer $d = 2\ell$ and a non-negative integer δ. Consider the locally closed subset $\mathcal{I}$ of $\mathcal{H} \times \mathcal{Q}_\ell^*$ of pairs $(\Gamma, \mathbb{D}^*)$ such that Γ does not coincide with the branch curve D of $\psi : \mathbb{D}^* \to \mathbb{P}^2$ and $\delta(D,\Gamma) = \delta$. Remember that we may equivalently interpret $\mathbb{D}^*$ as a surface in W_ℓ or in Z_ℓ. Each irreducible component of $\mathcal{I}$ is fixed by the obvious action of $\mathbb{G}_\ell$ on $\mathcal{H} \times \mathcal{Q}_\ell^*$.

For any $(\Gamma, \mathbb{D}^*) \in \mathcal{I}$, the pull–back $\Gamma^* \subset \mathbb{D}^*$ of Γ via ψ is a reduced curve in Z_ℓ. Hence there is a morphism $\mu : \mathcal{I} \to \mathcal{K}$, where $\mathcal{K}$ is the Hilbert scheme of curves of Z_ℓ. We let $\mathcal{V} \to \mathcal{K}$ be the corresponding universal family. The map μ is equivariant under the actions of $\mathbb{G}_\ell$ on both $\mathcal{I}$ and $\mathcal{K}$.

Let $\pi_1 : \mathcal{I} \to \mathcal{H}$ and $\pi_2 : \mathcal{I} \to \mathcal{Q}_\ell^*$ be the two projections. Under the hypotheses of Theorem 3.1 and with notation as in § 2.3, the following holds.

LEMMA 3.2. *There exists an irreducible component I of $\mathcal{I}$ which dominates $\mathcal{Q}_\ell$ via π_2. Hence I dominates also $\mathcal{U}_d \subset \mathcal{L}_d$ via $\beta \circ \pi_2$.*

Given I as in Lemma 3.2, we choose an irreducible, smooth subvariety $\mathcal{B}$ of I, such that π_2 restricts to an étale morphism of $\mathcal{B}$ onto its image, which is dense in $\mathcal{Q}_\ell$. To place our objects in the context of §1, consider the universal family $\mathbb{D}_\ell^* \to \mathcal{Q}_\ell^*$ (cf. § 2.3) of double planes $\mathbb{D}^*$ [resp. $\mathcal{V} \to \mathcal{K}$ of curves $\Gamma^* \subset \mathbb{D}^*$]. Up to possibly shrinking $\mathcal{B}$ and performing an étale cover of it, the morphisms $\mathcal{B} \xrightarrow{\pi_2} \mathcal{Q}_\ell^*$ and $\mathcal{B} \xrightarrow{\mu} \mathcal{K}$ give rise to families

$$\mathcal{D} := \pi_2^*(\mathbb{D}_\ell^*) \xrightarrow{p} \mathcal{B} \quad \text{and} \quad \mathfrak{C} := \mu^*(\mathcal{V}) \xrightarrow{q} \mathcal{B}$$

over $\mathcal{B}$ fitting in diagram (1.1). We may assume that there exists a simultaneous normalization ν and a family $\mathcal{C} \xrightarrow{q} \mathcal{B}$ as in (1.2), with $\mathcal{C}$ smooth fitting in (1.3), where $X = Z_\ell$.

3.1.2. *The odd degree case.* Fix now an odd positive integer $d = 2\ell - 1$ and a non-negative integer δ, and fix a line h in $\mathbb{P}^2$. We consider the locally closed subset $\mathcal{I}$ of $\mathcal{H} \times \mathcal{Q}_{\ell;h}$ consisting of pairs $(\Gamma, \mathbb{D}^*) \in \mathcal{H} \times \mathcal{Q}_\ell^*$ such that Γ is not contained in the branching divisor Δ of $\psi : \mathbb{D}^* \to \mathbb{P}^2$, $\delta(D, \Gamma) = \delta$, and Γ intersects h at m distinct points which are off D.

For any point $(\Gamma, \mathbb{D}^*) \in \mathcal{I}$, the pull–back $\Gamma^* \subset \mathbb{D}^*$ of Γ via ψ is an integral curve in Z_ℓ. So, we still have the morphisms $\mu : \mathcal{I} \to \mathcal{K}$, $\pi_1 : \mathcal{I} \to \mathcal{H}$ and $\pi_2 : \mathcal{I} \to \mathcal{Q}_{\ell;h}^*$ equivariant under actions of $\mathbb{G}_{\ell;h}$.

As before, we have the following

LEMMA 3.3. *There exists an irreducible component I of $\mathcal{I}$ which dominates $\mathcal{Q}_{\ell;h}^*$ via π_2.*

As in the even case, given I as in Lemma 3.3, we may construct a smooth $\mathcal{B}$ having an étale, dominant morphism to $\mathcal{Q}_{\ell;h}$, together with families

$$\mathcal{D} := \pi_2^*(\mathcal{D}_{\ell,h}^*) \xrightarrow{p} \mathcal{B}, \quad \mathfrak{C} := \mu^*(\mathcal{V}) \xrightarrow{q} \mathcal{B},$$

fitting in diagram (1.1). Consider a simultaneous normalization ν and a family $\mathcal{C} \xrightarrow{q} \mathcal{B}$ as in (1.2), with $\mathcal{C}$ smooth. In view of Lemmata 3.2 and 3.3, the constructed families fit in diagram (1.3), with $X = Z_\ell$.

In both cases, the next lemma allows to apply Proposition 1.3 in our setting.

LEMMA 3.4. *For any $d > 0$, $\mathcal{C} \xrightarrow{q} \mathcal{B}$ is a covering family, i.e. $z(\mathcal{C}) = 3$.*

PROOF. By the discussion in §3.1, for d even $\varphi(\mathcal{C})$ is stable under the action of $\mathbb{G}_\ell$ on Z_ℓ, which is transitive; for d odd $\varphi(\mathcal{C})$ is stable under the action of $\mathbb{G}_{\ell;h}$, which is transitive on the dense open subset of Z_ℓ whose complement is $\pi^{-1}(h) \cup E$. This proves the assertion. □

3.2. Proof of Theorem 3.1. Our proof follows the one of Theorem (1.2) in [**21**]. First we recall the following useful fact.

LEMMA 3.5 (See [**21**, Lemma (3.1)]). *Let $g : V \to W$ be a linear map of finite dimensional vector spaces. Suppose that* $\dim(g(V)) > k$. *Let $\{V_i\}_{i \in I}$ be a family of vector subspaces of V, such that $\bigcup_{i \in I} V_i$ spans V, and for any pair $(i, j) \in I \times I$, with $i \neq j$, there is a finite sequence $i_1 = i, i_2, \ldots, i_{t-1}, i_t = j$ of distinct elements of I with* $\dim(g(V_{i_h} \cap V_{i_{h+1}})) \geqslant k$, *for all $h \in \{1, \ldots, t-1\}$. Then there is an index $i \in I$ such that* $\dim(g(V_i)) > k$.

3.2.1. *The even degree case.* We need to construct a suitable subfamily of $\mathfrak{C} \to \mathcal{B}$ with the covering property.

Fix a general point $b_0 \in \mathcal{B}$, and let Γ_0^* and $\mathbb{D}_0^*$ be the corresponding elements of the families $\mathfrak{C} \to \mathcal{B}$ and $\mathcal{D} \to \mathcal{B}$, respectively.

Let $\mathcal{L}$ be the open subset of the linear system $\mathcal{L}_\ell(d-1)$ as in (2.2) consisting of the surfaces $F \in \mathcal{L}_\ell(d-1)$ which do not contain Γ_0^*. A general such surface F meets Γ_0^* transversely. By genericity, we may suppose that all surfaces F defined by the pull–back via $\pi : Z_\ell \to V_\ell \cong \mathbb{P}^2$ of degree $d-1$ monomials in the variables x_0, x_1, x_2 belong to $\mathcal{L}$. For a given $F \in \mathcal{L}$, we denote by $\mathcal{B}_F$ the subvariety of $\mathcal{B}$ parameterizing all double planes in $\mathcal{D} \to \mathcal{B}$ containing the complete intersection curve of F and $\mathbb{D}_0^*$. In addition, for a general point $\xi \in \Gamma_0^*$ we let $\mathcal{B}_{F,\xi}$ denote the subvariety of $\mathcal{B}_F$ parameterizing all surfaces in $\mathcal{D} \to \mathcal{B}_F$ which pass through ξ.

LEMMA 3.6. *For $F \in \mathcal{L}$ and $\xi \in \Gamma_0^*$ as above one has*

$$\dim(\mathcal{B}_F) = 3 \text{ and } \dim(\mathcal{B}_{F,\xi}) = 2\,.$$

Furthermore, b_0 is a smooth point of both $\mathcal{B}_F$ and $\mathcal{B}_{F,\xi}$.

PROOF. Consider the sublinear system Λ_F of $\widetilde{\mathcal{Q}}_\ell = \rho^*(\mathcal{Q}_\ell)$ on Z_ℓ consisting of all surfaces containing the complete intersection curve $F \cap \mathbb{D}_0^*$. Imposing to the surfaces in Λ_F the condition to contain a general point of F, the divisor $F + 2E$ splits off, and the residual surface sits in $\mathcal{L}_\ell(1)$. Hence Λ_F contains a codimension 1 sublinear system consisting of surfaces of the form $2E + F + L$, with L varying in $\mathcal{L}_\ell(1)$, which has dimension 2. Hence $\dim(\Lambda_F) = 3$. Since $\mathcal{B}$ dominates $\mathcal{Q}_\ell$ via π_2, which is finite on $\mathcal{B}$, and $\mathcal{B}_F$ is the inverse image of Λ_F, one has $\dim(\mathcal{B}_F) = \dim(\Lambda_F) = 3$. The proof is similar for $\mathcal{B}_{F,\xi}$. The final assertion follows by the genericity assumptions. □

We denote by T_0 the tangent space to $\mathcal{B}$ at b_0, and by T_F and $T_{F,\xi}$ the 3 and 2-dimensional subspaces of T_0 tangent to $\mathcal{B}_F$ and to $\mathcal{B}_{F,\xi}$ at b_0, respectively.

LEMMA 3.7. *One has:*

(a) $\bigcup_{F\in\mathcal{L}} T_F$ *spans* T_0*;*
(b) *given $F \in \mathcal{L}$, the union* $\bigcup_{\xi\in\Gamma_0^*} T_{F,\xi}$ *spans* T_F.

PROOF. (a) Since π_2 is étale on $\mathcal{B}$, T_0 is isomorphic to the tangent space to $\mathcal{Q}_\ell$ at $\mathbb{D}_0^*$. Remember that, by §2.3, the double plane $\mathbb{D}_0^*$, considered in W_ℓ, is cut out by a quadric with equation $z^2 = Q(x^I)_{|I|=\ell}$. So T_0 can be identified with the vector space of homogeneous quadratic polynomials of the form $az^2 - G(x^I)_{|I|=\ell}$ modulo the one-dimensional linear space spanned by $z^2 - Q(x^I)_{|I|=\ell}$ and by the linear space of quadratic polynomials in $\{x^I\}_{|I|=\ell}$ defining V_ℓ. [1] Hence T_0 can be identified with the vector space of quadratic polynomials in $\{x^I\}_{|I|=\ell}$, modulo the vector space of quadratic polynomials in $\{x^I\}_{|I|=\ell}$ defining V_ℓ. This, in turn, can be identified with the vector space S_d of homogeneous polynomials of degree d in x_0, x_1, x_2.

Now T_F can be identified with the vector subspace $S_d(f)$ of $T_0 \cong S_d$ of polynomials of the form fh, where f is a fixed homogeneous polynomial of degree $d-1$ (determined by F), and h is any linear form. By assumption on $\mathcal{L}$, $\bigcup_{F\in\mathcal{L}} T_F$ contains all monomials of degree d, which do span S_d.

(b) Given F, $T_{F,\xi}$ can be identified with the vector space of homogeneous polynomials of the form fh, where h vanishes at $\pi(\xi) \in \mathbb{P}^2$. These polynomials do span $T_F \cong S_d(f)$. □

Next we consider the restrictions

$$\mathcal{D}_F \xrightarrow{p} \mathcal{B}_F, \quad \mathfrak{C}_F \xrightarrow{\mathfrak{q}} \mathcal{B}_F, \text{ and } \mathcal{D}_{F,\xi} \xrightarrow{p} \mathcal{B}_{F,\xi}, \quad \mathfrak{C}_{F,\xi} \xrightarrow{\mathfrak{q}} \mathcal{B}_{F,\xi},$$

respectively, of the families

$$\mathcal{D} \xrightarrow{p} \mathcal{B} \text{ and } \mathfrak{C} \xrightarrow{\mathfrak{q}} \mathcal{B}.$$

[1] An explanation is in order. Consider a vector space V and a nonzero vector $v \in V$, along with the associated projective space $\mathbb{P}(V)$ and the corresponding point $[v] \in \mathbb{P}(V)$. Then the tangent space $T_{[v]}\mathbb{P}(V)$ can be canonically identified with $\mathrm{Hom}(\langle v\rangle, V/\langle v\rangle) \cong V/\langle v\rangle$.

PROPOSITION 3.8. *For general $F \in \mathcal{L}$ and $\xi \in \Gamma_0^*$, the families*

$$\mathfrak{C}_F \xrightarrow{\mathfrak{q}} \mathcal{B}_F \text{ and } \mathfrak{C}_{F,\xi} \xrightarrow{\mathfrak{q}} \mathcal{B}_{F,\xi}$$

have the covering property.

PROOF. We prove the assertion for $\mathfrak{C}_F \xrightarrow{\mathfrak{q}} \mathcal{B}_F$. The proof for $\mathfrak{C}_{F,\xi} \xrightarrow{\mathfrak{q}} \mathcal{B}_{F,\xi}$ is similar (and analogous to the proof of the corresponding statement in [**21**, Theorem (1.2)]), hence it can be left to the reader.

Let $\mathbb{M}$ be the set of all monomials of degree $d-1$ in x_0, x_1, x_2. Consider the family $\{F_M\}_{M\in\mathbb{M}}$, where $F_M \in \mathcal{L}$ is defined by the pull–back via $\pi\colon Z_\ell \to V_\ell \cong \mathbb{P}^2$ of the monomial M. Take two monomials M', M'' which differ only in degree 1, i.e., their lowest common multiple U has degree d. Then $\mathcal{B}_{F_{M'}} \cap \mathcal{B}_{F_{M''}}$ contains the pull–back via π_2 of an open, dense subset of the pencil $\langle \mathbb{D}_t^* \rangle$ spanned by $\mathbb{D}_0^*$ and F_U, where F_U is the pull–back via π of the monomial U. The base locus of this pencil does not contain Γ_0^*. Therefore, Γ_0^* varies in a non-trivial one-parameter family $\langle \Gamma_t^* \rangle$ together with members $\mathbb{D}_t^*$ varying in the pencil $\langle \mathbb{D}_t^* \rangle$.

Next we apply Lemma 3.5 with

- $V = T_0$;
- $W = H^0(\Gamma_0^*, N_{\Gamma_0^*|Z_\ell})$;
- the linear map g induced by the characteristic map (see Proposition 1.1 (a));
- the family of subspaces $\{V_i\}_{i\in I}$ given by $\{T_{F_M}\}_{M\in\mathbb{M}}$.

For each pair of monomials M', M'', there is a sequence of monomials $M_1 = M', M_2, \ldots, M_{t-1}, M_t = M''$, such that for all $i = 1, \ldots, t-1$, the lowest common multiple of M_i and M_{i+1} has degree d. The above argument implies that $g(T_{F_{M_i}} \cap T_{F_{M_{i+1}}})$ has dimension at least 1, for all $i = 1, \ldots, m-1$. Furthermore, one has $\dim(g(T_0)) \geqslant 2$, because $\mathfrak{C} \to \mathcal{B}$ is a covering family (see (b) of Proposition 1.1 and Lemma 3.4). By Lemma 3.5 there is a monomial $M \in \mathbb{M}$ such that $\dim(g(T_{F_M})) \geqslant 2$; by virtue of Lemma 3.6, this implies that $\mathfrak{C}_{F_M} \to \mathfrak{B}_{F_M}$ is a covering family. This proves the assertion. □

To finish the proof of Theorem 3.1 in this case, consider the covering family $\mathfrak{C}_{F,\xi} \xrightarrow{\mathfrak{q}} \mathcal{B}_{F,\xi}$, with $F \in \mathcal{L}$ and $\xi \in \Gamma_0^*$ general. Using (1.5), (2.6), and (2.8), for $b = (\Gamma_b^*, \mathbb{D}_b^*) \in \mathcal{B}_{F,\xi}$ general (see §3.1.1) we deduce

$$\deg(\Phi_b^{\mathrm{sm}}) \leqslant 4(g-1) + \delta + 2m(\ell+3) = 4(g-1) + \delta + m(d+6)\,. \tag{3.2}$$

On the other hand, by construction and by (a) of Proposition 1.3,

$$\deg(\Phi_b^{\mathrm{sm}}) \geqslant 1 + \deg(\Gamma_b^* \cap F) = 1 + 2(d-1)m\,. \tag{3.3}$$

Comparing (3.2) and (3.3) gives (3.1).

3.2.2. *The odd degree case.* The proof runs exactly as in the case of even d, so we will be brief and leave the details to the reader.

Fix again $b_0 \in \mathcal{B}$, Γ_0^* and $\mathbb{D}_0^*$ as in the even degree case. Following what we did in §3.1.2, we replace D_b by $D_b + h$, where $h \subset \mathbb{P}^2$ is a general line. In the present setting we let $\mathcal{L}$ be the open subset of $\mathcal{L}_\ell(d)$ consisting of the surfaces $F \in \mathcal{L}_\ell(d)$ which do not contain Γ_0^*. Again we may assume that all surfaces F defined by the pull–back via π of degree d monomials in the variables x_0, x_1, x_2 belong to $\mathcal{L}$. Given $F \in \mathcal{L}$, we define $\mathcal{B}_F$ and $\mathcal{B}_{F,\xi}$ as in the even degree case, and the analogue of Lemma 3.6 still holds. Then, with the usual meaning for T_0, T_F and $T_{F,\xi}$, the analogue of Lemma 3.7 holds. Similarly as in Proposition 3.8, the covering property holds for the restricted families

$$\mathcal{D}_F \xrightarrow{p} \mathcal{B}_F, \quad \mathfrak{C}_F \xrightarrow{q} \mathcal{B}_F, \quad \text{and} \quad \mathcal{D}_{F,\xi} \xrightarrow{p} \mathcal{B}_{F,\xi}, \quad \mathfrak{C}_{F,\xi} \xrightarrow{q} \mathcal{B}_{F,\xi}\,.$$

We conclude finally as in the even degree case: (3.2) holds with no change, whereas (3.3) has to be replaced by

$$\deg(\Phi_b^{\mathrm{sm}}) \geqslant 1 + \deg(\Gamma_b \cap F) = 1 + 2dm\,,$$

and again, (3.1) follows.

4. Genera of subvarieties: a survey

As we mentioned in the Introduction, inequality (0.2) allows to bound the genera of curves in double planes from below. In this section we provide a brief survey on genera bounds for subvarieties in different type of varieties, and discuss several conjectures. All varieties are supposed to be projective, reduced, irreducible, and defined over $\mathbb{C}$. The geometric genus $p_g(Y)$ of a variety Y is the geometric genus of a smooth model of Y.

Two important sources of interest in bounding genera are: the Clemens Conjecture on count of rational curves in Calabi-Yau varieties ([**27**]), and the celebrated Kobayashi Conjecture on hyperbolicity of general hypersurfaces in $\mathbb{P}^n$ of sufficiently large degree ([**60**]). Recall ([**43**], [**60**]) that the Kobayashi hyperbolicity of a variety X implies the algebraic hyperbolicity, and in particular, absence of rational and elliptic curves in X. A part concerning the Clemens Conjecture started with the following theorem.

THEOREM 4.1. (H. Clemens [**26**]) *The geometric genera of curves in a very general hypersurface X of degree $d \geqslant 2n-1$ in $\mathbb{P}^n$ satisfy the inequality $g \geqslant \frac{1}{2}(d-2n+1)+1$.*

This shows, in particular, the absence of rational curves in very general surfaces of degree $d \geqslant 5$ in $\mathbb{P}^3$. One of the subsequent results in higher dimensions was

THEOREM 4.2. (E. Ballico [**3**]) *There is an effective bound $\varphi(n)$ such that a very general hypersurface of degree $d \geqslant \varphi(n)$ in $\mathbb{P}^n$ is algebraically hyperbolic.*

A better effective bound was provided by Geng Xu ([**88**]). For instance ([**25**]), it follows from the results in [**88**] that a general sextic threefold in $\mathbb{P}^4$ is algebraically hyperbolic.

The Demailly algebraic hyperbolicity theorem states the following.

THEOREM 4.3. (J.-P. Demailly [**38**]) *For any hyperbolic subvariety $X \subset \mathbb{P}^n$ there exists $\varepsilon > 0$ such that, for any curve $C \subset X$, the geometric genus g of C is bounded below in terms of the degree: $g \geqslant \varepsilon \deg(C) + 1$. Therefore, the curves of bounded genera in X form bounded families.*

Due to a recent proof of the Kobayashi Conjecture, Theorem 4.3 can be applied to general (in Zariski sense) hypersurfaces in $\mathbb{P}^n$.

THEOREM 4.4. (D. Brotbek [**12**], Y.-T. Siu [**79**]) *A general hypersurface of sufficiently large degree in $\mathbb{P}^n$ is Kobayashi hyperbolic.*

For effective estimates of degrees of hyperbolic hypersurfaces, see Y. Deng ([**41**, **42**]); see also J.-P. Demailly [**40**] for a survey and a simplified proof.

L. Ein obtained some analogs of Clemens' estimate in higher dimensions.

THEOREM 4.5. (L. Ein [**44**, **45**]) *Let M be a smooth projective variety of dimension $n \geqslant 3$, let $L \to M$ be an ample line bundle, and let $X \in |dL|$ be a very general member. Then for $d \geqslant 2n - \ell$ any subvariety $Y \subset X$ of dimension ℓ has positive geometric genus, and for $d \geqslant 2n - \ell + 1$, Y is of general type.*

In the case $M = \mathbb{P}^n$ there is a sharper bound.

THEOREM 4.6. (C. Voisin [**81**, **82**]) *Let X be a very general hypersurface of degree $d \geqslant 2n - \ell - 1$ in $\mathbb{P}^n$, $n \geqslant 3$. Then for $d \geqslant 2n - l + 1$ any subvariety $Y \subset X$ of dimension $l \leq n - 3$ has positive geometric genus, and for $d \geqslant 2n - l$, Y is of general type.*

Sharper bounds are known also for certain toric varieties (A. Ikeda [**54**]). For $M = \mathbb{P}^n$ a further improvement is due to G. Pacienza.

THEOREM 4.7. (G. Pacienza [**73**]) *For $n \geqslant 6$ and for a very general hypersurface $X \subset \mathbb{P}^n$ of degree $d \geqslant 2n - 2$, any subvariety $Y \subset X$ is of general type.*

Geng Xu improved Ein's theorem as follows.

THEOREM 4.8. (G. Xu [**92**]) *Let X be a very general complete intersection of $m \leqslant n-3$ hypersurfaces of degrees $d_1, \ldots, d_m$ in $\mathbb{P}^n$, where $d_i \geqslant 2\ \forall i$, and let $Y \subset X$ be a reduced and irreducible divisor. Let $d = d_1 + \cdots + d_m$. Then $p_g(Y) \geqslant n - 1$ if $d \geqslant n + 2$, and Y is of general type if $d > n + 2$.*

See also Geng Xu ([**91**]), C. Chang and Z. Ran ([**15**]), L. Chiantini, A.-F. Lopez, and Z. Ran ([**22**]), H. Clemens and Z. Ran ([**28**]), S.S.-T. Lu and Y. Miyaoka ([**67**]), and L.-C. Wang ([**86**, **87**]). The results in [**86**] include some classes of divisors in Calabi-Yau hypersurfaces of degree $d = n + 1$ in $\mathbb{P}^n$.

Let us mention several sporadic results. See also, e.g., R. Beheshti ([**6**]), M. Bernardara ([**8**]), L. Bonavero and A. Hoering ([**10**]), T.D. Browning and P. Vishe ([**13**]), I. Coskun and E. Reidl ([**31**]), O. Debarre ([**37**]), K. Furukawa ([**48**, **49**]), J. Harris, M. Roth, and J. Starr ([**52**]), J. Kollar ([**61**]). Concerning the Clemens Conjecture on rational curves in quintic threefolds and Mirror Symmetry, see, e.g., M. Kontsevich ([**62**]), A. Libgober and J. Teitelbaum ([**64**]), D.A. Cox and S. Katz ([**35**]), T. Coates and A. Givental ([**29**]) and the references therein.

THEOREM 4.9.

- (G. Pacienza [**72**], E. Riedl and D. Yang [**74**]) *Let $X \subset \mathbb{P}^n$ be a very general hypersurface of degree d. If either $n = 6$ and $d = 2n - 3$, or $n \geqslant 7$ and $\frac{3n+1}{2} \leq d \leq 2n - 3$, then X contains lines but no other rational curves.*
- (D. Shin [**78**]) *A general hypersurface of degree $d > \frac{3}{2}n - 1$ in $\mathbb{P}^n$ does not contain any smooth conic; however* (S. Katz [**57**]), *a general quintic threefold in $\mathbb{P}^4$ does.*
- (S. Katz [**57**], P. Nijsse [**71**], T. Johnsen and S.L. Kleiman [**55**], J. D'Almeida [**36**], E. Cotterill [**32**, **33**], E. Ballico and C. Fontanari [**5**]; cf. also E. Ballico [**4**] and A. L. Knutsen [**58**, **59**]) *A general quintic threefold X in $\mathbb{P}^4$ contains only finitely many rational curves of degree ≤ 12, and*

each rational curve C of degree ≤ 11 either is smooth and embedded in X with a balanced normal bundle $\mathcal{O}(-1) \oplus \mathcal{O}(-1)$, or is a plane 6-nodal quintic.

- (D. Shin [**77**], G. Mostad Hana and T. Johnsen [**51**], E. Cotterill [**34**]) *A general hypersurface of degree 7 in $\mathbb{P}^5$ does not contain any rational curve of degree $d \in \{2, \ldots, 16\}$.*
- (G. Ferrarese and D. Romagnoli [**47**]) *The degree of an elliptic curve on a very general hypersurface X of degree 7 in $\mathbb{P}^4$ is a multiple of 7.*
- (B. Wang [**85**]) *A general hypersurface of degree 54 in $\mathbb{P}^{30}$ does not contain any rational quartic curve.*
- (B. Wang [**84**]) *A very general hypersurface of degree $d \geqslant 2n-1$ in $\mathbb{P}^n$ does not contain any smooth elliptic curve.*

For $n = 3$, Geng Xu replaced Clemens' initial genus bound in Theorem 4.1 by the optimal one. See also L. Chiantini and A.F. Lopez ([**21**]) for an alternative proof and some generalizations.

THEOREM 4.10. (G. Xu [**90**]) *The genera of curves on a very general surface of degree $d \geqslant 5$ in $\mathbb{P}^3$ satisfy the inequality $g \geqslant \frac{1}{2}d(d-3) - 2$, and this bound is sharp. For $d \geqslant 6$ this sharp bound can be achieved only by a tritangent hyperplane section.*

Let $\mathrm{Gaps}(d)$ be the set of all non-negative integers which cannot be realized as geometric genera of irreducible curves on a very general surface of degree d in $\mathbb{P}^3$. This set is union of finitely many disjoint and separated integer intervals. By Xu's Theorem 4.10, the first gap interval is $\mathrm{Gaps}_0(d) = [0, d(d-3)/2 - 3]$. For $d = 5$, this is the only gap interval. For $d \geqslant 6$, the next gap interval is $\mathrm{Gaps}_1(d) = [d(d-3)/2+2, d^2-2d-9]$ ([**23**]). One can show ([**24**]) that $\max(\mathrm{Gaps}(d)) = O(d^{8/3})$. The latter is based on certain existence results. For arbitrary smooth (not necessarily general) surfaces in $\mathbb{P}^3$, we have the following existence result.

THEOREM 4.11. ([**16**], [**24**]) *There exists a function $c(d) \sim d^3$ such that, for any smooth surface S in $\mathbb{P}^3$ of degree d and any $g \geqslant c(d)$, S carries a reduced, irreducible nodal curve of geometric genus g, whose nodes can be prescribed generically on S.*

To formulate an analog in higher dimensions, we recall the following notion. Let Y be an irreducible variety of dimension s. A singular point $y \in Y$ is called an *ordinary singularity of multiplicity m* ($m > 1$), if the Zariski tangent space of Y at y has dimension $s+1$, and the (affine) tangent cone to Y at y is a cone with vertex y over a smooth hypersurface of degree m in $\mathbb{P}^s$.

The next result was first established by J.A. Chen ([**16**]) for curves in n-dimensional varieties, and then in [**24**] for subvarieties of arbitrary dimension $s \leqslant n-1$ [2].

THEOREM 4.12. ([**16**], [**24**]) *Let X be an irreducible, smooth, projective variety of dimension $n > 1$, let L be a very ample divisor on X, and let $s \in \{1, \ldots, n-1\}$. Then there is an integer $p_{X,L,s}$ such that for any $p \geqslant p_{X,L,s}$ one can find an irreducible complete intersection $Y = D_1 \cap \ldots \cap D_{n-s} \subset X$ of dimension s with at most ordinary points of multiplicity $s+1$ as singularities such that $p_g(Y) = p$,*

---

[2] We are grateful to J.A. Chen for pointing out his nice paper [**16**] that we ignored when writing [**24**]. We apologize for our ignorance.

where $D_i \in |L|$ *for* $i = 1, \ldots, n-s-1$ *are smooth and transversal and* $D_{n-s} \in |mL|$ *for some* $m \geqslant 1$. *Moreover, for* $n \geqslant 3$ *and* $s = 1$ *one can find a* smooth *curve* Y *in* X *of a given genus* $g(Y) = p \geqslant p_{X,L,1}$.

Recall the famous:

Green-Griffith-Lang Conjecture. ([**50, 63**]; see also [**7**], [**39**]) *Let* X *be a projective variety of general type. Then there exists a proper closed subset* $Z \subset X$ *such that any subvariety* $Y \subset X$ *not of general type is contained in* Z.

The following conjecture is inspired by the previous results and by the Green-Griffiths-Lang Conjecture in the surface case.

CONJECTURE. *There exists a strictly growing function* $\varphi(d)$*, with natural values, such that the set of curves of geometric genus* $g \leqslant \varphi(d)$ *in any smooth surface* S *of degree* $d \geqslant 5$ *in* $\mathbb{P}^3$ *is finite.*

Notice ([**24**]) that for any $g \geqslant 0$ and $d \geqslant 1$ one can find a smooth surface $S \subset \mathbb{P}^3$ of degree d carrying a nodal curve of genus g. Notice also that a very general quartic surface in $\mathbb{P}^3$ contains an infinite countable set of rational curves, hence the restriction $d \geqslant 5$ is necessary. Let us mention a few facts supporting the conjecture. According to B. Segre ([**76**]) the number of lines on a smooth surface of degree $d \geqslant 3$ does not exceed $(d-2)(11d-6)$. The celebrated Bogomolov theorem ([**9**]) says that the number of rational and elliptic curves on a surface of general type with $c_1^2 > c_2$ is finite. Moreover, due to Y. Miyaoka, this number admits a uniform estimate:

THEOREM 4.13. (Y. Miyaoka [**69**]) *Let* S *be a minimal smooth projective surface of general type satisfying the inequality for Chern numbers* $c_1^2 > c_2$. *Then the number of irreducible curves of genus* 0 *and* 1 *on* S *is bounded by a function of* c_1 *and* c_2.

Analogous facts are true under certain weaker assumptions on Chern numbers (Y. Miyaoka [**70**]), or on the singularities of rational and elliptic curves in S (S.S.-Y. Lu and Y. Miyaoka [**68**]). It is plausible that the number of curves of genus $g \leqslant \varphi(d)$ on a smooth surface of degree d in $\mathbb{P}^3$ can be uniformly bounded above by a function of d. The conjecture above is coherent with the following ones.

CONJECTURE (C. Voisin [**83**]). *Let* $X \subset \mathbb{P}^n$ *be a very general hypersurface of degree* $d \geqslant n+2$. *Then the degrees of rational curves in* X *are bounded.*

CONJECTURE (P. Autissier, A. Chambert-Loir, and C. Gasbarri [**2**]). *Let* X *be a smooth projective variety of general type with the canonical line bundle* K_X. *Then there exist real numbers* A *and* B, *and a proper Zariski closed subset* $Z \subset X$ *such that for any curve* C *of geometric genus* g *in* X *not contained in* Z, *one has* $\deg_C(K_X) \leqslant A(2g-2) + B$.

References

[1] C. Araujo and J. Kollár, *Rational curves on varieties*, Higher dimensional varieties and rational points (Budapest, 2001), Bolyai Soc. Math. Stud., vol. 12, Springer, Berlin, 2003, pp. 13–68, DOI 10.1007/978-3-662-05123-8_3. MR2011743

[2] P. Autissier, A. Chambert-Loir, and C. Gasbarri, *On the canonical degrees of curves in varieties of general type*, Geom. Funct. Anal. **22** (2012), no. 5, 1051–1061, DOI 10.1007/s00039-012-0188-1. MR2989429

[3] E. Ballico, *Algebraic hyperbolicity of generic high degree hypersurfaces*, Arch. Math. (Basel) **63** (1994), no. 3, 282–283, DOI 10.1007/BF01189831. MR1287258

[4] E. Ballico, *Non-existence of smooth rational curves of degree* $d = 13, 14, 15$ *contained in a general quintic hypersurface of* $\mathbb{P}^4$ *and in some quadric hypersurface*, Note Mat. **36** (2016), no. 2, 77–88. MR3601833

[5] E. Ballico and C. Fontanari, *Finiteness of rational curves of degree 12 on a general quintic threefold*, Pure Appl. Math. Q. **11** (2015), no. 4, 537–557, DOI 10.4310/PAMQ.2015.v11.n4.a1. MR3613121

[6] R. Beheshti, *Hypersurfaces with too many rational curves*, Math. Ann. **360** (2014), no. 3-4, 753–768, DOI 10.1007/s00208-014-1024-8. MR3273644

[7] G. Berczi. *Towards the Green-Griffiths-Lang Conjecture via equivariant localization.* arXiv:1509.03406 (2015), 26p.

[8] M. Bernardara, *Calabi-Yau complete intersections with infinitely many lines*, Rend. Semin. Mat. Univ. Politec. Torino **66** (2008), no. 2, 87–97. MR2455452

[9] F. A. Bogomolov, *Families of curves on a surface of general type* (Russian), Dokl. Akad. Nauk SSSR **236** (1977), no. 5, 1041–1044. MR0457450

[10] L. Bonavero and A. Höring, *Counting conics in complete intersections*, Acta Math. Vietnam. **35** (2010), no. 1, 23–30. MR2642160

[11] D. Brotbek, *Hyperbolicity related problems for complete intersection varieties*, Compos. Math. **150** (2014), no. 3, 369–395, DOI 10.1112/S0010437X13007458. MR3187623

[12] D. Brotbek, *On the hyperbolicity of general hypersurfaces*, Publ. Math. Inst. Hautes Études Sci. **126** (2017), 1–34, DOI 10.1007/s10240-017-0090-3. MR3735863

[13] T. Browning and P. Vishe, *Rational curves on smooth hypersurfaces of low degree*, Algebra Number Theory **11** (2017), no. 7, 1657–1675, DOI 10.2140/ant.2017.11.1657. MR3697151

[14] J. Cao and A. Hoering. *Rational curves on compact Kaehler manifolds.* Avalaible at: http://math.unice.fr/ hoering/articles/a27-rat-curves.pdf

[15] M.-C. Chang and Z. Ran, *Divisors on some generic hypersurfaces*, J. Differential Geom. **38** (1993), no. 3, 671–678. MR1243790

[16] J. A. Chen, *On genera of smooth curves in higher-dimensional varieties*, Proc. Amer. Math. Soc. **125** (1997), no. 8, 2221–2225, DOI 10.1090/S0002-9939-97-03908-7. MR1401729

[17] X. Chen, *On the intersection of two plane curves*, Math. Res. Lett. **7** (2000), no. 5-6, 631–641, DOI 10.4310/MRL.2000.v7.n5.a9. MR1809289

[18] X. Chen, *On algebraic hyperbolicity of log varieties*, Commun. Contemp. Math. **6** (2004), no. 4, 513–559, DOI 10.1142/S0219199704001422. MR2078413

[19] X. Chen and Y. Zhu, $\mathbb{A}^1$*-curves on affine complete intersections*, Selecta Math. (N.S.) **24** (2018), no. 4, 3823–3834, DOI 10.1007/s00029-018-0421-3. MR3848034

[20] L. Chiantini and C. Ciliberto, *A few remarks on the lifting problem*, Astérisque **218** (1993), 95–109. Journées de Géométrie Algébrique d'Orsay (Orsay, 1992). MR1265310

[21] L. Chiantini and A. F. Lopez, *Focal loci of families and the genus of curves on surfaces*, Proc. Amer. Math. Soc. **127** (1999), no. 12, 3451–3459, DOI 10.1090/S0002-9939-99-05407-6. MR1676295

[22] L. Chiantini, A. F. Lopez, and Z. Ran, *Subvarieties of generic hypersurfaces in any variety*, Math. Proc. Cambridge Philos. Soc. **130** (2001), no. 2, 259–268, DOI 10.1017/S0305004100004904. MR1806776

[23] C. Ciliberto, F. Flamini, and M. Zaidenberg, *Genera of curves on a very general surface in* $\mathbb{P}^3$, Int. Math. Res. Not. IMRN **22** (2015), 12177–12205. MR3456718

[24] C. Ciliberto, F. Flamini, and M. Zaidenberg, *Gaps for geometric genera*, Arch. Math. (Basel) **106** (2016), no. 6, 531–541, DOI 10.1007/s00013-016-0908-0. MR3504187

[25] C. Ciliberto and M. Zaidenberg, *Scrolls and hyperbolicity*, Internat. J. Math. **24** (2013), no. 4, 1350026, 25, DOI 10.1142/S0129167X13500262. MR3062966

[26] H. Clemens, *Curves on generic hypersurfaces*, Ann. Sci. École Norm. Sup. (4) **19** (1986), no. 4, 629–636. MR875091

[27] H. Clemens, *Curves on higher-dimensional complex projective manifolds*, Proceedings of the International Congress of Mathematicians, Vol. 1, 2 (Berkeley, Calif., 1986), Amer. Math. Soc., Providence, RI, 1987, pp. 634–640. MR934266

[28] H. Clemens and Z. Ran, *Erratum to: "Twisted genus bounds for subvarieties of generic hypersurfaces" [Amer. J. Math.* ***126*** *(2004), no. 1, 89–120; MR2033565]*, Amer. J. Math. **127** (2005), no. 1, 241–242. MR2115667

[29] T. Coates and A. Givental, *Quantum Riemann-Roch, Lefschetz and Serre*, Ann. of Math. (2) **165** (2007), no. 1, 15–53, DOI 10.4007/annals.2007.165.15. MR2276766

[30] I. Coskun and J. Starr, *Rational curves on smooth cubic hypersurfaces*, Int. Math. Res. Not. IMRN **24** (2009), 4626–4641, DOI 10.1093/imrn/rnp102. MR2564370

[31] I. Coskun and E. Riedl, *Normal bundles of rational curves in projective space*, Math. Z. **288** (2018), no. 3-4, 803–827, DOI 10.1007/s00209-017-1914-z. MR3778979

[32] E. Cotterill, *Rational curves of degree 10 on a general quintic threefold*, Comm. Algebra **33** (2005), no. 6, 1833–1872, DOI 10.1081/AGB-200063325. MR2150847

[33] E. Cotterill, *Rational curves of degree 11 on a general quintic 3-fold*, Q. J. Math. **63** (2012), no. 3, 539–568, DOI 10.1093/qmath/har001. MR2967162

[34] E. Cotterill, *Rational curves of degree 16 on a general heptic fourfold*, J. Pure Appl. Algebra **218** (2014), no. 1, 121–129, DOI 10.1016/j.jpaa.2013.04.017. MR3120614

[35] D. A. Cox and S. Katz, *Mirror symmetry and algebraic geometry*, Mathematical Surveys and Monographs, vol. 68, American Mathematical Society, Providence, RI, 1999. MR1677117

[36] J. D'Almeida, *Courbes rationnelles de degré 11 sur une hypersurface quintique générale de* $\mathbf{P}^4$ (French, with French summary), Bull. Sci. Math. **136** (2012), no. 8, 899–903, DOI 10.1016/j.bulsci.2012.06.001. MR2995009

[37] O. Debarre. *Rational curves on hypersurfaces.* Lecture notes for the II Latin American School of Algebraic Geometry and Applications, 1–12 June 2015. Cabo Frio, Brazil. January 4, 2016, 48p.

[38] J.-P. Demailly, *Algebraic criteria for Kobayashi hyperbolic projective varieties and jet differentials*, Algebraic geometry—Santa Cruz 1995, Proc. Sympos. Pure Math., vol. 62, Amer. Math. Soc., Providence, RI, 1997, pp. 285–360, DOI 10.1090/pspum/062.2/1492539. MR1492539

[39] J. P. Demailly. *Towards the Green-Griffiths-Lang Conjecture.* Baklouti, Ali (ed.) et al., Analysis and geometry. MIMS-GGTM, Tunis, Tunisia, March 24–27, 2014. Proceedings of the international conference. In honour of Mohammed Salah Baouendi. Springer Proceedings in Mathematics and Statistics 127, 141–159 (2015).

[40] J. P. Demailly. *Recent results on the Kobayashi and Green-Griffiths-Lang conjectures.* arXiv:1801.04765 (2018), 76p.

[41] Y. Deng. *Effectivity in the hyperbolicity-related problems.* arXiv:1606.03831 (2016), 13p.

[42] Y. Deng. *On the Diverio–Trapani conjecture.* arXiv:1703.07560 (2017), 31p. To appear in Ann. Sci. École Norm. Sup.

[43] S. Diverio and E. Rousseau, *Hyperbolicity of projective hypersurfaces*, IMPA Monographs, vol. 5, Springer, [Cham], 2016. [Second edition]. MR3524135

[44] L. Ein, *Subvarieties of generic complete intersections*, Invent. Math. **94** (1988), no. 1, 163–169, DOI 10.1007/BF01394349. MR958594

[45] L. Ein, *Subvarieties of generic complete intersections. II*, Math. Ann. **289** (1991), no. 3, 465–471, DOI 10.1007/BF01446583. MR1096182

[46] D. Eklund, *Curves on Heisenberg invariant quartic surfaces in projective 3-space*, Eur. J. Math. **4** (2018), no. 3, 931–952, DOI 10.1007/s40879-018-0216-2. MR3851124

[47] G. Ferrarese and D. Romagnoli, *Elliptic curves on the general hypersurface of degree* 7 *of* $\mathbf{P}^4$, Nederl. Akad. Wetensch. Indag. Math. **50** (1988), no. 3, 249–252. MR964831

[48] K. Furukawa, *Rational curves on hypersurfaces*, J. Reine Angew. Math. **665** (2012), 157–188, DOI 10.1515/CRELLE.2011.114. MR2908742

[49] K. Furukawa. *Dimension of the space of conics on Fano hypersurfaces.* arXiv:1702.08890 (2017), 14p.

[50] M. Green and P. Griffiths, *Two applications of algebraic geometry to entire holomorphic mappings*, The Chern Symposium 1979 (Proc. Internat. Sympos., Berkeley, Calif., 1979), Springer, New York-Berlin, 1980, pp. 41–74. MR609557

[51] G. M. Hana and T. Johnsen, *Rational curves on a general heptic fourfold*, Bull. Belg. Math. Soc. Simon Stevin **16** (2009), no. 5, Linear systems and subschemes, 861–885. MR2574367

[52] J. Harris, M. Roth, and J. Starr, *Rational curves on hypersurfaces of low degree*, J. Reine Angew. Math. **571** (2004), 73–106, DOI 10.1515/crll.2004.045. MR2070144

[53] J. Harris and J. Starr, *Rational curves on hypersurfaces of low degree. II*, Compos. Math. **141** (2005), no. 1, 35–92, DOI 10.1112/S0010437X04001253. MR2099769

[54] A. Ikeda, *Subvarieties of generic hypersurfaces in a nonsingular projective toric variety*, Math. Z. **263** (2009), no. 4, 923–937, DOI 10.1007/s00209-008-0446-y. MR2551605

[55] T. Johnsen and S. L. Kleiman, *Rational curves of degree at most* 9 *on a general quintic threefold*, Comm. Algebra **24** (1996), no. 8, 2721–2753, DOI 10.1080/00927879608542652. MR1393281

[56] T. Johnsen and S. L. Kleiman, *Toward Clemens' conjecture in degrees between* 10 *and* 24, Serdica Math. J. **23** (1997), no. 2, 131–142. Conference on Geometry and Mathematical Physics (Zlatograd, 1995). MR1627602

[57] S. Katz, *On the finiteness of rational curves on quintic threefolds*, Compositio Math. **60** (1986), no. 2, 151–162. MR868135

[58] A. L. Knutsen, *On isolated smooth curves of low genera in Calabi-Yau complete intersection threefolds*, Trans. Amer. Math. Soc. **364** (2012), no. 10, 5243–5264, DOI 10.1090/S0002-9947-2012-05461-4. MR2931328

[59] A. L. Knutsen, *Smooth, isolated curves in families of Calabi-Yau threefolds in homogeneous spaces*, J. Korean Math. Soc. **50** (2013), no. 5, 1033–1050, DOI 10.4134/JKMS.2013.50.5.1033. MR3104098

[60] S. Kobayashi, *Hyperbolic manifolds and holomorphic mappings: An introduction*, 2nd ed., World Scientific Publishing Co. Pte. Ltd., Hackensack, NJ, 2005. MR2194466

[61] J. Kollár, *Rational curves on algebraic varieties*, Ergebnisse der Mathematik und ihrer Grenzgebiete. 3. Folge. A Series of Modern Surveys in Mathematics [Results in Mathematics and Related Areas. 3rd Series. A Series of Modern Surveys in Mathematics], vol. 32, Springer-Verlag, Berlin, 1996. MR1440180

[62] M. Kontsevich, *Enumeration of rational curves via torus actions*, The moduli space of curves (Texel Island, 1994), Progr. Math., vol. 129, Birkhäuser Boston, Boston, MA, 1995, pp. 335–368, DOI 10.1007/978-1-4612-4264-2_12. MR1363062

[63] S. Lang, *Hyperbolic and Diophantine analysis*, Bull. Amer. Math. Soc. (N.S.) **14** (1986), no. 2, 159–205, DOI 10.1090/S0273-0979-1986-15426-1. MR828820

[64] A. Libgober and J. Teitelbaum, *Lines on Calabi-Yau complete intersections, mirror symmetry, and Picard-Fuchs equations*, Internat. Math. Res. Notices **1** (1993), 29–39, DOI 10.1155/S1073792893000030. MR1201748

[65] Y. Liu, *Hyperbolicity of cyclic covers and complements*, Trans. Amer. Math. Soc. **370** (2018), no. 8, 5341–5357, DOI 10.1090/tran/7097. MR3812107

[66] Y. Liu, *Construction of hyperbolic Horikawa surfaces* (English, with English and French summaries), Ann. Inst. Fourier (Grenoble) **68** (2018), no. 2, 541–561. MR3803110

[67] S. S.-Y. Lu and Y. Miyaoka, *Bounding codimension-one subvarieties and a general inequality between Chern numbers*, Amer. J. Math. **119** (1997), no. 3, 487–502. MR1448213

[68] S. S.-Y. Lu and Y. Miyaoka, *Bounding curves in algebraic surfaces by genus and Chern numbers*, Math. Res. Lett. **2** (1995), no. 6, 663–676, DOI 10.4310/MRL.1995.v2.n6.a1. MR1362961

[69] Y. Miyaoka, *The orbibundle Miyaoka-Yau-Sakai inequality and an effective Bogomolov-McQuillan theorem*, Publ. Res. Inst. Math. Sci. **44** (2008), no. 2, 403–417, DOI 10.2977/prims/1210167331. MR2426352

[70] Y. Miyaoka, *Counting lines and conics on a surface*, Publ. Res. Inst. Math. Sci. **45** (2009), no. 3, 919–923, DOI 10.2977/prims/1249478969. MR2569571

[71] P. G. J. Nijsse, *Clemens' conjecture for octic and nonic curves*, Indag. Math. (N.S.) **6** (1995), no. 2, 213–221, DOI 10.1016/0019-3577(95)91244-P. MR1338327

[72] G. Pacienza, *Rational curves on general projective hypersurfaces*, J. Algebraic Geom. **12** (2003), no. 2, 245–267, DOI 10.1090/S1056-3911-02-00328-4. MR1949643

[73] G. Pacienza, *Subvarieties of general type on a general projective hypersurface*, Trans. Amer. Math. Soc. **356** (2004), no. 7, 2649–2661, DOI 10.1090/S0002-9947-03-03250-1. MR2052191

[74] E. Riedl, *Rational Curves on Hypersurfaces*, ProQuest LLC, Ann Arbor, MI, 2015. Thesis (Ph.D.)–Harvard University. MR3450467

[75] X. Roulleau and E. Rousseau, *On the hyperbolicity of surfaces of general type with small* c_1^2, J. Lond. Math. Soc. (2) **87** (2013), no. 2, 453–477, DOI 10.1112/jlms/jds053. MR3046280

[76] B. Segre, *The maximum number of lines lying on a quartic surface*, Quart. J. Math., Oxford Ser. **14** (1943), 86–96, DOI 10.1093/qmath/os-14.1.86. MR0010431

[77] D. Shin, *Rational curves on general hypersurfaces of degree 7 in* $\mathbb{P}^5$, Osaka J. Math. **44** (2007), no. 1, 1–10. MR2313025

[78] D. Shin, *Conics on a general hypersurface in complex projective spaces*, Bull. Korean Math. Soc. **50** (2013), no. 6, 2071–2077, DOI 10.4134/BKMS.2013.50.6.2071. MR3149585

[79] Y.-T. Siu, *Hyperbolicity of generic high-degree hypersurfaces in complex projective space*, Invent. Math. **202** (2015), no. 3, 1069–1166, DOI 10.1007/s00222-015-0584-x. MR3425387

[80] B. Teissier. *Résolution simultanée : I - Familles de courbes.* Séminaire sur les singularités des surfaces (1976-1977), 1-10.

[81] C. Voisin, *On a conjecture of Clemens on rational curves on hypersurfaces*, J. Differential Geom. **44** (1996), no. 1, 200–213. MR1420353

[82] C. Voisin, *A correction: "On a conjecture of Clemens on rational curves on hypersurfaces" [J. Differential Geom.* **44** *(1996), no. 1, 200–213; MR1420353 (97j:14047)]*, J. Differential Geom. **49** (1998), no. 3, 601–611. MR1669712

[83] C. Voisin, *On some problems of Kobayashi and Lang; algebraic approaches*, Current Developments in Mathematics, 2003, Int. Press, Somerville, MA, 2003, pp. 53–125. MR2132645

[84] B. Wang. *Genus of curves in generic hypersurfaces.* arXiv:1110.0185v3 (2011), 14p.

[85] B. Wang, *First order deformations of pairs and non-existence of rational curves*, Rocky Mountain J. Math. **46** (2016), no. 2, 663–678, DOI 10.1216/RMJ-2016-46-2-663. MR3529086

[86] L.-C. Wang, *A remark on divisors of Calabi-Yau hypersurfaces*, Asian J. Math. **4** (2000), no. 2, 369–372, DOI 10.4310/AJM.2000.v4.n2.a7. MR1797588

[87] L.-C. Wang, *Divisors of generic hypersurfaces of general type*, Taiwanese J. Math. **6** (2002), no. 4, 507–513, DOI 10.11650/twjm/1500407474. MR1937475

[88] G. Xu, *On the complement of a generic curve in the projective plane*, Amer. J. Math. **118** (1996), no. 3, 611–620. MR1393262

[89] G. Xu, *On the intersection of rational curves with cubic plane curves*, J. Algebraic Geom. **7** (1998), no. 1, 1–13. MR1620678

[90] G. Xu, *Subvarieties of general hypersurfaces in projective space*, J. Differential Geom. **39** (1994), no. 1, 139–172. MR1258918

[91] G. Xu, *Divisors on hypersurfaces*, Math. Z. **219** (1995), no. 4, 581–589, DOI 10.1007/BF02572382. MR1343663

[92] G. Xu, *Divisors on generic complete intersections in projective space*, Trans. Amer. Math. Soc. **348** (1996), no. 7, 2725–2736, DOI 10.1090/S0002-9947-96-01613-3. MR1348870

Dipartimento di Matematica, Università degli Studi di Roma "Tor Vergata", Via della Ricerca Scientifica, 00133 Roma, Italy

Email address: `cilibert@mat.uniroma2.it`

Dipartimento di Matematica, Università degli Studi di Roma "Tor Vergata", Via della Ricerca Scientifica, 00133 Roma, Italy

Email address: `flamini@mat.uniroma2.it`

Université Grenoble Alpes, Institut Fourier, CS 40700, 38058 Grenoble cedex 09, France

Email address: `mikhail.zaidenberg@univ-grenoble-alpes.fr`

Contemporary Mathematics
Volume **733**, 2019
https://doi.org/10.1090/conm/733/14738

Topological billiards, conservation laws and classification of trajectories

A. T. Fomenko and V. V. Vedyushkina

The authors dedicate this paper to the birthday of the prominent and remarkable mathematician and teacher Professor Selim Krein.

ABSTRACT. We survey the recent results on classification of topological billiards. In particular topological billiards model the bifurcations of Liouville tori of integrable Hamiltonian system. The paper contains the new results about topology of integrable billiards in the Minkowski space and the billiards with nonconvex angles.

1. Introduction

Integrable Hamiltonian systems occur in many problems of geometry, mechanics, and physics and have been of growing interest within the past years. At the end of the past century the theory of invariants based on the topological approach was suggested by A. T. Fomenko, H. Zieschang, A. V. Bolsinov, and others for the study of integrable Hamiltonian systems with two degrees of freedom. This theory allows to investigate various qualitative properties of such systems and to conclude whether two systems are Liouville equivalent [**6**, **8**, **9**]. For this type of equivalence, a non-degenerate integrable system restricted to a 3-dimensional isoenergy surface is assigned Fomenko-Zieschang invariant (molecule) which is a graph with some numerical marks. The main result of the Fomenko's theory can be formulated as follows: two integrable Hamiltonian systems considered on non-degenerate isoenergy 3-surfaces are equivalent in Liouville sense if and only if their corresponding invariants are the same. Possibility of establishing the fact of non-equivalence without using the theory of invariants seems doubtful.

In the present paper we will discuss the classification results, obtained using of Fomenko's theory for the integrable topological billiards.

2. Outline of the general theory

In this section we recall some basic definitions and main results of the theory of invariants for integrable Hamiltonian systems. For details we address the readers to the papers [**2**, **5**, **6**, **8**, **9**] and the book [**1**].

2010 *Mathematics Subject Classification.* Primary 37J15, 37J35.

The authors were supported by the program "Leading Scientific Schools" (grant no. NSh-6399.2018.1, Agreement No. 075-02-2018-867). They also thank the reviewer for useful remarks.

DEFINITION 1. A *Hamiltonian system* with n degrees of freedom is a triple (M^{2n}, ω, H) where M^{2n} is a $2n$-dimensional smooth manifold endowed with a symplectic form ω and H is a smooth function on M^{2n} called the *Hamiltonian function*. A *Hamiltonian vector field* is defined as $v := \operatorname{sgrad} H := \omega^{-1} dH$. We shall also denote a Hamiltonian system by (M^{2n}, v).

Let us recall the following classical result.

THEOREM 1 (Liouville). *Suppose that the smooth functions $f_1, \dots, f_n$ are the first integrals of the Hamiltonian system (M^{2n}, ω, H) satisfying the following conditions:*

1) *$f_1, \dots, f_n$ are functionally independent, i.e. their gradients are linearly independent almost everywhere on M;*
2) *$\{f_i, f_j\} = 0$ for $i, j = 1, \dots, n$, i.e. the integrals commute with respect to the Poisson bracket determined by the symplectic structure;*
3) *the vector fields* $\operatorname{sgrad} f_1, \dots, \operatorname{sgrad} f_n$ *are complete, i.e. the natural parameter on their integral trajectories is defined on the whole real axis.*

Let also $T_\xi = \{x \in M \mid f_i(x) = \xi_i,\ i = 1, \dots, n\}$ be a regular joint level set for the functions $f_1, \dots, f_n$ (regularity means that the differentials $df_1, \dots, df_n$ are linearly independent for all $x \in T_\xi$).

Then,

a) *T_ξ is a smooth Lagrangian submanifold which in case it is connected and compact is diffeomorphic to the n-dimensional torus T^n (the Liouville torus);*
b) *the Liouville foliation is trivial in some neighborhood U of the Liouville torus, that is, a neighborhood U of the torus T_ξ is the direct product of the torus T^n and the disc D^n;*
c) *in the neighbourhood U there exists a coordinate system (the action-angle variables) $s_1, \dots, s_n, \varphi_1, \dots, \varphi_n$ where $s_1, \dots, s_n$ are coordinates on the disk D^n depending only on the integrals $f_1, \dots, f_n$ and $\varphi_1, \dots, \varphi_n$ are standard angle coordinates on the torus, such that*
   (i) $\omega = \sum_{i=1}^{n} d\varphi_i \wedge ds_i$,
   (ii) *the Hamiltonian vector field takes the form $\dot{s}_i = 0$, $\dot{\varphi}_i = q_i(s_1, \dots, s_n)$, $i = 1, \dots, n$, and therefore determines a rectilinear winding (rational or irrational) on each of the Liouville tori.*

DEFINITION 2. A Hamiltonian system (M^{2n}, ω, H) is called *Liouville integrable* if there exists a set of smooth functions $f_1, \dots, f_n$ satisfying the conditions 1–3 of Theorem 1.

DEFINITION 3. The decomposition of the manifold M^{2n} into connected components of joint level surfaces of the integrals $f_1, \dots, f_n$ is called the *Liouville foliation* corresponding to the Liouville integrable Hamiltonian system (M^{2n}, ω, H).

The Liouville foliation consists of regular leaves (they are Liouville tori in the compact case), which fill M almost the whole and singular ones filling a set of zero measure on M.

We introduce now an equivalence relation between integrable Hamiltonian systems. Let $X_1 = (M_1^{2n}, \omega_1, H_1)$ and $X_2 = (M_2^{2n}, \omega_2, H_2)$ be two such systems.

DEFINITION 4. The systems X_1 and X_2 are called *Liouville equivalent* if there exists a diffeomorphism $\tau\colon M_1 \to M_2$ that transforms the Liouville foliation of the first system to that of the second one and satisfies two orientation conditions:

a: it preserves the orientation on the M_1 and M_2,

b: it also preserves the orientation on the critical trajectories given by the Hamiltonian flows and homeomorphic to the circles.

DEFINITION 5. The systems X_1 and X_2 are called *coarsely Liouville equivalent* if there exists a homeomorphism between the bases of the corresponding Liouville foliations that can locally (i.e. in the neighbourhood of each point of the base) be lifted up to a fiber homeomorphism of the Liouville foliations.

REMARK 1. In these definitions we consider integrable systems restricted to their invariant isoenergy surfaces $\{H = h = const\}$.

Suppose we are given two integrable Hamiltonian systems (M_1^4, v_1) and (M_2^4, v_2) with two degrees of freedom. Can we check whether they are equivalent? The Fomenko's theory of invariants gives the answer to this question in some cases. Namely, for systems restricted to their regular isoenergy surfaces there exist invariants which enable us to classify such systems up: coarse Liouville and (fine) Liouville equivalence.

Let $X = (Q, v = \operatorname{sgrad} H)$ be an integrable Hamiltonian system with 2 degrees of freedom restricted to the compact and regular 3-dimensional surface of constant energy $Q = \{H = h = const\}$ (regularity means that $dH|_Q \neq 0$). Suppose that an additional integral f of X is a Bott function on Q [**1**, Section 1.8] and that X is non-resonant on Q (i.e. the winding determined by X on the Liouville tori is irrational for almost all tori). We shall start with constructing the invariant for the coarse Liouville equivalence. First, we consider a graph that is the base of the Liouville foliation of Q. The points of this graph correspond to the connected components of the level surfaces of the function f on Q, i.e. to the leaves of the Liouville foliation. The graph obtains is sometimes called the Reeb graph of the function f on Q. The edges of this graph represent regular one-parameter families of Liouville tori. The vertices correspond to the singular leaves.

Now we endow each vertex with an invariant classifying the Liouville foliation in the neighbourhood of the corresponding singular leaf. Such invariant is a *3-atom* defined as a small neighbourhood of a singular leaf of the Liouville foliation considered up to the fiber diffeomorphism. Thus, a 3-atom is a compact 3-dimensional surface with boundary foliated into Liouville tori and one singular leaf. Its boundary consists of several (possibly one) Liouville tori. Each boundary torus corresponds to an edge of the graph adjacent to the given vertex.

There exists a one-to-one correspondence between 3-atoms and 2-atoms. By definition, a *2-atom* is a small invariant neighbourhood of a critical level of a Morse function on a 2-dimensional compact manifold considered up to a fiber diffeomorphism. A 2-atom P^2 is a compact 2-dimensional surface with boundary foliated by circles and one singular leaf K. Its boundary consists of one or several circles.

The connection between 3-atoms $U(L)$ (where L is the singular leaf) and 2-atoms is that each 3-atom is a Seifert fibration over a certain 2-atom, as described below.

Consider a 3-atom $U(L)$ with the structure of a Seifert fibration on it. Let $\pi : U(L) \to P^2$ denote its projection onto a two-dimensional base P^2 with the embedded graph $K = \pi(L)$. The vertices of the graph K are called the vertices of two-atom (P^2, K). Let us mark those points on the base P^2, into which the singular fibers of the Seifert fibration (i.e., the fibers of type (2,1)) are projected. We call this points a star-vertices of the corresponding two-atom. Recall that the base P^2 has a canonical orientation.

THEOREM 2. *(Fomenko,* [**1**, Th. 3.4]*)*

a): *Under the projection $\pi : U(L) \to P^2$, the 3-atom $U(L)$ turns into the 2-atom (P^2, K), and moreover, the singular fibers of the Seifert fibration on the 3-atom are in one-to-one correspondence with the star-vertices of the 2-atom.*

b): *This correspondence between 3-atoms and 2-atoms is a bijection.*

So we may assume that the vertices of the graph under consideration are 2-atoms, and there is a one-to-one correspondence between the boundary circles of each 2-atom and the edges of the graph adjacent to the given vertex.

In Figure 1 we show examples of 2-atoms. The corresponding 3-atoms can be obtained from them by multiplying by a circle.

FIGURE 1. Examples of 2-atoms

DEFINITION 6. The constructed graph W is called the *molecule* (or the *Fomenko invariant*) of the integrable system on the given isoenergy surface Q.

THEOREM 3 (Fomenko's Th. 3.5 in [**1**]). *Two integrable systems (Q^3_1, v_1) and (Q^3_2, v_2) are coarsely Liouville equivalent if and only if their molecules W_1 and W_2 are the same. This means that the corresponding graphs are homeomorphic, and this homeomorphism preserves the atoms at the vertices of the graphs.*

To construct the invariant for the (fine) Liouville equivalence we need to provide the molecule W of the integrable system (Q^3, v) with some additional information. Molecule W tells us which atoms should be glued together to obtain the Liouville foliation on Q^3. On each edge of W we need to glue together the adjacent atoms along their boundary tori corresponding to the given edge. Thus, we need to fix somehow the gluing diffeomorphism of two Liouville tori which are contained in the boundary of adjacent 3-atoms. For each 3-atom there exists a class of admissible coordinate systems on its boundary tori (see [**1**, Section 4.1]). Thus we consider on every edge of graph a natural integer-valued gluing matrix determined by the corresponding the isomorphism of the fundamental groups of the boundary tori. This isomorphism determines the diffeomorphism of the gluing tori up to isotopy. Of course, this matrix depends on the choice of coordinate systems on the tori. These coordinate systems cannot be chosen arbitrarily: see for details [**1**, Section 4.1].

Now we endow the molecule W with numerical marks r_i, ε_i, and n_k being functions of the gluing matrices. The marks $r_i \in \mathbb{R}/\mathbb{Z} \cup \{\infty\}$ and $\varepsilon_i = \pm 1$ are assigned to each edge of the molecule and the marks $n_k \in \mathbb{Z}$ are assigned to groups of atoms called families. If we cut the molecule along all the edges with finite r-mark, it will split into several connected pieces. The pieces that do not contain atoms of type A (the atoms corresponding the minimum or the maximum bifurcations of the level set of the additional integral) are said to be *families*. For explicit definitions of the marks r_i, ε_i, n_k see [**1**, Section 4.3].

DEFINITION 7. The molecule W^* endowed with the marks r_i, ε_i, n_k is called the *marked molecule* (or *Fomenko-Zieschang invariant*) of the integrable system on the isoenergy surface Q.

THEOREM 4 (Fomenko-Zieschang Th. 1.6 in [**9**]). *Two integrable systems (Q_1^3, v_1) and (Q_2^3, v_2) are Liouville equivalent if and only if their marked molecules $W_1^* = (W_1, r_i, \varepsilon_i, n_k)$ and $W_2^* = (W_2, r_i, \varepsilon_i, n_k)$ coincide.*

3. Topological integrable billiards

The integrability of a billiard in a domain bounded by an ellipse was noted by D.Birkhoff [**15**]. The integrability of the billiard is preserved if we consider the flat domain bounded by arcs of confocal ellipses and hyperbolas whose boundary does not contain angles equal to $\frac{3\pi}{2}$. In this case, all the angles of the boundary are equal to $\frac{\pi}{2}$, since confocal quadrics always intersect at a right angle. In the book by V.V.Kozlov, D.V.Treschev [**16**] it is noted, that these dynamical systems are completely Liouville integrable. For the flat billiard in an ellipse, there are coordinates such that the motion is represented as a periodic motion along tori. Up to Liouville equivalence, such systems have been studied in details in [**17, 18**] by V.Dragovic, M.Radnovic, and in [**24, 25**] by V.V.Fokicheva. In an interesting paper [**19**] V.Dragovic and M.Radnovic studied the Liouville foliation for the flat billiard in an ellipse, as well as geodesic flows on the ellipsoid in the Minkowski space, giving an answer in terms of the Fomenko-Zieschang invariants.

V.V. Fokicheva classified all locally flat billiards, bounded by the arcs of confocal ellipses and hyperbolas (not necessarily isometrically embedded in the plane), and also the billiards that are not necessarily flat, and are obtained by gluing together several elementary domains along convex boundary segments. Further, V.V. Fokicheva investigated the topology of Liouville foliations on isoenergy surfaces of such billiards by calculating the Fomenko-Zieschang invariants of these systems. Note that the billiard in the domain bounded by the arcs of confocal parabolas is also integrable. Confocal parabolas can be considered as a family of confocal ellipses and hyperbolas, where one focus is at infinity. The topology of the Liouville foliation of a billiard in domains bounded by the arcs of confocal parabolas, was investigated in [**26**] by V.V. Fokicheva.

Let us remind that two smooth integrable systems are called Liouville equivalent if and only if there exists a diffeomorphism sending the Liouville foliation of the first system to the Liouville foliation of the second system. In the "majority" of nondegenerate classical integrability cases the smooth Liouville tori with irrational windings form the everywhere dense subset in the ambient manifold. Such Liouville tori are the closures of the non-resonant trajectories of the system. In this case Liouville equivalence of the systems means that they have "identical" closures of solutions (i.e. integral trajectories) on the three-dimensional levels of constant energy.

In the case of billiards, the Liouville tori, the Liouville foliation, and the integral trajectories of the system are almost all piecewise smooth. Nevertheless, the Liouville foliation and the Liouville equivalence are well defined. The topological type of the Liouville foliation is completely determined by the Fomenko-Zieschang invariant, which is the graph with numeric marks, as described above. Analyzing a large number of calculated marked molecules for various billiards and the other integrable systems with two degrees of freedom, A.T. Fomenko formulated the following conjecture. Namely, many complicated integrable systems (for example, in the dynamics of a rigid body) can be "modeled" by much more visual topological billiards. In particular, this makes it possible to present and effectively classify the stable and unstable periodic solutions (trajectories) of integrable systems, in particular, in physics and mechanics. This conjecture was confirmed in the paper of V.V. Fokicheva and A.T. Fomenko. Namely, for many integrable cases of the rigid body dynamics the calculation of the Fomenko-Zieschang invariants made it possible to detect the Liouville equivalence of these systems to some topological billiards by comparing marked molecules (see [**28**]). Thus, roughly speaking, locally flat integrable billiards "visually model" many fairly complicated integrable cases in the dynamics of the rigid body.

In the paper by A.T.Fomenko and V.V.Vedyushkina [**30**] the investigation of integrable billiards was continued. New equivalences of billiards and systems of the rigid body dynamics were found. New classes of integrable billiards, bounded by the arcs of confocal quadrics, were constructed. Namely, noncompact billiards and topological billiards obtained by gluing together flat billiards along nonconvex arcs of the boundary (nonconvex topological billiards) are constructed. The noncompact billiards that are bounded by the arcs of confocal quadrics are completely classified and their topology is studied based upon Fomenko invariants describing the bifurcations of singular fibers of an additional integral.

The topology of the isoenergy surfaces for some nonconvex topological billiards was also studied. It turned out that they have exotic Liouville foliations. More precisely, at some special levels of additional integral the trajectories of the billiard do not admit continuous continuation. It turned out that such billiards are Liouville equivalent to the billiard in the ellipse in the Minkowski space, and also to an geodesic flows on the ellipsoid in the Minkowski space.

4. Fomenko's conjecture. Modeling of any nondegenerate integrable system of general form with two degrees of freedom by an integrable topological billiard

In this section, we discuss further Fomenko's conjecture and present some positive result in this direction. Namely, we simulate the arbitrary atoms (i.e., bifurcations of Liouville tori) and, more generally, we simulate the arbitrary Liouville foliation (of integrable systems with two degrees of freedom) by means of special integrable billiards. We call this billiards "book's billiards" (see below).

As it was noted above, the Fomenko-Zieschang invariants allows us to study and classify a large number of specific integrable systems. As a result, a lot of experimental material has been accumulated. Let us note the following important fact. For nontrivial gluing of locally flat billiard domains along the arcs of confocal quadrics, the additional (second) integral is preserved. In other words, the constructed topological billiard remains integrable. In this case, the billiard is represented as a two-dimensional cell complex. By gluing new cells (i.e. locally flat billiards) to the boundary, we complicate the topology of the cell complex. V.V.Vedyushkina constructed the class of "billiard's books" obtained by gluing together along the same arc of the quadric several elementary confocal billiards-cells. Roughly speaking, we get a "book", where several sheets are glued to the "spine". Different "books" can also be glued to each other along the different arcs. More complicated "books" can be obtained with many "spines" and many sheets. This operation extends essentially the class of topological integrable billiards (cell complexes). The natural problem arises: to describe (classify) all general integrable billiards (also with potentials). A.T.Fomenko noted that this construction has analogs in the knots theory.

A.T. Fomenko has also formulated the following five-part conjecture. Let us consider non-degenerate (Bott) integrable systems with two degrees of freedom, the isoenergy three-dimensional surfaces and the corresponding Liouville foliations.

- Conjecture A (Atoms). All bifurcation of the two-dimensional Liouville tori in such integrable dynamical systems can be modeled by integrable billiard's books. In other words, any non-degenerate 2-atom (and corresponding 3-atom) can be modeled by a suitable integrable billiard's book.
- Conjecture B (Coarse molecules). Any "coarse molecule" (Fomenko invariant) can be modeled by a suitable integrable billiard's book.
- Conjecture C (Marked molecules). Any "marked molecule" (Fomenko-Zieschang invariant) can be modeled by general integrable billiards. In other words, all Liouville foliations of non-degenerate (that is, Bott) integrable systems on isoenergy 3-surfaces are modeled by general integrable billiards (i.e., the corresponding foliations are Liouville equivalent).
- Conjecture D. Any three-dimensional closed isoenergy surface of arbitrary integrable nondegenerate system with two degrees of freedom can be realized by a general integrable billiard (maybe with a potential). This conjecture can be considered as a special case of conjecture C. If C is true, then conjecture D obviously follows.
- Conjecture E. According to the Conjecture C there exists the large class of Liouville foliations (i.e. the marked molecules), which are realized by integrable topological billiards. For this class the Fomenko-Zieschang invariant is "equivalent" to the corresponding billiard itself. More exactly, let us consider the integrable billiard as 2-dimensional cell-complex, where the boundary of 2-dimensional cells are formed by edges of the elementary billiards and by the segments of focal lines. Then there is one-to-one correspondence between the marked molecules (mentioned above and up to fiber equivalence of Liouville foliations) and integrable billiards (up to cell homeomorphism of cell complexes).

REMARK 2. Here is the comment to the conjecture D. Recall that this class of 3-manifolds coincides, according to the theorem by A.V. Brailov and A.T. Fomenko, with the class of graph-manifolds (Waldhausen manifolds). In addition, this remarkable class has four another unexpected representations, " faces ", see Section 6.8 [**11**].

Any answer to the Fomenko's conjecture would be of interest. For example, if it turns out that not all "marked molecules" (that is, Fomenko-Zieschang invariants) are realized by billiards, then it would be very useful to describe those molecules that are actually being realized. In this case, some topological obstructions could be discovered and that would distinguish realizable and non-realizable Liouville foliations (integrable systems). That would make clear - which non-degenerate integrable systems with two degrees of freedom are Liouville equivalent to billiards, and which are not, and why.

5. The part A of Fomenko's conjecture is correct. Kharcheva-Vedyushkina's theorem

DEFINITION 8. The family of the confocal quadrics on the plane $\mathbb{R}^2$ with Euclidean coordinates (x, y) is the set of curves described by the equation

$$\frac{x^2}{a-\lambda} + \frac{y^2}{b-\lambda} = 1. \tag{5.1}$$

Here $a > 0$ and $b > 0$ are called the parameters of this family[1]. Let us call the value $\lambda \in [0, max(a, b)]$ the parameter of the curve or the quadric's parameter.

REMARK 3. we say that the case $\lambda = a$ corresponds to the line $x = 0$, and $\lambda = b$ corresponds to the focal line $y = 0$. This curve from the family of confocal quadrics is an ellipse for $\lambda \in [0, b)$, a hyperbola for $\lambda \in (b, a)$, and a straight line for $\lambda = a$ and $\lambda = b$. The ellipses and hyperbolas from this family have the same foci and intersect at right angles.

DEFINITION 9. An elementary domain is a compact, connected subset of the plane whose boundary consists of arcs of confocal quadrics of the family (5.1) and does not contain internal angles greater than π.

It turns out that the billiards in elementary domains are integrable (see the book by V.V. Kozlov and D.V. Treschev [**16**, Ch. 4]). The straight lines containing the segments of the polygonal billiard trajectory are tangent to a certain quadric (ellipse or hyperbola). This quadric belongs to the same class of confocal quadrics as the quadrics whose arcs form the boundary of the domain (billiard). The quadric, which is tangent to the trajectory, is called the caustic. Thus, along the trajectory the caustic parameter is conserved. In what follows, we denote this parameter by Λ and consider it as an additional integral. The first integral of the system is the scalar square of the velocity vector. The level of the integral $\Lambda = b$ corresponds to the trajectories passing through the foci of the ellipse (or whose extensions pass through the foci if the billiard does not contain foci).

In the future we will use the word "billiard" both in relation to the domain and to the dynamic billiard system in it.

[1] Without loss of generality, we can always assume that $b < a$.

In [**27**] V.V. Fokicheva introduced an operation of isometric gluing of two elementary billiards along a convex arc of the boundary lying on the same quadric. The trajectory, passing along the first billiard, hits the common boundary arc, then reflects and continues the motion along the second billiard. It is also possible to determine the motion of the material point in a billiard glued from a pair of domains along two convex boundaries with a common point. Locally, such gluing is called gluing with a conical point. The motion is determined in the following way: when the material particle hits a conical point, it continues the motion along the same sheet as before the impact passing back along the same segment of trajectory, which fell into this corner. On Figure 2 below these trajectories are shown.

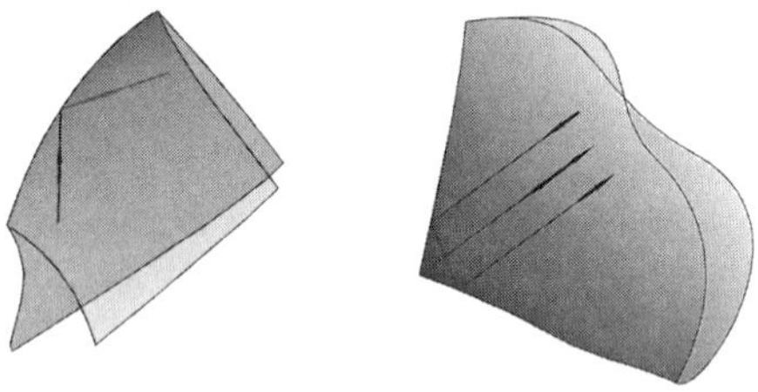

FIGURE 2. Trajectories of topological billiards.

DEFINITION 10. Let us fix an elementary billiard Ω, bounded by arcs of confocal quadrics, and a number $n \in \mathbb{N}$. Let us assign to each arc of the boundary Ω a permutation of order n satisfying the following conditions:

(1) if two arcs from the boundary have a common point, that is, they are adjacent, then the corresponding permutations σ_1 and σ_2 commute: $\sigma_1 \circ \sigma_2 = \sigma_2 \circ \sigma_1$;
(2) permutations assigned to the nonconvex arcs of the boundary, are always identical.

The pair $\eta = (\Omega, n)$ together with the described permutations is called *book's gluing*.

DEFINITION 11. Let us fix the book's gluing $\eta = (\Omega, n)$. Let us take a disjoint union of n elementary billiards (*sheets*) Ω. We mark them by integer numbers in arbitrary way and fix the numbering. Now we have the n elementary billiards Ω_i. Decompose the permutation on the arcs of the boundary into a product of independent cycles. Let us factorize the disjoint union $\bigsqcup_{i=1}^{n} \Omega_i$ by the following equivalence relation depending on the book's gluing: the edges of the boundaries $\Omega_i(\eta)$ and $\Omega_j(\eta)$ will be considered as equivalent if they correspond to these tho conditions:

a: they corresponds to the same arc of the elementary billiard domain,
b: the permutation assigned to this arc the $i-$th and $j-$th elements are in the same cycle.

The resulting cell complex $\Delta B(\eta) := (\bigsqcup_{i=1}^{n(\eta)} \Omega_i(\eta))/\sim$ will be called the *billiard's book* (or briefly *book*) corresponding to the book's gluing η. We require that the billiard's book should be connected.

REMARK 4. The construction of the book can be extended naturally to gluing arbitrary locally flat integrable topological billiards.

DEFINITION 12. We define the *billiard motion of the material particle on the book* $\Delta B(\eta)$, corresponding to the book's gluing η, as a motion without friction of a material point such that:

(1) inside all sheets, forming the billiard's book, the material point moves along a straight line without friction;
(2) at the boundary arc of an elementary billiard the material point reflects elastically, and continues its motion along another sheet determined by the permutation assigned to this arc;
(3) at a corner point, which is adjacent to two arcs with assigned permutations σ_1 and σ_2, material point goes from the sheet i to the sheet $(\sigma_1 \circ \sigma_2)(i)$ and passing back along the same segment of trajectory, which fell into this corner.

REMARK 5. Generally speaking, the glued billiard sheets do not have to be identical, the only necessary condition for gluing is the existence of a common arc of the boundary. But in the future we will use only billiard's book, glued from identical billiards of a special kind, namely bounded by an arc of an ellipse, two arcs of hyperbolas (convex and non-convex), and a focal straight line (see Figure 3). In [**27**] this billiard is denoted by A'_0. In further consideration we will use this specific billiard as the sheets Ω of the constructing billiard's book.

It turns out that the billiard's book is always integrable. This follows from the integrability of the elementary billiards. Since the parameter of the caustic is preserved along the trajectory of elementary billiards, this parameter will also be preserved after isometric gluing of elementary billiards with each other. Let us denote by M^4 the phase space of the billiard's book. Let us fix the length of the velocity vector (i.e. the energy of the system) and consider the corresponding isoenergy manifold Q. It can be shown that Q^3 is stratified into the tori and singular fibers. The singular value of the integral $\Lambda = b$ (see beginning of Section 5) as already mentioned above, corresponds to the trajectories lying on the straight lines passing through the foci of the family of the boundary.

THEOREM 5 (I.Kharcheva, V.Vedyushkina). *The conjecture of the Fomenko A holds true. Namely, for any 3-atom (with or without stars), we can algorithmically construct the billiard's book, glued from elementary billiards of type A'_0, such that the Liouville foliation on of the neighborhood of the singular value b for the integral Λ in the isoenergy surface Q^3 of this billiard is fiberwise homeomorphic to a given 3-atom.*

REMARK 6. Atoms without stars can be obtained by gluing along elliptical and horizontal segments. Billiard's book for atoms with stars are obtained with the help of conical points. The number of conical points is equal to the number of stars in the corresponding atom.

REMARK 7. Same 3-atom can be modeled (simulated) by several topologically non-equivalent billiards. An interesting problem is to estimate of the number of billiards modeling a given atom.

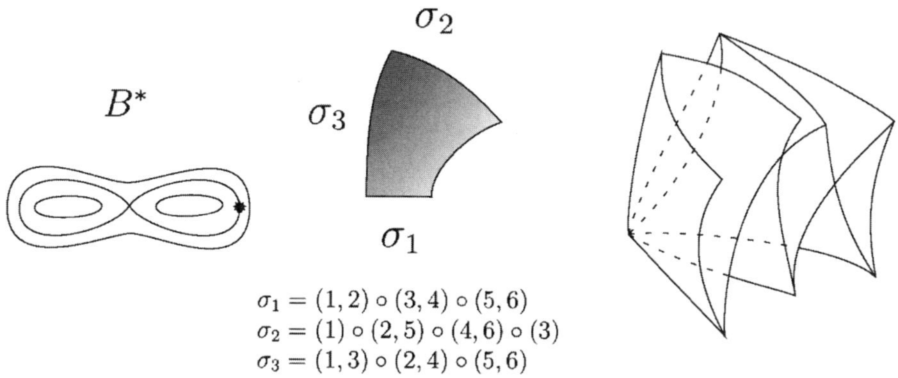

FIGURE 3. Modeling of the atom B^* (left) by a billiard's book. In the middle billiard-sheet of a book (elementary billiard A'_0) is shown with assigned permutations at the boundary. On the right is the resulting billiard's book (cell complex). The foliation of the preimage of a neighborhood of a singular value b for the integral Λ in the isoenergy surface Q^3 of this billiard is fiberwise homeomorphic to the atom B^*.

6. The topology of the Liouville foliation of the billiards with nonconvex angles.

Note that topological billiard in a domain with a piecewise smooth boundary and continuous flow of integral trajectories can be correctly defined in case when the internal angles of the boundary are equal to $\frac{\pi}{n}$, in particular $\frac{\pi}{2}$. When the material point hits corner $\frac{\pi}{2}$, then it continues its motion with the reverse velocity speed. This rule has a natural interpretation. This motion can be interpreted as the reflection from two boundaries forming the angle (see Figure 4). In the case when the angle is equal to $\frac{3\pi}{2}$, the trajectory can not be correctly determined by continuity (see Figure 4). The integral trajectory of the flow going into a singular point can not be extend for all values of time. In other words, the corresponding flow is incomplete in the sense of the theory of differential equations.

Let us recall that confocal quadrics intersects each other with the right angles.

DEFINITION 13. We call the internal boundary angle of a billiard nonconvex if it is equal to $\frac{3\pi}{2}$.

DEFINITION 14. Let us call the complexity of a billiard the number of nonconvex angles on its boundary.

Let us consider billiard bounded by the arcs of confocal quadrics and containing nonconvex angles. Note that such billiards are integrable (the second integral is simply the confocal quadric parameter).

DEFINITION 15. An elementary billiard with nonconvex angles is a compact, simply-connected subset of the plane whose boundary consists of arcs of confocal quadrics of the family (5.1) and contains both angles equal to $\frac{\pi}{2}$ and nonconvex angles equal to $\frac{3\pi}{2}$.

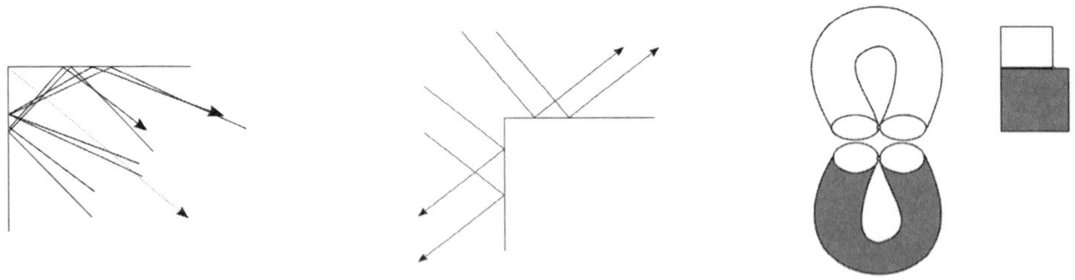

FIGURE 4. The internal angle $\frac{\pi}{2}$ (left). The trajectory that hits the right angle can be defined by continuity. The internal angle $\frac{3\pi}{2}$ (center). The trajectory that hits the nonconvex angle can not be evidently defined by continuity. Nonsingular leaf of the integral Λ for the rectangular billiard of the complexity 1 (right).

The next result describes the topology of regular fibers for integral Λ. This description was obtained in the papers by V.Dragovic and M.Radnovic [**20**–**23**] and then independently by V.Moskvin [**31**]. The analysis by V.Dragovic and M.Radnovic is based on the description of the dynamic of the integral trajectories on the regular fibers. The analysis by V.Moskvin describes the bifurcations of the level surfaces of the second integral Λ.

THEOREM 6. *For all nonsingular values of the integral, the connected component of the level surface of the integral Λ in the isoenergy surface Q^3 for the elementary billiard with nonconvex angles Ω is homeomorphic to the disjoint union of the two-dimensional spheres with $k+1$ handles and with k punctured points. Here k is a complexity (i.e. the number of boundary angles equal to $\frac{3\pi}{2}$) of the plane domain which is the projection of this connected component to the billiard Ω.*

In contrast to the classical case of complete flows, the regular leaves of the Liouville foliation are the spheres with handles and punctures, rather than Liouville tori. Nevertheless, here we can correctly define a coarse molecule – an analog of the Fomenko's invariant for complete flows. As before, the edges of this graph represent one-parametric families of regular fibers. Let us assign some additional information to the edges of the graph by putting on the edge a circle with indication of the genus of the surface of a one-parametric family, if the genus is not equal to one. That is, if the family of regular fibers is a family of tori, we will use the previous notation. The vertices of the molecule correspond to the bifurcations of these fibers at the critical levels of the integral. Thus, the molecules enable us describe the foliation of the isoenergy manifold Q^3 into the union of the levels of the additional integral. An analogue of numerical marks for such incomplete billiard systems will be introduced in future publications. The important step in the description of the global topology of Liouville foliation on the 3-dimensional isoenergy surfaces Q^3 is the result by V.Moskvin [**31**], which describes the bifurcations of regular fibers at the critical levels of the integral Λ. Below we formulate a special case of this general result.

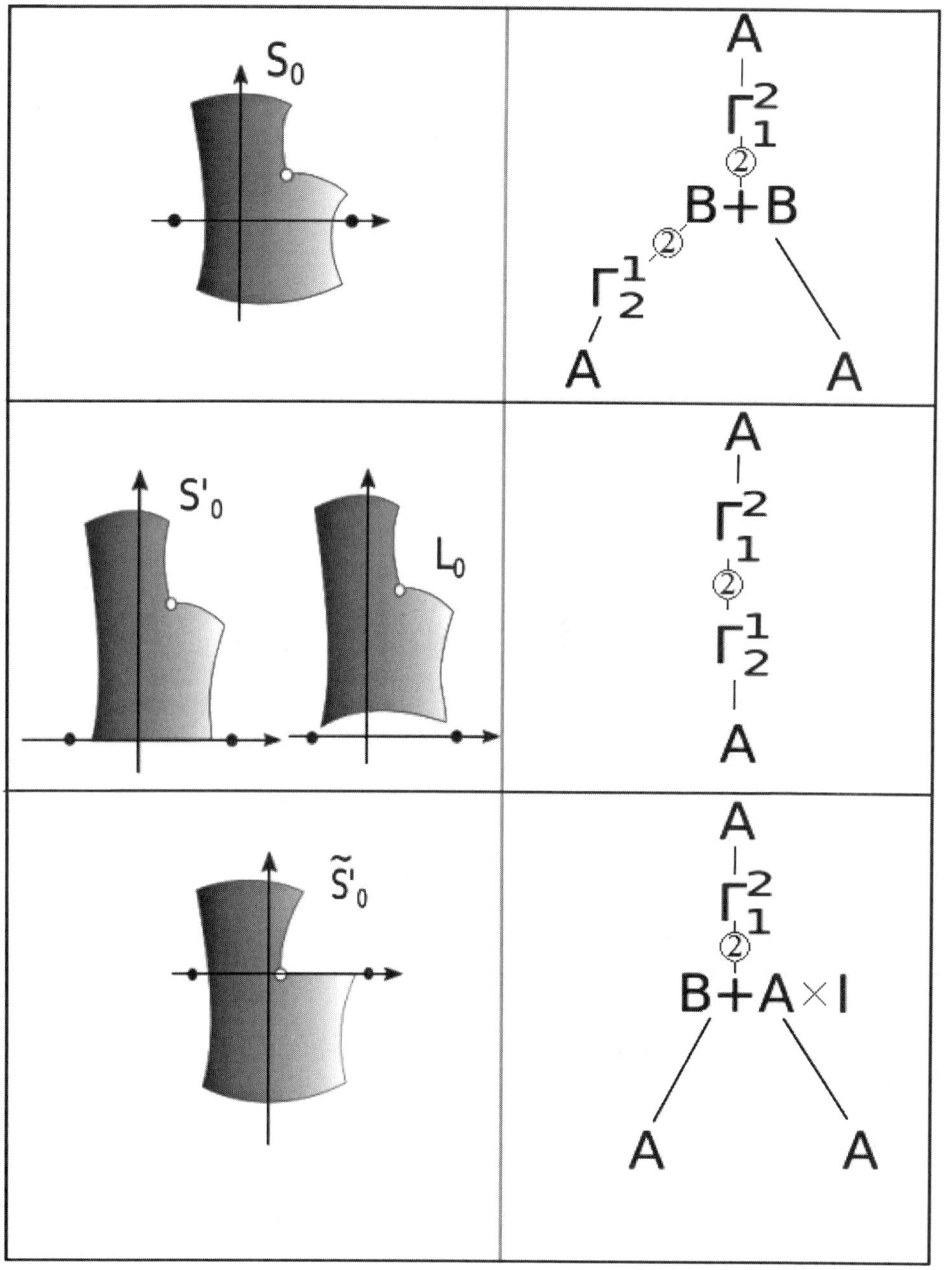

FIGURE 5. Description of the topology of the isoenergy surface Q^3 in terms of rough molecules.

THEOREM 7. [**31**] *Let S be the elementary billiard with nonconvex angles represented in the left column of Figure 5. Then the coarse molecules for this billiard have the form represented in the right column of Figure 5. The edges with the marks in the circles correspond to the two-dimensional spheres with two handles*

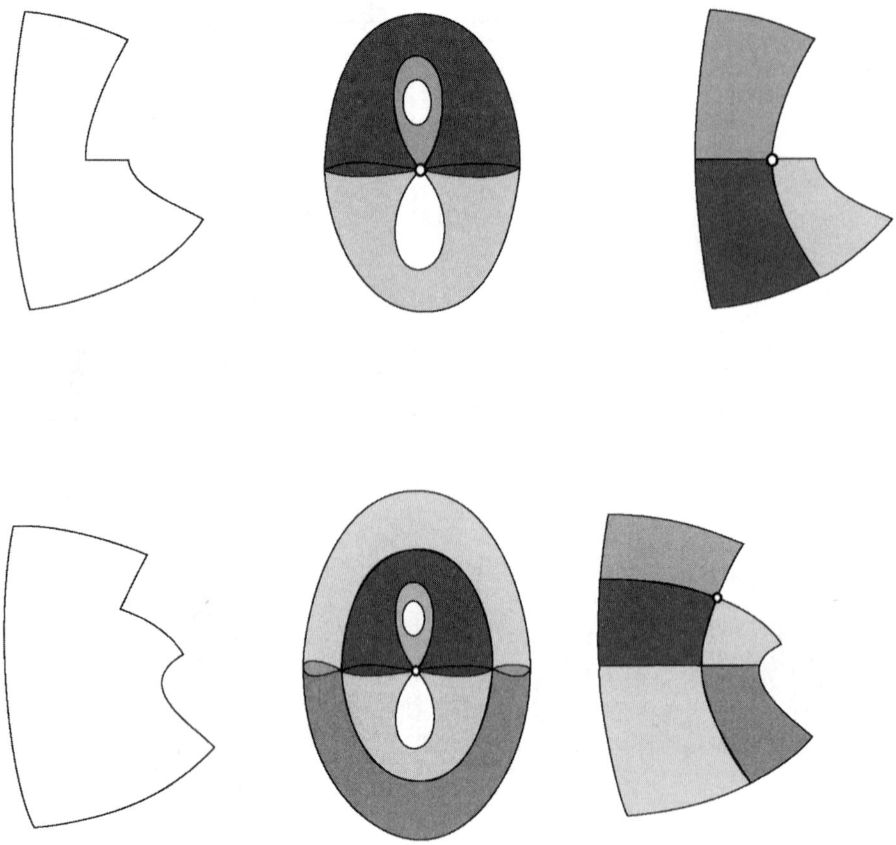

FIGURE 6. The saddle atom $B + A \times I$ for the billiard S_0' (above) and the saddle atom $B + B$ for the billiard S_0 (below), where the corresponding color shows the image of the critical fiber under the natural projection.

without a point. The atoms $B+B$ and $B+A\times I$ are shown in Figure 6. Complexes Γ_i^j, where $i \neq j$ and $i, j = \{1, 2\}$ are homeomorphic to a disjoint union of tori. To one of this tori the circle is glued at a point, whose projection onto the billiard domain is the vertex of a nonconvex angle.

7. Integrable billiards in the Minkowski space

In this section we develop the results of the interesting paper by V.Dragovic and M.Radnovic [**19**].

Let us suppose that the plane $\mathbb{R}^2$ is endowed with Minkowski metric with a scalar product $\langle x, y\rangle = x_1y_1 - x_2y_2$.

Since the scalar product can be negative, all vectors are divided into 3 non-intersecting subsets. The vector v is called

- *space-like*, if $\langle v, v\rangle > 0$;
- *time-like*, if $\langle v, v\rangle < 0$;
- *light-like*, if $\langle v, v\rangle = 0$.

The family of confocal quadrics on the Minkowski plane is given by

$$\frac{x^2}{a-\lambda}+\frac{y^2}{b+\lambda}=1.\lambda \leq a. \tag{7.1}$$

Here $a > b > 0$ are the real numbers. Among this family of quadrics we

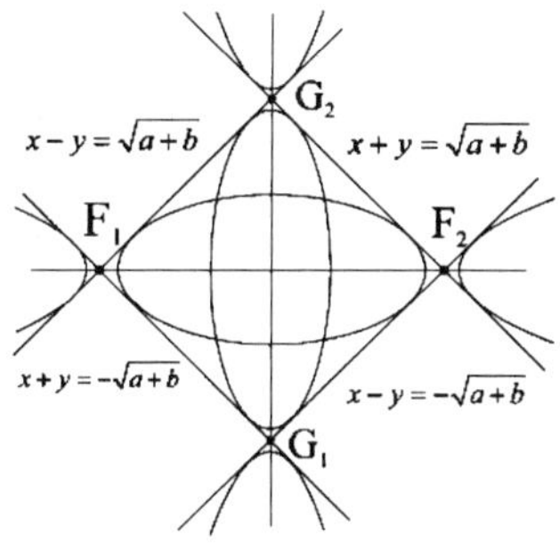

FIGURE 7. A family of confocal quadrics on the Minkowski plane. The bold dots indicate the four foci of the family of the quadrics.

distinguish 3 subfamilies

- by $\lambda \in (-\infty, -b)$ the quadric is a hyperbola with real axis x
- by $\lambda \in (a, \infty)$ the quadric is a hyperbola with real axis y
- by $\lambda \in (-b, a)$ the quadric is an ellipse

In addition, the values $\lambda = a, -b, \infty$ correspond to degenerate quadrics that are the axis y, the axis x and the line at infinity, respectively.

Let us define a billiard reflection in the Minkowski plane. Let v be a vector and ℓ a line. We represent a vector in the form $v = v_n + v_\ell$, where v_n is the normal component to the line ℓ (in the sense of the Minkowski metric) of the velocity vector, and v_ℓ belongs to ℓ. Let us call *the billiard reflection* v' of the vector v with respect to the line ℓ the vector $v' = -v_n + v_\ell$. Vector v is a billiard reflection of the vector v'. Note that if the vector v_n is light-like, the reflection is undefined.

Note that the reflection does not change the type of the vector. Indeed, it is not difficult to verify that $\langle v, v\rangle = \langle v', v'\rangle$, where v' and v are billiard reflections of each other.

It turns out that the flat billiard bounded by the arcs of confocal quadrics in the Minkowski space is also integrable. Along the trajectories of the billiard, the parameter of the confocal quadric is preserved (here we mean the billiard reflection in the Minkowski plane, see above).

Let us consider in details the billiard in the flat domain bounded by the ellipse. At four points, where the normal to the tangent is light-like (namely, the points of tangency of the lines $x \pm y = \pm\sqrt{a+b}$ and the ellipse), we can extend the billiard reflection by continuity. At these points the vector v' of the billiard reflection of the vector v can be defined by equation: $v' = -v$.

Let us demonstrate the evolution of trajectories induced by the change of the parameter Λ of the caustic.

For $\Lambda \in (-\infty, -b)$, the straight lines containing the trajectories of the billiard are tangent to the hyperbolas with real axis x to the family (7.1). Then all trajectories are time-like and completely fill the entire interior of the ellipse. Similarly, for $\Lambda \in (a, \infty)$, the caustics are the hyperbolas with real axis y, the trajectories are time-like and also fill the entire interior of the ellipse.

For $\Lambda = \infty$ the caustic is a line at infinity. Let us note that when Λ tends to $+\infty$ and $-\infty$ the pseudo-Euclidean length of the velocity vector can only decrease and its tilt angle tends to the tilt angle of the common tangents lines. Therefore, in the limit, we obtain light-like trajectories.

For $\Lambda \in (-b, 0)$ the trajectories are tangent to the confocal ellipse, and fill the two marked parts of the ellipse (see highlighted regions on Figure 8, *a.*). These trajectories are time-like.

For $\Lambda \in (0, a)$ the trajectories are also tangent to the confocal ellipse, but fill the other two marked parts of the ellipse (see Figure 8, *b.*). These trajectories are time-like.

For $\Lambda = a$, the tangent line to the trajectory is the y axis. The set of trajectories, which are time-like, consists of one periodic trajectory located along the axis y and two disjoint sets of homoclinic trajectories with tangent lines passing through pairs of foci located on the axis y.

For $\Lambda = -b$, the tangent line to the trajectory is the x axis.The set of trajectories, which are space-like, consists of one periodic trajectory located along the axis x and two disjoint sets of homoclinic trajectories with tangent lines passing through a pair of foci lying on the axis x.

For $\Lambda = -0, +0$ the caustic is the boundary ellipse. In the first case, the trajectories are space-like and the motion occurs along the segments marked in Figure 8, *c*. In the second case the trajectories are time-like and the motion occurs along the segments marked in Figure 8, *d*. There are no other trajectories for such Λ, since the segments are convex with respect to the interior of the ellipse. Hence, can not be reflected from the boundary in any other way.

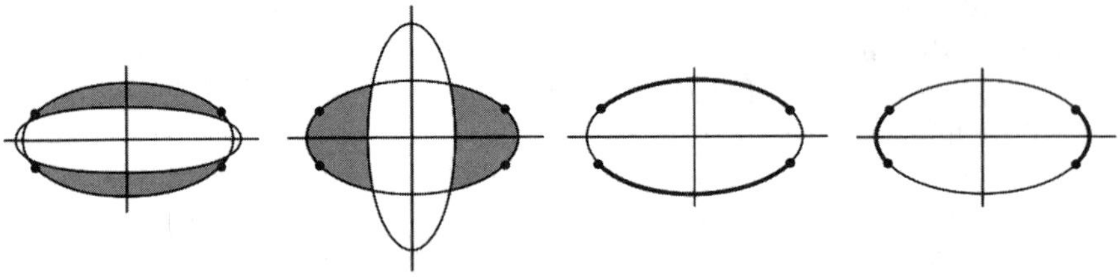

a. b. c. d.

FIGURE 8. Areas of possible motion for different values λ

DEFINITION 16. An elementary billiard in a Minkowski metric is a compact, connected subset of the Minkowski plane whose boundary consists of arcs of confocal quadrics for the family (7.1) and does not contain the angles greater than π.

REMARK 8. Karginova E.E. has obtained a complete classification of all elementary billiards on the Minkowski plane. For all such billiards E.Karginova calculated the Fomenko-Zieschang invariants. This classification will be presented in a future publication.

THEOREM 8 (E.Karginova). *Let Ψ be an elementary billiard in the Minkowski metric.*

- *Let the intersection of the interior of the billiard* Ψ *with the axes* Ox *and* Oy *be nonempty. Then the Fomenko-Zieschang invariant describing the topology of the Liouville foliation on the isoenergy surface for the billiard* Ψ *corresponds to the molecule presented in Figure* 9 *a.*
- *Let the interior of the billiard* Ψ *intersect only the one axis* Ox *or* Oy*. Then the Fomenko-Zieschang invariant corresponds to the molecule presented in Figure* 9 *b.*
- *Let us suppose that the interior of the billiard* Ψ *do not intersect with the axes* Ox *and* Oy*. Then the Fomenko-Zieschang invariant corresponds to the molecule presented in Figure* 9 *c.*

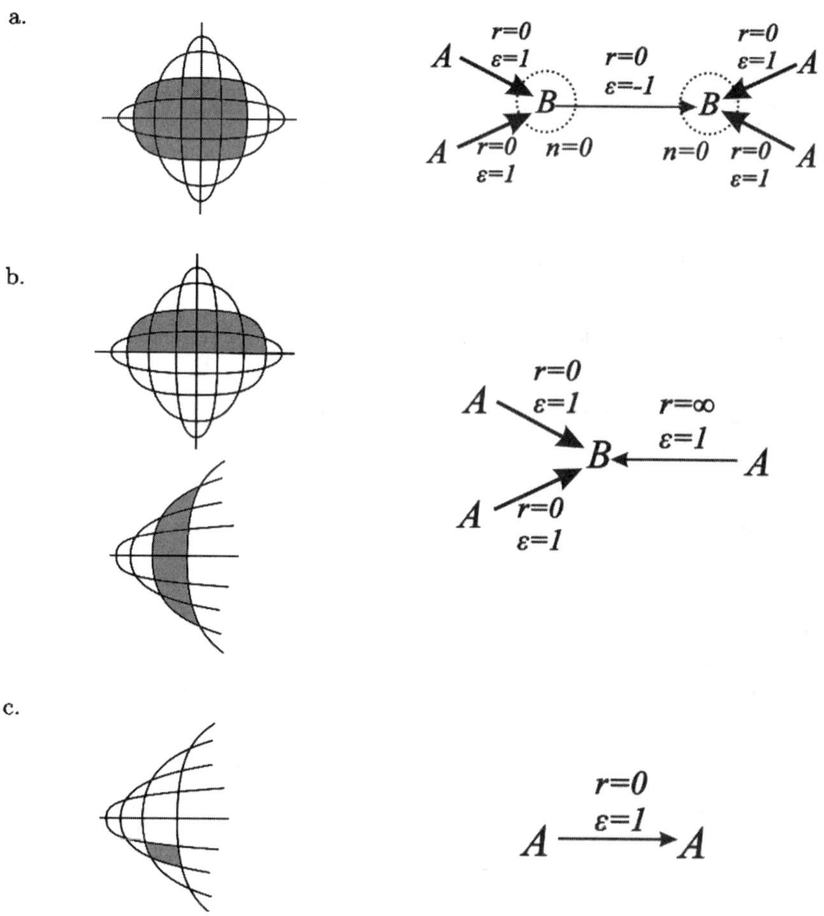

FIGURE 9. Fomenko-Zieschang invariants that classify the isoenergetic surfaces of billiards in the Minkowski space.

References

[1] A. V. Bolsinov and A. T. Fomenko, *Integrable Hamiltonian systems: Geometry, topology, classification*, Chapman & Hall/CRC, Boca Raton, FL, 2004. Translated from the 1999 Russian original. MR2036760

[2] A. V. Bolsinov and A. T. Fomenko, *Trajectory equivalence of integrable Hamiltonian systems with two degrees of freedom. A classification theorem. II* (Russian, with Russian summary), Mat. Sb. **185** (1994), no. 5, 27–78, DOI 10.1070/SM1995v082n01ABEH003551; English transl., Russian Acad. Sci. Sb. Math. **82** (1995), no. 1, 21–63. MR1275971

[3] A. V. Bolsinov and A. T. Fomenko, *Orbital classification of geodesic flows of two-dimensional ellipsoids. The Jacobi problem is orbitally equivalent to the integrable Euler case in the dynamics of a rigid body* (Russian, with Russian summary), Funktsional. Anal. i Prilozhen. **29** (1995), no. 3, 1–15, 96, DOI 10.1007/BF01077048; English transl., Funct. Anal. Appl. **29** (1995), no. 3, 149–160 (1996). MR1361950

[4] A. V. Bolsinov and A. T. Fomenko, *The geodesic flow of an ellipsoid is orbitally equivalent to the integrable Euler case in the dynamics of a rigid body* (Russian), Dokl. Akad. Nauk **339** (1994), no. 3, 253–296; English transl., Russian Acad. Sci. Dokl. Math. **50** (1995), no. 3, 412–417. MR1315180

[5] A. V. Bolsinov and A. T. Fomenko, Trajectory invariants of integrable Hamiltonian systems. The case of simple systems. Trajectory classification of Euler-type systems in rigid body dynamics. *Izvest. Akad. Nauk SSSR, Ser. Matem.*, **59** (1995), No. 1, P. 65–102.

[6] A. V. Bolsinov, S. V. Matveev, and A. T. Fomenko, *Topological classification of integrable Hamiltonian systems with two degrees of freedom. A list of systems of small complexity* (Russian), Uspekhi Mat. Nauk **45** (1990), no. 2(272), 49–77, 240, DOI 10.1070/RM1990v045n02ABEH002344; English transl., Russian Math. Surveys **45** (1990), no. 2, 59–94. MR1069348

[7] A. T. Fomenko, The topology of surfaces of constant energy in integrable Hamiltonian systems, and obstructions to integrability. Math. USSR Izvestija 29(3), 629-658 (1987)

[8] A. T. Fomenko, *Topological invariants of Hamiltonian systems that are integrable in the sense of Liouville* (Russian), Funktsional. Anal. i Prilozhen. **22** (1988), no. 4, 38–51, 96, DOI 10.1007/BF01077420; English transl., Funct. Anal. Appl. **22** (1988), no. 4, 286–296 (1989). MR976994

[9] A. T. Fomenko and Kh. Tsishang, *A topological invariant and a criterion for the equivalence of integrable Hamiltonian systems with two degrees of freedom* (Russian), Izv. Akad. Nauk SSSR Ser. Mat. **54** (1990), no. 3, 546–575, DOI 10.1070/IM1991v036n03ABEH002035; English transl., Math. USSR-Izv. **36** (1991), no. 3, 567–596. MR1072695

[10] A. T. Fomenko, H. Zieschang, On typical topological properties of integrable Hamiltonian systems. USSR-Izv. 32(2), 385-412 (1989)

[11] A. T. Fomenko, *Symplectic geometry*, 2nd ed., Advanced Studies in Contemporary Mathematics, vol. 5, Gordon and Breach Publishers, Luxembourg, 1995. Translated from the 1988 Russian original by R. S. Wadhwa. MR1673400

[12] A. A. Oshemkov, *Fomenko invariants for the main integrable cases of the rigid body motion equations*, Topological classification of integrable systems, Adv. Soviet Math., vol. 6, Amer. Math. Soc., Providence, RI, 1991, pp. 67–146. MR1141221

[13] P. Topalov, *Calculation of the fine Fomenko-Zieschang invariant for basic integrable cases of the motion of a rigid body* (Russian, with Russian summary), Mat. Sb. **187** (1996), no. 3, 143–160, DOI 10.1070/SM1996v187n03ABEH000120; English transl., Sb. Math. **187** (1996), no. 3, 451–468. MR1400349

[14] S. Tabachnikov, *Geometry and billiards*, Student Mathematical Library, vol. 30, American Mathematical Society, Providence, RI; published in cooperation with Mathematics Advanced Study Semesters, University Park, PA, 2005. MR2168892

[15] G. D. Birkhoff, *Dynamical systems*, American Mathematical Society Colloquium Publications, Vol. IX, American Mathematical Society, Providence, R.I., 1966. With an addendum by Jurgen Moser. MR0209095

[16] Kozlov, V. V.; Treshchev, D. V. (Russian) A genetic introduction to the dynamics of systems with impacts, Moskov. Gos. Univ., Moscow, 1991

[17] V. Dragović and M. Radnović, *Bifurcations of Liouville tori in elliptical billiards*, Regul. Chaotic Dyn. **14** (2009), no. 4-5, 479–494, DOI 10.1134/S1560354709040054. MR2551871

[18] V. Dragović and M. Radnović, *Poncelet porisms and beyond: Integrable billiards, hyperelliptic Jacobians and pencils of quadrics*, Frontiers in Mathematics, Birkhäuser/Springer Basel AG, Basel, 2011. MR2798784

[19] V. Dragovic, M. Radnovic, Topological invariants for elliptical billiards and geodesics on ellipsoids in the Minkowski space, Fundam. Sci. Appl. 20 (2015).

[20] V. Dragović and M. Radnović, *Bicentennial of the great Poncelet theorem (1813–2013): current advances*, Bull. Amer. Math. Soc. (N.S.) **51** (2014), no. 3, 373–445, DOI 10.1090/S0273-0979-2014-01437-5. MR3196793

[21] V. Dragović and M. Radnović, *Pseudo-integrable billiards and arithmetic dynamics*, J. Mod. Dyn. **8** (2014), no. 1, 109–132, DOI 10.3934/jmd.2014.8.109. MR3296569

[22] V. I. Dragovich and M. Radnovich, *Pseudo-integrable billiards and double reflection nets* (Russian, with Russian summary), Uspekhi Mat. Nauk **70** (2015), no. 1(421), 3–34, DOI 10.4213/rm9648; English transl., Russian Math. Surveys **70** (2015), no. 1, 1–31. MR3353115

[23] V. Dragović and M. Radnović, *Periods of pseudo-integrable billiards*, Arnold Math. J. **1** (2015), no. 1, 69–73, DOI 10.1007/s40598-014-0004-0. MR3331968

[24] V. V. Fokicheva, *Description of singularities for the "billiard in an ellipse" system* (Russian, with English and Russian summaries), Vestnik Moskov. Univ. Ser. I Mat. Mekh. **5** (2012), 31–34, DOI 10.3103/S0027132212050063; English transl., Moscow Univ. Math. Bull. **67** (2012), no. 5-6, 217–220. MR3076496

[25] V. V. Fokicheva, *Description of singularities for billiard systems bounded by confocal ellipses or hyperbolas*, Moscow Univ. Math. Bull. **69** (2014), no. 4, 148–158, DOI 10.3103/S0027132214040020. Translation of Vestnik Moskov. Univ. Ser. I Mat. Mekh. **2014**, no. 4, 18–27. MR3372782

[26] V. V. Fokicheva, *Classification of billiard motions in domains bounded by confocal parabolas* (Russian, with Russian summary), Mat. Sb. **205** (2014), no. 8, 139–160, DOI 10.1070/sm2014v205n08abeh004415; English transl., Sb. Math. **205** (2014), no. 7-8, 1201–1221. MR3288207

[27] V. V. Fokicheva, *Topological classification of billiards in locally planar domains bounded by arcs of confocal quadrics* (Russian, with Russian summary), Mat. Sb. **206** (2015), no. 10, 127–176, DOI 10.4213/sm8506; English transl., Sb. Math. **206** (2015), no. 9-10, 1463–1507. MR3438566

[28] V. V. Fokicheva and A. T. Fomenko, *Integrable billiards model important integrable cases of rigid body dynamics* (Russian, with Russian summary), Dokl. Akad. Nauk **465** (2015), no. 2, 150–153, DOI 10.1134/s1064562415060095; English transl., Dokl. Math. **92** (2015), no. 3, 682–684. MR3495614

[29] V. V. Fokicheva and A. T. Fomenko, *Billiard systems as the models for the rigid body dynamics*, Advances in dynamical systems and control, Stud. Syst. Decis. Control, vol. 69, Springer, [Cham], 2016, pp. 13–33. MR3616349

[30] A. T. Fomenko and V. V. Vedyushkina, *Integrable topological billiards and equivalent dynamical systems* (Russian, with Russian summary), Izv. Ross. Akad. Nauk Ser. Mat. **81** (2017), no. 4, 20–67, DOI 10.4213/im8602; English transl., Izv. Math. **81** (2017), no. 4, 688–733. MR3682783

[31] V. A. Moskvin, *Topology of Liouville foliations for integrable billiards in non-convex domains*, Moscow Univ. Math. Bull. **73** (2018), no. 3, 103–110. Translation of Vestnik Moskov. Univ. Ser. I Mat. Mekh. **2018**, no. 3, 21–29. MR3831453

[32] E. A. Kudryavtseva, Liouville integrable generalized billiard flows and Poncelet type theorems, Journal of Mathematical Sciences, 217 (2016)

[33] E. A. Kudryavtseva, I. M. Nikonov, and A. T. Fomenko, *Maximally symmetric cellular partitions of surfaces and their coverings* (Russian, with Russian summary), Mat. Sb. **199** (2008), no. 9, 3–96, DOI 10.1070/SM2008v199n09ABEH003962; English transl., Sb. Math. **199** (2008), no. 9-10, 1263–1353. MR2466854

[34] V. F. Lazutkin, *KAM theory and semiclassical approximations to eigenfunctions*, Ergebnisse der Mathematik und ihrer Grenzgebiete (3) [Results in Mathematics and Related Areas (3)], vol. 24, Springer-Verlag, Berlin, 1993. With an addendum by A. I. Shnirel′man. MR1239173

[35] A. A. Oshemkov, *Topology of isoenergetic surfaces, and bifurcation diagrams of integrable cases of the dynamics of a rigid body on* SO(4) (Russian), Uspekhi Mat. Nauk **42** (1987), no. 6(258), 199–200. MR934012

[36] A. A. Oshemkov, *Description of isoenergetic surfaces of integrable Hamiltonian systems with two degrees of freedom* (Russian), Trudy Sem. Vektor. Tenzor. Anal. **23** (1988), 122–132. MR1041275

[37] A. A. Oshemkov, *Fomenko invariants for the main integrable cases of the rigid body motion equations*, Topological classification of integrable systems, Adv. Soviet Math., vol. 6, Amer. Math. Soc., Providence, RI, 1991, pp. 67–146. MR1141221

[38] V. V. Kozlov, *Symmetries, topology and resonances in Hamiltonian mechanics*, Ergebnisse der Mathematik und ihrer Grenzgebiete (3) [Results in Mathematics and Related Areas (3)], vol. 31, Springer-Verlag, Berlin, 1996. Translated from the Russian manuscript by S. V. Bolotin, D. Treshchev and Yuri Fedorov. MR1411677

[39] D. Genin, B. Khesin, and S. Tabachnikov, *Geodesics on an ellipsoid in Minkowski space*, Enseign. Math. (2) **53** (2007), no. 3-4, 307–331. MR2455947

[40] E. Gutkin and S. Tabachnikov, *Billiards in Finsler and Minkowski geometries*, J. Geom. Phys. **40** (2002), no. 3-4, 277–301, DOI 10.1016/S0393-0440(01)00039-0. MR1866992

[41] A. V. Brailov and A. T. Fomenko, *Topology of integral submanifolds of completely integrable Hamiltonian systems* (Russian), Mat. Sb. (N.S.) **134(176)** (1987), no. 3, 375–385, 447–448, DOI 10.1070/SM1989v062n02ABEH003244; English transl., Math. USSR-Sb. **62** (1989), no. 2, 373–383. MR922630

[42] S. V. Matveev and A. T. Fomenko, *Isoenergetic surfaces of Hamiltonian systems, the enumeration of three-dimensional manifolds in order of growth of their complexity, and the calculation of the volumes of closed hyperbolic manifolds* (Russian), Uspekhi Mat. Nauk **43** (1988), no. 1(259), 5–22, 247, DOI 10.1070/RM1988v043n01ABEH001554; English transl., Russian Math. Surveys **43** (1988), no. 1, 3–24. MR937017

[43] A. V. Bolsinov, *Methods of calculation of the Fomenko-Zieschang invariant*, Topological classification of integrable systems, Adv. Soviet Math., vol. 6, Amer. Math. Soc., Providence, RI, 1991, pp. 147–183. MR1141222

[44] A. T. Fomenko and A. Konyaev, *Algebra and geometry through Hamiltonian systems*, Continuous and distributed systems, Solid Mech. Appl., vol. 211, Springer, Cham, 2014, pp. 3–21, DOI 10.1007/978-3-319-03146-0_1. MR3204085

[45] A. T. Fomenko and E. O. Kantonistova, *Topological classification of geodesic flows on revolution 2-surfaces with potential*, Continuous and distributed systems. II, Stud. Syst. Decis. Control, vol. 30, Springer, Cham, 2015, pp. 11–27, DOI 10.1007/978-3-319-19075-4_2. MR3381622

Email address: `atfomenko@mail.ru`

Moscow State University, Leninskiye Gory 1, Moscow, Russia Federation 119991

Moscow State University, Leninskiye Gory 1, Moscow, Russia Federation 119991
Email address: `aivenirra@gmail.com`

Contemporary Mathematics
Volume **733**, 2019
https://doi.org/10.1090/conm/733/14739

Hasse–Schmidt derivations and Cayley–Hamilton theorem for exterior algebras

Letterio Gatto and Inna Scherbak

In memory of those participants of the Voronezh Winter Mathematical School who have already passed into another world where all problems are solved.

ABSTRACT. Using the natural notion of *Hasse–Schmidt derivations on an exterior algebra*, we relate two classical and seemingly unrelated subjects. The first is the famous Cayley–Hamilton theorem of linear algebra, "*each endomorphism of a finite-dimensional vector space is a root of its own characteristic polynomial*", and the second concerns the expression of the bosonic vertex operators occurring in the representation theory of the (infinite-dimensional) Heisenberg algebra.

1. Introduction

In 1937, Hasse and Schmidt introduced the notion of higher derivations [**8**], nowadays called *Hasse–Schmidt (HS) derivations*. Let $(A, *)$ be an algebra over a ring B, not necessarily commutative or associative. A *HS–derivation on A* is a B-algebra homomorphism, $D(t) : A \to A[[t]]$, that is, a B-linear mapping satisfying

$$D(t)(a_1 * a_2) = D(t)a_1 * D(t)a_2, \quad \forall a_1, a_2 \in A.$$

A fundamental example of a HS–derivation is given by the map sending any function $f = f(z)$, holomorphic in some domain of the complex plane, to its *formal Taylor series*,

$$f(z) \mapsto T(t)[f(z)] = \left(\exp\left(t\frac{d}{dz}\right)\right) f(z).$$

The property $T(t)[f(z)g(z)] = T(t)[f(z)] \cdot T(t)[g(z)]$ encodes the full set of the Leibniz's rules,

$$\frac{d^i(fg)}{dz^i} = \sum_{j=0}^{i} \binom{i}{j} \frac{d^j f}{dz^i} \cdot \frac{d^{i-j} f}{dz^{i-j}} .$$

In general, if $(A, *)$ is any commutative $\mathbb{Q}$-algebra and $\delta(t) : A \to A[[t]]$ is a derivation in the Leibniz rule sense, i.e., $\delta(t)(a * b) = \delta(t)a * b + a * \delta(t)b$, then $\exp(\delta(t))$ is a HS–derivation.

2010 *Mathematics Subject Classification.* Primary 15A75, 17B69. 050E5.

Key words and phrases. Hasse-Schmidt Derivations on Grassmann Algebras, Theorem of Cayley and Hamilton, Vertex Operators.

Work parially sponsored by PRIN "Geometria sulle Varietà algebriche", INDAM-GNSAGA e "Finanziamento diffuso della Ricerca" del Politecnico di Torino.

The aim of Hasse and Schmidt was to find a counterpart to the Taylor series that would work in positive characteristic. Their definition does not require division by integers and is therefore particularly suitable for this purpose. Schmidt later applied the theory to investigate Weierstrass points and Wronskians on curves in positive characteristic [**15**].

In a number of papers motivated by Schubert Calculus [**3**, **6**] (see also the book [**5**]), one of us proposed to study HS–derivations for exterior algebras.

If A is a commutative ring with unit, M a module over A, and $\bigwedge M$ its exterior algebra, then a *HS–derivation on* $\bigwedge M$ is a $\wedge$-homomorphism

$$\mathcal{D}(t) : \bigwedge M \to \left(\bigwedge M\right)[[t]],$$

that is, a linear mapping satisfying

$$\mathcal{D}(t)(u \wedge v) = \mathcal{D}(t)u \wedge \mathcal{D}(t)v, \quad \forall u, v \in \bigwedge M. \tag{1.1}$$

In this paper we consider HS–derivations $\mathcal{D}(t) = \sum_{i\geq 0} \mathcal{D}_i \cdot t^i$ with $\mathcal{D}_0 = \mathbf{1}$, the identity on $\bigwedge M$. Such $\mathcal{D}(t)$ is invertible as an element of $\mathrm{End}(\bigwedge M)[[t]]$, that is, there exists $\overline{\mathcal{D}}(t) \in \mathrm{End}(\bigwedge M)[[t]]$ satisfying

$$\overline{\mathcal{D}}(t)D(t) = \mathcal{D}(t)\overline{\mathcal{D}}(t) = \mathbf{1}. \tag{1.2}$$

A straightforward calculation shows that $\overline{\mathcal{D}}(t)$ is also a HS–derivation on $\bigwedge M$. Hence,

$$\mathcal{D}(t)(\overline{\mathcal{D}}(t)u \wedge v) = u \wedge \mathcal{D}(t)v, \ \overline{\mathcal{D}}(t)(\mathcal{D}(t)u \wedge v) = u \wedge \overline{\mathcal{D}}(t)v, \quad \forall u, v \in \bigwedge M. \tag{1.3}$$

The coefficients of t in these equations give $u \wedge \mathcal{D}_1 v = \mathcal{D}_1(u \wedge v) - \mathcal{D}_1 u \wedge v$. That is why in [**3**] we call (1.3) the *integration by parts* formulas.

In the present article we show how these simple formulas link two classical and seemingly unrelated subjects (one finite-dimensional and the other infinite-dimensional), apparently leading to a unified interpretation.

One topic is the classical Cayley–Hamilton Theorem of Linear Algebra saying that *each endomorphism f of an r-dimensional vector space M is a root of its own characteristic polynomial* $\det(t\mathbf{1} - f)$. Let us reformulate this theorem as a linear recurrence relation on the sequence of endomorphisms $(f^j)_{j\geq 0}$,

$$f^{r+k} - e_1 f^{r+k-1} + \cdots + (-1)^k e_r f^k = \mathbf{0}, \quad \forall k \geq 0, \tag{1.4}$$

where $\det(t\mathbf{1} - f) = t^r - e_1 t^{r-1} + \cdots + (-1)^r e_r$, and $\mathbf{0}$ denotes the zero endomorphism. Consider $D(t) = \sum_{i\geq 0} D_i \cdot t^i$, the unique HS–derivation on the exterior algebra of M such that $D_{i|M} = f^i$, $i \geq 0$. It turns out that the sequence $(D_i)_{i\geq 0}$ of endomorphisms of $\bigwedge M$ satisfies relations similar to (1.4), see Theorem 2.3 in Section 2.1 for the exact formulation, and Section 3 for the proof.

The other topic concerns bosonic vertex operators arising in the representation theory of the (infinite-dimensional) Heisenberg algebra (see, for example, [**9**]). As we observe in Section 2.2, any countably generated vector space over the rationals can be equipped with the structure of a free module of finite rank r over a ring of polynomials in r variables with rational coefficients, for any integer $r > 0$. We present the construction in Section 4, and in Section 5 we apply it to obtain the "finite-dimensional approximation" to the well-known expressions of the vertex operators $\Gamma(t)$ and $\Gamma^*(t)$ generating the bosonic Heisenberg vertex algebra (see [**9**, p. 56]). We interpret $\Gamma(t)$ and $\Gamma^*(t)$ as the limit, when $r \to \infty$, of the ratio of two characteristic polynomials associated to the shift endomorphisms of steps $+1$ and -1, respectively. The precise formulation can be found in Section 2.3.

Our work is based on interpreting (1.3) as a sort of abstract Cayley–Hamilton theorem, holding for general invertible HS–derivations on exterior algebras of arbitrary modules (not necessarily free). If the module is countably generated, then (1.3) produces a sequence of Cayley-Hamilton relations (3.5) which specialize to the classical Cayley–Hamilton formulas (2.7) when considering Hasse–Schmidt derivations associated to endomorphisms of finitely generated free modules.

Plan of the paper. In Section 2 we formulate the main statements. Section 2.1 is devoted to our extension of the Cayley–Hamilton theorem on an exterior algebra of a finite rank free module, which is Theorem 2.3. The proof can be found in Section 3, which also includes the necessary information on the Hasse–Schmidt derivations, and the discussion concerning the case when the ring A contains the rationals.

In Section 2.2, we equip a countably infinite-dimensional $\mathbb{Q}$-vector space with a natural structure of a free module of rank r over the ring of polynomials of r variables with rational coefficients, for any integer $r > 0$. The construction is based on a Giambelli's type formula. The details are explained in Section 4.

In Section 2.3 we apply our construction to the bosonic Heisenberg vertex algebra and interpret the truncation of bosonic vertex operators as the ratio of two characteristic polynomials, respectively associated to the shift endomorphisms of step ± 1; see Section 5 for detailed explanation.

2. Formulation of the results

2.1. Cayley–Hamilton theorem for exterior algebras. Let M be a free A-module of at most countable (i.e., either finite or countable) rank. If $b_1, b_2, \ldots$ is a basis of M, then $\{b_{i_1} \wedge \ldots \wedge b_{i_j}\}_{1 \le i_1 < \ldots < i_j}$ is a basis of $\bigwedge^j M$, $j \ge 1$. Following a referee's suggestion, we observe that any A-endomorphism $f : M \to M$ naturally extends to the endomorphism $\bigwedge f$ of $\bigwedge M$ as follows.

Define $\bigwedge f = \bigoplus_{j \ge 0} \bigwedge^j f$ by setting $\bigwedge^0 f$ to be the identity on $\bigwedge^0 M = A$, and by defining the action of $\bigwedge^j f$ on the basis of $\bigwedge^j M$ as follows,

$$b_{i_1} \wedge \ldots \wedge b_{i_j} \;\mapsto\; fb_{i_1} \wedge \ldots \wedge fb_{i_j}, \quad j \ge 1. \tag{2.1}$$

As a consequence, any formal power series $P(t) = \sum_{i \ge 0} p_i t^i$, where $p_i \in \mathrm{End}_A M$, clearly defines a HS–derivation $\bigwedge P(t)$ on $\bigwedge M$. If, in addition, $p_0 = \mathbf{1}$, the identity on M, then the formal inverse $P^{-1}(t) \in (\mathrm{End}_A M)[[t]]$ defines the inverse HS–derivation,

$$\overline{\bigwedge P(t)} = \bigwedge P^{-1}(t).$$

PROPOSITION 2.1. *Let M be an A-module of at most countable rank, and $f \in \mathrm{End}_A M$. Set $f^0 = \mathbf{1}$. Then*

$$D(t) = \bigwedge \left(\sum_{i \ge 0} f^i t^i \right) \tag{2.2}$$

is an invertible HS–derivation, and its inverse is

$$\overline{D}(t) = \bigwedge (\mathbf{1} - ft)\,. \tag{2.3}$$

□

Our extension of the Cayley–Hamilton theorem concerns modules of finite rank. If M has rank r over A, then $\bigwedge^r M$ has rank 1, and so $D(t)$ and $\overline{D}(t)$ act on $\bigwedge^r M$ by multiplication by some formal power series. Indeed, $\bigwedge^r M = \mathrm{Span}_A\{b_1 \wedge \ldots \wedge b_r\}$, and, by definition (2.3),

$$\overline{D}(t)\,(b_1 \wedge \ldots \wedge b_r) = (\mathbf{1} - ft)b_1 \wedge \ldots \wedge (\mathbf{1} - ft)b_r = \det(\mathbf{1} - ft)\,(b_1 \wedge \ldots \wedge b_r)\,.$$

Thus, the "eigenvalue" of $\overline{D}(t)$ on $\bigwedge^r M$ is $\det(\mathbf{1} - ft)$, considered as an element of $A[[t]]$. Let us write $\det(\mathbf{1} - ft) = E_r(t)$, where

$$E_r(t) = 1 - e_1 t + \cdots + (-1)^r t^r. \tag{2.4}$$

REMARK 2.2. (1) Clearly "the eigenvalue" $H_r(t)$ of $D(t)$ on $\bigwedge^r M$ is the formal inverse of $E_r(t)$, i.e.,

$$H_r(t)E_r(t) = 1. \tag{2.5}$$

If we write $H_r(t) = \sum_{j\geq 0} h_j t^j$, then (2.4) and (2.5) determine each h_j as a polynomial of $e_1, \ldots, e_r$. For example, $h_0 = 1$, $h_1 = e_1$, $h_2 = e_1^2 - e_2$ etc.

(2) It is worth emphasizing that $\{e_i\}$ and $\{h_j\}$ are related in exactly the same way as *the elementary* and *the complete symmetric functions* of r variables, since $E_r(t)$ and $H_r(t)$ are their generating functions, respectively (see, for example, [**13**, I 2]).

For $f \in \mathrm{End}_A M$, one can write HS–derivations $D(t)$ and $\overline{D}(t)$, defined by (2.2) and (2.3), respectively, in the form

$$D(t) = \sum_{i\geq 0} D_i \cdot t^i, \quad \overline{D}(t) = \sum_{i\geq 0} (-1)^i \overline{D}_i \cdot t^i.$$

We have $D_0 = \mathbf{1} = \overline{D}_0$, $D_1 = \overline{D}_1$, and $D_i\,|_M = f^i$, $i \geq 0$.

Denote $\bigwedge^{>i} M = \bigoplus_{j=i+1}^r \bigwedge^j M$.

THEOREM 2.3. *For $1 \leq k < r$, the endomorphism*

$$D_k - e_1 D_{k-1} + \cdots + (-1)^k e_k \mathbf{1} \tag{2.6}$$

vanishes on $\bigwedge^{>(r-k)} M$, and for $i \geq r$ the endomorphism

$$D_i - e_1 D_{i-1} + \cdots + (-1)^r e_r D_{i-r} \tag{2.7}$$

vanishes on the whole of $\bigwedge M$.

According to (2.4), the characteristic polynomial of f is

$$\det(t\mathbf{1} - f) = t^r E_r(1/t) = t^r - e_1 t^{r-1} + \cdots + (-1)^r e_r. \tag{2.8}$$

Hence, for $i = r+k$, the restriction of (2.7) to M gives the classical Cayley–Hamilton theorem (1.4).

2.2. A look at the infinite-dimensional case. Let M_0 be a $\mathbb{Q}$-vector space with a countable basis, and $\bigwedge M_0 = \bigoplus_{j\geq 0} \bigwedge^j M_0$ its exterior algebra.

We equip M_0 with a structure of a free module of rank r over the ring of polynomials of r variables with rational coefficients, for any integer $r > 0$. See Section 4 for details.

We fix a basis $(b_j)_{j\geq 1}$ of M_0, and define the shift operators σ_{+1}, σ_{-1} on M_0 by their action on the basis,

$$\sigma_{+1}(b_j) = b_{j+1},\ j \geq 1, \text{ and } \sigma_{-1}(b_1) = 0,\ \sigma_{-1}(b_j) = b_{j-1},\ j > 1.$$

One can attach to each of the endomorphisms σ_{+1}, σ_{-1} a unique HS–derivation and its inverse, as in (2.2), (2.3). In this subsection we need only the HS–derivations

$$\sigma_+(t), \overline{\sigma}_+(t) : \bigwedge M_0 \to \left(\bigwedge M_0\right)[[t]]$$

generated by σ_{+1}. We shall denote by $\sigma_{+i}, \overline{\sigma}_{+i} : \bigwedge M_0 \to \bigwedge M_0$ the coefficients of t^i in $\sigma_+(t)$ and $\overline{\sigma}_+(t)$ respectively. In the next subsection, the HS–derivations corresponding to σ_{-1} will also appear.

Let us fix $r > 0$. It is convenient to enumerate the basis of $\bigwedge^r M_0$, which corresponds to $(b_j)_{j\geq 1}$, by partitions $\boldsymbol{\lambda} = (\lambda_1 \geq \cdots \geq \lambda_r \geq 0)$ of length at most r. We write $\mathcal{P}_r$ for the set of all such partitions, and denote the basis vectors as follows,

$$[\mathbf{b}]^r_{\boldsymbol{\lambda}} = b_{1+\lambda_r} \wedge b_{2+\lambda_{r-1}} \wedge \cdots \wedge b_{r+\lambda_1}\,, \quad \boldsymbol{\lambda} \in \mathcal{P}_r\,. \tag{2.9}$$

In particular, the zero partition $\mathbf{0} = (\lambda_1 = 0)$ gives $[\mathbf{b}]^r_{\mathbf{0}} = b_1 \wedge \cdots \wedge b_r$.

Now consider $(e_i)_{1\leq i\leq r}$ as indeterminates, and the polynomial ring

$$B_r = \mathbb{Q}[e_1, e_2, \ldots, e_r]. \tag{2.10}$$

Let us equip $\bigwedge^r M_0$ with a B_r-module structure via

$$E_r(t)[\mathbf{b}]^r_{\boldsymbol{\lambda}} = \overline{\sigma}_+(t)[\mathbf{b}]^r_{\boldsymbol{\lambda}}\,, \tag{2.11}$$

where $E_r(t)$ is given by (2.4). In terms of the inverses, see (2.5), the same structure is given by

$$H_r(t)[\mathbf{b}]^r_{\boldsymbol{\lambda}} = \sigma_+(t)[\mathbf{b}]^r_{\boldsymbol{\lambda}}\,. \tag{2.12}$$

The interpretation of e_j's and h_j's as the elementary and the complete symmetric functions of r variables, see Remark 2.2 (2), suggests to consider B_r as a $\mathbb{Q}$-vector spaces generated by the Schur polynomials (see, for example, [**13**, I3]),

$$\Delta_{\boldsymbol{\lambda}}(H_r) = \det(h_{\lambda_j-j+i})_{1\leq i,j\leq r}\,, \quad \boldsymbol{\lambda} \in \mathcal{P}_r. \tag{2.13}$$

Here h_j's are defined as in Remark 2.2 (1) for $j \geq 0$, and $h_j = 0$ for $j < 0$.

According to *Giambelli's formula* as in [**3**, p. 321]),

$$[\mathbf{b}]^r_{\boldsymbol{\lambda}} = \Delta_{\boldsymbol{\lambda}}(H_r)[\mathbf{b}]^r_{\mathbf{0}}, \tag{2.14}$$

that is, $\bigwedge^r M_0$ is a free B_r-module of rank 1 generated by $[\mathbf{b}]^r_{\mathbf{0}}$. This allows us to equip M_0 with a multiplicative structure over B_r, see Proposition 4.3. Denote M_0, endowed with this multiplicative structure, by M_r. In Section 4, we check that

- M_r is a B_r-module of rank r freely generated by $b_1, \ldots, b_r$.
- $\bigwedge^r M_r$ is $\bigwedge^r M_0$ with the B_r-module structure defined by (2.11) or (2.12).
- e_i is the eigenvalue of $\overline{\sigma}_{+i}$ restricted to $\bigwedge^r M_r$, $1 \leq i \leq r$.
- h_j is the eigenvalue of the restriction of σ_{+j} to $\bigwedge^r M_r$, $j \geq 0$.

REMARK 2.4. The notion of HS–derivation on an exterior algebra enables one to extend some finite-dimensional linear algebra concepts (like eigenvalues and characteristic polynomials) to an infinite-dimensional situation. Indeed, an endomorphism of an infinite-dimensional vector space does not have a characteristic polynomial, whereas the corresponding HS–derivation is still defined.

2.3. Finite-dimensional approximations of bosonic vertex operators. We apply the construction of the previous subsection in order to get a "finite-dimensional approximation" of the well-known expression of the vertex operators occurring in the boson-fermion correspondence. We interpret this approximation as the ratio of certain characteristic polynomials.

Details are in Section 5, see also [**5**].

Take the polynomial ring of countably many indeterminates, $B = \mathbb{Q}[x_1, x_2, \ldots]$, and define the *bosonic vertex operators*, following [**9**, p. 56],

$$\Gamma(t) = \exp\left(\sum_{i\geq 1} x_i t^i\right) \cdot \exp\left(-\sum_{i\geq 1} \frac{1}{it^i}\frac{\partial}{\partial x_i}\right) : B \to B[t^{-1}, t]],$$

$$\Gamma^*(t) = \exp\left(-\sum_{i\geq 1} x_i t^i\right) \cdot \exp\left(\sum_{i\geq 1} \frac{1}{it^i}\frac{\partial}{\partial x_i}\right) : B \to B[t^{-1}, t]].$$

We find finite-dimensional counterparts of these operators using the symmetric functions interpretation. Namely, similarly to the finite-dimensional case, define $E_\infty(t)$ and $H_\infty(t)$,

$$E_\infty(t) = 1 - e_1 t + e_2 t^2 + \cdots + (-1)^k e_k t^k + \ldots\ , \qquad H_\infty(t) = 1/E_\infty(t),$$

as the generating functions of the elementary and the complete symmetric functions of a countable set of variables, say, $(\xi_k)_{k\geq 1}$. Consider also $x_j = j\sum_{k\geq 1} \xi_k^j$, *the power sum symmetric functions*, see [**13**, I 3]. We have $\mathbb{Q}[x_1, x_2, \ldots] = \mathbb{Q}[e_1, e_2, \ldots]$. Moreover, $X_\infty(t) = \sum_{i\geq 1} x_i t^i$, the generating function of $(x_i)_{i\geq 1}$, satisfies

$$\exp\left(\sum_{i\geq 1} x_i t^i\right) = \sum_{i\geq 0} h_i t^i.$$

Clearly, $E_r(t)$, $H_r(t)$, $X_r(t)$ are obtained from $E_\infty(t)$, $H_\infty(t)$, $X_\infty(t)$ by setting $\tau_k = 0$ for $k > r$.

In order to define $\Gamma_r(t)$ and $\Gamma_r^*(t)$ for $r > 0$, use the notation of Section 2.2. In particular, the ring (2.10) is freely generated by the Schur polynomials (2.13), and $\bigwedge^r M_r$ is spanned over B_r by $[\mathbf{b}]^r_{\mathbf{0}}$, according to (2.14). Juxtaposing (2.11) or (2.12) and (2.14) we get, respectively,

$$\overline{\sigma}_+(t)[\mathbf{b}]^r_{\boldsymbol{\lambda}} = \overline{\sigma}_+(t)\left(\Delta_{\boldsymbol{\lambda}}(H_r)[\mathbf{b}]^r_{\mathbf{0}}\right) = E_r(t)\Delta_{\boldsymbol{\lambda}}(H_r)[\mathbf{b}]^r_{\mathbf{0}}, \tag{2.15}$$

$$\sigma_+(t)[\mathbf{b}]^r_{\boldsymbol{\lambda}} = \sigma_+(t)\left(\Delta_{\boldsymbol{\lambda}}(H_r)[\mathbf{b}]^r_{\mathbf{0}}\right) = H_r(t)\Delta_{\boldsymbol{\lambda}}(H_r)[\mathbf{b}]^r_{\mathbf{0}}\,. \tag{2.16}$$

Thus each of $\overline{\sigma}_+(t)$, $\sigma_+(t)$ defines certain homomorphism $B_r \to B_r[[t]]$, which we denote in the same way.

For the HS–derivations generated by the shift operator σ_{-1} of Section 2.2, we use the indeterminate t^{-1} instead of t, and denote them by $\sigma_-(t^{-1})$, $\overline{\sigma}_-(t^{-1})$. The corresponding homomorphisms are defined via

$$(\sigma_-(t^{-1})\Delta_{\boldsymbol{\lambda}}(H_r))[\mathbf{b}]^r_0 = \sigma_-(t^{-1})[\mathbf{b}]^r_{\boldsymbol{\lambda}}, \quad (\overline{\sigma}_-(t^{-1})\Delta_{\boldsymbol{\lambda}}(H_r))[\mathbf{b}]^r_0 = \overline{\sigma}_-(t^{-1})[\mathbf{b}]^r_{\boldsymbol{\lambda}}.$$

Definition of σ_{-1} implies that $\sigma_-(t^{-1})b_i = b_i + b_{i-1}t^{-1} + b_{i-2}t^{-2} + \cdots + b_1 t^{1-i}$ is a polynomial of t^{-1} for each $i > 0$. It follows that $\sigma_-(t^{-1})$, $\overline{\sigma}_-(t^{-1})$ also send all $\Delta_{\boldsymbol{\lambda}}(H_r)$'s to B_r-polynomials of t^{-1}.

Now we are ready to define the homomorphisms $\Gamma_r(t), \Gamma_r^*(t) : B_r \to B_r[t^{-1}, t]]$ by their values on $\Delta_{\boldsymbol{\lambda}}(H_r)$'s, as follows,

$$\begin{aligned}\Gamma_r(t)\left(\Delta_{\boldsymbol{\lambda}}(H_r)\right) &= \frac{1}{E_r(t)}\left(\overline{\sigma}_-(t^{-1})\Delta_{\boldsymbol{\lambda}}(H_r)\right),\\ \Gamma_r^*(t)\left(\Delta_{\boldsymbol{\lambda}}(H_r)\right) &= E_r(t)\cdot\left(\sigma_-(t^{-1})\Delta_{\boldsymbol{\lambda}}(H_r)\right).\end{aligned}$$

For $r_1 < r_2$, the natural projection $B_{r_2} \to B_{r_1}$ sending each of $e_{r_1+1}, \ldots, e_{r_2}$ to zero, sends $E_{r_2}(t)$ to $E_{r_1}(t)$, $H_{r_2}(t)$ to $H_{r_1}(t)$, and $X_{r_2}(t)$ to $X_{r_1}(t)$. In this sense, $E_r(t) \to E_\infty(t)$, $H_r(t) \to H_\infty(t)$, $X_r(t) \to X_\infty(t)$ as $r \to \infty$.

Thus, $\Gamma_r(t)$ and $\Gamma_r^*(t)$ tend to $\Gamma(t)$ and $\Gamma^*(t)$ when $r \to \infty$.

3. Cayley–Hamilton Theorem revisited

3.1. Hasse-Schmidt derivations on exterior algebras [3, 6]. Let A be a commutative ring with unit, M a free A-module of rank r, and $b_1, \ldots, b_r$ some A-basis of M.

Set $\bigwedge^0 M = A$. For $1 \le j \le r$, denote by $\bigwedge^j M$ the A-module generated by all $b_{i_1} \wedge \ldots \wedge b_{i_j}$ modulo permutations,

$$b_{i_{\tau(1)}} \wedge \ldots \wedge b_{i_{\tau(j)}} = \operatorname{sgn}(\tau) b_{i_1} \wedge \ldots \wedge b_{i_j},$$

where $\operatorname{sgn}(\tau)$ is the sign of permutation τ. In particular, $\bigwedge^1 M = M$.

The exterior algebra $\bigwedge M = \bigoplus_{j=0}^r \bigwedge^j M$ possesses the natural graduation and the byproduct is given by juxtaposition $\wedge : \bigwedge^i M \times \bigwedge^j M \to \bigwedge^{i+j} M$.

We denote by $(\bigwedge M)[[t]]$ the ring of formal power series of t with coefficients in $\bigwedge M$, and by $(\operatorname{End}_A(\bigwedge M))[[t]]$ the ring of formal power series of t with coefficients in $\operatorname{End}_A(\bigwedge M)$.

For $\mathcal{D}(t) = \sum_{i\ge 0} \mathcal{D}_i t^i$, $\mathcal{D}(t) = \sum_{j\ge 0} \widetilde{\mathcal{D}}_j t^j \in (\operatorname{End}_A(\bigwedge M))[[t]]$, their product is defined as follows,

$$\mathcal{D}(t)\widetilde{\mathcal{D}}(t)u = \mathcal{D}(t)\sum_{j\ge 0}\widetilde{\mathcal{D}}_j u\cdot t^j = \sum_{j\ge 0}(\mathcal{D}(t)\tilde{\mathcal{D}}_j u)\cdot t^j, \quad \forall u \in \bigwedge M.$$

Given series $\mathcal{D}(t)$, we use the same notation for the induced A-homomorphism,

$$\mathcal{D}(t) : \bigwedge M \to \bigwedge M[[t]], \quad u \mapsto \mathcal{D}(t)u = \sum_{i\ge 0}\mathcal{D}_i u\cdot t^i, \quad \forall u \in \bigwedge M.$$

The series $\mathcal{D}(t) = \sum_{i\ge 0}\mathcal{D}_i t^i$ is *invertible* in $(\operatorname{End}_A(\bigwedge M))[[t]]$, if there exists $\overline{\mathcal{D}}(t) \in (\operatorname{End}_A(\bigwedge M))[[t]]$ such that

$$\mathcal{D}(t)\overline{\mathcal{D}}(t) = \overline{\mathcal{D}}(t)\mathcal{D}(t) = \mathbf{1}_{\bigwedge M}. \tag{3.1}$$

We call $\overline{\mathcal{D}}(t)$ the *inverse* series and write it in the form $\overline{\mathcal{D}}(t) = \sum_{i\ge 0}(-1)^i\overline{\mathcal{D}}_i t^i$. Then (3.1) is equivalent to

$$\mathcal{D}_j - \overline{\mathcal{D}}_1\mathcal{D}_{j-1} + \ldots + (-1)^j\overline{\mathcal{D}}_j = 0, \ \forall j \ge 1. \tag{3.2}$$

One can check that $\mathcal{D}(t)$ is invertible if and only if $\mathcal{D}_0$ is an automorphism of $\bigwedge M$.

PROPOSITION 3.1. The following two statements are equivalent:

i) $\mathcal{D}(t)(u \wedge v) = \mathcal{D}(t)u \wedge \mathcal{D}(t)v, \ \ \forall u, v \in \bigwedge M;$

ii) $\mathcal{D}_i(u \wedge v) = \sum_{j=0}^i \mathcal{D}_j u \wedge \mathcal{D}_{i-j} v, \ \ \forall u, v \in \bigwedge M, \ \ \forall i \ge 0.$

PROOF. $i) \Rightarrow ii)$ By definition of $\mathcal{D}(t)$, one can write $i)$ as

$$\sum_{i\geq 0}\mathcal{D}_i(u\wedge v)t^i = \sum_{j_1\geq 0}\mathcal{D}_{j_1}u\cdot t^{j_1}\wedge\sum_{j_2\geq 0}\mathcal{D}_{j_2}v\cdot t^{j_2}. \tag{3.3}$$

Hence $\mathcal{D}_i(u\wedge v)$ is the coefficient of t^i on the right hand side of (3.3), which is $\sum_{j_1+j_2=i}\mathcal{D}_{j_1}u\wedge\mathcal{D}_{j_2}v = \sum_{j=0}^{i}\mathcal{D}_j u\wedge\mathcal{D}_{i-j}v$.

$ii) \Rightarrow i)$ We have

$$\begin{aligned}\mathcal{D}(t)(u\wedge v) &= \sum_{i\geq 0}\mathcal{D}_i(u\wedge v)t^i = \sum_{i\geq 0}\left(\sum_{i_1+i_2=j}\mathcal{D}_{i_1}u\wedge\mathcal{D}_{i_2}v\right)t^i\\ &= \sum_{i\geq 0}\left(\sum_{i_1}\mathcal{D}_{i_1}u\cdot t^{i_1}\wedge\sum_{i_2}\mathcal{D}_{i_2}v\cdot t^{i_2}\right) = \mathcal{D}(t)u\wedge\mathcal{D}(t)v. \quad\square\end{aligned}$$

DEFINITION 3.2. (Cf. [**3**]) Let $\mathcal{D}(t)\in(\mathrm{End}_A(\bigwedge M))[[t]]$. The induced map $\mathcal{D}(t):\bigwedge M\to(\bigwedge M)[[t]]$ is called a *Hasse–Schmidt derivation* (or, for brevity, a *HS–derivation*) on $\bigwedge M$, if it satisfies the (equivalent) conditions of Proposition 3.1.

PROPOSITION 3.3. *(Cf.* [**3**], [**6**]*)* The product of two *HS*–derivations on $\bigwedge M$ is a HS–derivation. The inverse of a HS–derivation is a HS–derivation.

PROOF. For the product of HS–derivations $\mathcal{D}(t)$ and $\tilde{\mathcal{D}}(t)$, the statement $i)$ of Proposition 3.1 holds. Indeed, $\forall u,v\in\bigwedge M$,

$$\begin{aligned}\mathcal{D}(t)\tilde{\mathcal{D}}(t)(u\wedge v) &= \mathcal{D}(t)\left(\sum_{j\geq 0}\sum_{j_1+j_2=j}\tilde{\mathcal{D}}_{j_1}u\wedge\tilde{\mathcal{D}}_{j_2}v\right)t^j\\ &= \sum_{j\geq 0}\sum_{j_1+j_2=j}\mathcal{D}(t)D_{j_1}u\cdot t^{j_1}\wedge\mathcal{D}(t)\mathcal{D}_{j_2}v\cdot t^{j_2}\\ &= \mathcal{D}(t)\tilde{\mathcal{D}}(t)u\wedge\mathcal{D}(t)\tilde{\mathcal{D}}(t)v.\end{aligned}$$

Similarly, if $\overline{\mathcal{D}}(t)$ is the inverse of the HS–derivation $\mathcal{D}(t)$, then $\forall u,v\in\bigwedge M$,

$$\begin{aligned}\overline{\mathcal{D}}(t)(u\wedge v) &= \overline{\mathcal{D}}(t)(\mathcal{D}(t)\overline{\mathcal{D}}(t)u\wedge\mathcal{D}(t)\overline{\mathcal{D}}(t)v)\\ &= (\overline{\mathcal{D}}(t)\mathcal{D}(t))(\overline{\mathcal{D}}(t)u\wedge\overline{\mathcal{D}}(t)v)\\ &= \overline{\mathcal{D}}(t)u\wedge\overline{\mathcal{D}}(t)v. \quad\square\end{aligned}$$

COROLLARY 3.4. [**6**] *If $\overline{\mathcal{D}}(t)$ is the inverse of a HS–derivation $\mathcal{D}(t)$, then*

$$\begin{aligned}\mathcal{D}(t)u\wedge v &= \mathcal{D}(t)u\wedge\mathcal{D}(t)\overline{\mathcal{D}}(t)v = \mathcal{D}(t)(u\wedge\overline{\mathcal{D}}(t)v),\\ u\wedge\overline{\mathcal{D}}(t)v &= \overline{\mathcal{D}}(t)\mathcal{D}(t)u\wedge\overline{\mathcal{D}}(t)v = \overline{\mathcal{D}}(t)(\mathcal{D}(t)u\wedge v)\end{aligned} \tag{3.4}$$

for all $u,v\in\bigwedge M$. Equivalently, for any $k\geq 1$,

$$\begin{aligned}\mathcal{D}_k u\wedge v &= \mathcal{D}_k(u\wedge v)-\mathcal{D}_{k-1}(u\wedge\overline{\mathcal{D}}_1 v)+\ldots+(-1)^k u\wedge\overline{\mathcal{D}}_k v,\\ u\wedge\overline{D}_k v &= \overline{D}_k(u\wedge v)-\overline{D}_{k-1}(D_1 u\wedge v)+\ldots+(-1)^k D_k u\wedge v\end{aligned} \tag{3.5}$$

3.2. Proof of the Theorem 2.3. As we have seen in Proposition 2.1, any endomorphism $f\in\mathrm{End}_A(M)$ defines two graded mutually inverse HS–derivations,

$$\overline{D}(t) = \bigwedge(\mathbf{1}-ft) \quad\text{and}\quad D(t) = \bigwedge\left(\sum_{i\geq 0}f^i t^i\right),$$

where $\mathbf{1}$ denotes the identity endomorphism. Write $\overline{D}(t)$ and $D(t)$ in the form

$$\overline{D}(t) = \sum_{i\geq 0}(-1)^i\overline{D}_i t^i \quad \text{and} \quad D(t) = \sum_{i\geq 0} D_i t^i,$$

then these HS–derivations satisfy the following properties.

LEMMA 3.5. *We have*
(i) $\overline{D}_0 = D_0 = \mathbf{1}_{\bigwedge M}$ *and* $D_1 = \overline{D}_1$,
(ii) $D_i\,|_M = f^i$, $i \geq 0$,
(iii) $\overline{D}_k u = 0$, *for all* $u \in \bigwedge^i M$ *with* $i < k$.

Indeed, $\overline{D}(t)\,|_{\bigwedge^k M}$ is a polynomial of t of degree k, $1 \leq k \leq r$. □

As before, we assume that our A-module M is freely generated by $(b_j)_{1\leq j\leq r}$. Thus $\bigwedge^r M$ has rank 1 and is spanned by $[\mathbf{b}]^r_{\mathbf{0}} = b_1 \wedge \cdots \wedge b_r$. The restriction of each $\overline{D}_i$ to $\bigwedge^r M$ is a multiplication by some scalar $e_i \in A$,

$$\overline{D}_i\left([\mathbf{b}]^r_{\mathbf{0}}\right) = e_i[\mathbf{b}]^r_{\mathbf{0}}, \quad 1 \leq i \leq r. \tag{3.6}$$

Take now $u \in \bigwedge^i M$ and $v \in \bigwedge^{r-i} M$. Then $D_j u \wedge v \in \bigwedge^r M$ for $1 \leq j \leq k$. Applying (3.5) to our situation, we can write

$$D_k u \wedge v - e_1(D_{k-1}u \wedge v) + \cdots + (-1)^k e_k(u \wedge v) = (-1)^k u \wedge \overline{D}_k v$$

for $1 \leq k \leq r$, and

$$D_k u \wedge v - e_1(D_{k-1}u \wedge v) + \cdots + (-1)^r e_r(D_{k-r}u \wedge v) = (-1)^k u \wedge \overline{D}_k v$$

for $k > r$.

Equivalently, we have

$$\left(D_k u - e_1 D_{k-1}u + \cdots + (-1)^k e_k u\right) \wedge v = (-1)^k u \wedge \overline{D}_k v, \quad 1 \leq k \leq r, \tag{3.7}$$

and

$$(D_k u - e_1 D_{k-1}u + \cdots + (-1)^r e_r D_{k-r}u) \wedge v = (-1)^k u \wedge \overline{D}_k v, \quad k > r. \tag{3.8}$$

Of course, one can set $e_k = 0$ for $k > r$, in order do not distinguish between the two cases. However, we prefer a division into cases.

Assume now $i > r - k > 0$. Then, according to Remark 3.5, *(iii)*, the right hand side of (3.7) vanishes $\forall v \in \bigwedge^{r-i} M$, as $r - i < k$. This means that

$$D_k u - e_1 D_{k-1}u + \cdots + (-1)e_k u = 0$$

for any $u \in \bigwedge^i M$ with $i > r - k > 0$. This proves the first part of Theorem 2.3.

If $k > r$, then the left hand side of (3.8) vanishes for each $i \geq 0$, and this proves the second part. □

REMARK 3.6. Thus we understand (1.3) as an abstract Cayley–Hamilton theorem valid for general invertible HS–derivations on exterior algebras of arbitrary (not necessarily free) modules. If the module is free and at most countably generated, then (1.3) produces a sequence of Cayley–Hamilton relations (3.5). This sequence turns into the classical Cayley–Hamilton formulas (2.7) when the HS–derivation corresponds to an endomorphism of a finitely generated free module.

3.3. Example. Take $M = \mathbb{R}^3$. Let $f : \mathbb{R}^3 \to \mathbb{R}^3$ have an eigenbasis,

$$f\mathbf{u} = a\mathbf{u}, \quad f\mathbf{v} = b\mathbf{v}, \quad f\mathbf{w} = c\mathbf{w}.$$

Then $\det(\mathbf{1}t - f) = t^3 - e_1t^2 + e_2t - e_3$, where

$$e_1 = a + b + c, \quad e_2 = ab + ac + bc, \quad e_3 = abc.$$

In notation of Theorem 2.3, we have $r = 3$.

1) Let us take $k = 2$ and check that $D_2 - e_1D_1 + e_2\mathbf{1}$ vanishes on $\mathbb{R}^3 \wedge \mathbb{R}^3$, as it should be, according to (2.6). We show the calculation of $(D_2 - e_1D_1 + e_2\mathbf{1})(\mathbf{u} \wedge \mathbf{v})$; for two other basis vectors $\mathbf{u} \wedge \mathbf{w}$ and $\mathbf{v} \wedge \mathbf{w}$ it is completely similar.

First, we find the action of D_1, D_2 on $\mathbf{u} \wedge \mathbf{v}$. We have

$$(\mathbf{1} + ft + f^2t^2 + \text{o}(t^2))\mathbf{u} \wedge (\mathbf{1} + ft + f^2t^2 + \text{o}(t^2))\mathbf{v}$$

$$= \mathbf{u} \wedge \mathbf{v} + (f\mathbf{u} \wedge \mathbf{v} + \mathbf{u} \wedge f\mathbf{v})t + (f^2\mathbf{u} \wedge \mathbf{v} + f\mathbf{u} \wedge f\mathbf{v} + \mathbf{u} \wedge f^2\mathbf{v})t^2 + \text{o}(t^2).$$

Thus $D_1(\mathbf{u} \wedge \mathbf{v}) = (a + b)(\mathbf{u} \wedge \mathbf{v})$, $D_2(\mathbf{u} \wedge \mathbf{v}) = (a^2 + ab + b^2)(\mathbf{u} \wedge \mathbf{v})$, and

$$(D_2 - e_1D_1 + e_2\mathbf{1})(\mathbf{u} \wedge \mathbf{v}) = (a^2 + ab + b^2 - e_1a - e_1b + e_2)(\mathbf{u} \wedge \mathbf{v}).$$

Now, we substitute the expressions for e_1, e_2, and get

$$a^2 + ab + b^2 - e_1a - e_1b + e_2 = a^2 + ab + b^2 - a^2 - ab - ac - ab - b^2 - bc + ab + ac + bc = 0.$$

2) Take $k = 4$ and check that $D_4 - e_1D_3 + e_2D_2 - e_3D_1$ vanishes on $\mathbb{R}^3 \wedge \mathbb{R}^3$. According to (2.7), this endomorphism vanishes on the whole of $\bigwedge \mathbb{R}^3$, and in fact the verification for the rest of the direct summands is simpler. Again we calculate the image of $\mathbf{u} \wedge \mathbf{v}$.

First, we obtain $D_3(\mathbf{u} \wedge \mathbf{v})$ and $D_4(\mathbf{u} \wedge \mathbf{v})$ in the standard way, writing

$$(\mathbf{1} + ft + f^2t^2 + f^3t^3 + f^4t^4 + \text{o}(t^5))\mathbf{u} \wedge (\mathbf{1} + ft + f^2t^2 + f^3t^3 + f^4t^4 + \text{o}(t^5))\mathbf{v},$$

and collecting the coefficients of t^3, t^4, respectively. We get

$$D_3(\mathbf{u}\wedge\mathbf{v}) = (a^3 + a^2b + ab^2 + b^3)(\mathbf{u}\wedge\mathbf{v}), D_4(\mathbf{u}\wedge\mathbf{v}) = (a^4 + a^3b + a^2b^2 + ab^3 + b^4)(\mathbf{u}\wedge\mathbf{v}),$$

substitute all the expressions for D_4, D_3, D_2, D_1, e_1, e_2, e_3 in terms of a, b, c into $D_4 - e_1D_3 + e_2D_2 - e_3D_1$, and safely get 0.

REMARK 3.7. In general, if $f \in \text{End}(M)$ is diagonalizable, and if $(\mathbf{v}_i)_{1 \le i \le r}$ is an eigenbasis, $f\mathbf{v}_i = x_i\mathbf{v}_i$, $1 \le i \le r$, then the vector $\mathbf{v}_1 \wedge \ldots \wedge \mathbf{v}_l \in \bigwedge^l M$ is an eigenvector of D_k with the eigenvalue which is the complete symmetric polynomial of $x_1, \ldots, x_l$ of degree k. Therefore, for a diagonalizable endomorphism our Theorem 2.3 is reduced to the following identity. Denote by $h_i(\mathbf{x}_j)$ the complete symmetric polynomial of degree i in $x_1, \ldots, x_j$, and by $e_k(\mathbf{x}_n)$ the elementary symmetric polynomial of degree k in $x_1, \ldots, x_n$. Then, for $n \ge 1$ and all $1 \le j \le n$, we have

$$h_n(\mathbf{x}_j) - e_1(\mathbf{x}_n)h_{n-1}(\mathbf{x}_j) + \ldots + (-1)^n e_n(\mathbf{x}_n) = 0.$$

One can deduce the identity, for example, from the formula (*) of [**13**, I 3 28].

This remark can be turned into a rigorous general proof, using a standard (though rather long) reasoning. Another possible way, which was suggested by our referee, is based on the Frobenius proof of the classical Cayley–Hamilton theorem for the complex matrices, [**2**]. We were not aware of that 1896 paper by Frobenius. One could translate the previous arguments into the language of matrix minors. However, our approach, through the relationship to symmetric functions, is short, easy, and, in addition, allows us to concern with the infinite-dimensional case.

3.4. The case of a $\mathbb{Q}$-algebra. If A is a $\mathbb{Q}$-algebra, then for $f \in \mathrm{End}_A M$ the exponential

$$\exp(ft) := \sum_{k\geq 0} \frac{f^k t^k}{k!} \in (\mathrm{End}_A M)[[t]]$$

is well-defined. In the ring $(\mathrm{End}_A M)[[t]]$, there is the formal derivative with respect to t,

$$y(t) = \sum_{k\geq 0} g_k t^k \;\Rightarrow\; y'(t) = \sum_{k\geq 0} k g_k t^{k-1}, \quad g_k \in \mathrm{End}_A M.$$

Recall the notation $E_r(t)$ given by (2.4), and write the characteristic polynomial of f as in (2.8). In [**4**] for any commutative ring R containing the rational numbers, *the formal Laplace transform* $L : R[[t]] \to R[[t]]$ and its inverse L^{-1} are defined as follows,

$$L \sum_{n\geq 0} a_n t^n = \sum_{n\geq 0} n! a_n t^n, \quad L^{-1} \sum_{n\geq 0} c_n t^n = \sum_{n\geq 0} c_n \frac{t^n}{n!}, \quad a_n, c_n \in R.$$

Take the *inverse formal Laplace transform of the HS–derivation* $D(t) = \sum_{i\geq 0} D_i t^i$ corresponding to $f \in \mathrm{End}_A M$,

$$D^*(t) = L^{-1} \sum_{k\geq 0} D_k t^k = \sum_{k\geq 0} \frac{D_k t^k}{k!} \in \left(\mathrm{End}_A(\textstyle\bigwedge M)\right)[[t]]. \tag{3.9}$$

Define $\mathtt{p}_k(D)$ as the coefficient of t^k in $E_r(t)D(t)$. We have

$$\mathtt{p}_0(D) = \mathbf{1},\ \mathtt{p}_1 = D_1 - e_1\mathbf{1},\ \mathtt{p}_j = D_j - e_1 D_{j-1} + \ldots + (-1)^j e_j \mathbf{1},\ 1 < j < r,$$

and

$$\mathtt{p}_{r+j}(D) = D_{r+j} - e_1 D_{r+j-1} + \ldots + (-1)^r e_r D_j = \mathbf{0}, \quad j \geq 0,$$

according to Theorem (2.3). Therefore,

PROPOSITION 3.8. *We have*

$$D(t) = \frac{\mathbf{1} + \mathtt{p}_1(D)t + \ldots + \mathtt{p}_{r-1}(D)t^{r-1}}{E_r(t)}. \tag{3.10}$$

□

COROLLARY 3.9. *Let $\mathbb{Q} \subseteq A$ and the characteristic polynomial of $f \in \mathrm{End}_A M$ be given by (2.8). Then the series $D^*(t)$ defined in (3.9) solves the ordinary differential equation*

$$y^{(r)}(t) - e_1 y^{(r-1)}(t) + \ldots + (-1)^r e_r y(t) = 0 \tag{3.11}$$

in $(\mathrm{End}_A(\bigwedge M))[[t]]$.

PROOF. Take the inverse formal Laplace transform of (3.10). We obtain

$$D^*(t) = u_0 + \mathtt{p}_1(D)u_{-1} + \ldots + \mathtt{p}_{r-1}(D)u_{-r+1},$$

where

$$u_{-j} = u_{-j}(t) = L^{-1}\left(\frac{t^j}{E_r(t)}\right), \quad 0 \leq j \leq r-1.$$

Let us re-write the series $u_0, u_{-1}, \ldots, u_{-r+1}$ in terms of $H_r(t) = 1/E_r(t)$, see Remark 2.2(1),

$$u_{-j} = L^{-1}(t^j H_r(t)) = \sum_{n\geq j} h_{n-j} \frac{t^n}{n!}, \quad 0 \leq j \leq r-1.$$

In [**7**], we proved that these series form an A-basis of solutions to the ODE (3.11) in $R[[t]]$. For $R = \mathrm{End}_A(\bigwedge M)$ we get the claim. □

3.5. Elementary remarks. We finish this section with a few remarks relevant to the case when A is a $\mathbb{Q}$-algebra.

(1) The characteristic polynomial of $f \in \mathrm{End}_A M$ is given by (2.8) if and only if $y(t) = \exp(ft)$ satisfies the linear ordinary differential equation (3.11). This is our Corollary 3.9 restricted to M.

In particular,

$$\exp(ft) = v_0(t)\mathbf{1}_M + v_1(t)f + \cdots + v_{r-1}(t)f^{r-1},$$

where $(v_j(t))_{0\le j\le r-1}$ is the *standard* A-basis of solutions to (3.11) in $A[[t]]$, that is, $v_j^{(i)}(t) = \delta_{ij}$, $0 \le i,j \le r-1$. Indeed, $\mathbf{1}_M, f, \dots, f^{r-1}$ are the initial conditions of the solution $\exp(ft)$.

In the context of endomorphisms of complex vector spaces, the formula for $\exp(ft)$ was obtained in 1966 by Putzer [**14**], and then re-obtained in 1998 by Leonard and Liz, [**11, 12**], in a different way.

(2) The relation between the standard fundamental system $v_j(t))_{0\le j\le r-1}$ and the fundamental system $u_{-j}(t)_{0\le j\le r-1}$ appeared in the proof of Corollary 3.9 is as follows. Consider the linear system of first order differential equations equivalent to our ODE (3.11),

$$y_1' = y_2,\ \ y_2' = y_3,\ \dots,\ y_{r-2}' = y_{r-1},\ \ y_{r-1}' = e_1 y_{(r-1)} - \dots + (-1)^{r-1}e_r y_1.$$

Denote the matrix of this system by P_r. Then $Q = \exp(P_r t)$ is the Wronski matrix of $v_1(t), \dots,\ v_{r-1}(t)$,

$$(Q)_{ij} = v_j^{(i)}(t),\quad 0 \le i,j \le r-1,$$

and $u_0(t), u_{-1}(t), \dots,\ u_{1-r}(t)$ is the last column of Q,

$$v_{r-1}(t) = u_{1-r}(t),\ v_{r-1}'(t) = u_{2-r}(t), \dots,\ v_{r-1}^{(r-1)}(t) = u_0(t).$$

(3) As another elementary corollary of our considerations, we get formulas for the coefficients e_k of the characteristic polynomial of $f \in \mathrm{End}_A M$ in terms of its matrix elements. If $C = (c_{ij})$ is the $r\times r$ matrix of f in some A-basis of M, denote by $D(i_1,\dots,i_k)$ the determinant of the $(r-k)\times(r-k)$-matrix obtained from the matrix C by deleting the i_1-th,…,i_k-th rows and columns. Then

$$(-1)^k e_k = \sum_{1\le i_1<\dots<i_k\le r} D(i_1,\dots,i_k),\ 1\le k\le r.$$

This formula was obtained differently by Brooks in [**1**].

4. Countably generated $\mathbb{Q}$-vector spaces

Let M_0 be a $\mathbb{Q}$-vector space generated by $(b_j)_{j\ge1}$ and $\bigwedge M_0 = \bigoplus_{r\ge0}\bigwedge^r M_0$ be its exterior algebra.

As in Section 2.2, we take shift operators $\sigma_{\pm1} \in \mathrm{End}_{\mathbb{Q}} M_0$, and denote by $\sigma_\pm(t^{\pm1}), \overline{\sigma}_\pm(t^{\pm1})$ the corresponding HS-derivations. Below we will write

$$\sigma_+(t) = \sum_{j\ge0}\sigma_j t^j,\quad \overline{\sigma}_+(t) = \sum_{j\ge0}\overline{\sigma}_j t^j, \tag{4.1}$$

skipping sign + in the subscript.

In this section, we treat $e_1, \ldots, e_r$ as indeterminates. As we already pointed out in Section 2.2, the ring B_r, given by (2.10), has a basis formed by Schur polynomials $\Delta_{\boldsymbol{\lambda}}(H_r)$, see (2.13). The structure of a principal B_r-module on $\bigwedge^r M_0$ is defined via any of the two equivalent equalities (2.11) and (2.12).

Let $(\beta_i)_{i\geq 1}$ be linear forms on M_0 defined by $\beta_i(b_j) = \delta_{ij}$. Their linear span is, by definition, the restricted dual M_0^*. Each β_j induces a $\mathbb{Q}$-linear contraction map $\beta_j : \bigwedge^r M_0 \to \bigwedge^{r-1} M_0$ defined by $\beta_j \lrcorner m = \beta_j(m)$ for all $m \in M_0$ and

$$\beta_j \lrcorner (m \wedge \eta) = \beta_j(m)\eta - m \wedge \beta_j \lrcorner \eta. \tag{4.2}$$

As each $\zeta \in \bigwedge M_0$ is a sum of homogeneous elements of the form $m \wedge \eta$, equation (4.2) defines the contraction operator over the entire exterior algebra $\bigwedge M_0$.

LEMMA 4.1. *Let $m, m' \in M_0$ satisfy*

$$m \wedge \eta = m' \wedge \eta \tag{4.3}$$

for all $\eta \in \bigwedge^{r-1} M_0$. Then $m = m'$.

PROOF. Under the hypothesis (4.3), suppose first that $m' = am$ for some $a \neq 1$. If $m \neq 0$, then there is $\mu \in M_0^*$ such that $\mu(m) \neq 0$. Because of the isomorphism $\bigwedge^{r-1} M_0 \cong M_0^*$, there is then $\eta \in \bigwedge^{r-1} M_0$ such that $m \wedge \eta \neq 0$. We get $m' \wedge \eta = a(m \wedge \eta)$, hence $m \wedge \eta \neq m' \wedge \eta$.

If m and m' are not proportional, take their duals $\mu, \mu' \in M_0^*$ and choose $\zeta_{m,m'} = \mu' \lrcorner (\mu \lrcorner [\mathbf{b}]_0^r) \in \bigwedge^{r-1} M_0$. Then $\mu(m) = \mu'(m') = 1$ and so $m \wedge m \wedge \zeta_{m,m'} = 0$ while $m' \wedge m \wedge \zeta_{m,m'} \neq 0$. $\square$

COROLLARY 4.2. *For each e_i, $1 \leq i \leq r$, there exists the unique mapping $M_0 \to M_0$, called the multiplication by e_i, $m \mapsto e_i m$, such that*

$$(e_i m) \wedge \eta = e_i(m \wedge \eta), \tag{4.4}$$

for all $\eta \in \bigwedge^{r-1} M_0$.

The vector space M_0 endowed with the multiplications by e_i, $1 \leq i \leq r$, becomes a B_r-module, denoted by M_r.

PROPOSITION 4.3. *The B_r-module M_r is freely generated by $(b_i)_{1\leq i\leq r}$, and $\sigma_+(t) \in End_{B_r}(M_r)[[t]]$.*

PROOF. Let us check first that $b_1, b_2, \ldots, b_r$ are B_r-linearly independent. Denote by $(\beta_j)_{j\geq 0}$ the generators of M_0^* dual to $(b_i)_{i\geq 1}$. Notice that $a_1 b_1 + \cdots + a_r b_r = 0$ implies $0 = a_i b_i \wedge \eta_i$, where $\eta_i = \beta_i \lrcorner [\mathbf{b}]_0^r$, that is, $a_i = 0$ for all $1 \leq i \leq r$.

Now, let us show that $b_{i+r} - e_1 b_{i+r-1} + \cdots + e_r b_i = 0$ for all $i \geq 0$. This will prove, by induction, that M_r is generated over B_r by $b_1, b_2, \ldots, b_r$. It is enough to observe that

$$\sum_{i=1}^{\infty} (b_i - e_1 b_{i-1} + (-1)^r e_r b_{r-1+i}) t^i = E_r(t)\sigma_+(t) b_1$$

is a polynomial of degree r (here we set $b_j = 0$ for $j < 1$). By definition of the module structure, for each $\eta \in \bigwedge^{r-1} M_0$ we have

$$\begin{aligned}
E_r(t)(\sigma_+(t)b_0)\wedge\eta &= E_r(t)(\sigma_+(t)b_0\wedge\eta) && \text{(definition of the } B_r\text{-module structure)}\\
&= E_r(t)\sigma_+(t)(b_0\wedge\overline{\sigma}_+(t)\eta) && \text{(Proposition 3.4)}\\
&= E_r(t)\frac{1}{E_r(t)}(b_0\wedge\overline{\sigma}_+(t)\eta) && \text{(definition of the } B_r\text{-module structure)}\\
&= b_0\wedge\overline{\sigma}_+(t)\eta.
\end{aligned}$$

We use now the agreement (4.1). As in Remark 3.5, we see that $\overline{\sigma}_r$ vanishes on $\bigwedge^{r-1} M_0$, hence the expression obtained above is a polynomial in t of degree $r-1$. We have so proven that M_r is a B_r-module of rank r. Moreover σ_1 is B_r-linear. In fact,

$$\sigma_1(e_im)\wedge\eta = \sigma_1(e_im\wedge\eta)-e_im\wedge\sigma_1\eta = e_i\sigma_1(m\wedge\eta)-e_im\wedge\sigma_1\eta = e_1\sigma_1 m\wedge\eta. \qquad \square$$

COROLLARY 4.4. *The elements e_i, $1\le i\le r$, and h_j, $j\ge 0$, of B_r are the eigenvalues of $\overline{\sigma}_i$ and σ_j, respectively, thought of as endomorphisms of $\bigwedge^r M_r$.* $\square$

Similarly to (2.15) and (2.16), the mappings $\sigma_-(t^{-1})$ and $\overline{\sigma}_-(t^{-1})$ define two homomorphisms $B_r\to B_r[t^{-1}]$, via the equalities

$$(\sigma_-(t^{-1})\Delta_{\boldsymbol\lambda}(H_r))[\mathbf{b}]_0^r := \sigma_-(t^{-1})[\mathbf{b}]_{\boldsymbol\lambda}^r, \tag{4.5}$$

and

$$(\overline{\sigma}_-(t^{-1})\Delta_{\boldsymbol\lambda}(H_r))[\mathbf{b}]_0^r := \overline{\sigma}_-(t^{-1})[\mathbf{b}]_{\boldsymbol\lambda}^r. \tag{4.6}$$

By abuse of notation, we denote the homomorphisms in the same way.

5. Bosonic vertex operators

5.1. Let $B:=\mathbb{Q}[x_1,x_2,\ldots]$ be the polynomial ring in infinitely many indeterminates and $M_0:=\bigoplus_{i\ge 0}\mathbb{Q}b_i$. The aim of this section is to show that the *bosonic vertex operators*

$$\Gamma(t):=\exp(\sum_{i\ge1}x_it^i)\cdot\exp\left(-\sum_{i\ge1}\frac{1}{it^i}\frac{\partial}{\partial x_i}\right): B\to B[t^{-1},z]]$$

and

$$\Gamma^*(t):=\exp(-\sum_{i\ge1}x_it^i)\cdot\exp\left(\sum_{i\ge1}\frac{1}{it^i}\frac{\partial}{\partial x_i}\right): B\to B[t^{-1},z]]$$

may be identified with ratios of characteristic series operators associated to the shift endomorphisms of step ±1 of M_0.

Let $\Gamma_r(t),\Gamma_r^*(t): B_r\to B_r[t^{-1},t]]$ be defined by

$$\Gamma_r(t)\Delta_{\boldsymbol\lambda}(H_r)[\mathbf{b}]_0^r := \sigma_+(t)\overline{\sigma}_-(t^{-1})\Delta_{\boldsymbol\lambda}(H_r))[\mathbf{b}]_0^r$$

and

$$\Gamma_r^*(t)\Delta_{\boldsymbol\lambda}(H_r)[\mathbf{b}]_0^r := \overline{\sigma}_+(t)\sigma_-(t^{-1})\Delta_{\boldsymbol\lambda}(H_r)[\mathbf{b}]_0^r.$$

Then, due to (2.15) and (2.16) one can write:

$$\Gamma_r(t)=\frac{1}{E_r(t)}\cdot\overline{\sigma}_-(t^{-1}) \qquad \text{and} \qquad \Gamma_r^*(t)=E_r(t)\cdot\frac{1}{\overline{\sigma}_-(t^{-1})}\,.$$

REMARK 5.1. Notice that $E_r(t)$ is indeed the characteristic polynomial of σ_1, thought of as endomorphism of M_r, and $\overline{\sigma}_-(t^{-1})$ is the characteristic series operator associated to σ_{-1}.

PROPOSITION 5.2. [**5**] *The operators* $\Gamma_r(t), \Gamma_r^*(t)$ *tend to* $\Gamma(t), \Gamma^*(t)$ *as* r *goes to infinity.*

PROOF. We sketch the arguments of [**5**]. First of all, notice that for all $r \geq 1$

$$\overline{\sigma}_-(t^{-1})h_n = h_n - \frac{h_{n-1}}{t} \quad \text{and} \quad \sigma_-(t^{-1})h_n = \sum_{i\geq 0} \frac{h_{n-i}}{t^i}. \tag{5.1}$$

Let $\overline{\sigma}_-(t^{-1})H_r = (\sigma_-(t^{-1})h_n)_{n\in\mathbb{Z}}$ and $\sigma_-(t^{-1})H_r = (\sigma_-(t^{-1})h_n)_{n\in\mathbb{Z}}$. Then

$$\overline{\sigma}_-(t^{-1})\Delta_{\boldsymbol{\lambda}}(H_r) = \Delta_{\boldsymbol{\lambda}}(\overline{\sigma}_-(t^{-1})H_r), \quad \sigma_-(t^{-1})\Delta_{\boldsymbol{\lambda}}(H_r) = \Delta_{\boldsymbol{\lambda}}(\sigma_-(t^{-1})H_r), \tag{5.2}$$

by [**5**, Propositions 6.2.10 and 6.2.13]. Now, according to [**5**, Corollaries 6.2.11 and 6.2.14],

$$\sigma_-(t^{-1})(h_{i_1} \cdot \ldots \cdot h_{i_r}) = \sigma_-(t^{-1})h_{i_1} \cdot \ldots \cdot \sigma_-(t^{-1})h_{i_r} \tag{5.3}$$

and

$$\overline{\sigma}_-(t^{-1})(h_{i_1} \cdot \ldots \cdot h_{i_r}) = \overline{\sigma}_-(t^{-1})h_{i_1} \cdot \ldots \cdot \overline{\sigma}_-(t^{-1})h_{i_r}. \tag{5.4}$$

Clearly formulas (5.1) and (5.2) do not depend on r when r is big enough (that is, at least the length of the partition $\boldsymbol{\lambda}$). Thus these formulas hold for $r = \infty$ as well. We set

$$E_\infty(t) = 1 - e_1 z + e_2 t^2 + \cdots \quad \text{and} \quad H_\infty(t) = 1/E_\infty(t).$$

Define now

$$\exp\left(\sum_{i\geq 1} x_i t^i\right) := \sum_{n\geq 0} h_n t^n.$$

Then we have

$$B = \mathbb{Q}[e_1, e_2, \ldots] = \mathbb{Q}[h_1, h_2, \ldots] = \mathbb{Q}[x_1, x_2, \ldots],$$

see, for example, [**13**, I 3]. Moreover,

$$\frac{\partial h_n}{\partial x_i} = h_{n-i} \quad \text{and} \quad \frac{\partial^j h_n}{\partial x_1^j} = \frac{\partial h_n}{\partial x_j}. \tag{5.5}$$

As it follows from (5.3) and (5.4), $\overline{\sigma}_-(t^{-1})$ and $\sigma_-(t^{-1})$, for $r = \infty$, become ring homomorphisms $B \to B[t^{-1}]$. Thus

$$\begin{aligned}
\overline{\sigma}_-(t^{-1})h_n &= h_n - \frac{h_{n-1}}{t} && \text{(first formula in (5.1))}\\
&= \left(1 - \frac{1}{z}\frac{\partial}{\partial x_1}\right) h_n\\
&= \exp\left(-\sum_{i\geq 1}\frac{1}{it}\frac{\partial^i}{\partial x_1^i}\right) h_n && \text{(definition of the logarithm}\\
& && \text{of a formal power series)}\\
&= \exp\left(-\sum_{i\geq 1}\frac{1}{it}\frac{\partial}{\partial x_i}\right) h_n && \text{(second equality in (5.5)).}
\end{aligned}$$

Notice that $\exp\left(-\sum_{i\geq 1}\frac{1}{it}\frac{\partial}{\partial x_i}\right)$, being the exponent of a first order differential operator, is a ring homomorphism whose value at h_n coincides with $\overline{\sigma}_-(t^{-1})h_n$. This means that

$$\overline{\sigma}_-(t^{-1}) = \exp\left(-\sum_{i\geq 1}\frac{1}{it}\frac{\partial}{\partial x_i}\right).$$

Similarly one shows that

$$\sigma_-(t^{-1}) = \exp\left(\sum_{i\geq 1}\frac{1}{it}\frac{\partial}{\partial x_i}\right).$$

Thus

$$\Gamma_\infty(t) = \frac{1}{E_\infty(t)}\overline{\sigma}_-(t) = \exp\left(\sum_{i\geq 1} x_i t^i\right)\exp\left(-\sum_{i\geq 1}\frac{1}{it}\frac{\partial}{\partial x_i}\right) = \Gamma(t)$$

and

$$\Gamma^*_\infty(t) = E_\infty(t)\sigma_-(t^{-1}) = \exp\left(-\sum_{i\geq 1} x_i t^i\right)\exp\left(\sum_{i\geq 1}\frac{1}{it}\frac{\partial}{\partial x_i}\right) = \Gamma^*(t)$$

as claimed. □

Acknowledgments

We thank the referees for their efforts reading the first version of the paper. Their sharp criticism really contributed to the improvement of the article. In particular, we appreciate the help of the referee who found a mistake in our text. Her/his suggestions allowed us to simplify our proof. We are also grateful to the other referee who demanded more motivations.

References

[1] B. P. Brooks, *The coefficients of the characteristic polynomial in terms of the eigenvalues and the elements of an $n \times n$ matrix*, Appl. Math. Lett. **19** (2006), no. 6, 511–515, DOI 10.1016/j.aml.2005.07.007. MR2221507

[2] G. Frobenius, *ÜvertauschbareMatrizen*, Sitzungsberichte der Preussischen Akademie der Wissenschaften zu Berlin, Reichsdr., 1896. Available at `http://www.e-rara.ch/zut/content/titleinfo/5929398`

[3] L. Gatto, *Schubert calculus via Hasse-Schmidt derivations*, Asian J. Math. **9** (2005), no. 3, 315–321, DOI 10.4310/AJM.2005.v9.n3.a2. MR2214955

[4] L. Gatto and D. Laksov, *From linear recurrence relations to linear ODEs with constant coefficients*, J. Algebra Appl. **15** (2016), no. 6, 1650109, 23, DOI 10.1142/S0219498816501097. MR3479813

[5] L. Gatto and P. Salehyan, *Hasse-Schmidt derivations on Grassmann algebras: With applications to vertex operators*, IMPA Monographs, vol. 4, Springer, [Cham], 2016. MR3524604

[6] L. Gatto and T. Santiago, *Schubert calculus on a Grassmann algebra*, Canad. Math. Bull. **52** (2009), no. 2, 200–212, DOI 10.4153/CMB-2009-023-x. MR2512308

[7] L. Gatto and I. Scherbak, *"On one property of one solution of one equation" or linear ODE's, Wronskians and Schubert calculus* (English, with English and Russian summaries), Mosc. Math. J. **12** (2012), no. 2, 275–291, 460. MR2978757

[8] F. K. Schmidt and H. Hasse, *Noch eine Begründung der Theorie der höheren Differentialquotienten in einem algebraischen Funktionenkörper einer Unbestimmten. (Nach einer brieflichen Mitteilung von F.K. Schmidt in Jena)* (German), J. Reine Angew. Math. **177** (1937), 215–237, DOI 10.1515/crll.1937.177.215. MR1581557

[9] V. G. Kac and A. K. Raina, *Bombay lectures on highest weight representations of infinite-dimensional Lie algebras*, Advanced Series in Mathematical Physics, vol. 2, World Scientific Publishing Co., Inc., Teaneck, NJ, 1987. MR1021978

[10] D. Laksov, *Linear maps and linear recursion* (Norwegian, with English summary), Normat **55** (2007), no. 2, 59–77, 96. MR2355290

[11] I. E. Leonard, *The matrix exponential*, SIAM Rev. **38** (1996), no. 3, 507–512, DOI 10.1137/S0036144595286488. MR1409596

[12] E. Liz, *A note on the matrix exponential*, SIAM Rev. **40** (1998), no. 3, 700–702, DOI 10.1137/S0036144596320752. MR1642760

[13] I. G. Macdonald, *Symmetric functions and Hall polynomials*, Oxford Mathematical Monographs, The Clarendon Press, Oxford University Press, New York, 1979. MR553598

[14] E. J. Putzer, *Avoiding the Jordan canonical form in the discussion of linear systems with constant coefficients*, Amer. Math. Monthly **73** (1966), 2–7, DOI 10.2307/2313914. MR0194636

[15] F. K. Schmidt, *Die Wronskische Determinante in beliebigen differenzierbaren Funktionenkörpern* (German), Math. Z. **45** (1939), no. 1, 62–74, DOI 10.1007/BF01580273. MR1545804

Dipartimento di Scienze Matematiche, Politecnico di Torino, C.so Duca degli Abruzzi, 24, 10129 - Torino, Italia

Email address: `letterio.gatto@polito.it`

School of Mathematical Sciences, Raymond & Beverly Sackler Faculty of Exact Sciences, Tel Aviv University, P.O. Box 39040, Tel Aviv 6997801, Israel

Email address: `scherbak@post.tau.ac.il`

Contemporary Mathematics
Volume **733**, 2019
https://doi.org/10.1090/conm/733/14740

Complex analysis on the real sphere, or variations on a Maxwell's theme

Simon Gindikin

In memoriam of Selim Grigorievich Krein, who loved life and mathematics

ABSTRACT. Maxwell has related spherical polynomials to homogeneous polynomials on the complex quadratic cone. We interpret this duality using an analogue of the Radon transform - the horospherical Cauchy transform on the sphere, which is different from the Minkowski-Funk transform on the sphere.

Introduction

I had known Selim Grigorievich for more than 50 years and recall with enormous pleasure his fantastic charm. S.G. was a very original mathematician, with very broad interests. He played an outstanding role in the phenomenon for which there does not exist a better name than "Soviet Mathematics". The mathematical life in the fSU was hyper centralized in the capitals, but thanks to S.G. and M.A.Krasnoselski Voronezh was a singular point on the mathematical map. The high point was Voronezh Winter Mathematical School. Every January many mathematicians would come there and enjoy Mathematics and warm friendly interactions, which were expertly lead by Selim Grigorievich, who reigned everywhere - from meetings to dances. I was happy to collaborate with him for several years organizing "Summer satellites" of the winter school - conferences on functional analysis and differential equations. I will never forget our last meeting at the ICM Zürich, how he was gladden by our unexpected encounter. I consider the short etude below as my virtual talk at the Voronezh School.

1. Dual spherical polynomials (following Maxwell)

Let $S^n \subset \mathbb{R}^{n+1}$ be the n-dimensional sphere:

$$\square(x) := x_1^2 + \cdots + x_{n+1}^2 = 1.$$

Spherical (harmonic) polynomials are eigenfunctions of the spherical Laplacian on S^n. The equivalent definition is that they are restrictions to S^n of harmonic homogeneous polynomials in $\mathbb{R}^{n+1}$. Degrees of these polynomials are defined by the eigenvalues corresponding to the spherical polynomials.

2010 *Mathematics Subject Classification.* Primary 32A45, 33C55, 43A75, 44A12.

Maxwell in his famous 1873 "Treatise on Electricity and Magnetism" related spherical polynomials to homogeneous polynomials on the (complex!) cone

$$\Xi := \{\zeta \in \mathbb{C}^{n+1} | \Box(\zeta) = 0\}.$$

One can find a detailed exposition of this construction in [**1**]. Of course, it is very impressive that in the middle of XIX century Maxwell appealed to a complex construction in a very classical real situation. I have not been able to figure out what was his background geometric picture that had led to such a relation. I suggest below such an interpretation.

Spherical polynomials can be extended to homogeneous harmonic polynomials on $\mathbb{R}^{n+1}$ and then on $\mathbb{C}^{n+1}$ as well. The restriction to the cone Ξ produces homogeneous polynomials, while condition of harmonicity disappears. Indeed, it is possible to see that homogeneous polynomials on Ξ admit unique harmonic extensions on $\mathbb{C}^{n+1}$. Such a cone exists only in the complex domain. Maxwell gave an algebraic formula for spherical polynomials through the homogeneous polynomials on the complex cone, which can be found in [**1**].

We can extend this interpretation. While S^n has no real boundary, we can interpret Ξ as a line bundle over the "complex boundary" of S^n , homogeneous polynomial on Ξ as sections of line bundles on this boundary, and Maxwell's dual spherical polynomials as "boundary values" of spherical polynomials on it. Correspondingly, it would be natural to replace the algebraic Maxwell formula with an analytic formula that one could interpret as a version of the Poisson type formula for the compact manifold S^n and complex "boundary values." It is the goal of this note.

Let us jump to the XX-th century and look at Maxwell's construction from the point of view of group representation theory. Irreducible representations of the orthogonal group $SO(n+1)$, as well as of its complexification $SO(n+1;\mathbb{C})$, can be realized for fixed eigenvalues in subspaces of spherical polynomials. The equivalent representations of $SO(n+1)$ or, equivalently, of its complexification are realized in corresponding subspaces of dual spherical polynomials. Maxwell's formula gives an intertwining operator between these equivalent representations.

We will construct an intertwining operator that not only replaces the algebraic construction of Maxwell with an analytic one, but acts on $C^\infty(S^n)$, automatically intertwining subspaces of spherical polynomials and the dual ones and is universal for all eigenvalues. And lo and behold, it happens to be a version of the Radon transform.

Let us mention that not only the group $SO(n;\mathbb{C})$ but also the (Abelian) multiplicative group $\mathbb{C}^\times$ act on the cone Ξ, and these two actions commute. The decomposition into Fourier series relative to the latter group, is the decomposition into subspaces of homogeneous polynomials realizing representations of the orthogonal group.

2. The Cauchy- Radon transform in $\mathbb{R}^n$

Before discussing integral geometry on the sphere, we will make a small modification of the usual Radon transform in $\mathbb{R}^n$. Let $\mathbb{R}^n_x, \mathbb{R}^n_\xi$ be dual copies of $\mathbb{R}^n$ with the duality form

$$< \xi, x >= \xi_1 x_1 + \cdots + \xi_n x_n$$

and $L(\xi, p)$ be the hyperplanes

$$< \xi, x >= p.$$

Let $f(x) \in C_0^\infty(\mathbb{R}_x^n)$. Then its Radon transform is

$$\mathcal{R}f(\xi, p) := \int_{\mathbb{R}^n} f(x)\delta(< \xi, x > -p)dx = \int_{L(\xi,p)} f(x)(< \xi, dx > \rfloor dx).$$

Here $< \xi, dx > \rfloor dx$ is the form α such that $(\xi, dx) \wedge \alpha = dx$. Here $\rfloor$ denotes the interior product; its restriction on $L(\xi, p)$ is unique. In the 2nd formula we take the residue on $L(\xi, p)$. The inversion formulas for $\mathcal{R}$ have different structures for odd and even n.

Let us consider a modification of this definition in which we replace the δ-function by the Cauchy kernel. Let us define the *Cauchy-Radon transform* of a function f as follows:

$$\mathcal{C}f(\xi, p - i\varepsilon) := \int_{\mathbb{R}^n} \frac{f(x)}{< \xi, x > -p + i\varepsilon} dx, \varepsilon > 0.$$

It is well defined, since for $\varepsilon > 0$ there are no singularities[1]. We can relate this to the absence of real points on (complex) hyperplanes

$$< \zeta, z >= p - i\varepsilon$$

in $\mathbb{C}_z^n$. We define the boundary values

$$\mathcal{C}f(\xi, p - i0) := \lim_{\varepsilon \to 0} \mathcal{C}f(\xi, p - i\varepsilon)$$

in the distribution sense. We can now express $\mathcal{C}f(\xi, p - i0)$ through $\mathcal{R}f$ by integrating first along parallel real hyperplanes. (The converse is also possible.)

Since

$$\lim_{\varepsilon \to 0} \frac{1}{p - i\varepsilon} = (p - i0)^{-1} = p^{-1} + i\pi\delta(p),$$

we have

$$2\mathcal{R}f(\xi, p) = \mathcal{C}f(\xi, p - i0) + \mathcal{C}f(\xi, p + i0). \tag{1}$$

It is remarkable that, if one deals with the Cauchy-Radon transform, the difference between even and odd dimensions disappears. Namely, there is an universal inversion formula that works in both cases:

$$f(x) = \left(\frac{i}{2\pi}\right)^n \int_\Gamma \mathcal{C}f_p^{(n-1)}(\xi, < \xi, x > -i0)\omega(\xi, d\xi), \tag{2}$$

where ω is the canonical projective form

$$\omega(\xi, d\xi) := \sum_{j \le n} (-1)^j \xi_j \bigwedge_{l \ne j} d\xi_j.$$

In (2) differentiation is of the order $n - 1$ with respect to p and integration is done along any cycle Γ in $\mathbb{R}_\xi^n$ around the origin that intersects once almost any ray emanating from it (e.g., a sphere centered at the origin).

The proof of (2) is a direct combination of (1) and the inversion formula for the Radon transform. One can also check that formally applying the even dimensional Radon inversion formula in odd dimensions (or vice versa), one gets zero.

[1] Of course, we could have chosen $\varepsilon < 0$ as well.

We can also write (2) in the following way:

$$f(x) = \left(\frac{i}{2\pi}\right)^n \int_\Gamma \int_{\mathbb{R}^n} f((y)(< \xi, (y-x) > -i0)^{-n} \omega(\xi, d\xi) \wedge dy.$$

The real form in this integral coincides with the form in Leray's Cauchy-Fantappie formula. This formal coincidence of Radon and Leray forms probably was not noticed by Leray.

3. The horospherical Cauchy transform on the sphere

We want now to relate the Maxwell's dual spherical polynomials to an analog of the Radon transform on the unit sphere. What can such a transform be? There is an illusion that the Minkowski-Funk transform - the integration along big spheres - is the natural analog of the Radon transform on the sphere. It looks natural, since big spheres are complete geodesics, but the experience of integral geometry on homogeneous spaces suggests differently. A great Gelfand's idea was to integrate on homogeneous non-flat spaces along horospheres - limits of spheres when centers and radii simultaneously tend to infinity. Besides, unlike the horospheric ones, the geodesic versions of the Radon transform have a rather weak relation to the geometry and harmonic analysis on homogeneous spaces.

The space $\mathbb{R}^n$ has two homogeneous curved "relatives:" hyperbolic space H^n of negative curvature and the sphere S^n of positive curvature. It is possible to consider the geodesic analog of the Radon transform on H^n (cf. with the Minkowski-Funk transform on S^n), which does not help much with harmonic analysis on H^n, but its horospherical analog does provide such a possibility. Namely, the spherical Fourier transform on H^n - spectral decomposition - is connected to the horospherical Radon on H^n by the Euclidean one-dimensional Fourier transform. This existence of "twins" - analogs of Fourier and Radon transforms connected by an Euclidean Fourier transform - is the principal aim of integral geometry on homogeneous spaces. This is achievable neither with the Minkowski-Funk transform on sphere, nor with the geodesic Radon transform on the hyperbolic space.

We want to achieve such a goal on the sphere, but ... oops! There are no (real) horospheres on S^n! So, let us consider complex ones. Following [**2**–**4**], we consider the horospherical Cauchy transform.

Consider the complexification $\mathbb{C}S^n \subset \mathbb{C}^n_z$ of S^n, given as follows:

$$\Box(z) = 1.$$

(Complex) horospheres $L(\zeta)$ are isotropic sections of $\mathbb{C}S^n$:

$$< \zeta, z >= 1, \zeta \in \Xi.$$

These sections are paraboloids.

Let us recall that on Ξ,

$$\Box(\zeta) = 0, \zeta \neq 0.$$

We are interested in horospheres without real points (i.e., points of S^n), in order to use them for the definition of the horospherical Cauchy transform.

Let Ξ_0 be the subset of such points $\zeta = \xi + i\eta \in \Xi$ that horospheres $L(\zeta)$ have no real points (i.e., do not intersect S^n). We have on Ξ:

$$\Box(\xi) = \Box(\eta), < \xi, \eta >= 0.$$

It turns out that Ξ_0 is given by the condition

$$\Box(\xi) = \Box(\eta) < 1.$$

The complex orthogonal group $SO(n+1;\mathbb{C})$ acts on Ξ and the conditions defining Ξ_0 are invariant with respect to its maximal compact subgroup $SO(n+1)$. Acting by this subgroup on Ξ, we can transform any ζ to the form $(a, ia, 0, \dots, 0)$. It is evident that such point lies in Ξ_0 if and only if $|a| < 1$. The $SO(n+1)$-invariant form of this condition is just $\Box(\xi) < 1$, and we get a description of Ξ_0.

The boundary $\partial\Xi_0$ corresponds to complex horospheres $L(\zeta)$ with

$$\Box(\xi) = \Box(\eta) = 1, < \xi, \eta >= 1.$$

Such horosphere intersects S^n at the unique point

$$x = \xi.$$

So we have a fibration

$$\partial\Xi \to S^n$$

by $(n-1)$-dimensional spheres, which can be identified with spheres at tangent planes to S^n. Correspondingly, the complex domain $\Xi_0 \subset \Xi$ is fibering over the ball B^n into $n-1$-dimensional spheres.

The group $\mathbb{C}^\times$ acts on Ξ, defining a cone structure on Ξ. Its part $\Lambda = \{\lambda \in \mathbb{R}; |\lambda| < 1\}$ acts on the domain Ξ_0 as contractions:

$$\zeta \mapsto \lambda\zeta.$$

It contracts the boundary $\partial\Xi_0$ into the interior of Ξ_0: $\Box(\lambda\xi) = \lambda^2 < 1$.

We thus have constructed a geometrical background for introducing a horospherical Cauchy transform on the sphere. Now we are ready to give its definition. Let $f \in C^\infty(S^n)$ and

$$\mathcal{C}f(\zeta) := \int_{S^n} \frac{f(x)}{< \zeta, x > -1} \omega(x; dx), \zeta \in \Xi_0.$$

This integral is well defined and represents a holomorphic function of $\zeta \in \Xi_0$. So, we have defined an operator

$$\mathcal{C} : C^\infty(S^n) \to \mathcal{O}(\Xi_0),$$

where $\mathcal{O}(\Xi_0)$ is the space of holomorphic functions on Ξ_0. This transform can be extended also to the boundary points $\zeta = \xi + i\eta \in \partial\Xi_0$, since then the integrand in $\mathcal{C}f(\zeta)$ will be singular only at the point $x = \xi$, and it is thus easy to regularize the integral. We can interpret it as boundary values of $\mathcal{C}f(\zeta), \zeta \in \Xi_0$:

$$\mathcal{C}f(\zeta) := \lim_{\varepsilon \to 0} \mathcal{C}f((1-\varepsilon)\zeta), \zeta \in \partial\Xi_0.$$

It would be interesting to obtain Paley - Wiener Theorems for various functional spaces on S^n (correspondingly, boundary values at Ξ_0).

Let us make a couple of remarks. Our functional spaces include in particular the space $\mathcal{O}(\mathbb{C}S^n)$ of analytic functions on the Stein manifold $\mathbb{C}S^n$). For $f \in \mathcal{O}(\mathbb{C}S^n)$, the transform can be holomorphically extended on the whole Ξ, since we can deform the sphere S^n of integration to another real form of $\mathbb{C}S^n$, which will change Ξ_0 correspondingly. As a result, we produce a holomorphic $SO(n+1;\mathbb{C})$-isomorphism between $\mathcal{O}(\mathbb{C}S^n)$ and $\mathcal{O}(\Xi)$ [**5**].

Let us consider another example. Let $Hyp(S^n)$ be the space of functionals on the space of functions holomorphic in a neighborhood of S^n in $\mathbb{C}S^n$. We can interpret elements of this space as hyperfunctions on the sphere, with the topology of the inductive limit. In particular, it contains distributions on S^n. For $F \in Hyp(S^n)$, we can interpret $\mathcal{C}F$ as values of this functional on the function $(<\zeta, x> -1)^{-1}$, which for any $\zeta \in \Xi_0$ is holomorphic in a neighborhood of S^n. So, we extend the horospherical Cauchy transform to the space of hyperfunctions.

4. The inversion of the horospherical Cauchy transform on the sphere

For the horospherical Cauchy transform on the sphere, there is an inversion formula similar to the inversion formula in $\mathbb{R}^n$. Namely, let

$$D_u F(\zeta) := \lim_{u \to 0} \frac{\partial F((1-u)\zeta)}{\partial u}.$$

Then

$$f(x) = c \int_{[\Box(\eta)=0. <\eta, x>=0]} (D_u)^{n-1} \mathcal{C} f(x + i\eta) \omega(\eta; d\eta),$$

where $c = (i/2\pi)^n$ and $x \in S^n$. Thus, we take the average over the $(n-1)$-dimensional sphere of (complex) horospheres that contain the unique real point $x \in S^n$. Let us emphasize that we reconstruct functions on the sphere through the "dilation" derivatives of the "boundary" values of their horospherical Cauchy transforms.

There are several ways to prove this formula: either using representations of the group $SO(n)$ [**4**], or specializing the general method of the reconstruction functions on the sphere from their integrals over hyperplane sections [**2, 3**]. This formula is not only similar to the inversion formula in $\mathbb{R}^n$, but in appropriate notations they are identical, as well as identical to the horospherical inversion Cauchy formula for the hyperbolic space. There are important reasons for this coincidence, which we cannot address here.

As we have mentioned already, dilations $\lambda \Xi_0$ for $0 < \lambda < 1$ of the boundary lie inside Ξ_0. We can formally apply the inversion formula to the restriction of $\mathcal{C}f$ on them to the surface $\Box(x) = \lambda$. As the result, we extend the function $f(x) \in C^\infty(S^n)$ to the unit ball B^{n+1}. One can verify that this extension is harmonic. Indeed, the direct differentiation with respect to x shows that $\Box(\partial/\partial x) f(x) = 0$, since $\Box(\zeta) = 0$ on Ξ.

5. Relation to Maxwell polynomials

We have considered a special case of a very general phenomenon of duality between symmetric spaces and manifolds of their horospheres [**4, 5**]. In our example, it is the pair $(\mathbb{C}S^n, \Xi)$ in the complex setting and (S^n, Ξ_0) in the real one. The crucial observation is that on the space of horospheres there is an action not only of the motion group of the symmetric space, but also of its extension by an Abelian group. We have mentioned already that the group $\mathbb{C}^\times$ acts on Ξ. It does not preserve Ξ_0, but the semigroup of contractions $\{0 < |\lambda| < 1\}$ preserves it, as well as multiplications $\zeta \mapsto \exp(i\theta)\zeta$. We thus have an action of the unit disc and the domain Ξ_0 fibers into unit discs D. The base of this fibration Ξ_0/D can be interpreted as the "complex boundary" of S^n.

Let us consider the Fourier series decomposition generated by the action of the circle:

$$\mathcal{O}(\Xi_0) = \bigoplus_k \mathcal{O}_k(\Xi_0).$$

It happens to be exactly the decomposition into subspaces of homogeneous polynomials - Maxwell polynomials. A similar decomposition of $\mathcal{O}(\Xi)$ also holds.

We have constructed on $\mathcal{O}(\Xi_0)$ the inverse horospherical Cauchy operator $(\mathcal{C})^{-1}$, which transforms this decomposition into the decomposition into spherical polynomials:

$$Hyp(S^n) = \bigoplus_k H_k(S^n).$$

It is possible to see, either by using the representation theory, or by directly investigating the integral $(\mathcal{C})^{-1}F(x)$ for a homogeneous polynomial $F(\zeta)$ on Ξ of a degree k, that it is holomorphically extendible on $\mathbb{C}^n$ as a homogeneous harmonic polynomial. Of course, the convergence of series expansions into spherical polynomials for hyperfunctions requires some clarifications.

Thus, the composition of the horospherical Cauchy transform and the one-dimensional discrete Fourier transform (Fourier series) gives the decomposition into irreducible representations of $SO(n)$. This is similar to the connection between the usual Fourier and Radon transforms through the one-dimensional Fourier transform (sometimes called Fourier-slice formula).

In particular, the horospherical Cauchy transform of any spherical polynomial is a dual spherical Maxwell polynomial and the inverse horospherical Cauchy transform of any Maxwell polynomial is a spherical polynomial. This gives the following integral representations of spherical polynomials:

$$f(x) = (\mathcal{C})^{-1}F(x) = ck^{n-1} \int_{\square(\eta)=0,<\eta,x>=0} F(x+i\eta)\omega(\eta;d\eta),$$

where $c = (i/2\pi)^n$ and F is an arbitrary homogeneous polynomial of degree k on Ξ. We notice that the differentiations in the case of homogeneous polynomials transform into multiplications. We can interpret this as an analog of the classical Poisson formula. Surely, significant modifications are necessary, since we work with the complex boundary and holomorphic boundary values. It is also an analytic analog of Maxwell's algebraic formula connecting homogeneous polynomials on the cone Ξ with spherical polynomials.

References

[1] A. Erdélyi, W. Magnus, F. Oberhettinger, and F. G. Tricomi, *Higher transcendental functions. Vols. I, II*, McGraw-Hill Book Company, Inc., New York-Toronto-London, 1953. Based, in part, on notes left by Harry Bateman. MR0058756

[2] S. Gindikin, *Complex horospherical transform on real sphere*, Geometric analysis of PDE and several complex variables, Contemp. Math., vol. 368, Amer. Math. Soc., Providence, RI, 2005, pp. 227–232, DOI 10.1090/conm/368/06781. MR2126472

[3] S. Gindikin, *Holomorphic horospherical duality "sphere-cone"*, Indag. Math. (N.S.) **16** (2005), no. 3-4, 487–497, DOI 10.1016/S0019-3577(05)80037-0. MR2313635

[4] S. Gindikin, *The horospherical Cauchy-Radon transform on compact symmetric spaces* (English, with English and Russian summaries), Mosc. Math. J. **6** (2006), no. 2, 299–305, 406. MR2270615

[5] S. Gindikin, *Harmonic analysis on symmetric Stein manifolds from the point of view of complex analysis*, Japanese J. Math. **1** (2006), no. 1, 87–105, DOI 10.1007/s11537-006-0503-4. MR2261062

Department of Mathematics, Hill Center, Rutgers University 110 Frelinghysen Road, Piscataway, NJ 08854, U.S.A.

Email address: sgindikin@gmail.com

Contemporary Mathematics
Volume **733**, 2019
https://doi.org/10.1090/conm/733/14741

The weighted Laplace transform

Gisele Ruiz Goldstein, Jerome A. Goldstein, Giorgio Metafune, and Luigi Negro

Dedicated to the memory of Selim Krein.

ABSTRACT. We consider for $1 \le p \le \infty$, $\alpha \in \mathbb{R}$ and $0 \le K \in L^\infty_{\text{loc}}\,(]0,\infty[)$ the operator

$$T_\alpha f(\lambda) := \int_0^\infty K(\lambda t) f(t) t^\alpha \, dt$$

and we investigate boundedness properties of T_α over the spaces $L^P\,(]0,\infty[, t^\alpha dt)$.

1. Introduction

The Laplace Transform $\mathcal{L}$ has long been an important linear operator. But, surprisingly, it has not been a popular operator in texts on functional analysis. We begin by summarizing a few of its properties. The definition is

$$\mathcal{L}f(x) = \int_0^\infty e^{-tx} f(t)\, dt.$$

We view $\mathcal{L}$ as an operator on $L^p := L^p\,(]0,\infty[)$, $1 \le p \le \infty$.

Here are some basic properties.

(i) $\mathcal{L}$ is a bounded operator on L^p if and only if $p = 2$.
(ii) $\mathcal{L}$ is bounded and selfadjoint on L^2 with norm $\sqrt{\pi}$.
(iii) $\mathcal{L}^2$ is bounded on L^p for all p, $1 < p < \infty$.

We begin by giving some quick proofs. Let $\alpha > 0$, $f_\alpha(x) := e^{-\alpha x}$ so that $f_\alpha \in \bigcap\limits_{1\le p\le\infty} L^p$, and

$$\begin{aligned}
\mathcal{L}f_\alpha(x) &= \int_0^\infty e^{-tx} e^{-\alpha t}\, dt = \frac{1}{\alpha + x}, \\
\|f_\alpha\|_p^p &= \int_0^\infty e^{-\alpha p x}\, dx = \frac{1}{\alpha p}, \qquad 1 \le p < \infty, \\
\|\mathcal{L}f_\alpha\|_p^p &= \int_0^\infty \frac{1}{(\alpha + x)^p}\, dx = \int_\alpha^\infty \frac{1}{y^p}\, dy = \frac{\alpha^{1-p}}{p-1}, \qquad 1 < p < \infty, \\
\frac{\|\mathcal{L}f_\alpha\|_p}{\|f_\alpha\|_p} &= \left(\frac{p}{p-1}\right)^{\frac{1}{p}} \alpha^{\frac{2}{p}-1} \to \infty
\end{aligned}$$

2010 *Mathematics Subject Classification.* Primary 44A10, 45P05, 47B38, 47G10.
Key words and phrases. Laplace transform, integral operator.

as $\alpha \to \infty$ for $1 < p < 2$ and as $\alpha \to 0$ for $2 < p < \infty$.

Thus $\mathcal{L} \notin \mathcal{B}(L^p)$ for $1 < p < \infty$ and $p \neq 2$; this also holds for $p = 1, \infty$. Clearly it is possible that $\mathcal{L} \in \mathcal{B}\left(L^2\right)$. Proving this is an exercise in the 1958 book of Dunford and Schwartz. Here is a classical proof. For $f \in L^2$,

$$(1.1) \qquad |\mathcal{L}f(x)|^2 = \left| \int_0^\infty \left(e^{-\frac{t}{2}x} t^{-\frac{1}{4}}\right) \left(e^{-\frac{t}{2}x} t^{\frac{1}{4}} f(t)\right) dt \right|^2 \le \left(\int_0^\infty e^{-tx} t^{-\frac{1}{2}}\, dt \right) \left(\int_0^\infty e^{-tx} t^{\frac{1}{2}} |f(t)|^2\, dt \right).$$

Let $s = tx$, $ds = xdt$. Recalling

$$\Gamma(\gamma) = \int_0^\infty e^{-s} s^{\gamma - 1}\, ds,$$

$$(1.2) \qquad \int_0^\infty e^{-tx} t^{-\frac{1}{2}}\, dt = \int_0^\infty e^{-s} s^{-\frac{1}{2}}\, ds\, x^{-\frac{1}{2}} = \Gamma\left(\frac{1}{2}\right) x^{-\frac{1}{2}} = \sqrt{\pi} x^{-\frac{1}{2}},$$

we deduce from (1.1), (1.2) that

$$\begin{aligned} \int_0^\infty |\mathcal{L}f(x)|^2\, dx &\le \int_0^\infty \sqrt{\pi} x^{-\frac{1}{2}} \int_0^\infty e^{-tx} t^{\frac{1}{2}} |f(t)|^2\, dt\, dx \\ &= \sqrt{\pi} \int_0^\infty \left(\int_0^\infty e^{-tx} x^{-\frac{1}{2}}\, dx \right) t^{\frac{1}{2}} |f(t)|^2\, dt \\ &= \pi \int_0^\infty t^{-\frac{1}{2}} t^{\frac{1}{2}} |f(t)|^2\, dt \qquad \text{by} \quad (1.2) \\ &= \pi \int_0^\infty |f(t)|^2\, dt, \end{aligned}$$

and so $\|\mathcal{L}\|_{\mathcal{B}(L^2)} \le \sqrt{\pi}$. In the above calculations, the only inequality used was the Cauchy-Schwarz inequality, and in it, $|\int_\Omega gh\, d\mu| \le \|g\|_2 \|h\|_2$, equality holds if and only if one of g, h is a multiple of the other. We cannot have $0 \neq f(t) = ce^{-tx} t^{-\frac{1}{2}}$ for some fixed $c > 0$ and all $t > 0$ since $t \mapsto e^{-tx} t^{-\frac{1}{2}} \notin L^2$. But by modifying f near $t = 0$, we see that the constant $\sqrt{\pi}$ is best possible but is not attained, i.e., there are not extremals other then 0.

Furthermore

$$\begin{aligned} \mathcal{L}^2 f(x) &= \int_0^\infty e^{-tx} \mathcal{L}f(t)\, dt = \int_0^\infty \int_0^\infty e^{-tx} e^{-st} f(s)\, ds\, dt \\ &= \int_0^\infty \left(\int_0^\infty e^{-(s+x)t}\, dt \right) f(s)\, ds = \int_0^\infty \frac{f(s)}{s+x}\, ds. \end{aligned}$$

$\mathcal{M}f(x) := \mathcal{L}^2 f(x)$ clearly exists for all $x > 0$ for, say, $f \in C_c([0, \infty[)$. It is related to the Hilbert transform $\mathcal{H}$,

$$\mathcal{H}g(x) = \text{P.V.}\ \frac{1}{\pi} \int_{-\infty}^\infty \frac{g(y)}{x - y}\, dy = \lim_{\epsilon \to 0+} \frac{1}{\pi} \int_{|x-y| \ge \epsilon} \frac{g(y)}{x - y}\, dy.$$

Extend f to be 0 on $(-\infty, 0)$, apply $\pi\mathcal{H}$ to this extension, replace x by $-x$ and multiply the result by -1; one thus factors $\mathcal{M}$ as a product of bounded operators on L^p for $1 < p < \infty$ since $\mathcal{H} \in \bigcap_{1<p<\infty} \mathcal{B}(L^p)$ by the Calderón-Zygmund Theorem.

The "folklore" results (i)-(iii) are interesting and deserve to be more widely known. Our goal in this paper is to embed these results into a richer, more substantial theory.

We close this introduction by stating the elegant theorem of Gilliam, Schulenberger, Lund [**2**].

THEOREM 1.1 ([**2**]). *There is a unitary operator*

$$U : L^2 \to L^2 \left((-\sqrt{\pi}, \sqrt{\pi}), |y|dy\right)$$

such that

$$U\mathcal{L}U^{-1}f(x) = xf(x)$$

for all $f \in L^2 ((-\sqrt{\pi}, \sqrt{\pi}), |y|dy)$ *and a.e.* $x \in (-\sqrt{\pi}, \sqrt{\pi})$. *The operator* $\mathcal{L}$ *has spectrum* $\sigma(\mathcal{L}) = [-\sqrt{\pi}, \sqrt{\pi}]$ *and is spectrally absolutely continuous of multiplicity one.* $\mathcal{M} = \mathcal{L}^2$ *is spectrally absolutely continuous on* L^2 *of multiplicity two with spectrum* $[0, \pi]$.

Theorem 1.1 is the final bit of "folklore" we want to mention. It is part of the deeper theory of spectra of selfadjoint operators. The Spectral Theorem of M. Stone and J. von Neumann says that a selfadjoint operator A on a complex Hilbert space H is unitarily equivalent to a real multiplication operator M_m on some concrete L^2 space $L^2(\Omega, \Sigma, \mu)$, so that $M_m f(\omega) = m(\omega)f(\omega)$ for some Σ-measurable function $m : \Omega \to \mathbb{R}$, and $D(M_m) = \{f : f, mf \in L^2(\Omega, \Sigma, \mu)\}$. Finding m and $L^2(\Omega, \Sigma, \mu)$ explicitly has long been a goal of spectral theory. If A on H has a compact resolvent, then $A\varphi_n = \lambda_n \varphi_n$ for $\{\varphi_n\}_{n\in J}$ an orthonormal basis for H, $\lambda \in \mathbb{R}$, $J \subset \mathbb{Z}$ and $Au = \sum_{n\in J} \lambda_n \langle u, \varphi_n\rangle \varphi_n$. Then $L^2(\Omega, \Sigma, \mu) = l^2_J = L^2(J, 2^J, \text{counting measure})$. For $A = -\Delta$ on $L^2(\mathbb{R}^N)$ and F the Fourier transform, $(Fu)(\xi) = (2\pi)^{-N/2} \int_{\mathbb{R}^N} u(x) \exp(-ix\cdot\xi)dx$, then $F : L^2(\mathbb{R}^N_x) \to L^2(\mathbb{R}^N_\xi)$ is unitary and $F^{-1}AF = M_m$ where $m(\xi) = |\xi|^2$, so that $F^{-1}AF(\xi) = M_{|\xi|^2}$. In the beginning of quantum theory in the late 1920s, P. Dirac [**1**] independently did something similar to what Stone and von Neumann did, but differently and less rigorously. That is, he "invented" Hilbert space and stated his version of the Spectral Theorem for selfadjoint operators (which, in hindsight, included Schrödinger and Dirac type operators but not all selfadjoint operators). But for Dirac [**1**], the Hilbert space was an L^2 space and the selfadjoint operator A satisfied $A = A_1 + A_2$ where

$$A_1 u = \sum_{n\in J} \lambda_n \langle u, \varphi_n\rangle \varphi_n,$$

$$A_2 u(x) = \int_S \lambda(\xi) \left(\int_S u(y)\varphi(y, \xi)dy\right) \overline{\varphi(x, \xi)}d\xi,$$

with $S \subset \mathbb{R}^N$. In the above example of the negative Laplacian,

$$-\Delta u(x) = \int_{\mathbb{R}^N} |\xi|^2 \left(\int_{\mathbb{R}^N} u(y)\varphi(y, \xi)dy\right) \overline{\varphi(x, \xi)}d\xi$$

where $\varphi(x, \xi) = \exp(-ix \cdot \xi)/(2\pi)^{\frac{N}{2}} \in L^\infty(\mathbb{R}^N) \setminus L^2(\mathbb{R}^N)$. These are generalized eigenfunction expansions, since $A\varphi(x, \xi) = \lambda(\xi)\varphi(x, \xi)$, but they are not bona fide eigenfunctions since $\varphi(x, \xi)$ are not in L^2. Many heroic efforts led to realizations of this goal in special cases. The PhD Thesis [**4**] of T. Ikebe (in Physics at Tokyo, under T. Kato) led to generalized eigenfunction expansions for many Schrödinger operators and Dirac type operators in $L^2(\mathbb{R}^N)$.

The Weyl-Stone-Titchmarsh-Kodaira Theorem [**5**] did essentially the same thing for Schrödinger operators on the half line. From a different perspective, these four researchers extended Sturm-Liouville theory for operators of the form $-\frac{d^2}{dx^2} + V(x)$

from $L^2(0,a)$ to $L^2(0,\infty)$. These selfadjoint operators in general have both discrete and continuous spectra.

Extending Theorem 1.1 to the more general context just discussed is a future project.

In the present paper we consider a family T_α of linear integral operators depending on a parameter $\alpha \in \mathbb{R}$ that defines the measure $d\mu = t^\alpha dt$ on the positive half line $]0,\infty[$.

The operator T_α is defined by

$$T_\alpha f(\lambda) := \int_0^\infty K(\lambda t)f(t)\,d\mu(t) = \int_0^\infty K(\lambda t)f(t)t^\alpha\,dt. \tag{1.3}$$

The function $K :]0,\infty[\to \mathbb{R}$ is measurable, $K \geq 0$, $K \neq 0$ and $K \in L^\infty_{\text{loc}}(]0,\infty[)$. The function $T_\alpha f$ is well defined if f has a compact support in $]0,\infty[$. If $\alpha = 0$ and $K(t) = e^{-t}$, the operator T_α reduces to the Laplace Transform $\mathcal{L}$ and so we generalize the above results.

Notation. The weighted Lebesgue spaces with respect to the measures $d\mu = t^\alpha\,dt$ are denoted by L^p_α but we write L^∞ for L^∞_α and L^p for L^p_0. When it is clear from the context, we also write $\|f\|_p$ to denote the norm of a function f in L^p_α. Finally, for the operators T_α defined above, $\|T_\alpha\|_{p,q}$ denotes the norm of T_α from L^p_α to L^q_α .

2. Boundedness of T_α

We start with some technical lemmas.

LEMMA 2.1. *For every $\alpha \in \mathbb{R}$ one has*

$$\|T_\alpha\|_{1,\infty} = \|K\|_\infty \leq \infty.$$

PROOF. For a fixed $\lambda > 0$,

$$\sup_{\|f\|_1=1} |T_\alpha f(\lambda)| = \sup_{\|f\|_1=1} \left| \int_0^\infty K(\lambda t)f(t)\,d\mu(t) \right| = \|K(\lambda\cdot)\|_\infty = \|K\|_\infty.$$

Taking the supremum over $\lambda > 0$ we get the result. □

LEMMA 2.2. *$\|T_\alpha\|_{p,1} = \infty$ holds for every $p \in [1,\infty]$ if $\alpha \neq -1$ and for every $p \in]1,\infty]$ if $\alpha = -1$.*

PROOF. Let us consider $f \geq 0$. We have

$$\begin{aligned}\|T_\alpha f\|_1 &= \int_0^\infty T_\alpha f(\lambda)\,d\mu(\lambda) = \int_0^\infty d\mu(\lambda) \int_0^\infty K(t\lambda)f(t)\,d\mu(t) \\ &= \int_0^\infty f(t) \left[\int_0^\infty K(t\lambda)\,d\mu(\lambda) \right] d\mu(t).\end{aligned}$$

Then

$$\|T_\alpha\|_{p,1} = \sup_{\substack{f\geq 0\\ \|f\|_p=1}} \|T_\alpha f\|_1 = \sup_{\substack{f\geq 0\\ \|f\|_p=1}} \int_0^\infty f(t)g(t)\,d\mu(t) = \|g\|_{p'},$$

where

$$g(t) = \int_0^\infty K(t\lambda)\,d\mu(\lambda) = \int_0^\infty K(s)\frac{1}{t^{\alpha+1}}s^\alpha\,ds = \frac{\|K\|_1}{t^{\alpha+1}}$$

with $0 < \|K\|_1 \leq \infty$.

This implies, in particular, that $\|g\|_{p'} = \infty$ for every $p \in [1, \infty]$ if $\alpha \neq -1$ and for every $p \in]1, \infty]$ if $\alpha = -1$. $\square$

LEMMA 2.3. *If $\alpha \neq -1$ and $\|T_\alpha\|_{p,q} < \infty$, then $q > 1$ and $q = p'$.*

PROOF. Lemma 2.2 implies immediately that $q > 1$. To prove the second property, let us fix $f \in L^p_\alpha$, $f \geq 0$ and define $f_s(t) := f(st)$. Then we have

$$\|f_s\|_p^p = \int_0^\infty |f(st)|^p t^\alpha \, dt = \int_0^\infty |f(x)|^p \frac{x^\alpha}{s^{\alpha+1}} \, dx = \frac{\|f\|_p^p}{s^{\alpha+1}}.$$

Moreover

$$\begin{aligned} T_\alpha f_s(\lambda) &= \int_0^\infty K(\lambda t) \, f(st) t^\alpha \, dt \\ &= \int_0^\infty K\left(\frac{\lambda}{s} x\right) f(x) \frac{x^\alpha}{s^{\alpha+1}} \, dx = \frac{1}{s^{\alpha+1}} \left(T_\alpha f\right)_{\frac{1}{s}}(\lambda). \end{aligned}$$

Therefore

$$\|T_\alpha f_s\|_q = \frac{1}{s^{\alpha+1}} \| \left(T_\alpha f\right)_{\frac{1}{s}} \|_q = \|T_\alpha f\|_q \frac{s^{\frac{\alpha+1}{q}}}{s^{\alpha+1}} = \|T_\alpha f\|_q s^{(\alpha+1)\left(\frac{1}{q} - 1\right)}.$$

Inserting the last relations into the inequality $\|T_\alpha f_s\|_q \leq C \|f_s\|_p$ yields the estimate

$$\|T_\alpha f\|_q \leq C s^{(\alpha+1)(1 - \frac{1}{q} - \frac{1}{p})} \|f\|_p.$$

If $\alpha \neq -1$, the arbitrariness of $s > 0$ implies $1 - \frac{1}{q} - \frac{1}{p} = 0$. $\square$

LEMMA 2.4. *If $\alpha \neq -1$ and $\|T_\alpha\|_{p,q} < \infty$ then $1 \leq p \leq 2$ and $q = p'$.*

PROOF. Lemma 2.3 implies $q = p'$. For every $\gamma \in \mathbb{R}$ with $\gamma p - \alpha > 1$, let us consider the function $f(t) = \frac{1}{t^\gamma} \chi_{]1,\infty[}$. Then

$$\|f\|_p = \left(\int_1^\infty \frac{1}{t^{\gamma p - \alpha}} \, dt \right)^{\frac{1}{p}} = \frac{1}{(\gamma p - \alpha - 1)^{\frac{1}{p}}}$$

and

$$\left(T_\alpha f\right)(\lambda) = \int_1^\infty \frac{K(\lambda t)}{t^{\gamma - \alpha}} \, dt \overset{s = \lambda t}{=} \int_\lambda^\infty \frac{K(s)}{s^{\gamma - \alpha}} \, ds \, \lambda^{-\alpha + \gamma - 1}.$$

Let us fix $\lambda_0 > 0$ and a constant $c > 0$ such that

$$\int_{\lambda_0}^\infty \frac{K(s)}{s^{\gamma - \alpha}} \, ds \geq c > 0$$

for every γ in a right neighborhood of $\frac{1+\alpha}{p}$. Then

$$\left(T_\alpha f\right)(\lambda) \geq \frac{c}{\lambda^{1 + \alpha - \gamma}}$$

for every $\lambda \leq \lambda_0$ and therefore

$$\|T_\alpha f\|_{p'}^{p'} \geq c^{p'} \int_0^{\lambda_0} \frac{1}{\lambda^{p'(1+\alpha-\gamma)}} \lambda^\alpha \, d\lambda = c^{p'} \frac{\lambda_0^{\alpha + 1 - p'(1 + \alpha - \gamma)}}{\alpha + 1 - p'(1 + \alpha - \gamma)}.$$

Since

$$\alpha + 1 - p'(1 + \alpha - \gamma) = \frac{\gamma p - \alpha - 1}{p - 1}$$

it follows that

$$\|T_\alpha\|_{p,p'} \geq C\,(\gamma p - \alpha - 1)^{\frac{1}{p} - \frac{1}{p'}}\,. \tag{2.1}$$

In the case $p > 2$, since $\frac{1}{p} - \frac{1}{p'} < 0$, (2.1) immediately implies $\|T_\alpha\|_{p,p'} \to +\infty$ as $\gamma \to \frac{1+\alpha}{p}$. This proves that $1 \leq p \leq 2$. □

REMARK 2.5. When $\alpha = -1$, the conclusion of Lemma 2.3 ceases to hold and the argument of the proof of Lemma 2.4 fails since we do not have $q = p'$.

LEMMA 2.6. *If $\alpha = -1$ and $\|T_{-1}\|_{p,q} < \infty$ then $q \geq p$.*

PROOF. Fixing $\epsilon > 0$, we adapt the reasoning of Lemma 2.4 for $\alpha = -1$ and we consider the function

$$f(t) = \frac{1}{t^\gamma}\chi_{]1,\infty[}$$

for $0 < \gamma < \epsilon$. Then

$$\|f\|_p = \left(\frac{1}{\gamma p}\right)^{\frac{1}{p}}$$

and $(T_\alpha f)(\lambda) \geq c\lambda^\gamma$ for every $\lambda \leq \lambda_0$ and $0 < \gamma < \epsilon$. This yields

$$\|T_{-1}f\|_q^q \geq c^q \int_0^{\lambda_0} \lambda^{q\gamma - 1}\, d\lambda = c^q \frac{(\lambda_0)^{q\gamma}}{q\gamma}$$

and therefore

$$\|T_{-1}\|_{p,q} \geq C\gamma^{\frac{1}{p} - \frac{1}{q}}.$$

This yields, if $q < p$, that $\|T_{-1}\|_{p,q} \to \infty$ as $\gamma \to 0$, proving the Lemma. □

In the following proposition we compute the norm of T_α in L^2. In the special case of the Laplace transform, we reprove property (ii) of the Introduction.

PROPOSITION 2.7. *For every $\alpha \in \mathbb{R}$, one has*

$$\|T_\alpha\|_{2,2} = \int_0^\infty K(s)s^{\frac{\alpha-1}{2}}\, ds \leq \infty.$$

PROOF. Let us first assume that $\alpha = 0$. Since

$$T_0 f(\lambda) = \int_0^\infty K(\lambda t) f(t)\, dt \overset{s=\lambda t}{=} \int_0^\infty K(s) f\left(\frac{s}{\lambda}\right)\frac{1}{\lambda}\, ds,$$

the Minkowski inequality yields for every $f \in C_c^\infty(]0,\infty[)$,

$$\|T_0 f\|_2 \leq \int_0^\infty K(s)\left[\int_0^\infty \frac{|f\left(\frac{s}{\lambda}\right)|^2}{\lambda^2}\, d\lambda\right]^{\frac{1}{2}} ds$$

$$\overset{x=\frac{s}{\lambda}}{=} \int_0^\infty K(s)\left[\int_0^\infty \frac{|f(x)|^2}{s}\, dx\right]^{\frac{1}{2}} ds = \|f\|_2 \int_0^\infty \frac{K(s)}{\sqrt{s}}\, ds.$$

Hence

$$\|T_0\|_{2,2} \leq \int_0^\infty \frac{K(s)}{\sqrt{s}}\, ds.$$

To prove the converse inequality, let us consider

$$f_n(t) := \frac{1}{\sqrt{t}}\chi_{]\frac{1}{n},n[}.$$

A simple computation shows that $\|f_n\|_2^2 = 2\log n$ and

$$T_0 f_n(\lambda) = \frac{1}{\sqrt{\lambda}} \int_{\frac{\lambda}{n}}^{\lambda n} \frac{K(s)}{\sqrt{s}}\, ds.$$

Let us fix, now, $0 < a < b < \infty$ and $n^2 \leq \frac{b}{a}$ i.e. $\frac{b}{n} \leq an$. If $a \leq \lambda \leq b$ we have

$$|T_0 f_n(\lambda)|^2 \geq \frac{1}{\lambda} \left[\int_{\frac{b}{n}}^{an} \frac{K(s)}{\sqrt{s}}\, ds \right]^2.$$

Then

$$\frac{\|T_0 f_n\|_2^2}{\|f_n\|_2^2} \geq \frac{\left(\int_a^b \frac{1}{\lambda}\, d\lambda\right) \left(\int_{\frac{b}{n}}^{an} \frac{K(s)}{\sqrt{s}}\, ds\right)^2}{2\log n} = \frac{\log(\frac{b}{a})}{2\log n} \left(\int_{\frac{b}{n}}^{an} \frac{K(s)}{\sqrt{s}}\, ds \right)^2.$$

For a fixed $\epsilon > 0$ we specialize a and b to $a = n^{-1+\frac{\epsilon}{2}}$ and $b = n^{1-\frac{\epsilon}{2}}$. With this choice of parameters, the above inequality implies

$$\frac{\|T_0 f_n\|_2^2}{\|f_n\|_2^2} \geq \frac{2-\epsilon}{2} \left(\int_{n^{-\frac{\epsilon}{2}}}^{n^{\frac{\epsilon}{2}}} \frac{K(s)}{\sqrt{s}}\, ds \right)^2.$$

Taking $n \to \infty$ and then $\epsilon \to 0$ in the last estimate, we get

$$\|T_0\|_{2,2} \geq \int_0^\infty \frac{K(s)}{\sqrt{s}}\, ds,$$

which proves the Proposition in the case $\alpha = 0$.

The case $\alpha \neq 0$ can be deduced from the first one by making use of the isometry $J : L^2 \to L^2_\alpha$, $Jf \mapsto ft^{-\frac{\alpha}{2}}$. Indeed, we have

$$\begin{aligned} J^{-1} T_\alpha J f(\lambda) &= \lambda^{\frac{\alpha}{2}} \int_0^\infty K(\lambda t) f(t) t^{-\frac{\alpha}{2}} t^\alpha\, dt \\ &= \lambda^{\frac{\alpha}{2}} \int_0^\infty K(\lambda t) f(t) t^{\frac{\alpha}{2}}\, dt = \int_0^\infty H(\lambda t) f(t)\, dt \end{aligned}$$

where we set $H(s) = s^{\frac{\alpha}{2}} K(s)$.

Define the operator T_H by

$$T_H f(\lambda) = \int_0^\infty H(\lambda t) f(t) t^\alpha\, dt$$

(compare with (1.3)).

The previous computation shows that $J^{-1} T_\alpha J = T_H$. Applying now to T_H the results already proved for $\alpha = 0$, we conclude that

$$\|T_\alpha\|_{L^2_\alpha} = \|J^{-1} T_\alpha J\|_{L^2} = \|T_H\|_{L^2} = \int_0^\infty K(s) s^{\frac{\alpha-1}{2}}\, ds.$$

□

We can now prove the main results of this section.

THEOREM 2.8. *Let $1 \leq p, q \leq \infty$, $\alpha \in \mathbb{R}$ and $K \in L^\infty$ such that*

$$\int_0^\infty K(s) s^{\frac{\alpha-1}{2}}\, ds < \infty.$$

(i) *If $1 \leq p \leq 2$ and $q = p'$, then the operator T_α defined by (1.3) is bounded from L^p_α to L^q_α.*

(ii) *If $\alpha \neq -1$ the conditions on p, q are also necessary for the boundedness.*

PROOF. The second claim is a direct consequence of Lemmas 2.3 and 2.4. To prove (i), we observe that by Lemma 2.1 and Proposition 2.7,

$$\|T_\alpha\|_{1,\infty} = \|K\|_\infty, \qquad \|T_\alpha\|_{2,2} = \int_0^\infty K(s)s^{\frac{\alpha-1}{2}}\, ds < \infty.$$

Applying the Riesz-Thorin Theorem, we get the boundedness for $1 \le p \le 2$ and $q = p'$. □

REMARK 2.9. Let us suppose for simplicity that $\alpha = 0$. Using Hölder's inequality we have

$$\begin{aligned}|T_0 f(\lambda)| &\le \left(\int_0^\infty |K(\lambda t)|^{p'}\, dt\right)^{\frac{1}{p'}} \|f\|_p \\ &= \left(\int_0^\infty |K(s)|^{p'} \frac{1}{\lambda}\, ds\right)^{\frac{1}{p'}} \|f\|_p = \|K\|_{p'} \frac{\|f\|_p}{\lambda^{\frac{1}{p'}}}.\end{aligned}$$

Then, if $K \in L^{p'}$, T_0 is of weak type (p, p') for every $1 \le p \le \infty$. However, the Marcinkiewicz theorem cannot be applied for $p > 2$ since it requires $p \le p'$.

The case $\alpha = -1$ is special, since then

$$\int_0^\infty \frac{K(\lambda t)}{t}\, dt$$

is independent of λ.

THEOREM 2.10. *Let $1 \le p, q \le \infty$, $\alpha = -1$ and $K \in L^\infty \cap L^1_{-1}$. Then T_{-1} is a bounded operator from L^p_{-1} to L^q_{-1} if and only if $q \ge p$.*

PROOF. We already know from Lemma 2.6 that the condition $q \ge p$ is necessary for boundedness. Since

$$T_{-1} f(\lambda) = \int_0^\infty K(\lambda t) \frac{f(t)}{t}\, dt$$

we immediately deduce

$$\|T_{-1} f\|_\infty \le \|f\|_\infty \int_0^\infty \frac{K(\lambda t)}{t}\, dt = \|f\|_\infty \int_0^\infty \frac{K(s)}{s}\, ds.$$

Hence, T_{-1} is bounded from L^∞ onto L^∞. Since the operator T_α is selfadjoint for every α, it follows by duality that T_{-1} is bounded from L^1_{-1} onto itself. Moreover, Proposition 2.7 ensures the boundedness of T_{-1} on L^2_{-1}. The Riesz-Thorin Theorem shows that T_{-1} is bounded from L^p_{-1} to itself for every $p \ge 2$, and then also for $p < 2$ by duality.

Next, $K \in L^\infty \cap L^1_{-1} \subseteq L^{p'}_{-1}$. Therefore, using the Hölder inequality, we get

$$|T_{-1} f(\lambda)| \le \left(\int_0^\infty \frac{K^{p'}(\lambda t)}{t}\, dt\right)^{\frac{1}{p'}} \|f\|_p = \left(\int_0^\infty \frac{K^{p'}(s)}{s}\, ds\right)^{\frac{1}{p'}} \|f\|_p,$$

and so T_{-1} is bounded from L^p_{-1} to L^∞.

Finally, since T_{-1} is bounded from L^p_{-1} to either of L^p_{-1} and L^∞, another application of the Riesz-Thorin theorem proves that T_{-1} is bounded from L^p_{-1} to L^q_{-1} for every $q \ge p$. □

3. The operator T_α^2

In this section we consider the operator T_α^2 and find sufficient summability conditions on K that ensure that T_α^2 is bounded in L_α^p for every $1 < p < \infty$.

For every $\lambda, s > 0$ we define the function

$$H(\lambda, s) = \int_0^\infty K(\lambda t) K(ts) t^\alpha \, dt$$

and begin by observing that T_α^2 can be written as an integral operator whose kernel is given by H. Indeed,

$$\begin{aligned} T_\alpha^2 f(\lambda) &= \int_0^\infty K(\lambda t) T_\alpha f(t) t^\alpha \, dt \\ &= \int_0^\infty K(\lambda t) \left(\int_0^\infty K(ts) f(s) s^\alpha \, ds \right) t^\alpha \, dt \\ &= \int_0^\infty H(\lambda, s) f(s) s^\alpha \, ds. \end{aligned}$$

LEMMA 3.1. *H is positive and homogeneous of degree $-(\alpha+1)$.*

PROOF. The positivity of K implies $H > 0$. For the homogeneity property it is sufficient to observe that for every $\lambda, r, s > 0$,

$$\begin{aligned} H(r\lambda, rs) &= \int_0^\infty K(r\lambda t) K(rts) t^\alpha \, dt \\ &\overset{x=rt}{=} \int_0^\infty K(\lambda x) K(xs) \frac{x^\alpha}{r^{\alpha+1}} \, dx = \frac{H(\lambda, s)}{r^{\alpha+1}}. \end{aligned}$$

□

PROPOSITION 3.2. *For every $1 < p < \infty$ one has*

$$\|T_\alpha^2 f\|_p \le A_p \|f\|_p$$

where

$$A_p = \left(\int_0^\infty K(t) t^{\frac{\alpha-1}{p'}} \, dt \right) \left(\int_0^\infty K(s) s^{\frac{\alpha-1}{p}} \, ds \right).$$

In particular, if $A_p < \infty$, then T_α^2 is a bounded operator on L_α^p.

PROOF. Let us consider the isometry $J : L^p \to L_\alpha^p$, $Jf \mapsto f s^{-\frac{\alpha}{p}}$. Then

$$J^{-1} T_\alpha^2 J f = \lambda^{\frac{\alpha}{p}} \int_0^\infty H(\lambda, s) f(s) s^{\frac{\alpha}{p'}} \, ds.$$

The operator $\Gamma_\alpha = J^{-1} T_\alpha^2 J$ acts in L^p with kernel $\lambda^{\frac{\alpha}{p}} H(\lambda, s) s^{\frac{\alpha}{p'}}$, which is homogeneous of degree -1. The Proposition follows from [**3**, Theorem 319, page 229], once we observe that

$$\begin{aligned} A_p &= \int_0^\infty \lambda^{\frac{\alpha}{p}} H(\lambda, 1) \lambda^{-\frac{1}{p}} \, d\lambda = \int_0^\infty \lambda^{\frac{\alpha}{p}} \int_0^\infty K(\lambda t) K(t) \lambda^{-\frac{1}{p}} t^\alpha \, dt \, d\lambda \overset{\lambda t = s}{=} \\ &= \int_0^\infty K(t) t^{\frac{\alpha-1}{p'}} \, dt \int_0^\infty K(s) s^{\frac{\alpha-1}{p}} \, ds. \end{aligned}$$

□

References

[1] P. A. M. Dirac, *The Principles of Quantum Mechanics*, Oxford, at the Clarendon Press, 1947. 3d ed. MR0023198

[2] D. S. Gilliam, J. R. Schulenberger, and J. L. Lund, *Spectral representation of the Laplace and Stieltjes transforms* (English, with Portuguese summary), Mat. Apl. Comput. **7** (1988), no. 2, 101–107. MR987878

[3] G. H. Hardy, J. E. Littlewood, and G. Pólya, *Inequalities*, Cambridge Mathematical Library, Cambridge University Press, Cambridge, 1988. Reprint of the 1952 edition. MR944909

[4] T. Ikebe, *Eigenfunction expansions associated with the Schroedinger operators and their applications to scattering theory*, Arch. Rational Mech. Anal. **5** (1960), 1–34 (1960), DOI 10.1007/BF00252896. MR0128355

[5] K. Yosida, *Lectures on differential and integral equations*, Pure and Applied Mathematics, Vol. X, Interscience Publishers, New York-London, 1960. MR0118869

Department of Mathematical Sciences, 373 Dunn Hall, The University of Memphis, Memphis, Tennessee 38152
Email address: ggoldste@memphis.edu

Department of Mathematical Sciences, 373 Dunn Hall, The University of Memphis, Memphis, Tennessee 38152
Email address: jgoldste@memphis.edu

Dipartimento di Matematica "Ennio De Giorgi", Università del Salento, C.P.193, 73100, Lecce, Italy
Email address: giorgio.metafune@unisalento.it

Dipartimento di Matematica "Ennio De Giorgi", Università del Salento, C.P.193, 73100, Lecce, Italy
Email address: luigi.negro@unisalento.it

Contemporary Mathematics
Volume **733**, 2019
https://doi.org/10.1090/conm/733/14742

Surfaces with big automorphism groups

Shulim Kaliman

In memory of Professor Selim Krein, whose Voronezh winter mathematical schools were of immense benefit to the author

ABSTRACT. We describe a classification of $\mathbb{G}_m$-actions on quasi-homogeneous (Gizatullin) surfaces and recent results about $\mathbb{A}^1$-fibrations on these surfaces and their algebraic automorphism groups. We present also a classification of generalized quasi-homogeneous surfaces, where a normal complex affine surface V is generalized quasi-homogeneous if the group of holomorphic automorphisms of V acts transitively on the complement to a finite subset in V.

Introduction

The aim of this survey is to present recent developments in the theory of affine algebraic surfaces with big automorphism groups, i.e. automorphism groups that possess open orbits with finite complements. Unless it is stated otherwise, we consider these surfaces over the field of complex numbers $\mathbb{C}$.

The classical example of such a surface is the plane $\mathbb{C}^2$ which we equip with a coordinate system (x, y). Let us recall some facts about the group of algebraic automorphisms $\mathrm{Aut}(\mathbb{C}^2)$ of $\mathbb{C}^2$. It contains the subgroup of affine transformations $\mathcal{A}_2 = \{\varphi = (q_1, q_2) \in \mathrm{Aut}(\mathbb{C}^2) | \deg q_1 = \deg q_2 = 1\}$ and the subgroup $\mathcal{E}_2$ of elementary automorphisms that are of the form $(x, y) \to (a_1x + b, a_2y + p(x))$, where $q_1, q_2 \in \mathbb{C}[x, y]$, $p \in \mathbb{C}[x]$, $a_1, a_2 \in \mathbb{C}^*$, and $b \in \mathbb{C}$. The Jung-van der Kulk theorem states that $\mathrm{Aut}(\mathbb{C}^2)$ is the amalgamated product of $\mathcal{A}_2$ and $\mathcal{E}_2$, i.e. every element α of $\mathrm{Aut}(\mathbb{C}^2)$ can be presented as a composition $\alpha = \varphi_1 \circ \psi_1 \circ \ldots \circ \varphi_m \circ \psi_m$, where $\varphi_i \in \mathcal{A}_2$ and $\psi_i \in \mathcal{E}_2$ are determined uniquely modulo the subgroup $\mathcal{A}_2 \cap \mathcal{E}_2$.[1] The group $\mathrm{Aut}(\mathbb{C}^2)$ itself is not algebraic and the theorem of Wright [**26**] states that every algebraic subgroup G of $\mathrm{Aut}(\mathbb{C}^2)$ is of bounded length. That is, there exists $m_0 > 0$ such that every $\alpha \in G$ can be presented as a composition $\varphi_1 \circ \psi_1 \circ \ldots \circ \varphi_m \circ \psi_m$ with $m \le m_0$. By Serre's theorem [**22**], such a subgroup is isomorphic to a subgroup of either $\mathcal{A}_2$ or $\mathcal{E}_2$. This fact yields the earlier results of Gutwirth [**16, 17**] and Renschler [**21**] that state that in suitable polynomial coordinate systems on $\mathbb{C}^2$

2010 *Mathematics Subject Classification.* Primary 14R20, 32M17.
Key words and phrases. Affine algebraic surface, automorphism group, action, open orbit.

[1]This theorem is valid over any field (not necessarily algebraically closed or of characteristic zero).

every effective $\mathbb{G}_m$-action[2] is given by

(1) $$(x,y) \to (\lambda^k x, \lambda^l y), \text{ where } \lambda \in \mathbb{C}^* \text{ and } k,l \in \mathbb{Z} \text{ are coprime},$$

and every $\mathbb{G}_a$-action is given by

(2) $$(x,y) \to (x, y + tp(x)), \text{ where } t \in \mathbb{C}_+.$$

Actually, Gutwirth formulated his result (which implies (2)) in terms of $\mathbb{A}^1$-fibrations. Recall that an $\mathbb{A}^1$-fibration is a dominant morphism $\varrho : X \to B$ of algebraic varieties such that general fibers of ϱ are isomorphic to $\mathbb{C}$. If X and B are affine then there is a $\mathbb{G}_a$-action whose general orbits are fibers of ϱ. Vice versa, every nontrivial $\mathbb{G}_a$-action on an affine algebraic variety X of dimension at most 3 generates an $\mathbb{A}^1$-fibration into an affine base B (the so-called categorical quotient morphism [**11**]). In this terminology the Gurwirth's result says that every polynomial on $\mathbb{C}^2$ which yields an $\mathbb{A}^1$-fibration is a variable (more precisely, a variable in a suitable coordinate system). A much more delicate fact is the Abhyankar-Moh-Suzuki theorem that says that an irreducible polynomial whose zero locus is isomorphic to $\mathbb{C}$ is a variable or, equivalently, every plane curve isomorphic to $\mathbb{C}$ can be sent to a coordinate line by an automorphism of $\mathbb{C}^2$. It is worth mentioning that the analytic analogue of this statement does not hold - there are proper holomorphic embeddings of $\mathbb{C}$ into $\mathbb{C}^2$ such that their images cannot be sent to a coordinate line by holomorphic automorphisms of $\mathbb{C}^2$ [**10**]. However, there is a positive result in the analytic setting - M. Suzuki proved in [**25**] that every holomorphic $\mathbb{C}^*$-action on $\mathbb{C}^2$ can be presented in a suitable holomorphic coordinate system by Formula (1). Nothing like this can be true in the case of the holomorphic $\mathbb{G}_a$-actions.

The author expresses his gratitude to the referee for useful comments.

1. Danielewski's Surfaces

In this section we consider one of the simplest nontrivial examples of a surface with a big automorphism group - a Danielewski's surface $S_{1,p}$. Recall that a Danielewski's surface is a hypersurface in $\mathbb{C}^3_{x,y,z}$ [3] such that it is given by $S_{n,p} = \{(x,y,z) \in \mathbb{C}^3 : x^n y = p(z)\}$ where p is a polynomial of degree at least 2 with simple roots. The remarkable fact established by Danielewski is that $S_{n,p} \times \mathbb{C}$ is isomorphic to $S_{k,p} \times \mathbb{C}$ while for different n and $k \in \mathbb{N}$ the surfaces $S_{n,p}$ and $S_{k,p}$ are not isomorphic (i.e. these surfaces yield a counterexample to the general version of the Zariski cancellation conjecture). Actually, in the original approach of Danielewski and Fieseler a stronger fact was established: the fundamental groups of $S_{n,p}$ and $S_{k,p}$ at infinity are different which implies that they are not even homeomorphic. However, there are other methods to prove the absence of isomorphism between $S_{1,p}$ and $S_{n,p}$ for $n \geq 2$, which are more interesting for the purpose of this survey. Namely, one can see that $S_{1,p}$ admits two obvious $\mathbb{A}^1$-fibrations $x : S_{1,p} \to \mathbb{C}$ and $y : S_{1,p} \to \mathbb{C}$ and, therefore, it admits two $\mathbb{G}_a$-actions with different general orbits. On the other hand, for $n \geq 2$ there is no $\mathbb{A}^1$-fibration on $S_{n,p}$ but the obvious one $x : S_{n,p} \to \mathbb{C}$. It is worth studying underlying reasons for this fact.

Consider $\mathbb{C}^2_{x,z}$ as the complement in $S' = \mathbb{P}^1 \times \mathbb{P}^1$ to the lines $C_0' = \{x = \infty\}$ and $C_1' = \{z = \infty\}$. Let $C_2' = \{x = 0\}$. Suppose for simplicity that $p(z) = z(z-1)$. Let $\pi : \tilde{S} \to S'$ be the blowing up of S' at the two points in C_2' given by $x = z = 0$

[2]Recall that a $\mathbb{G}_m$-action (resp. $\mathbb{G}_a$-action) on an algebraic variety X is an algebraic action of the group $\mathbb{C}^*$ (resp. $\mathbb{C}_+$).

[3]The lower indices mean that $\mathbb{C}^3$ is equipped with a coordinate system (x,y,z).

and $x = z - 1 = 0$ and let C_i be the proper transform of C_i'. Then one can see that the Danielewski surface $S_{1,p}$ is obtained from $\tilde{S}$ by removing the divisor $D = C_0 + C_1 + C_2$, i.e. $\tilde{S}$ is a simple normal crossing (SNC) completion of $S_{1,p}$ with D as the boundary divisor.

Recall that the weighted dual graph Γ of a closed SNC curve in a complete surface is the graph whose vertices are the irreducible components of the curve, the edges between the vertices are the intersection points of these components, and the weight of every vertex C is the selfintersection number $C \cdot C$ of this component C in the surface. Since after blowing the surface up at a smooth point of C one has to reduce the selfintersection number of the proper transform of C by 1 and $C_i' \cdot C_i' = 0$ (for C_i' in S'), we have the following linear dual graph of D:

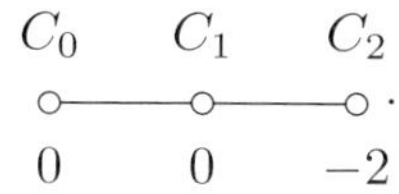

It is a classical fact [**2**] that any smooth rational curve C in a smooth complete surface Y with selfintersection $C \cdot C = 0$ is a fiber of a $\mathbb{P}^1$-fibration $Y \to B$ onto a curve B. In particular, C_0 induces a fibration $\tau : \tilde{S} \to \mathbb{P}^1$. Note that C_1 is a section of this fibration since it meets C_0 transversely. Therefore, the restriction of τ to $S_{1,p}$ is an $\mathbb{A}^1$-fibration (that is nothing but the one provided by the function x). The only fiber of τ different from $\mathbb{P}^1$ is the fiber over $0 \in \mathbb{P}^1$ containing C_2. It contains also two exceptional (-1)-curves E_0 and E_1 of π, where each E_i meets C_2 transversely and $E_i \cap S_{1,p}$ corresponds to the line $x = z - i = 0$ in $S_{1,p}$.

One can see that the dual graph of the SNC-curve $D + E_0 + E_1$ has the following form:

(3)

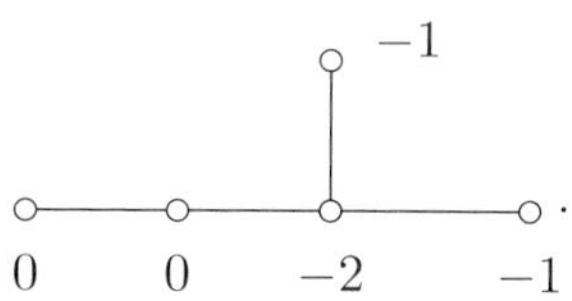

In order to obtain $S_{2,p}$, one has to consider the blowing up of $\tilde{S}$ at the two points $x = z = y = 0$ and $x = z - 1 = y = 0$ and remove the proper transforms of D and also of E_0 and E_1 (whose weights in the resulting surface are equal to -2). This leads to the following dual graph for the boundary divisor of the SNC-completion of $S_{2,p}$

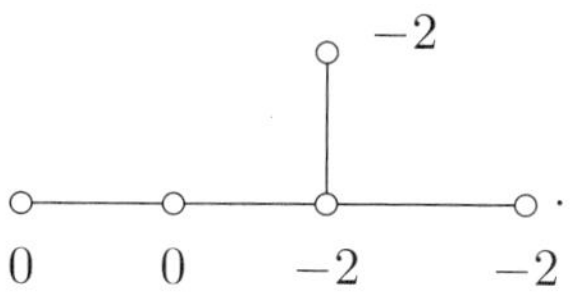

We see that this graph is not linear and this is the crucial difference.

2. Gizatullin surfaces

THEOREM 2.1. ([**12**]) *Let V be a normal affine surface different from $\mathbb{C}^* \times \mathbb{C}$ or the torus $\mathbb{T} = \mathbb{C}^* \times \mathbb{C}^*$. Then the following are equivalent:*

(i) *V admits two different $\mathbb{A}^1$-fibrations with affine bases;*

(ii) $\mathrm{Aut}(V)$ *has an open orbit whose complement is at most finite;*

(iii) *V admits an SNC-completion $\tilde{V}$ such that every component of $D = \tilde{V} \setminus V$ is rational and the dual graph $\Gamma(D)$ of D is*

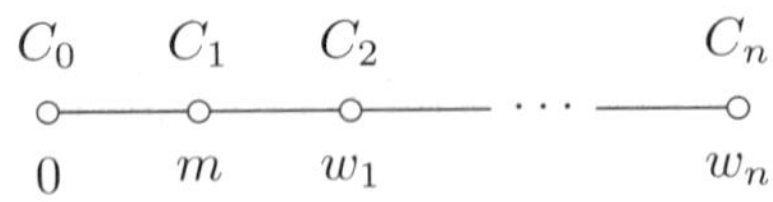

where $n \geq 0$, $m \in \mathbb{Z}$, and every $w_i \leq -2$.

Furthermore, for the given surface V the ordered sequence of weights $[[w_1, w_2, \ldots, w_n]]$ is unique up to the reversion $[[w_n, \ldots, w_2, w_1]]$, while m can be chosen arbitrarily.

REMARK 2.2. One can easily see that the automorphism group $\mathrm{Aut}(V)$ acts transitively on V when $V = \mathbb{C}^* \times \mathbb{C}$ or $V = \mathbb{T}$, i.e. statement (ii) is valid for each of these surfaces. However, they do not satisfy (i) and (iii).

DEFINITION 2.3. (1) A surface with such an SNC-completion is called a Gizatullin (or quasi-homogeneous) surface.

(2) The graph in Theorem 2.1 (iii) is called a zigzag. In the case of $m = 0$ we call such a zigzag standard.

(3) The $\mathbb{P}^1$-fibration defined by C_0 on $\tilde{V}$ has only one non-general fiber and it contains $C_2, \ldots, C_n$ and some other irreducible curves called feathers. The union of D and feathers is an SNC-curve and its dual graph is called the extended graph of the zigzag.

REMARK 2.4. In a smooth Gizatullin surface the complement to the open orbit of $\mathrm{Aut}(V)$ may be non-empty (Danilov-Gizatullin and Kovalenko [**19**]).

Recall that given a (-1)-vertex in a dual graph with at most two neighbors (a so-called linear vertex), we can contract it, which leads to a change of the graph as below:

Using these transformations, we can expand the Gizatullin theorem further.

THEOREM 2.5. ([**12**]) *A normal affine surface is a Gizatullin surface or $\mathbb{C}^* \times \mathbb{C}$ if and only if any SNC-completion of this surface has rational components and its dual graph is contractible to a linear one.*

Hence one can see that the surface $S_{2,p}$ in the previous section is not a Gizatullin surface and thus it has only one $\mathbb{A}^1$-fibration.

3. Danilov-Gizatullin surfaces

DEFINITION 3.1. Let a smooth surface V with the Picard group different from $\mathbb{Z}_2$ have an SNC-completion with rational components and a dual graph of the form

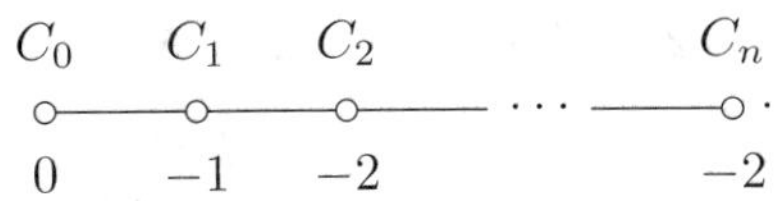

Then it is called a Danilov-Gizatulin surface (e.g. the Danielewski surface $S = \{xy = z(z-1)\}$ is a Danilov-Gizatullin surface).

Using transformations

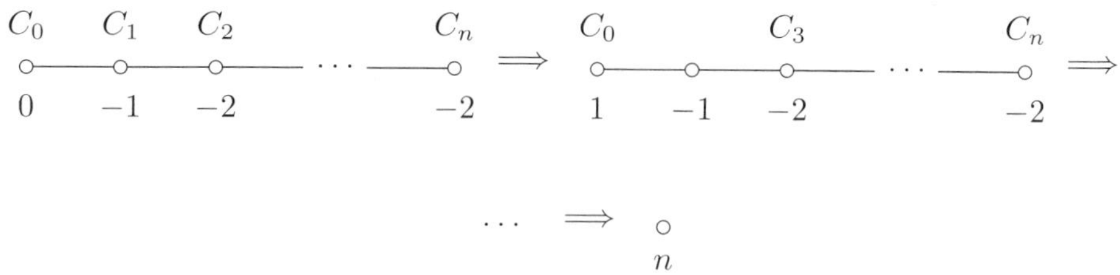

we get an SNC-completion of this surface by an irreducible curve of selfintersection index n. One can check that this completion is a Hirzebruch surface Σ_d over $\mathbb{P}^1$ and the curve is a section L with $L \cdot L = n$, where $d \leq n$.

THEOREM 3.2. ([**5**]) *The isomorphism class of a Danilov-Gizatullin surface is completely determined by the number $L \cdot L$ (which is called the index of the surface).*

REMARK 3.3. Though the standard zigzag of such a surface is unique, the extended graph is not. Here is its general form in which the index t can take any value from 2 to n

(4)

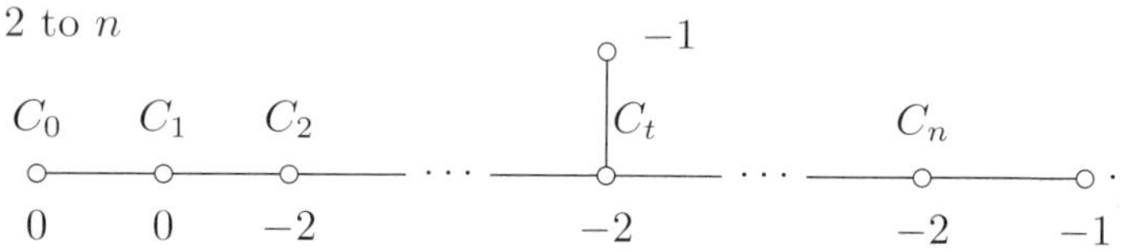

4. $\mathbb{G}_m$-actions on normal surfaces

NOTATION 4.1. We abbreviate by $A_k = [[(-2)_k]]$ the weighted linear graph with k vertices of weight -2 each and by a box $\square$ with a rational weight e/m the weighted linear graph

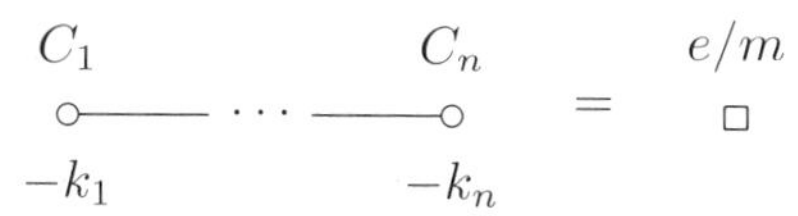

with $k_1, \ldots, k_n \geq 2$, where m/e coincides with the continuous fraction $[k_1, \ldots, k_n]$[4], $0 < e < m$ and $\gcd(m, e) = 1$. A box with the weight $(e/m)^*$ will denote the similar linear chain of vertices with the reverse order of the weights.

THEOREM 4.2. ([**6**]) *A standard zigzag occurs as the boundary divisor D of a smooth Gizatullin surface $V \neq \mathbb{C}^2$ or $\mathbb{C} \times \mathbb{C}^*$ with a nontrivial $\mathbb{G}_m$-action if and only if it can be written in one of the forms*

(A)

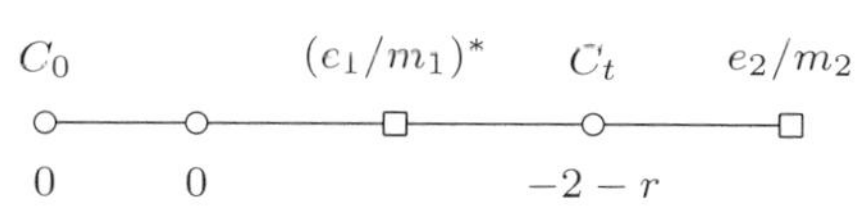

or (B)

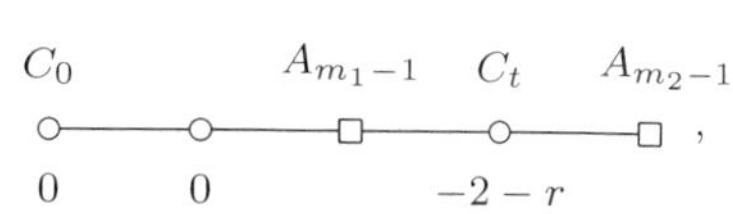

where $r \geq 0$, $m_i \geq 1$, $\gcd(e_i, m_i) = 1$ for $i = 1, 2$, and either

$$\frac{e_1}{m_1} + \frac{e_2}{m_2} = 1, \quad or \quad \frac{e_1}{m_1} + \frac{e_2}{m_2} = 1 - \frac{1}{m_1 m_2}.$$

Furthermore, the extended graph $\Gamma_{\text{ext}}(D)$ has only one branch vertex C_t.

[4]For a sequence of real numbers $k_0, \ldots, k_n$ with $k_0, \ldots, k_{n-1} \geq 2$ and $k_n \geq 1$ we let $[k_0, \ldots, k_n]$ be the continued fraction defined inductively via $[k_0] = k_0$ and $[k_0, \ldots, k_n] = k_0 - \frac{1}{[k_1, \ldots, k_n]}$.

A few words about the proof of this result. By the Sumihiro theorem [**23**], there exists a $\mathbb{G}_m$-equivariant completion $\tilde{V}$ of V. One can show that such a completion can be chosen as a zigzag. Hence the $\mathbb{G}_m$-action preserves the 0-vertex C_0 and the $\mathbb{P}^1$-fibration associated with it. In particular, it preserves the feathers (since they are components of the singular fiber of this fibration) and the points where the feathers meet the components of the boundary. Therefore, if C_i is a branch point of $\Gamma_{\text{ext}}(D)$, then the restriction of the $\mathbb{G}_m$-action to it is trivial, since it has at least three fixed points on $C_i \simeq \mathbb{P}^1$. Another easy claim is that when this $\mathbb{G}_m$-action is trivial on two components of the boundary, then it is trivial on V. This yields the last claim of the theorem, from which it is not difficult to extract the rest.

REMARK 4.3. When $r = 0$ in class (B) from Theorem 4.2, then V is just a Danilov-Gizatullin surface with the extended graph as in (4). When $r \geq 1$ and $t \neq n$, then there are $r+1$ feathers $F, F_1, \ldots, F_r$ that are neighbors of C_t (in the case of $t = n$ there is one extra feather neighboring C_t). Furthermore, $F_1, \ldots, F_r$ are (-1)-vertices in $\Gamma_{\text{ext}}(D)$, while the weight of F is $1 - t$.

DEFINITION 4.4. (1) We call a Gizatullin surface special if it has a dual graph as in class (B) from Theorem 4.2 with $r \geq 1$. It is special of type I if either $t \in \{2, n\}$ or $r = 1$. Otherwise it is special of type II.

(2) We say that two $\mathbb{G}_m$-actions (resp. $\mathbb{A}^1$-fibrations) on an algebraic surface V are equivalent if there exists an automorphism of V that transforms one into the other.

THEOREM 4.5. ([**7**, **9**]) *Let V be a smooth affine surface with a nontrivial $\mathbb{G}_m$-action. Then this action is unique up to equivalence and taking the inverse action $\lambda \to \lambda^{-1}$ if and only if V is not a toric surface* [5]*, a Danilov-Gizatullin surface of index $n \geq 4$, or a special Gizatullin surface.*

In the case of a Danilov-Gizatullin surface V with an extended graph $\Gamma_{\text{ext}}(D)$ as in (4), the multiplicities of the two feathers in the singular fiber of the $\mathbb{P}^1$-fibration $\psi_t : \tilde{V} \to \mathbb{P}^1$ associated with C_0 are 1 and $t - 1$ respectively. As we mentioned, when $\tilde{V}$ is a $\mathbb{G}_m$-equivariant completion, then the $\mathbb{G}_m$-action preserves ψ_t and, in particular, the $\mathbb{A}^1$-fibration $\psi_t|_V : V \to \mathbb{C}$. This enabled P. Russell to conclude that for different values of t such actions are not equivalent, i.e. there are at least $n - 1$ non-equivalent $\mathbb{G}_m$-actions on V. In fact, there are no more [**6**]. In the case of special smooth Gizatullin surfaces, one has the following.

THEOREM 4.6. [**8**]) *Let V be a special smooth Gizatullin surface. If V is of type I, then the equivalence classes of $\mathbb{G}_m$-actions on V form in a natural way a 1-parameter family, while in case of type II they form a 2-parameter family.*

The proof of this result is based on a difficult generalization of Theorem 3.2 [**8**, Corollary 6.1.4] that can be formulated in the following form:
Up to an affine transformation of $C_t \setminus C_{t-1} \simeq \mathbb{C}$, the isomorphism class of the special Gizatullin surface V from Theorem 4.2 (B) is uniquely determined by the set of points $\{F_i \cap C_t\}_{i=1}^r$.

[5]Recall that a surface is called toric if it admits an effective algebraic action of the torus $\mathbb{T}$. The list of smooth affine toric surfaces consists of $\mathbb{T}$, $\mathbb{C}^* \times \mathbb{C}$ and $\mathbb{C}^2$ while every other affine normal toric surface can be obtained as the quotient of $\mathbb{C}^2$ with respect to a linear action of the group of d-roots of unity for some $d \in \mathbb{N}$.

5. $\mathbb{A}^1$-fibrations

There is an analogue of Theorem 4.5 for $\mathbb{A}^1$-fibrations.

THEOREM 5.1. ([**7**]) *Let V be a smooth affine surface with a nontrivial $\mathbb{G}_m$-action such that V is not a Danilov-Gizatullin surface of index $n \geq 4$ or a special Gizatullin surface. Then V admits at most two non-equivalent $\mathbb{A}^1$-fibrations $V \to \mathbb{A}^1$.*

In the case of a Danilov-Gizatullin surface, the number of non-equivalent $\mathbb{A}^1$-fibrations is finite when the index $n \leq 5$. If $n \geq 6$, then there exist $\mathbb{A}^1$-fibrations that are not invariant under any $\mathbb{G}_m$-action (unlike the ones discussed in the previous section). It is unknown if the number of such fibrations is finite for $n = 6$.

THEOREM 5.2. ([**8**, Example 6.3.21]) *If V is a Danilov-Gizatullin surface of index $n \geq 7$, then V carries continuous families of pairwise non-equivalent $\mathbb{A}^1$-fibrations with the number of parameters being an increasing function of n.*

The similar fact holds for special Gizatullin surfaces.

THEOREM 5.3. ([**8**, Corollary 6.3.22]) *Let V be a special Gizatullin surface with a zigzag of length n. Then there are families of pairwise non-equivalent $\mathbb{A}^1$-fibrations $V \to \mathbb{A}^1$ depending on $r(n) \geq 1$ parameters with $\lim_{n\to\infty} r(n) = \infty$.*

There is no analogue of the Abhyankar-Moh-Suzuki Theorem for Gizatullin surfaces. Indeed, consider the Danielewski surface $S = \{xy = z(z-1)\}$ and the curves L_1 and L_2 in it given by $x = 1$ and by $x = z = 0$ respectively. Both of them are isomorphic to $\mathbb{C}$, but there is no automorphism of S that transforms one into the other, since L_1 is a principal divisor while L_2 is not.

A more delicate question was studied in [**15**]. It turns out that, if the Picard group $\mathrm{Pic}(V)$ of a Gizatullin surface V is not a torsion group, then there may exist an affine line $L \simeq \mathbb{C}$ in V that is not a component of a fiber of any $\mathbb{A}^1$-fibration $V \to \mathbb{C}$. Actually, the authors found such an L for which the logarithmic Kodaira dimension of $V \setminus L$ is nonnegative, while for any component of a fiber of an $\mathbb{A}^1$-fibration the similar dimension is $-\infty$.

6. Amalgams

Recall that a tree is a connected (non-weighted) graph without cycles. Consider a pair $(T, \mathcal{G})$ that consists of a tree T and a collection $\mathcal{G}$ of vertex groups $(G_P)_{P\in \mathrm{vert}T}$ and edge groups $(G_\nu)_{\nu\in \mathrm{edge}T}$ such that for every edge $\nu = [P, Q]$ of T, there are monomorphisms $G_\nu \to G_P$ and $G_\nu \to G_Q$ identifying G_ν with proper subgroups of the vertex groups G_P and G_Q. Then one can construct a unique group $G = \lim_\to(T, \mathcal{G})$ called the free amalgamated product of the vertex groups, where G is freely generated by the subgroups (G_P) and (G_ν) with unified subgroups $G_P \cap G_Q = G_\nu$ for each $\nu = [P, Q] \in \mathrm{edge}T$ [**22**, Ch. I, Sections 4,5].

The following analogue of the Jung-Van der Kulk theorem was proved in [**5**].

THEOREM 6.1. *Let V be a Danilov -Gizatullin surface of index $n \leq 5$ and $\mathrm{Aut}_o(V)$ be the connected component of identity in $\mathrm{Aut}(V)$. Then $\mathrm{Aut}_o(V)$ is an amalgamated product of at most three algebraic groups.*

On the other hand, the following result holds [**20**].

PROPOSITION 6.2. *Let Y be a normal affine variety that does not admit a unipotent group action with a general orbit of dimension ≥ 2. Suppose that the connected component* $\mathrm{Aut}_o(Y)$ *of identity in* $\mathrm{Aut}(Y)$ *is an amalgamated product of a countable number of algebraic groups. Then the set of non-equivalent $\mathbb{A}^1$-fibrations on Y over affine bases is countable.*

In combination with Theorems 5.2 and 5.3, this Proposition yields the next fact.

THEOREM 6.3. ([**20**]) *Let V be either a Danilov-Gizatullin surface of index $n \geq 7$ or a special Gizatullin surface. Then* $\mathrm{Aut}_o(V)$ *is not an amalgamated product of a countable number of algebraic groups.*

Another characterization of the complexity of the automorphism groups of Gizatullin surfaces was found by Blanc and Dubouloz [**3**]. Namely, there exists a surface V such that for the normal subgroup $N(V) \subset \mathrm{Aut}(V)$ generated by all algebraic subgroups of $\mathrm{Aut}(V)$ the quotient $\mathrm{Aut}(V)/N(V)$ contains a free group on an uncountable set of generators.

7. Generalized Gizatullin surfaces

Recall that a holomorphic vector field ν on a complex space X is called complete if the ODE $\frac{d}{dt}\Phi(x,t) = \nu(\Phi(x,t))$ with initial data $\Phi(x,0) = x$ has a solution $\Phi(x,t)$ defined for all values of complex time $t \in \mathbb{C}$ and every starting point $x \in X$, i.e. the flow $\Phi(*,t)$ yields a complex one-parameter subgroup of the holomorphic automorphism group $\mathrm{Aut}_{\mathrm{hol}}(X)$.

Every $\mathbb{G}_a$-action (resp. $\mathbb{G}_m$-action) on an algebraic variety X is the flow of a complete algebraic vector field which is called locally nilpotent (resp. semi-simple). However, there are complete algebraic vector fields that are neither locally nilpotent nor semi-simple and their flows are not algebraic but only holomorphic actions of $\mathbb{C}_+$. Up to conjugation by elements of $\mathrm{Aut}(\mathbb{C}^2)$, such complete algebraic fields on $\mathbb{C}^2$ were classified by Brunella [**4**]. Consider some examples of these fields.

EXAMPLE 7.1. The vector field $yx\frac{\partial}{\partial x}$ is complete, but its flow $(x,y) \to (e^{ty}x, y)$ is not algebraic, i.e. elements of this flow are only holomorphic automorphisms. Of course, the canonical forms of complete algebraic vector fields found by Brunella are more complicated and here is one of them: $ax\frac{\partial}{\partial x} + p(x^n y^m)[ny\frac{\partial}{\partial y} - mx\frac{\partial}{\partial x}]$ where $a \in \mathbb{C}$, $n, m \geq 1$ are coprime and $p(z) \in \mathbb{C}[z]$.

DEFINITION 7.2. Denote by $\mathrm{AAut}(X)$ the subgroup of the group $\mathrm{Aut}_{\mathrm{hol}}(X)$ of holomorphic automorphisms of X generated by the elements of the flows of complete algebraic vector fields on X. We call a normal affine surface V a generalized Gizatullin surface if $\mathrm{AAut}(V)$ admits an open orbit whose complement is at most finite.[6]

The technique of Brunella and some fundamental results of McQuillan were originally used in the proof of Theorem 7.4 below that describes generalized Gizatullin surfaces. It was later extracted in a much easier way from the next very general result of Guillot and Rebelo on semi-complete vector fields [**14**].

---

[6]It is worth mentioning that in all Kovalenko's examples [**19**] the group $\mathrm{AAut}(V)$ acts transitively on V while $\mathrm{Aut}(V)$ does not.

Theorem 7.3. *Let V be a normal affine algebraic surface that admits a nonzero complete algebraic vector field. Then either:*

(1) all complete algebraic fields share the same rational first integral (i.e. there is a rational map $f : V \dashrightarrow B$ such that all complete algebraic vector fields on V are tangent to the fibers of f), or

(2) V is a rational surface with an open orbit of $\mathrm{AAut}(V)$ and, furthermore, for every complete algebraic vector field ν on V there is a regular function $f : V \to \mathbb{C}$ (depending on ν) with general fibers isomorphic to $\mathbb{C}$ or $\mathbb{C}^*$ such that the flow of ν sends fibers of f to fibers of f.

As we have mentioned, Theorem 7.3 is an essential ingredient in the proof of the following main result of [**18**].

Theorem 7.4. *A normal complex affine algebraic surface V is generalized Gizatullin if and only if it admits an SNC-completion $\bar{V}$ for which the boundary $\bar{V} \setminus V$ is connected, consists of rational curves, and has a dual graph that belongs to one of the following types*

(1) *a standard zigzag or a linear chain of three 0-vertices (i.e. Gizatullin surfaces and $\mathbb{C} \times \mathbb{C}^*$),*

(2) *a circular graph with the following possibilities for weights*

(2a) $((0, 0, w_1, \ldots, w_n))$ *where* $n \geq 0$ *and* $w_i \leq -2$,

(2b) $((0, 0, w))$ *with* $-1 \leq w \leq 0$ *or* $((0, 0, 0, w))$ *with* $w \leq 0$,

(2c) $((0, 0, -1, -1))$;

(3) 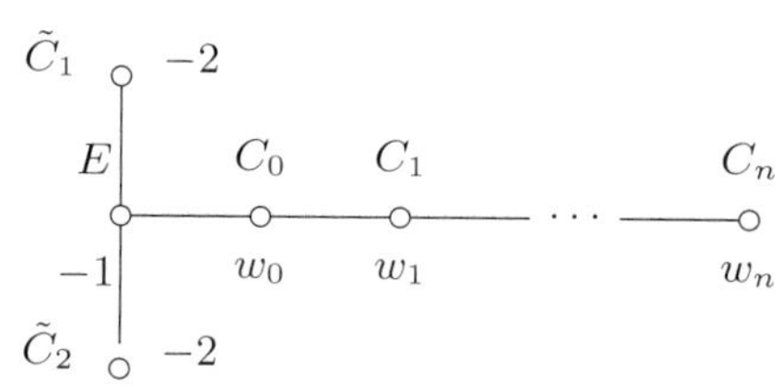

where $n \geq 0$, $w_0 \geq 0$ *and* $w_i \leq -2$ *for* $i \geq 1$,

(4) 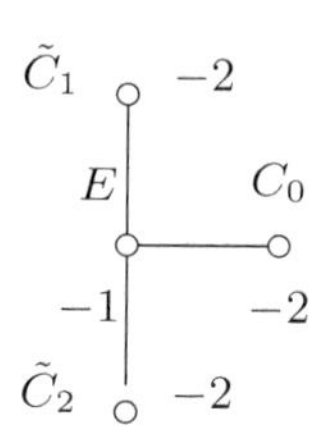

(5) 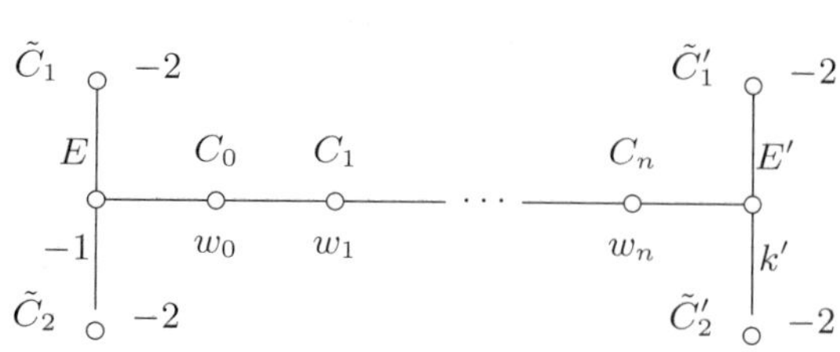

where $n \geq 0$, $w_0 \geq 0$ *and* $w_i \leq -2$ *for* $i \geq 1$; *moreover* $k' \leq -1$ *if* $n = 0$ *while for* $n > 0$ *one has* $k' \leq -2$,

(6) 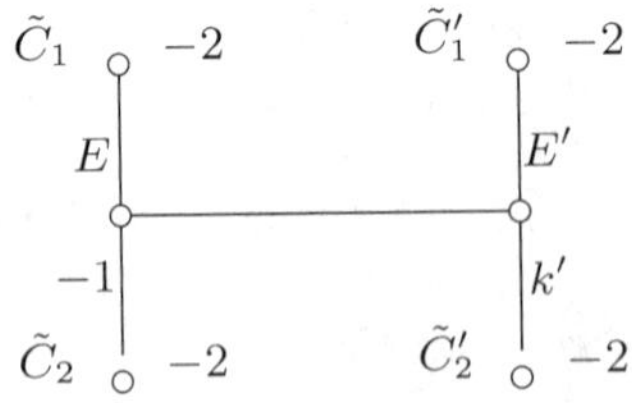

where $k' \geq -1$.

REMARK 7.5. With the exception of (2a), every graph in Theorem 7.4 can appear as the dual graph of an SNC-curve in a rational surface. In order to guarantee the same in (2a), one has to impose the following additional necessary and sufficient assumption: there is a linear chain $C_1 + \ldots + C_n$ of rational curves with weights $[[u_1, \ldots, u_n]]$ such that $u_i \geq w_i$, the curve $C_2 + \ldots + C_{n-1}$ is contractible and the weights of the proper transforms of C_1 and C_n become zero after this contraction.

Without going into a rather technical description of the proof of this result, let us consider some interesting examples.

EXAMPLE 7.6. (a) Consider hypersurfaces $\{xp(x) + yq(y) + xyz = 1\} \subset \mathbb{C}^3_{x,y,z}$, where the polynomials $1 - xp(x)$ and $1 - yq(y)$ have simple roots only (say, $x + y + xyz = 1$). None of them admits nontrivial algebraic $\mathbb{G}_a$ or $\mathbb{G}_m$-actions, but they are generalized Gizatullin surfaces and as in the case of the torus, for each such a surface V the dual graph of an SNC-completion $\bar{V}$ can be chosen as a cycle.

(b) The following special case of (5)

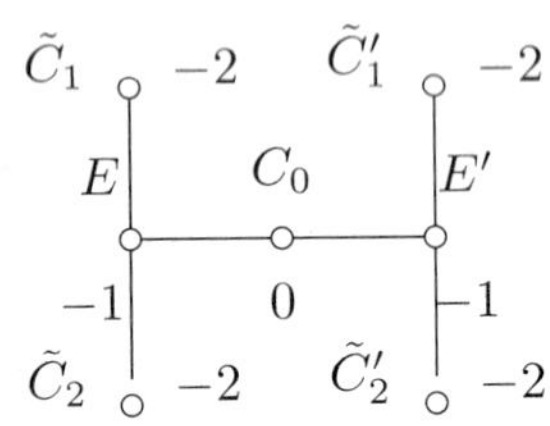

yields a surface which is a locally trivial twisted $\mathbb{C}^*$-fibration over $\mathbb{C}^*$ (i.e. it is nothing but a complexification of the Klein bottle).

We conclude the paper with a theorem describing some singular generalized Gizatullin surfaces, which include a unique non-toric surface of this type.

THEOREM 7.7. ([**18**]) *Let V_0 be a normal generalized Gizatullin surface such that for a complete algebraic vector field ν_0 on V_0 there is a surjective rational first integral $f_0 : V_0 \dashrightarrow B$ into a complete curve B. Then*

(1) *either V_0 is toric (and, in particular, quasi-homogeneous), or V_0 is isomorphic to the hypersurface $y(x^2 + y^2) + z^2 = 0$ and, in particular, it has $(-D_4)$-singularity*[7]*;*

[7]A singularity of type $-D_{n+1}$ is locally isomorphic to the hypersurface $yx^2 + y^n + z^2 = 0$ in $\mathbb{C}^3_{x,y,z}$.

(2) *up to a constant nonzero factor* ν_0 *is semi-simple.*

References

[1] S. S. Abhyankar and T. T. Moh, *Embeddings of the line in the plane*, J. Reine Angew. Math. **276** (1975), 148–166. MR0379502

[2] W. P. Barth, K. Hulek, C. A. M. Peters, and A. Van de Ven, *Compact complex surfaces*, 2nd ed., Ergebnisse der Mathematik und ihrer Grenzgebiete. 3. Folge. A Series of Modern Surveys in Mathematics [Results in Mathematics and Related Areas. 3rd Series. A Series of Modern Surveys in Mathematics], vol. 4, Springer-Verlag, Berlin, 2004. MR2030225

[3] J. Blanc and A. Dubouloz, *Affine surfaces with a huge group of automorphisms*, Int. Math. Res. Not. IMRN **2** (2015), 422–459, DOI 10.1093/imrn/rnt202. MR3340326

[4] M. Brunella, *Complete polynomial vector fields on the complex plane*, Topology **43** (2004), no. 2, 433–445, DOI 10.1016/S0040-9383(03)00050-8. MR2052971

[5] V. I. Danilov, M. H. Gizatullin, *Automorphisms of affine surfaces*. I. Math. USSR Izv. 9 (1975), 493–534; II. *ibid.* **11** (1977), 51–98.

[6] H. Flenner, S. Kaliman, and M. Zaidenberg, *Completions of* $\mathbb{C}^*$*-surfaces*, Affine algebraic geometry, Osaka Univ. Press, Osaka, 2007, pp. 149–201. MR2327238

[7] H. Flenner, S. Kaliman, and M. Zaidenberg, *Uniqueness of* $\mathbb{C}^*$*- and* $\mathbb{C}_+$*-actions on Gizatullin surfaces*, Transform. Groups **13** (2008), no. 2, 305–354, DOI 10.1007/s00031-008-9014-0. MR2426134

[8] H. Flenner, S. Kaliman, and M. Zaidenberg, *Smooth affine surfaces with non-unique* $\mathbb{C}^*$*-actions*, J. Algebraic Geom. **20** (2011), no. 2, 329–398, DOI 10.1090/S1056-3911-2010-00533-4. MR2762994

[9] H. Flenner and M. Zaidenberg, *On the uniqueness of* $\mathbb{C}^*$*-actions on affine surfaces*, Affine algebraic geometry, Contemp. Math., vol. 369, Amer. Math. Soc., Providence, RI, 2005, pp. 97–111, DOI 10.1090/conm/369/06807. MR2126657

[10] F. Forstneric, J. Globevnik, and J.-P. Rosay, *Nonstraightenable complex lines in* $\mathbf{C}^2$, Ark. Mat. **34** (1996), no. 1, 97–101, DOI 10.1007/BF02559509. MR1396625

[11] G. Freudenburg, *Algebraic theory of locally nilpotent derivations*, Encyclopaedia of Mathematical Sciences, vol. 136, Springer-Verlag, Berlin, 2006. Invariant Theory and Algebraic Transformation Groups, VII. MR2259515

[12] M. H. Gizatullin, I. *Affine surfaces that are quasihomogeneous with respect to an algebraic group*, Math. USSR Izv. 5 (1971), 754–769; II. *Quasihomogeneous affine surfaces*, *ibid.* 1057–1081.

[13] M. H. Gizatullin, *Quasihomogeneous affine surfaces* (Russian), Izv. Akad. Nauk SSSR Ser. Mat. **35** (1971), 1047–1071. MR0286791

[14] A. Guillot and J. Rebelo, *Semicomplete meromorphic vector fields on complex surfaces*, J. Reine Angew. Math. **667** (2012), 27–65. MR2929671

[15] R. V. Gurjar, K. Masuda, M. Miyanishi, and P. Russell, *Affine lines on affine surfaces and the Makar-Limanov invariant*, Canad. J. Math. **60** (2008), no. 1, 109–139, DOI 10.4153/CJM-2008-005-8. MR2381169

[16] A. Gutwirth, *An inequality for certain pencils of plane curves*, Proc. Amer. Math. Soc. **12** (1961), 631–638, DOI 10.2307/2034258. MR0126759

[17] A. Gutwirth, *The action of an algebraic torus on the affine plane*, Trans. Amer. Math. Soc. **105** (1962), 407–414, DOI 10.2307/1993728. MR0141664

[18] S. Kaliman, F. Kutzschebauch, M. Leuenberger, *Complete algebraic vector fields on affine surfaces*, preprint, 42 p., arXiv:1411.5484.

[19] S. Kovalenko, *Transitivity of automorphism groups of Gizatullin surfaces*, Int. Math. Res. Not. IMRN **21** (2015), 11433–11484, DOI 10.1093/imrn/rnv025. MR3456050

[20] S. Kovalenko, A. Perepechko, and M. Zaidenberg, *On automorphism groups of affine surfaces*, Algebraic varieties and automorphism groups, Adv. Stud. Pure Math., vol. 75, Math. Soc. Japan, Tokyo, 2017, pp. 207–286. MR3793368

[21] R. Rentschler, *Opérations du groupe additif sur le plan affine* (French), C. R. Acad. Sci. Paris Sér. A-B **267** (1968), A384–A387. MR0232770

[22] J.-P. Serre, *Trees*, Springer-Verlag, Berlin-New York, 1980. Translated from the French by John Stillwell. MR607504

[23] H. Sumihiro, *Equivariant completion*, J. Math. Kyoto Univ. **14** (1974), 1–28, DOI 10.1215/kjm/1250523277. MR0337963

[24] M. Suzuki, *Propriétés topologiques des polynômes de deux variables complexes, et automorphismes algébriques de l'espace* $\mathbf{C}^2$ (French), J. Math. Soc. Japan **26** (1974), 241–257, DOI 10.2969/jmsj/02620241. MR0338423

[25] M. Suzuki, *Sur les opérations holomorphes du groupe additif complexe sur l'espace de deux variables complexes* (French), Ann. Sci. École Norm. Sup. (4) **10** (1977), no. 4, 517–546. MR0590938

[26] D. Wright, *Abelian subgroups of* $\mathrm{Aut}_k(k[X, Y])$ *and applications to actions on the affine plane*, Illinois J. Math. **23** (1979), no. 4, 579–634. MR540400

University of Miami, Coral Gables, FL 33124, USA
Email address: kaliman@math.miami.edu

Contemporary Mathematics
Volume **733**, 2019
https://doi.org/10.1090/conm/733/14743

Some binomial formulas for non-commuting operators

Peter Kuchment and Sergey Lvin

Dedicated to the memory of our beloved teacher, colleague, and co-author Selim Krein

ABSTRACT. Let D and U be linear operators in a vector space (or more generally, elements of an associative algebra with a unit). We establish binomial-type identities for D and U assuming that either their commutator $[D, U]$ or the second commutator $[D, [D, U]]$ is proportional to U.

Operators $D = d/dx$ (differentiation) and U- multiplication by $e^{\lambda x}$ or by $\sin \lambda x$ are basic examples, for which some of these relations appeared unexpectedly as byproducts of an authors' medical imaging research [**2**–**5**].

Introduction

While working on range conditions for a Radon type transform arising in emission medical imaging, the authors [**3**, **4**] (see also [**5**]) discovered that one of their theorems was equivalent to an infinite series of puzzling nonlinear combinatorial-differential identities for the classical exponential, linear, and some trigonometric functions (sic!). Here are the examples:

- For any non-negative integer n and $u = e^{\lambda x}$, one has

$$\sum_{k=0}^{n} \binom{n}{k} \left[\left(\frac{d}{dx} - u + 0\lambda \right) \circ \cdots \circ \left(\frac{d}{dx} - u + (k-1)\lambda \right) \right] u^{n-k} = 0. \tag{0.1}$$

- For $u = \sin \lambda x$, similar identities hold for any odd natural n:

$$\sum_{k=0}^{n} \binom{n}{k} \left[\left(\frac{d}{dx} - u + 0\lambda \right) \circ \cdots \circ \left(\frac{d}{dx} - u + i(k-1)\lambda \right) \right] u^{n-k} = 0. \tag{0.2}$$

In order to avoid quite possible misunderstanding, let us explain briefly the meaning of various terms of these identities.

REMARK 0.1. The factors in the $\circ$-products are understood as operators on smooth functions on the line. In particular, u there means the multiplication operator by the function u (exponential or sine). The circle $\circ$ means the composition of operators, and the order of factors is important, due to them non-commuting. On the other hand, u^{n-k} at the end is considered as a function, to which the operator [...] is applied. If one tries to understand u^{n-k} also as an operator and thus

---

2010 *Mathematics Subject Classification.* Primary 11B65, 16B99, 20F12, 47B47; Secondary 33B9, 44A12; 92C55; 81S05.

The authors thank the reviewer for very helpful comments and literature references. P. K. expresses his gratitude to the support from DMS NSF Grants #1211463 and #1816430.

considers $\sum \binom{n}{k} [\cdots] \circ u^{n-k}$, the resulting **operator** is NOT identically equal to zero. It needs to be applied to the function identically equal to 1 to preserve the identity. This understanding will be important throughout the text.

It is also why, to avoid misinterpretation, **we use later the notation $\widetilde{f}$ for the operator of multiplication by the function f.**

In the paper, we significantly extend the results of [**3**–**5**]), as well as generalize them to a much wider algebraic situation. Namely, the setting in which we obtained these results before was of a commutative algebra (where u belongs to) with differentiation D and u satisfying "differential equations" $Du = \lambda u$ or $D^2u = \lambda^2 u$. Now we show that the results have generalization to elements D and U of any associative algebra with a unit, with appropriate conditions on their first and second commutators of D and U.

When these identities appeared in medical imaging, they have attracted quite a lot of attention, especially after discovering their relations to unusual Hartogs type analytic continuation theorems in several complex variables [**1**, **7**, **8**, **10**]. It is fair to notice that, in spite of a variety of different proofs and generalizations of the identities available, the authors still feel that they do not have good understanding of the origins of such formulas.

Here is the structure of the paper: The main notions are introduced and results stated in Section 1. Section 2 is devoted to particular cases of elementary functions. The proofs are delegated to Section 3, followed by a final remarks section.

1. Formulation of main results

Let D and U be elements of an associative algebra A over a field $\mathbb{Q}$ with identity I (for instance, the algebra of all linear operators in a vector space $\mathbb{F}$).

Let us introduce, lead by (0.1)-(0.2), an nth order binomial-type combination of D and U

$$B(n, \lambda, U, D) := \sum_{k=0}^{n} \binom{n}{k} \left(\prod_{j=0}^{k-1} (D - U + j\lambda I) \right) U^{n-k}, \tag{1.1}$$

where $n \geq 0$ is an integer, $\lambda \in \mathbb{Q}$, and $\binom{n}{k}$ is the binomial coefficient. When $k = 0$, the product is understood as I.

Due to non-commutativity of A, we will adhere to the following agreement: The products $\prod_j \cdots$ are understood in the order of the index j increasing from the left to the right.

For instance,

$$B(0, \lambda, U, D) = I,$$

$$B(1, \lambda, U, D) = U + (D - U) = D,$$

$$B(2, \lambda, U, D) = U^2 + 2(D - U)U + (D - U)(D - U + \lambda I), \text{ etc.},$$

with formulas getting more complex with n increasing.

1.1. First order commutator. In the following theorems we make certain assumptions about the commutator $[D, U] = DU - UD$.

The following rather surprising result holds:

THEOREM 1.1. *Suppose that* $[D, U] = \lambda U$. *Then* $B(n, \lambda, U, D)$ *does not depend on* U. *Moreover,*

$$B(n, \lambda, U, D) = B(n, \lambda, 0, D) = \prod_{j=0}^{n-1} (D + j\lambda I) . \tag{1.2}$$

REMARK 1.2.

- If D and U commute ($\lambda = 0$), then the theorem states that

  $$B(n, 0, U, D) = D^n,$$

  which is an obvious consequence of the standard binomial formula. Indeed, when D and U commute,

  $$B(n, 0, U, D) = \sum_{k=0}^{n} \binom{n}{k} (D - U)^k U^{n-k} = (D - U + U)^n = D^n.$$

- The equality $[D, U] = \lambda U$ is homogeneous of degree one with respect to (D, λ). Homogeneity is not obvious for the originally defined $B(n, \lambda, U, D)$, however the statement of the theorem implies that the homogeneity does hold:

  $$B(n, \lambda, U, D) = \lambda^n B(n, 1, U, D/\lambda), \tag{1.3}$$

  This shows that essentially the study boils down to only the cases when $\lambda = 0$ (considered above) and $\lambda = 1$, which simplifies considerations.

Here are some immediate consequences of the Theorem:

COROLLARY 1.3. *Suppose* $[D, U] = \lambda U$. *Then,*

(1) *If* $V \in A$ *and for some* $j \in \{0, ..., n - 1\}$, *one has* $(D + j\lambda I)V = 0$, *then*

$$(B(n, \lambda, U, D)) V = 0. \tag{1.4}$$

If A *is the algebra of linear operators on a vector space, the equivalent reformulation, under the same assumptions, is:*

$$Ker(D + j\lambda I) \subset Ker(B(n, \lambda, U, D)).$$

(2) *If* $[V, W] = 0$ *and* $[D, W] = \lambda W$, *then* $B(n, \lambda, V + W, D)$ *does not depend on* W:

$$B(n, \lambda, V + W, D) = B(n, \lambda, V, D).$$

Indeed, just substitute D *in the Theorem with* $D - V$.

For what follows, it is interesting to understand what happens with $B(n, \lambda, U, D)$ when $[D, U] = -\lambda U$ (notice the wrong sign in the commutation relation, and thus Theorem 1.1 describes $B(n, -\lambda, U, D)$, rather than $B(n, \lambda, U, D)$). Things get more complicated here.

In the following results we denote by $\mathbb{F}_0$ the set of all $V \in A$ is such that $DV = 0$. We use the standard notation $(2m - 1)!!$ for $1 \cdot 3 \cdot ... \cdot (2m - 1)$.

THEOREM 1.4. *Let* $[D, U] = -\lambda U$, *then*

- $B(n, \lambda, U, D)|_{\mathbb{F}_0} = 0$ *for any odd* n,
- $B(n, \lambda, U, D)|_{\mathbb{F}_0} = (n-1)!!(-2\lambda U)^{n/2}|_{\mathbb{F}_0}$ *for even* $n > 0$, *and*
- $(2D + \lambda n I)B(n, \lambda, U, D)|_{\mathbb{F}_0} = 0$ *for all* $n \geq 0$.

The reason why we interested in this case will be clear in the proof of Theorem 1.5 below.

1.2. Second order commutators. Here we will be interested in using conditions on the second order commutators $[D, [D, U]]$ and $[U, [D, U]]$. Here, the condition $[D, U] = \lambda U$ is replaced with

$$[D, [D, U]] = \lambda^2 U$$

(notice that we preserve the (D, λ)-homogeneity)[1]. We are interested in behavior of the polynomials $B(n, \lambda, U, D)$ in this situation.

THEOREM 1.5. *Suppose* $[D, [D, U]] = \lambda^2 U$ *and* $[U, [D, U]] = 0$. *Then*

- $B(n, \lambda, U, D)|_{\mathbb{F}_0} = 0$ *for all odd* n,
- $B(n, \lambda, U, D)|_{\mathbb{F}_0} = (n-1)!!([D, U] - \lambda U)^{n/2}|_{\mathbb{F}_0}$ *for all even* $n > 0$, *and*
- $(2D + \lambda n I)B(n, \lambda, U, D)|_{\mathbb{F}_0} = 0$ *for all* $n \geq 0$.

REMARK 1.6. Theorem 1.5 (unlike Theorem 1.1) remains nontrivial even when $\lambda = 0$.

2. Differential identities for some elementary functions

Here we apply the above results to the algebra of linear operators acting in the vector space $\mathbb{F} := C^\infty(\mathbb{R})$ of all smooth functions on the real line.

DEFINITION 2.1.

- We **denote by $\widetilde{f}$ the operator of multiplication by such a function** $f(x)$.
- We use the notation **1** for the function that is identically equal to 1.
- We also denote $D := d/dx$.

2.1. Exponential functions. It is clear that $e^{-j\lambda x}$ belongs the kernel of $D + j\lambda I$ and, in particular, **1** is an element in the kernel of D.

Now one clearly has

$$[D, \widetilde{e^{\pm\lambda x}}] = \pm\lambda\widetilde{e^{\pm\lambda x}}.$$

Thus, the results of Section 1.1 apply to produce the following formulas:

THEOREM 2.2.

$$B(n, \lambda, d/dx, \widetilde{e^{\lambda x}})e^{-j\lambda x} = 0 \text{ for all } n > 0 \text{ and } 0 \leq j \leq n-1, \tag{2.1}$$

$$B(n, \lambda, d/dx, \widetilde{e^{-\lambda x}})\mathbf{1} = 0 \text{ for all odd } n, \tag{2.2}$$

$$B(n, \lambda, d/dx, \widetilde{e^{-\lambda x}})\mathbf{1} = (n-1)!!(-2\lambda)^{n/2}e^{-n\lambda x/2} \text{ for all even } n > 0, \tag{2.3}$$

$$(2d/dx + \lambda n I)B(n, \lambda, d/dx, \widetilde{e^{-\lambda x}})\mathbf{1} = 0 \text{ for all } n > 0. \tag{2.4}$$

[1]For elementary functions example (see the next section), this means switching from exponential to trigonometric functions.

2.2. Trigonometric functions. Since multiplications by functions commute, we have in this case the condition $[\widetilde{f}, [D, \widetilde{f}]] = 0$, needed in Section 1.2, automatically satisfied for any smooth function f.

Let us now check the condition $[D, [D, \widetilde{f}]] = \lambda^2 \widetilde{f}$ for the natural candidates: trigonometric and hyperbolic sine and cosine.

It is an easy computation that when $U = \widetilde{\sin \lambda x}$ is the operator of multiplication by $\sin \lambda x$, then $[D, U] = \lambda \widetilde{\cos \lambda x}$, $[D, [D, \widetilde{\sin \lambda x}]] = (i\lambda)^2 \widetilde{\sin \lambda x}$, and as we have mentioned above, $[U, [D, U]] = 0$ is automatic.

Thus, the results of section 1.2 provide the following set of differential identities for sine functions:

THEOREM 2.3.

- $$B(n, i\lambda, d/dx, \widetilde{\sin \lambda x})\mathbf{1} = 0 \textit{ for all odd } n, \tag{2.5}$$

- $$B(n, i\lambda, d/x, \widetilde{\sin \lambda x})\mathbf{1} = (n-1)!!\lambda^{n/2} e^{-in\lambda x/2} \textit{ for all even } n > 0, \tag{2.6}$$

- $$(2d/dx + i\lambda n I)B(n, i\lambda, d/x, \widetilde{\sin \lambda x})\mathbf{1} = 0 \textit{ for all } n > 0. \tag{2.7}$$

REMARK 2.4. Similar identities hold for any solutions of the differential equation $\frac{d^2u}{du^2} = \lambda^2 u$, including cosines, hyperbolic sine and cosine, and linear functions. In particular, here are the binomial-type identities for linear functions:

- $$B(n, 0, d/dx, (\widetilde{ax+b}))\mathbf{1} = 0 \text{ for all odd } n, \tag{2.8}$$

- $$B(n, 0, d/dx, (\widetilde{ax+b}))\mathbf{1} = (n-1)!!a^{n/2} \text{ for all even } n > 0, \tag{2.9}$$

- $$(d/dx)B(n, 0, d/dx, (\widetilde{ax+b}))\mathbf{1} = 0 \text{ for all } n > 0. \tag{2.10}$$

2.3. Change of variables. One can play with changes of variables in the formulas of Theorems 2.2 and 2.3, to get a variety of new identities. For instance, the change $x \to x^2/2$ and correspondingly $d/dx \to x^{-1} d/dx$ gives

$$B(n, \lambda, \frac{1}{x}\frac{d}{dx}, \widetilde{e^{\lambda x^2/2}})e^{-j\lambda x^2/2} = 0 \text{ for all } n > 0 \text{ and } 0 \le j \le n-1, \tag{2.11}$$

while $x \to \ln x, d/dx \to x\, d/dx$ produces

$$B(n, \lambda, x\frac{d}{dx}, \widetilde{\lambda x})x^{-\lambda j} = 0 \text{ for all } n > 0 \text{ and } 0 \le j \le n-1. \tag{2.12}$$

2.4. Vector functions. The results easily translate to vector-valued functions. Let $\mathbb{F}$ is the space of all smooth $\mathbb{C}^m$-valued functions. Then the kernel of $D = d/dx$ consists of constant column vectors $\overrightarrow{c}$. Let A be an $m \times m$ constant matrix. Then the following set of statements hold:

(1)

$$B(n, \lambda, d/dx, \widetilde{e^{\lambda x} A})e^{-j\lambda x} I = 0 \text{ for all } n > 0 \text{ and } 0 \le j \le n-1,$$

(2)
$$B(n,\lambda,d/dx,\widetilde{e^{-\lambda x}A})\overrightarrow{c}=\overrightarrow{0} \text{ for all odd } n,$$
(3)
$$B(n,\lambda,d/dx,\widetilde{e^{-\lambda x}A})\overrightarrow{c}=(n-1)!!(-2\lambda)^{n/2}e^{-n\lambda x/2}A^{n/2}\overrightarrow{c} \text{ for all even } n>0,$$
(4)
$$(2d/dx-\lambda nI)B(n,\lambda,d/dx,\widetilde{e^{-\lambda x}A})\overrightarrow{c}=\overrightarrow{0} \text{ for all } n>0,$$
(5)
$$B(n,i\lambda,d/dx,\widetilde{\sin\lambda xA})\overrightarrow{c}=\overrightarrow{0} \text{ for all odd } n,$$
(6)
$$B(n,i\lambda,d/x,\widetilde{\sin\lambda xA})\overrightarrow{c}=(n-1)!!\lambda^{n/2}e^{-in\lambda x/2}A^{n/2}\overrightarrow{c} \text{ for all even } n>0,$$
(7)
$$(2d/dx+i\lambda nI)B(n,i\lambda,d/x,\widetilde{\sin\lambda xA})\overrightarrow{c}=\overrightarrow{0} \text{ for all } n>0.$$
(8) If A_1 and A_2 are commuting $m\times m$ matrices, then
$$B(n,0,d/dx,\widetilde{A_1x+A_2})\overrightarrow{c}=\overrightarrow{0} \text{ for all odd } n,$$
$$B(n,0,d/dx,\widetilde{A_1x+A_2})\overrightarrow{c}=(n-1)!!A_1^{n/2}\overrightarrow{c} \text{ for all even } n>0,$$
$$(d/dx)B(n,0,d/dx,\widetilde{A_1x+A_2})\overrightarrow{c}=\overrightarrow{0} \text{ for all } n>0.$$

3. Proofs

3.1. Proof of Theorem 1.1. We start with the following lemma, which can be easily proved by induction.

LEMMA 3.1.
(1) *If $[D,U]=\lambda U$, then $[D,U^m]=\lambda mU^m$ for all integer $m\geq 0$.*
(2) *If $[D,U]=V$, $[U,V]=0$, and $[D,V]=0$, then $[D,U^m]=mVU^{m-1}$ for all natural m.*

We now provide a different representation for $B(n,\lambda,U,D)$.

LEMMA 3.2. *For all $n>0$*

$$B(n,\lambda,U,D)=\sum_{k=0}^{n-1}\binom{n-1}{k}\left(\prod_{j=0}^{k-1}(D-U+\lambda j)\right)(D+\lambda kI)U^{n-1-k}. \tag{3.1}$$

Proof Using $\binom{n}{k}=\binom{n-1}{k}+\binom{n-1}{k-1}$, we can rewrite $B(n,\lambda,U,D)$ as follows:

$$\sum_{k=0}^{n-1}\binom{n-1}{k}\left(\prod_{j=0}^{k-1}(D-U+j\lambda I)\right)U^{n-k}$$
$$+\sum_{k=1}^{n}\binom{n-1}{k-1}\left(\prod_{j=0}^{k-1}(D-U+j\lambda I)\right)U^{n-k}.$$

(**We remind the reader that in the products the order is in increasing j (or k) from the left to the right!**) Changing k to $k+1$ in the second sum and combining both sums proves the lemma. □.

Proof of Theorem 1.1. By definition, $B(0,\lambda,U,D)=I$ and $B(1,\lambda,U,D)=D$.

Let now $n>1$ and $[D,U]=\lambda U$. Using Lemma 3.1, we get

$$(D+\lambda kI)U^{n-1-k}=U^{n-1-k}(D+\lambda(n-1)I).$$

Substituting this into (3.1), we obtain the recurrent formula

$$B(n,\lambda,u,D)=B(n-1,\lambda,u,D)(D+\lambda(n-1)I). \tag{3.2}$$

By induction, (3.2) implies Theorem 1.1 for all n. □

3.2. Proof of Theorem 1.4. The following lemma (appeared in a somewhat more restricted and implicit form in [**4**]) provides an important insight on the nature of $B(n,\lambda,U,D)$, and the technique of its proof will be used below. It uses an additional assumption that U is invertible. It is easy to see that if $[D,U]=\lambda U$, then $\left[D,U^{-1}\right]=-\lambda U^{-1}$, and the statement of Lemma 3.1 holds now for all integers m, including $m<0$.

The following lemma provides a useful recurrent relation for $B(n,\lambda,U,D)$ in the case when $[D,U]=-\lambda U$.

LEMMA 3.3. *If $[D,U]=-\lambda U$, then*

$$\begin{aligned}B(n,\lambda,U,D)&=B(n-1,\lambda,U,D)(D+\lambda(n-1)I)-2(n-1)\lambda UB(n-2,\lambda,U,D)\\&+2(n-1)(n-2)\lambda^2UB(n-3,\lambda,U,D)\end{aligned} \tag{3.3}$$

for all $n>1$, with the last term absent when $n=2$.

Proof of Lemma 3.3 Using Lemma 3.1, we get

$$\begin{aligned}&(D+\lambda kI)U^{n-1-k}=U^{n-1-k}(D+\lambda(2k-n+1)I)\\&=U^{n-1-k}(D+\lambda(n-1)I-2\lambda(n-1-k)I).\end{aligned}$$

Substituting it into (3.3), we obtain

$$\begin{aligned}B(n,\lambda,U,D)&=\sum_{k=0}^{n-1}\binom{n-1}{k}\left(\prod_{j=0}^{k-1}(D-U+\lambda jI)\right)U^{n-1-k}(D+\lambda(n-1)I)\\&-2\lambda\sum_{k=0}^{n-1}\binom{n-1}{k}(n-1-k)\left(\prod_{j=0}^{k-1}(D-U+\lambda jI)\right)U^{n-1-k}.\end{aligned}$$

Here the first sum is equal to

$$B(n-1,\lambda,U,D)(D+\lambda(n-1)I),$$

in the second sum the term with $k=n-1$ is zero, and

$$\binom{n-1}{k}(n-1-k)=(n-1)\binom{n-2}{k}.$$

Thus,

$$\begin{aligned}Z&:=B(n,\lambda,U,D)-B(n-1,\lambda,U,D)(D+\lambda(n-1)I)\\&=-2(n-1)\lambda\sum_{k=0}^{n-2}\binom{n-2}{k}\left(\prod_{j=0}^{k-1}(D-U+\lambda jI)\right)U^{n-1-k}.\end{aligned}$$

Since $(D - U + \lambda jI)U = U(D - U + \lambda(j-1)I)$, we have

$$Z = -2(n-1)\lambda U \sum_{k=0}^{n-2} \binom{n-2}{k} \left(\prod_{j=-1}^{k-2} (D - U + \lambda jI) \right) U^{n-2-k}.$$

When $j = -1$, $D - U + \lambda jI = (D - U + (k-1)\lambda I) - k\lambda I$. Thus,

$$Z = -2(n-1)\lambda U \sum_{k=0}^{n-2} \binom{n-2}{k} \left(\prod_{j=0}^{k-1} (D - U + \lambda jI) \right) U^{n-2-k}$$
$$+2(n-1)\lambda^2 U \sum_{k=1}^{n-2} \binom{n-2}{k} k \left(\prod_{j=0}^{k-2} (D - U + \lambda jI) \right) U^{n-2-k}.$$

The first sum here is $B(n-2, \lambda, U, D)$. Using the identity

$$\binom{n-2}{k} k = (n-2)\binom{n-3}{k-1}$$

and changing $k-1$ to k, we conclude that the second sum equals $(n-2)B(n-3, \lambda, U, D)$. Therefore,

$$Z = -2(n-1)\lambda U B(n-2, \lambda, U, D) + 2(n-1)(n-2)\lambda^2 U B(n-3, \lambda, U, D),$$

which coincides with (3.3) and completes the proof. □

Proof of Theorem 1.4 We know that $B(0, \lambda, U, D) = I$ and $B(1, \lambda, U, D) = D$, so $B(1, \lambda, U, D) = 0$ on $\mathbb{F}_0$ and the first statement of the Theorem holds true for $n = 0$ and $n = 1$. Let $[D, U] = -\lambda U$, then on $\mathbb{F}_0$, the recurrence (3.3) is reduced to

$$B(n, \lambda, U, D) = \lambda(n-1)I)B(n-1, \lambda, U, D) - 2(n-1)\lambda U B(n-2, \lambda, U, D)$$
$$+2(n-1)(n-2)\lambda^2 U B(n-3, \lambda, U, D).$$

From here, the statement of the Theorem follows by induction for all $n > 0$, both odd and even. □

3.3. Proof of Theorem 1.5. The proof will be different for $\lambda \neq 0$ and for $\lambda = 0$. This is not that surprising, since the claim, if considered on elementary functions, leads to differential binomial for quite different functions: sine functions when $\lambda \neq 0$ and linear functions when $\lambda = 0$.

The case $\lambda \neq 0$. Let us introduce the elements $V := (\lambda U - [D, U])/2\lambda$ and $W := (\lambda U + [D, U])/2\lambda$. A direct calculation shows that $U = V + W$, and under the Theorem's assumptions $[D, V] = -\lambda V$, $[D, W] = \lambda W$, and $[V, W] = 0$. From Corollary 1.3 we obtain that $B(n, \lambda, U, D) = B(n, \lambda, V + W, D) = B(n, \lambda, V, D)$. Now one applies Theorem 1.4.

The case $\lambda = 0$. By definition, we have $B(1, 0, U, D) = D$ and $B(2, 0, U, D) = D^2 + [D, U]$. Let $V := [D, U]$. Then $B(1, 0, U, D) = 0$ and $B(2, 0, U, D) = V$ on $\mathbb{F}_0$ so the claim is true for $n = 1$ and $n = 2$. Now, under the assumption that $[U, V] = 0$ and $[D, V] = 0$, we will show that for $n > 2$

$$B(n, 0, U, D) = (n-1)VB(n-2, 0, U, D) \text{ on } \mathbb{F}_0. \tag{3.4}$$

Then induction proves the statement for all n.

Let $n > 2$. According to Lemma 3.2,

$$B(n, 0, U, D) = \sum_{k=0}^{n-1} \binom{n-1}{k} (D - U)^k D U^{n-1-k},$$

where the term with $k = n-1$ is equal to zero on $\mathbb{F}_0$. Also, according to Lemma 3.1, $DU^m = mVU^{m-1}$ on $\mathbb{F}_0$. Then on $\mathbb{F}_0$ we get

$$B(n,0,U,D) = \sum_{k=0}^{n-2} \binom{n-1}{k} (D-U)^k (n-1-k) V U^{n-2-k}.$$

From here we obtain (3.4) by using

$$\binom{n-1}{k}(n-1-k) = (n-1)\binom{n-2}{k}$$

and commuting V to the left. This completes the proof. □

4. Final remarks and conclusions

- It would be interesting to figure out what can be done under the condition of vanishing of the third commutator $[D,[D,[D,U]]] = \lambda^3 U$ (maybe plus some other restrictions). It has been checked by direct computation that the natural analog of (0.1)-(0.2) for the solutions of the third order equation $D^3 u = \lambda^3 u$ does not hold [**5**, **9**].
- As we have already mentioned, we do not truly understand the origin of such identities. It looks like this issue is in the realm of the techniques of [**6**], in which we are not experts, to say the least. In particular, one can compare with the identity in [**6**, Proposition 8.65], due to O. V. Viskov [**11**].
- There still might be interesting relations to SCV, as the ones to Hartogs' type theorems in [**1**, **7**, **8**, **10**]. One also wonders about such higher dimension analogs of Hartogs' theorems.
- We cannot help it providing a cute lemma used in [**4**]. An older version of this text used it, but we have managed to avoid this. A reader, however, could find it interesting:

  LEMMA 4.1. [**4**] *For any two elements A_1 and A_2 of algebra A, the following equality holds:*

  $$\sum_{k=0}^{n} \binom{n}{k} (A_1 - I)^k (A_2 + I)^{n-k} = \sum_{k=0}^{n} \binom{n}{k} A_1^k A_2^{n-k}. \tag{4.1}$$

  If A_1 and A_2 commute, then both sums in (4.1) are equal to $(A_1 + A_2)^n$ and thus to each other. The Lemma states that (4.1) still holds in the non-commutative case, when the binomial formula does not apply.
- Another sometimes useful observation is

  LEMMA 4.2. *Let $[D,U] = \lambda U$ and U^{-1} exist. Then*

  $$B(n,\lambda,U,D) = (DU^{-1})^n U^n.$$

  Proof Suppose U^{-1} exists. If $[D,U] = \lambda U$, then Lemma 3.1 implies that $(D - U + j\lambda I)U^{-j} = U^{-j}(D-U)$.

  So, when $k = 2$

  $$\begin{aligned} &(D-U)(D-U+\lambda I)U^{n-2} \\ &= (D-U)U^{-1}(D-U)U^{-1}U^n \\ &= \left((D-U)U^{-1}\right)^2 U^n = (DU^{-1} - I)^2 U^n, \end{aligned}$$

when $k = 3$

$$\begin{aligned}&(D-U)(D-U+\lambda I)(D-U+2\lambda I)U^{n-3}\\&=\left((D-U)U^{-1}\right)^3 U^n\\&=(DU^{-1}-I)^3U^n\end{aligned}$$

and so on. Thus, by induction

$$B(n,\lambda,U,D)=\left(\sum_{k=0}^{n}\binom{n}{k}(DU^{-1}-I)^k I^{n-k}\right)U^n. \tag{4.2}$$

Using the standard binomial formula for the commuting operators $(DU^{-1} - I)$ and I we get $B(n,\lambda,U,D) = ((DU^{-1}-I)+I)^n U^n$, or $B(n,\lambda,U,D) = (DU^{-1})^n U^n$. □

- Note that if U^{-1} exists and $[D,U] = -\lambda U$, then the technique used in the proof of Lemma 4.2 allows us to rewrite $B(n,\lambda,U,D)$ as

$$B(n,\lambda,U,D)=\left(\sum_{k=0}^{n}\binom{n}{k}(DU-U^2)^k U^{2(n-k)}\right)U^{-n},$$

however the standard binomial formula is not applicable to this sum, since the operators $DU-U^2$ and U^2 do not commute.

- The authors thank NSF for the support, as well as the reviewer for useful comments and references.

References

[1] V. Aguilar, L. Ehrenpreis, and P. Kuchment, *Range conditions for the exponential Radon transform*, J. Anal. Math. **68** (1996), 1–13, DOI 10.1007/BF02790201. MR1403248

[2] P. Kuchment, *The Radon transform and medical imaging*, CBMS-NSF Regional Conference Series in Applied Mathematics, vol. 85, Society for Industrial and Applied Mathematics (SIAM), Philadelphia, PA, 2014. MR3137435

[3] P. A. Kuchment and S. Ya. L′vin, *Range of the Radon exponential transform* (Russian), Dokl. Akad. Nauk SSSR **313** (1990), no. 6, 1329–1330; English transl., Soviet Math. Dokl. **42** (1991), no. 1, 183–184. MR1080033

[4] P. A. Kuchment and S. Ya. L′vin, *Paley-Wiener theorem for exponential Radon transform*, Acta Appl. Math. **18** (1990), no. 3, 251–260, DOI 10.1007/BF00049128. MR1065641

[5] P. Kuchment and S. Lvin, *Identities for* $\sin x$ *that came from medical imaging*, Amer. Math. Monthly **120** (2013), no. 7, 609–621, DOI 10.4169/amer.math.monthly.120.07.609. MR3096467

[6] T. Mansour and M. Schork, *Commutation relations, normal ordering, and Stirling numbers*, Discrete Mathematics and its Applications (Boca Raton), CRC Press, Boca Raton, FL, 2016. MR3408225

[7] O. Öktem, Comparing range characterizations of the exponential Radon transform, *Research reports in Math.*, Dept. Math., Stockholm University, Sweden, no. 17, 1996.

[8] O. Öktem, *Extension of separately analytic functions and applications to range characterization of the exponential Radon transform: Complex analysis and applications (Warsaw, 1997)*, Ann. Polon. Math. **70** (1998), 195–213, DOI 10.4064/ap-70-1-195-213. MR1668725

[9] E. Rodriguez, unpublished.

[10] A. Tumanov, *Analytic continuation from a family of lines*, J. Anal. Math. **105** (2008), 391–396, DOI 10.1007/s11854-008-0043-3. MR2438433

[11] O. V. Viskov, *On a theorem of R. A. Sack for shift operators* (Russian), Dokl. Akad. Nauk **340** (1995), no. 4, 463–466. MR1327828

Mathematics Department, Texas A&M University, College Station, Texas 77845
Email address: kuchment@math.tamu.edu

Mathematics and Statistic Department, University of Maine, Orono, Maine 04469
Email address: lvin@math.umaine.edu

Contemporary Mathematics
Volume **733**, 2019
https://doi.org/10.1090/conm/733/14744

Similarity of holomorphic matrices on 1-dimensional Stein spaces

Jürgen Leiterer

Dedicated to the memory of Selim Krein

ABSTRACT. R. Guralnick [Linear Algebra Appl. 99, 85-96 (1988)] proved that two holomorphic matrices on a noncompact connected Riemann surface, which are locally holomorphically similar, are globally holomorphically similar. In the preprints [arXiv:1703.09524] and [arXiv:1703.09530], a generalization of this to arbitrary (possibly, nonsmooth) 1-dimensional Stein spaces was obtained. The present paper contains a revised version of the proof from [arXiv:1703.09524]. The method of this revised proof can be used also in the higher dimensional case, which will be the subject of a forthcoming paper.

1. Introduction

Let X be a (reduced) complex space, e.g., a complex manifold or an analytic subset of a complex manifold. Let $\mathrm{Mat}(n \times n, \mathbb{C})$ be the algebra of complex $n \times n$ matrices, and $\mathrm{GL}(n, \mathbb{C})$ the group of invertible complex $n \times n$ matrices.

Two holomorphic maps $A, B : X \to \mathrm{Mat}(n \times n, \mathbb{C})$ are called (globally) **holomorphically similar on** X if there is a holomorphic map $H : X \to \mathrm{GL}(n, \mathbb{C})$ with $B = H^{-1}AH$ on X. They are called **locally holomorphically similar at** $\xi \in X$ if there is a neighborhood U of ξ such that $A|_U$ and $B|_U$ are holomorphically similar on U. Correspondingly, **continuous** and $\mathcal{C}^k$ **similarity** are defined.

R. Guralnick [**14**] proved the following theorem.

THEOREM 1.1. *Suppose X is a noncompact connected Riemann surface. Then any two holomorphic maps $A, B : X \to \mathrm{Mat}(n \times n, \mathbb{C})$, which are locally holomorphically similar at each point of X, are globally holomorphically similar on X.*

In [**17**], the following generalization was obtained.

THEOREM 1.2. *The claim of Theorem 1.1 remains true if X is an arbitrary 1-dimensional Stein space (for example, a 1-dimensional closed analytic subset of some $\mathbb{C}^N$, or, more general, of a domain of holomorphy in $\mathbb{C}^N$).*

Guralnick's proof of Theorem 1.1 consists in proving a theorem for matrices with elements in a Bezout ring (with some extra properties) and then applying

2010 *Mathematics Subject Classification.* Primary: 32E10, 47A56, 15A21.
Key words and phrases. Holomorphic matrices, similarity, Oka principle, 1-dimensional Stein spaces.

this to the ring of holomorphic functions on a noncompact connected Riemann surface. The ring of holomorphic functions on an arbitrary (possibly nonsmooth) 1-dimensional Stein space need not be Bezout. Therefore, it seems that this proof cannot be used to prove Theorem 1.2, at least not in a straightforward way.

The proof of Theorem 1.2 given in [**17**] proceeds as follows. First we use the Oka principle for Oka-pairs of Forster and Ramspott [**7**] to show that Theorem 1.2 is equivalent to a certain topological statement (see Theorem 3.2 below). Then we prove this topological statement. This proof is independent of Guralnick's result. Using Guralnick's result, this proof can be shortened, passing to the normalization of X (which is smooth). This is shown in [**19**].

The aim of the present paper is to give a revised version of the proof from [**17**].

Note added in proof. I want to thank to the referee for pointing out the following: Another proof of Theorem 1.1 is outlined by F. Forstnerič in the second edition of his book [**9**, Theorem 6.14.9]. He does not use the above mentioned Oka principle of Forster and Ramspott, but a new Oka principle originally established by him in [**8**].

2. Notations and our use of the language of sheaves

$\mathbb{N}$ is the set of natural numbers including 0. $\mathbb{N}^* = \mathbb{N} \setminus \{0\}$.

If $n, m \in \mathbb{N}^*$, then $\mathrm{Mat}(n \times m, \mathbb{C})$ is the space of complex $n \times m$ matrices (n rows, m columns), and $\mathrm{GL}(n, \mathbb{C})$ is the group of invertible complex $n \times n$ matrices.

The unit matrix in $\mathrm{Mat}(n \times n, \mathbb{C})$ will be denoted by I_n or simply by I.

If a matrix $\Phi \in \mathrm{Mat}(n \times m, \mathbb{C})$ is interpreted as a linear map from $\mathbb{C}^m$ to $\mathbb{C}^n$, then $\mathrm{Ker}\, \Phi$ denotes the kernel, $\mathrm{Im}\, \Phi$ the image and $\|\Phi\|$ the operator norm (induced by the Euclidean norm) of Φ.

By a complex space we always mean a *reduced* complex space in the sense of [**12**], which is the same as an *analytic space* in the terminology used in [**5**] and [**20**].

From now on, in this section, X is topological space, and G is a topological group (possibly non-abelian).

By $\mathcal{C}^G$, or more precisely by $\mathcal{C}^G_X$, we denote the sheaf of continuous G-valued maps on X, that is, the map which assigns to each nonempty open $U \subseteq X$ the group $\mathcal{C}^G(U)$ of all continuous maps $f : U \to G$. Also $\mathcal{C}^G(\emptyset) := \{1\}$ (1 being the neutral element of G).

A subsheaf of $\mathcal{C}^G_X$ is a map $\mathcal{F}$ which assigns to each open $U \subseteq X$ a subgroup $\mathcal{F}(U)$ of $\mathcal{C}^G(U)$ such that:

- If $V \subseteq U$ are nonempty open subsets of X, then, for each $f \in \mathcal{F}(U)$, the restriction of f to V, $f|_V$, belongs to $\mathcal{F}(V)$.
- If $U \subseteq X$ is open and $f \in \mathcal{C}^G(U)$ is such that, for each $\xi \in U$, there is an open neighborhood $V \subseteq U$ of ξ with $f|_V \in \mathcal{F}(V)$, then $f \in \mathcal{F}(U)$.

The elements of $\mathcal{F}(U)$ are called **sections** of $\mathcal{F}$ over U.

If $\mathcal{F}$ and $\mathcal{G}$ are two subsheaves of $\mathcal{C}^G_X$, then $\mathcal{F}$ is called a subsheaf of $\mathcal{G}$ if $\mathcal{F}(U) \subseteq \mathcal{G}(U)$ for all open $U \subseteq X$.

If X is a complex space and G a complex Lie group, then we denote by $\mathcal{O}^G_X$, or simply by $\mathcal{O}^G$, the subsheaf of $\mathcal{C}^G_X$ which assigns to each nonempty open $U \subseteq X$, the group $\mathcal{O}^G(U)$ of all holomorphic maps from U to G.

Let $\mathcal{F}$ be a subsheaf of $\mathcal{C}^G_X$.

Let $\mathcal{U} = \{U_i\}_{i \in I}$ an open covering of X.

A family $f_{ij} \in \mathcal{F}(U_i \cap U_j)$, $i, j \in I$, is called a $(\mathcal{U}, \mathcal{F})$**-cocycle**, or simply a **cocycle**, if (the group operation of G being a multiplication)

$$f_{ij} f_{jk} = f_{ik} \quad \text{on} \quad U_i \cap U_j \cap U_k \quad \text{for all} \quad i, j, k \in I. \text{ [1]}$$

Note that then always $f_{ij}^{-1} = f_{ji}$ and $f_{ii} \equiv 1$. The set of all $(\mathcal{U}, \mathcal{F})$-cocycles will be denoted by $Z^1(\mathcal{U}, \mathcal{F})$.

Let $f = \{f_{ij}\} \in Z^1(\mathcal{U}, \mathcal{F})$. We say that f **splits** (or **is trivial**) if there exists a family $h_i \in \mathcal{F}(U_i)$, $i \in I$, such that

$$f_{ij} = h_i h_j^{-1} \quad \text{on} \quad U_i \cap U_j \quad \text{for all} \quad i, j \in I.$$

We say that f is an $\mathcal{F}$**-cocycle** on X, if there exists an open covering $\mathcal{U}$ of X with $f \in Z^1(\mathcal{U}, \mathcal{F})$. Then $\mathcal{U}$ is called the **covering of** f. As usual, we write $H^1(X, \mathcal{F}) = 0$ to say that each $\mathcal{F}$-cocycle on X splits, and $H^1(X, \mathcal{F}) \neq 0$ to say that there exist non-splitting $\mathcal{F}$-cocycles on X.

Now let $\mathcal{U}^* = \{U_\alpha^*\}_{\alpha \in I^*}$ be a second open covering of X, which is a refinement of $\mathcal{U}$, i.e., there is a map $\tau : I^* \to I$ with $U_\alpha^* \subseteq U_{\tau(\alpha)}$ for all $\alpha \in I^*$. Then we say that a $(\mathcal{U}^*, \mathcal{F})$-cocycle $\{f_{\alpha\beta}^*\}_{\alpha,\beta \in I^*}$ is **induced** by a $(\mathcal{U}, \mathcal{F})$-cocycle $\{f_{ij}\}_{i,j \in I}$ if this map τ can be chosen so that

$$f_{\alpha\beta}^* = f_{\tau(\alpha)\tau(\beta)} \quad \text{on} \quad U_i^* \cap U_j^* \quad \text{for all} \quad \alpha, \beta \in I^*.$$

We need the following well-known and simple proposition, see [**15**, p. 41] for "only if" and [**5**, p. 101] for "if".

PROPOSITION 2.1. *Let $f \in Z^1(\mathcal{U}, \mathcal{F})$ and $f^* \in Z^1(\mathcal{U}^*, \mathcal{F})$ such that f^* is induced by f. Then f splits if and only if f^* splits.*

Let Y be a nonempty open subset of X.

Then we denote by $\mathcal{F}|_Y$ the subsheaf of $\mathcal{C}_Y^G$ defined by $\mathcal{F}|_Y(U) = \mathcal{F}(U)$ for each open $U \subseteq Y$. $\mathcal{F}|_Y$ is called the **restriction** of $\mathcal{F}$ to Y.

If $\mathcal{U} = \{U_i\}_{i \in I}$ is an open covering of X, then we define $\mathcal{U} \cap Y = \{U_i \cap Y\}_{i \in I}$, and, for each $f = \{f_{ij}\}_{i,j \in I} \in Z^1(\mathcal{U}, \mathcal{F})$, we denote by $f|_Y = \{(f|_Y)_{ij}\}_{i,j \in I}$ the $(\mathcal{U} \cap Y, \mathcal{F}|_Y)$-cocycle defined by

$$(f|_Y)_{ij} = f_{ij}\big|_{U_i \cap U_j \cap Y} \quad \text{for} \quad i, j \in I.$$

We call $f|_Y$ the **restriction** of f to Y.

REMARK 2.2. Let $\mathcal{U} = \{U_i\}_{i \in I}$ be an open covering of X, and $f = \{f_{ij}\}_{i,j \in I} \in Z^1(\mathcal{U}, \mathcal{F})$. Then $f|_{U_i}$ splits for each $i \in I$ with $U_i \neq \emptyset$.

Indeed, the one-set family $\{U_i\}$ is an open covering of U_i which is a refinement of $\mathcal{U} \cap U_i$, and there is precisely one $(\{U_i\}, \mathcal{F})$-cocycle which is induced by $f|_{U_i}$, namely $\{f_{ii}\}$. Since $f_{ii} \equiv 1$, it is trivial that $\{f_{ii}\}$ splits. Therefore it follows from Proposition 2.1 that $f|_{U_i}$ splits.

3. Topological criteria for global holomorphic similarity

DEFINITION 3.1. Let $\Phi \in \mathrm{Mat}(n \times n, \mathbb{C})$. Then we denote by $\mathrm{Com}\,\Phi$ the algebra of all $\Theta \in \mathrm{Mat}(n \times n, \mathbb{C})$ with $\Phi\Theta = \Theta\Phi$, and by $\mathrm{GCom}\,\Phi$ we denote the group of

---

[1] We use the convention that statements like "$f = g$ on $\emptyset$" or "$f := g$ on $\emptyset$" have to be omitted.

invertible elements of $\operatorname{Com}\Phi$. It is easy to see that

$$\operatorname{GCom}\Phi = \operatorname{GL}(n,\mathbb{C}) \cap \operatorname{Com}\Phi, \tag{3.1}$$

$$\operatorname{Com}(\Gamma^{-1}\Phi\Gamma) = \Gamma^{-1}(\operatorname{Com}\Phi)\Gamma \quad \text{for all} \quad \Gamma \in \operatorname{GL}(n,\mathbb{C}). \tag{3.2}$$

Now let X be a complex space (of arbitrary dimension), and $A : X \to \operatorname{Mat}(n \times n, \mathbb{C})$ a holomorphic map.

We introduce the families

$$\operatorname{Com} A := \{\operatorname{Com} A(\zeta)\}_{\zeta\in X} \quad \text{and} \quad \operatorname{GCom} A := \{\operatorname{GCom} A(\zeta)\}_{\zeta\in X}.$$

If the dimension of $\operatorname{Com} A(\zeta)$ does not depend on ζ, then it is well-known (and easy to see) that $\operatorname{Com} A$ is a holomorphic vector bundle. But also in this special case, the family of groups $\operatorname{GCom} A$ need not be locally trivial. It is possible that $\operatorname{Com} A$ is a holomorphic vector bundle, but $\operatorname{GCom} A$ is even not locally trivial as a family of topological spaces. For an example, see [**19**, Sec. 4].

Nevertheless the sheaves of holomorphic and continuous sections of $\operatorname{Com} A$ and $\operatorname{GCom} A$ are well-defined. We denote them by $\mathcal{O}^{\operatorname{Com} A}$, $\mathcal{O}^{\operatorname{GCom} A}$, $\mathcal{C}^{\operatorname{Com} A}$ and $\mathcal{C}^{\operatorname{GCom} A}$, respectively.

We define a subsheaf $\widehat{\mathcal{O}}^{\operatorname{Com} A}$ of $\mathcal{C}^{\operatorname{Com} A}$ as follows: if U is a nonempty open subset of X, then $\widehat{\mathcal{O}}^{\operatorname{Com} A}(U)$ is the subalgebra of all $f \in \mathcal{C}^{\operatorname{Com} A}(U)$ such that, for each $\xi \in U$,

$$\left\{ \begin{array}{l} \text{there exist a neighborhood } V_\xi \text{ of } \xi \\ \text{and } h_\xi \in \mathcal{O}^{\operatorname{Com} A}(V_\xi) \text{ with } h(\xi) = f(\xi). \end{array} \right. \tag{3.3}$$

Further, we define a subsheaf $\widehat{\mathcal{O}}^{\operatorname{GCom} A}$ of $\mathcal{C}^{\operatorname{GCom} A}$ setting $\widehat{\mathcal{O}}^{\operatorname{GCom} A}(U) = \mathcal{C}^{\operatorname{GL}(n,\mathbb{C})}(U) \cap \widehat{\mathcal{O}}^{\operatorname{Com} A}(U)$ for each nonempty open $U \subseteq X$.

The Oka principle for Oka pairs of Forster and Ramspott [**7**, Satz 1], yields the following

PROPOSITION 3.2. *Let X be a Stein space, and let $A : X \to \operatorname{Mat}(n \times n, \mathbb{C})$ be holomorphic. Then each $\mathcal{O}^{\operatorname{GCom} A}$-cocycle, which splits as an $\widehat{\mathcal{O}}^{\operatorname{GCom} A}$-cocycle, splits also as an $\mathcal{O}^{\operatorname{GCom} A}$-cocycle.*

Indeed, it is easy to see that, for each nonempty open $U \subseteq X$ we have: If $h \in \mathcal{O}^{\operatorname{Com} A}(U)$, then $e^h \in \mathcal{O}^{\operatorname{GCom} A}(U)$, and, if $H \in \mathcal{O}^{\operatorname{GCom} A}(U)$ with $\sup_{\zeta\in U} \|H(\zeta) - I\| < 1$, then

$$\log H := -\sum_{\mu=1}^{\infty} \frac{(I-H)^\mu}{\mu}$$

belongs to $\mathcal{O}^{\operatorname{Com} A}(U)$. This shows that $\mathcal{O}^{\operatorname{GCom} A}$ is a *coherent $\mathcal{O}$-subsheaf* of $\mathcal{O}^{\operatorname{GL}(n,\mathbb{C})}$ in the sense of [**7**, §2], where $\mathcal{O}^{\operatorname{Com} A}$ is the *generating sheaf of Lie algebras*. Moreover, as observed in [**7**, §2.3, example b], the pair $\left(\mathcal{O}^{\operatorname{GCom} A}, \widehat{\mathcal{O}}^{\operatorname{GCom} A}\right)$ is an *admissible pair* in the sense of [**7**], which, trivially, satisfies condition (PH) in Satz 1 of [**7**]). Therefore Proposition 3.2 is one of the statements of that Satz 1.

THEOREM 3.3. *Let X be a Stein space, and let $A : X \to \operatorname{Mat}(n \times n, \mathbb{C})$ be holomorphic. Then the following are equivalent.*

(i) *Each holomorphic $B : X \to \operatorname{Mat}(n \times n, \mathbb{C})$, which is locally holomorphically similar to A at each point of X, is globally holomorphically similar to A on X.*

(ii) *Each* $\mathcal{O}^{\mathrm{GCom}\,A}$*-cocycle on* X*, which splits as a* $\mathcal{C}^{\mathrm{GL}\,(n,\mathbb{C})}$*-cocycle, splits also as an* $\widehat{\mathcal{O}}^{\mathrm{GCom}\,A}$*-cocycle.*

PROOF. (ii) $\Longrightarrow$ (i): Let $B : X \to \mathrm{Mat}(n \times n, \mathbb{C})$ be holomorphic and locally holomorphically similar to A at each point of X. Then we can find an open covering $\{U_i\}_{i\in I}$ of X and holomorphic maps $H_i : U_i \to \mathrm{GL}(n, \mathbb{C})$, $i \in I$, such that

$$B = H_i^{-1} A H_i \quad \text{on} \quad U_i. \tag{3.4}$$

Hence, for all $i, j \in I$ with $U_i \cap U_j \neq \emptyset$, $AH_iH_j^{-1} = H_iH_j^{-1}A$ on $U_i \cap U_j$, i.e., $H_iH_j^{-1} \in \mathcal{O}^{\mathrm{GCom}\,A}(U_i \cap U_j)$. Clearly,

$$(H_iH_j^{-1})(H_jH_k^{-1}) = H_iH_k^{-1} \quad \text{on} \quad U_i \cap U_j \cap U_k, \qquad i, j, k \in I.$$

Therefore, the family $\{H_iH_j^{-1}\}_{i,j\in I}$ is a well-defined $O^{\mathrm{GCom}\,A}$-cocycle. It is clear that this cocycle splits as an $\mathcal{O}^{\mathrm{GL}\,(n,\mathbb{C})}$-cocycle. In particular it splits as a $\mathcal{C}^{\mathrm{GL}\,(n,\mathbb{C})}$-cocycle. Since condition (ii) is satisfied, this implies that $\{h_{ij}\}_{i,j\in I}$ splits as an $\widehat{\mathcal{O}}^{\mathrm{GCom}\,A}$-cocycle. By Proposition 3.2, this further implies that $\{h_{ij}\}_{i,j\in I}$ splits as an $\mathcal{O}^{\mathrm{GCom}\,A}$-cocycle, i.e., there is a family $h_i \in \mathcal{O}^{\mathrm{GCom}\,A}(U_i)$ with $H_iH_j^{-1} = h_ih_j^{-1}$ on $U_i \cap U_j$. Therefore $h_i^{-1}H_i = h_j^{-1}H_j$ on $U_i \cap U_j$, and we have a well-defined global holomorphic map $H : X \to \mathrm{GL}(n, \mathbb{C})$ with $H = h_i^{-1}H_i$ on U_i for all $i \in I$. By (3.4) and since $h_iAh_i^{-1} = A$, we get $H^{-1}AH = B$ on X.

(i) $\Longrightarrow$ (ii): Let an open covering $\mathcal{U} = \{U_i\}_{i\in I}$ of X and a $(\mathcal{U}, \mathcal{O}^{\mathrm{GCom}\,A})$-cocycle $f = \{f_{ij}\}_{i,j\in I}$ be given such that f splits as a $\mathcal{C}^{\mathrm{GL}\,(n,\mathbb{C})}$-cocycle. Then, by Grauert's Oka principle [**11**, Satz I] (see also [**9**, Theorem 5.3.1]), f even splits as an $\mathcal{O}^{\mathrm{GL}\,(n,\mathbb{C})}$-cocycle. This means (we may assume that $U_i \neq \emptyset$ for all i) that there exists a family of holomorphic maps $f_i : U_i \to \mathrm{GL}\,(n, \mathbb{C})$ such that

$$f_{ij} = f_if_j^{-1} \quad \text{on} \quad U_i \cap U_j, \quad i, j \in I. \tag{3.5}$$

Since $f_{ij} \in \mathcal{O}^{\mathrm{GCom}\,A}(U_i \cap U_j)$, this implies that

$$f_if_j^{-1}A = Af_if_j^{-1} \quad \text{and, hence,} \quad f_j^{-1}Af_j = f_i^{-1}Af_i \quad \text{on} \quad U_i \cap U_j, \quad i, j \in I.$$

Hence, there is a well-defined holomorphic map $B : X \to \mathrm{Mat}(n \times n, \mathbb{C})$ with

$$B = f_i^{-1}Af_i \quad \text{on} \quad U_i, \quad i \in I. \tag{3.6}$$

From its definition it is clear that B is locally holomorphically similar to A at each point of X. Since condition (i) is satisfied, it follows that there is a holomorphic map $T : X \to \mathrm{GL}\,(n, \mathbb{C})$ such that

$$B = T^{-1}AT \quad \text{on} \quad X. \tag{3.7}$$

Set $h_i = f_iT^{-1}$ on U_i. By (3.7) and (3.6), then, on each U_i,

$$h_iA = f_iT^{-1}A = f_iBT^{-1} = f_if_i^{-1}Af_iT^{-1} = Af_iT^{-1} = Ah_i,$$

i.e., $h_i \in \mathcal{O}^{\mathrm{GCom}\,A}(U_i)$. Moreover, by (3.5),

$$h_ih_j^{-1} = f_iT^{-1}Tf_j^{-1} = f_if_j^{-1} = f_{ij} \quad \text{on} \quad U_i \cap U_j.$$

So, f splits as an $\mathcal{O}^{\mathrm{GCom}\,A}$-cocycle and, above all, as an $\widehat{\mathcal{O}}^{\mathrm{GCom}\,A}$-cocycle. □

For us, the following two immediate corollaries of Theorem 3.3 are important.

COROLLARY 3.4. *Let X be a Stein space, and let $A, B : X \to \mathrm{Mat}(n \times n, \mathbb{C})$ be holomorphic maps, which are locally holomorphically similar at each point of X. If*

$$H^1(X, \widehat{\mathcal{O}}^{\mathrm{GCom}\,A}) = 0,$$

then A and B are globally holomorphically similar on X.

COROLLARY 3.5. *Let X be a Stein space such that $H^1(X, \mathcal{O}^{\mathrm{GL}(n,\mathbb{C})}) = 0$, and let $A : X \to \mathrm{Mat}(n \times n, \mathbb{C})$ be a holomorphic map such that*

$$H^1(X, \widehat{\mathcal{O}}^{\mathrm{GCom}\,A}) \neq 0.$$

Then there exists a holomorphic $B : X \to \mathrm{Mat}(n \times n, \mathbb{C})$, which is locally holomorphically similar to A at each point of X, but which is not globally holomorphically similar to A on X.

Corollary 3.4 will be used to prove Theorem 1.2. Corollary 3.5 is useful for counterexamples, as, e.g., in [**19**, Sec. 8].

4. Bumps on Riemann surfaces

DEFINITION 4.1. Let X be a Riemann surface. Denote by Δ the closed unit disk centered at the origin in $\mathbb{C}$, and set

$$\Delta_I = \{u \in \Delta \mid \operatorname{Im} u \leq 0 \text{ and } 1/2 \leq |u| \leq 1\},$$
$$\Delta_{II} = \{u \in \Delta \mid |\operatorname{Im} u| \leq |\operatorname{Re} u| \text{ and } 1/2 \leq |u| \leq 1\}.$$

A pair (B_1, B_2) will be called a **bump in** X if B_1, B_2 are compact subsets of X such that either

$$B_1 \cap B_2 = \emptyset, \tag{4.1}$$

or there exists a neighborhood U of B_2 and a $\mathcal{C}^\infty$-diffeomorphism, z, from U onto a neighborhood of Δ such that[2]

$$B_2 \subseteq \{|z| \leq 1\}, \tag{4.2}$$
$$B_1 \cap \{|z| \leq 1\} \subseteq B_2, \tag{4.3}$$

and one of the following two conditions is satisfied.

$$B_1 \cap \{1/2 \leq |z| \leq 1\} = B_2 \cap \{1/2 \leq |z| \leq 1\} = \{z \in \Delta_I), \tag{4.4}$$
$$B_1 \cap \{1/2 \leq |z| \leq 1\} = B_2 \cap \{1/2 \leq |z| \leq 1\} = \{z \in \Delta_{II}\}. \tag{4.5}$$

An m-tuple $(B_1, \dots, B_m)$, $m \geq 2$, of compact subsets of X is called a **bump extension in** X if, for each $1 \leq \mu \leq m-1$, $(B_1 \cup \ldots \cup B_\mu, B_{\mu+1})$ is a bump in X.

LEMMA 4.2. *Let X be a Riemann surface, and let $\rho : X \to \mathbb{R}$ be a $\mathcal{C}^\infty$ function such that, for some real numbers $\alpha < \beta$, the set $\{\rho \leq \beta\}$ is compact and ρ has no critical points on $\{\alpha \leq \rho \leq \beta\}$. Moreover, let $\mathcal{U}$ be an open covering of X. Then there exist $B_1, \dots, B_m$ such that*

(i) *$(\{\rho \leq \alpha\}, B_1, \ldots B_m)$ is a bump extension in X;*
(ii) *$\{\rho \leq \alpha\} \cup B_1 \cup \ldots \cup B_m = \{\rho \leq \beta\}$;*
(iii) *for each $1 \leq \mu \leq m$, $\{\rho \leq \alpha - 1\} \cap B_\mu = \emptyset$ and B_μ is contained in at least one set of $\mathcal{U}$.*

[2] $\{|z| \leq 1\} := \{z \in \Delta\} := z^{-1}(\Delta) := \{\zeta \in U \mid |z(\zeta)| \leq 1\}$ etc.

PROOF. It is sufficient to prove that, for each $\alpha \le t \le \beta$, there exists $\varepsilon > 0$ such that, if $t-\varepsilon \le t_1 \le t \le t_2 \le t+\varepsilon$, then there exist $B_1, \dots, B_m$ such that

(i') $\big(\{\rho \le t_1\}, B_1, \dots, B_m\big)$ is a bump extension in X;

(ii') $\{\rho \le t_1\} \cup B_1 \cup \dots \cup B_m = \{\rho \le t_2\}$;

(iii') for each $2 \le \mu \le m$, $\{\rho \le \alpha - 1\} \cap B_\mu = \emptyset$ and B_μ is contained in at least one set of $\mathcal{U}$.

Let $\alpha \le t \le \beta$ be given.

Since ρ has no critical points on $\{\rho = t\}$ and $\{\rho = t\}$ is compact, then we can find $\widetilde{\varepsilon} > 0$, open subsets $\widetilde{U}_1, \dots, \widetilde{U}_m$ of X, and $\mathcal{C}^\infty$ diffeomorphisms $\widetilde{z}_\mu$ from $\widetilde{U}_\mu$ onto a neighborhood of Δ such that

(a) $\big\{t - \widetilde{\varepsilon} \le \rho \le t + \widetilde{\varepsilon}\big\} \subseteq \big\{|\widetilde{z}_1| < 1/8\} \cup \dots \cup \{|\widetilde{z}_m| < 1/8\big\}$;

(b) $\rho = \operatorname{Im} \widetilde{z}_\mu + t$ on $\widetilde{U}_\mu$ for $1 \le \mu \le m$;

(c) for each $1 \le \mu \le m$, $\widetilde{U}_\mu \cap \{\rho \le \alpha - 1\} = \emptyset$ and $\widetilde{U}_\mu$ is contained in at least one set of $\mathcal{U}$.

By (a) we can choose $\mathcal{C}^\infty$-functions $\chi_1, \dots, \chi_m : X \to [0,1]$ such that

(d) $\operatorname{supp} \chi_\mu \subseteq \{|\widetilde{z}_\mu| < 1/4\}$ for $1 \le \mu \le m$;

(e) $\sum_{\mu=1}^m \chi_\mu = 1$ on $\big\{t - \widetilde{\varepsilon} \le \rho \le t + \widetilde{\varepsilon}\big\}$, and $0 \le \sum_{\mu=1}^m \chi_\mu \le 1$ on X.

Further, for each $1 \le \mu \le m$, take an open set $U_\mu \subseteq \widetilde{U}_\mu$, which is relatively compact in $\widetilde{U}_\mu$ and such that $\widetilde{z}_\mu(U_\mu)$ is still a neighborhood of Δ. Then we can find $\varepsilon > 0$ so small that, for all $v, w \in \mathbb{C}$ with $|v|, |w| \le 2\varepsilon$, and, for $1 \le \mu \le m$,

(f) the function $\widetilde{z}_\mu + v + w \sum_{\nu=1}^\mu \chi_\nu$ restricted to U_μ is a $\mathcal{C}^\infty$ diffeomorphism from U_μ onto a neighborhood of Δ;

(g) $\big\{|\widetilde{z}_\mu| \le 1/4\big\} \subseteq \big\{\zeta \in U_\mu \;\big|\; |\widetilde{z}_\mu(\zeta) + v + w \sum_{\nu=1}^\mu \chi_\nu(\zeta)| \le 1/2\big\}$.

Moreover, we may assume that

(h) $\varepsilon < \widetilde{\varepsilon}/4$.

To prove that this ε has the required property, let t_1, t_2 with $t - \varepsilon \le t_1 \le t \le t_2 \le t + \varepsilon$ be given. Define

on U_μ: $z_\mu = \widetilde{z}_\mu + i(t - t_1) - i(t_2 - t_1) \sum_{\nu=1}^\mu \chi_\nu$ for $1 \le \mu \le m$;
on X: $\rho_0 = \rho - t_1$ and $\rho_\mu = \rho - t_1 - (t_2 - t_1) \sum_{\nu=1}^\mu \chi_\nu$ for $1 \le \mu \le m$;
$B_\mu = \big\{\rho_\mu \le 0\big\} \cap \big\{|z_\mu| \le 1\big\}$ for $1 \le \mu \le m$.

Then, by (e) and (h), $\sum_{\nu=1}^m \chi_\nu = 1$ on $\big\{t - 4\varepsilon \le \rho \le t + 4\varepsilon\big\}$, and $0 \le \sum_{\nu=1}^m \chi_\mu \le 1$ everywhere on X. Since $0 \le t_2 - t_1 \le 2\varepsilon$ and $0 \le t - t_1 \le \varepsilon$, it follows that

$$\rho_m \begin{cases} = \rho - t_2 & \text{on} \quad \big\{t - 4\varepsilon \le \rho \le t + 4\varepsilon\big\}, \\ \ge \rho - t_1 - (t_2 - t_1) > 0 & \text{on} \quad \big\{\rho \ge t + 4\varepsilon\big\}, \\ \le \rho - t_1 < 0 & \text{on} \quad \big\{\rho \le t - 4\varepsilon\big\}. \end{cases}$$

Therefore

$$\big\{\rho_m \le 0\big\} = \big\{\rho \le t_2\big\}. \tag{4.6}$$

By (g), $\big\{|\widetilde{z}_\mu| \le 1/4\big\} \subseteq \big\{|z_\mu| \le 1/2\big\}$ for $1 \le \mu \le m$. Together with (d), this gives

$$\rho_\mu = \rho_{\mu-1} \quad \text{on} \quad X \setminus \big\{|z_\mu| \le 1/2\big\} \quad \text{for} \quad 1 \le \mu \le m. \tag{4.7}$$

Let $1 \le \mu \le m$. Then, by (4.7),

$$\{\rho_{\mu-1} \le 0\} = \Big(\{\rho_{\mu-1} \le 0\} \cap \{|z_\mu| \le 1\}\Big) \cup \Big(\{\rho_\mu \le 0\} \cap \big(X \setminus \{|z_\mu| \le 1\}\big)\Big)$$

and, further,

$$\begin{aligned}\{\rho_{\mu-1} \le 0\} \cup B_\mu =&\Big(\{\rho_{\mu-1} \le 0\} \cap \{|z_\mu| \le 1\}\Big)\\ &\cup \Big(\{\rho_\mu \le 0\} \cap \big(X \setminus \{|z_\mu| \le 1\}\big)\Big) \cup \Big(\{\rho_\mu \le 0\} \cap \{|z_\mu| \le 1\}\Big)\\ =&\Big(\{\rho_{\mu-1} \le 0\} \cap \{|z_\mu| \le 1\}\Big) \cup \{\rho_\mu \le 0\}.\end{aligned}$$

Since $\rho_{\mu-1} \ge \rho_\mu$ and therefore $\{\rho_{\mu-1} \le 0\} \cap \{|z_\mu| \le 1\} \subseteq \{\rho_\mu \le 0\}$, it follows that

$$\{\rho_{\mu-1} \le 0\} \cup B_\mu = \{\rho_\mu \le 0\},$$

and, hence,

$$\{\rho_0 \le 0\} \cup B_1 \cup \ldots \cup B_\mu = \{\rho_\mu \le 0\}.$$

Since $\rho_0 = \rho - t_1$, so we have proved that

$$\{\rho \le t_1\} \cup B_1 \cup \ldots \cup B_\mu = \{\rho_\mu \le 0\} \quad \text{for} \quad 1 \le \mu \le m. \tag{4.8}$$

Since $\rho_1 \ge \rho_0 = \rho - t_1$, we have $\{\rho \le t_1\} = \{\rho_0 \le 0\} \subseteq \{\rho_1 \le 0\}$, which implies by definition of B_1 that

$$\{\rho \le t_1\} \cap \{|z_1| \le 1\} \subseteq B_1. \tag{4.9}$$

For $2 \le \mu \le m$, it follows from (4.8) that

$$\Big(\{\rho \le t_1\} \cup B_1 \cup \ldots \cup B_{\mu-1}\Big) \cap \{|z_\mu| \le 1\} = \{\rho_{\mu-1} \le 0\} \cap \{|z_\mu| \le 1\}.$$

Sine $\rho_{\mu-1} \le \rho_\mu$, this implies by definition of B_μ that

$$\Big(\{\rho \le t_1\} \cup B_1 \cup \ldots \cup B_{\mu-1}\Big) \cap \{|z_\mu| \le 1\} \subseteq B_\mu \quad \text{for} \quad 2 \le \mu \le m. \tag{4.10}$$

Further it follows from (4.7) that

$$\begin{aligned}\{\rho_{\mu-1} \le 0\} \cap \big\{1/2 \le |z_\mu| \le 1\big\} &= \{\rho_\mu \le 0\} \cap \big\{1/2 \le |z_\mu| \le 1\big\}\\ &= B_\mu \cap \big\{1/2 \le |z_\mu| \le 1\big\} \quad \text{for} \quad 1 \le \mu \le m.\end{aligned} \tag{4.11}$$

Since $\{\rho_0 \le 0\} = \{\rho \le t_1\}$, this implies that

$$\{\rho \le t_1\} \cap \big\{1/2 \le |z_1| \le 1\big\} = B_1 \cap \big\{1/2 \le |z_1| \le 1\big\}, \tag{4.12}$$

Moreover, by (4.8), from (4.11) we get

$$\begin{aligned}&\Big(\{\rho \le t_1\} \cup B_1 \cup \ldots \cup B_{\mu-1}\Big) \cap \big\{1/2 \le |z_\mu| \le 1\big\}\\ &\qquad = B_\mu \cap \big\{1/2 \le |z_\mu| \le 1\big\} \quad \text{for} \quad 2 \le \mu \le m.\end{aligned} \tag{4.13}$$

By (b), $\operatorname{Im} z_\mu = \rho_\mu\big|_{U_\mu}$. Therefore $\{\rho_\mu \le 0\} \cap \big\{1/2 \le |z_\mu| \le 1\big\} = \{z_\mu \in \Delta_I\}$. Since $B_\mu \cap \big\{1/2 \le |z_\mu| \le 1\big\} = \{\rho_\mu \le 0\} \cap \big\{1/2 \le |z_\mu| \le 1\big\}$, this means

$$B_\mu \cap \big\{1/2 \le |z_\mu| \le 1\big\} = \{z_\mu \in \Delta_I\} \quad \text{for} \quad 1 \le \mu \le m. \tag{4.14}$$

We summarize:

By (f), z_1 is a $\mathcal{C}^\infty$ diffeomorphism from U_1 onto a neighborhood of Δ and, by definition of B_1, we have $B_1 \subseteq \{|z_1| \le 1\}$. Together with (4.9), (4.12) and (4.14) (for $\mu = 1$), this shows that $(\{\rho \le t_1\}, B_1)$ is a bump in X ((4.4) is satisfied).

For $2 \le \mu \le m$, by (f), z_μ is a $\mathcal{C}^\infty$ diffeomorphism from U_μ onto a neighborhood of Δ and, by definition of B_μ, we have $B_\mu \subseteq \{|z_\mu| \le 1\}$. Together with (4.10), (4.13) and (4.14), this shows that $\big(\{\rho \le t_1\} \cup B_1 \cup \ldots \cup B_{\mu-1}, B_\mu\big)$ is a bump in X.

Therefore (i') holds. (ii') follows from (4.8) and (4.6). (iii') follows from (c). □

THEOREM 4.3. *Let X be noncompact and connected Riemann surface, and let $\mathcal{U}$ be an open covering of X. Then there exists a sequence $(B_\mu)_{\mu\in\mathbb{N}}$ of compact subsets of X such that*

(a) *for each $\mu \in \mathbb{N}$, B_μ is contained in at least one set of $\mathcal{U}$;*
(b) *for each $\mu \in \mathbb{N}^*$, $(B_0 \cup \ldots \cup B_{\mu-1}, B_\mu)$ is a bump in X;*
(c) *$X = \bigcup_{\mu\in\mathbb{N}} B_\mu$.*
(d) *for each compact set $\Gamma \subseteq X$, there exists $N(\Gamma) \in \mathbb{N}$ such that $B_\mu \cap \Gamma = \emptyset$ if $\mu \geq N(\Gamma)$.*

PROOF. Since X is a Stein manifold (see, e.g., [**6**, Corollary 26.8]), we can find a strictly subharmonic $\mathcal{C}^\infty$ function $\rho : X \to \mathbb{R}$ such that, for all $\alpha \in \mathbb{R}$, $\{\rho \leq \alpha\}$ is compact (see, e.g., [**16**, Theorem 5.1.6]). By Morse theory (see, e.g., [**13**, §7 and §19, Exercise 19]), we may assume that all critical points of ρ are non-degenerate, and, for each $t \in \mathbb{R}$, at most one critical point of ρ lies on $\{\rho = t\}$. Since ρ is strictly subharmonic (which implies that ρ has no local maxima), then, for each critical value t of ρ, there are only two possibilities: either ρ has precisely one critical point on $\{\rho = t\}$, and this is the point of a strong local minimum of ρ, or ρ has precisely one critical point on $\{\rho = t\}$, and this is the point of a strong saddle point of ρ.

In particular, then there is precisely one point in X, $\xi_{\min}$, where ρ assumes its absolute minimum, and this minimum is strong. Therefore, we can find $\varepsilon_0 > 0$ such that $\{\rho \leq \rho(\xi_{\min}) + \varepsilon_0\}$ is contained in at least one of the sets of $\mathcal{U}$, and ρ has no critical points in $\{\rho(\xi_{\min}) < \rho \leq \rho(\xi_{\min}) + \varepsilon_0\}$. Set $B_0 = \{\rho(\xi_{\min}) \leq \rho \leq \rho(\xi_{\min}) + \varepsilon_0\}$.

If $\xi_{\min}$ is the only critical point of ρ, then the proof of the theorem can be completed inductively, applying Lemma 4.2 with $\alpha = \rho(\xi_{\min}) + \varepsilon_0 + N$ and $\beta = \rho(\xi_{\min}) + \varepsilon_0 + N + 1$ for $N = 0, 1, 2, \ldots$.

If there are further critical points of ρ, it remains to complete Lemma 4.2 by the following statement.

(*) Let ξ be a critical point of ρ with $t := \rho(\xi) > \rho(\xi_{\min}) + \varepsilon_0$. Then there exists $\varepsilon > 0$ such that, for each $0 < \delta \leq \varepsilon$, we can find an m-tuple $(A_1, \ldots, A_m)$ such that $\big(\{\rho \leq t - \delta\}, A_1, \ldots, A_m\big)$ is a bump extension in X, $\{\rho \leq t - \delta\} \cup A_1 \cup \ldots \cup A_m = \{\rho \leq t + \delta\}$, and, for each $1 \leq \mu \leq m$, $A_\mu \cap \{\rho \leq t - \delta - 1\} = \emptyset$ and A_μ is contained in at least one set of $\mathcal{U}$.

Proof of (*) if ξ is the point of a strong local minimum of ρ:

Then ξ is an isolated point of $\{\rho \leq t\}$. Therefore we can find an open neighborhood U of ξ and an open neighborhood V of $\{\rho \leq t\} \setminus \{\xi\}$ such that $U \cap V = \emptyset$ and U is contained in at least one set of $\mathcal{U}$. Choose $\varepsilon > 0$ such that ρ has no critical points on $V \cap \{t - \varepsilon \leq \rho \leq t + \varepsilon\}$ and

$$\{\rho \leq t + \varepsilon\} \subseteq U \cup V.$$

To prove that this ε has the required property, let $0 < \delta \leq \varepsilon$ be given. Then, by Lemma 4.2, we can find $A_1, \ldots, A_{m-1}$ such that $(\{\rho \leq t - \delta\}, A_1, \ldots, A_{m-1})$ is a bump extension in V, $\{\rho \leq t - \delta\} \cup A_1 \cup \ldots \cup A_{m-1} = V \cap \{\rho \leq t + \delta\}$, and, for $1 \leq \mu \leq m - 1$, $A_\mu \cap \{\rho \leq t - \delta - 1\} = \emptyset$ and A_μ is contained in at least one set of $\mathcal{U}$. Since $U \cap V = \emptyset$, it remains to set $A_m = U \cap \{\rho \leq t + \delta\}$.

Proof of (*) if ξ is the point of a strong saddle point of ρ:

By a lemma of Morse (see, e.g., [**21**, Lemma 2.2]), then we can find $R > 0$, an open neighborhood W of ξ and a $\mathcal{C}^\infty$ diffeomorphism w from W onto a neighborhood of $\Delta_R := \{\lambda \in \mathbb{C} \mid |\lambda| \le R\}$ such that $w(\xi) = 0$ and

$$\rho = t + (\operatorname{Re} w)^2 - (\operatorname{Im} w)^2 \quad \text{on} \quad W. \tag{4.15}$$

Moreover, we may choose W so small that

$$W \cap \{\rho \le t - 1\} = \emptyset \text{ and } W \text{ is contained in at least one set of } \mathcal{U}. \tag{4.16}$$

Take $\varepsilon > 0$ so small that ξ is the only critical point of ρ on $\{t - \varepsilon \le \rho \le t + \varepsilon\}$. To prove that this ε has the required property, let $0 < \delta \le \varepsilon$ be given.

Choose $0 < r \le R/2$ with

$$\{|w| \le r\} \subseteq \{t - \delta < \rho < t + \delta\}, \tag{4.17}$$

and take a $\mathcal{C}^\infty$ function $\chi : X \to [0, 1]$ with $\chi = 1$ on $\{|z| \le r/2\}$ and $\chi = 0$ on $X \setminus \{|w| \le r\}$. By (4.15), then we can find $c > 0$ so small that the functions $\rho_+, \rho_- : X \to \mathbb{R}$ defined by

$$\rho_+ = \rho + c\chi \quad \text{and} \quad \rho_- = \rho - c\chi$$

have the same critical points as ρ. Then $\rho_+ = \rho_- = \rho$ on $X \setminus \{|w| \le r\}$ and $\rho_+ \ge \rho \ge \rho_-$ on X. Therefore and by (4.17),

$$\{\rho_+ \le t - \delta\} = \{\rho \le t - \delta\}, \tag{4.18}$$

$$\{t - \delta \le \rho_+ \le t\} \subseteq \{t - \delta \le \rho \le t\} \subseteq \{t - \varepsilon \le \rho \le t + \varepsilon\}, \tag{4.19}$$

$$\{\rho_+ \le t - \delta - 1\} = \{\rho \le t - \delta - 1\}, \tag{4.20}$$

$$\{\rho_- \le t + \delta\} = \{\rho \le t + \delta\}, \tag{4.21}$$

$$\{t \le \rho_- \le t + \delta\} \subseteq \{t \le \rho \le t + \delta\} \subseteq \{t - \varepsilon \le \rho \le t + \varepsilon\}, \tag{4.22}$$

$$\{\rho_- \le t - 1\} \supseteq \{\rho \le t - \delta - 1\}. \tag{4.23}$$

Since $\rho_+(\xi) = \rho(\xi) + c > t$, we have $\xi \notin \{t - \delta \le \rho_+ \le t\}$. As ξ is the only critical point of ρ in $\{t - \varepsilon \le \rho \le t + \varepsilon\}$ and by (4.19), this implies that ρ has no critical point in $\{t - \delta \le \rho_+ \le t\}$. Since ρ_+ has the same critical points as ρ, this means that ρ_+ has no critical point in $\{t - \delta \le \rho_+ \le t\}$. Therefore, by Lemma 4.2, by (4.18) and by (4.20), we can find $A_1, \dots, A_k$ such that

(a) $(\{\rho \le t - \delta\}, A_1, \dots, A_k)$ is a bump extension in X, $\{\rho \le t - \delta\} \cup A_1 \cup \dots \cup A_k = \{\rho_+ \le t\}$ and, for each $1 \le \mu \le k$, $A_\mu \cap \{\rho \le t - \delta - 1\} = \emptyset$ and A_μ is contained in at least one set of $\mathcal{U}$.

Since $\rho_-(\xi) = \rho(\xi) - c < t$, we have $\xi \notin \{t \le \rho_- \le t + \delta\}$. As ξ is the only critical point of ρ in $\{t - \varepsilon \le \rho \le t + \varepsilon\}$ and by (4.22), this implies that ρ has no critical points in $\{t \le \rho_- \le t + \delta\}$. Since ρ_- has the same critical points as ρ, this means that ρ_- has no critical points in $\{t \le \rho_- \le t + \delta\}$. Therefore, by Lemma 4.2, by (4.21) and by (4.23), we can find $A_{k+2}, \dots, A_m$ such that

(b) $(\{\rho_- \le t\}, A_{k+2}, \dots, A_m)$ is a bump extension in X, $\{\rho_- \le t\} \cup A_{k+2} \cup \dots \cup A_m = \{\rho \le t + \delta\}$ and, for each $k+2 \le \mu \le m$, $A_\mu \cap \{\rho \le t - \delta - 1\} = \emptyset$ and A_μ is contained in at least one set of $\mathcal{U}$.

Set $z = w/2r$ on W. Since $r \le R/2$, then z is a diffeomorphism from W onto a neighborhood of Δ. Since

$$\rho_+ = \rho_- = \rho \quad \text{on} \quad X \setminus \{|z| \le 1/2\} = X \setminus \{|w| \le r\}, \tag{4.24}$$

it follows from (4.15) that

$$\rho_+ = \rho_- = \rho = t + 4r^2\Big((\operatorname{Re} z)^2 - (\operatorname{Im} z)^2\Big) \quad \text{on} \quad \big\{1/2 \le |z| \le 1\big\}. \tag{4.25}$$

Set $A_{k+1} = \big\{\rho_- \le t\big\} \cap \big\{|z| \le 1\big\}$. Then

$$A_{k+1} \subseteq \big\{|z| \le 1\big\}. \tag{4.26}$$

Since $\rho_- \le \rho_+$, we have

$$\{\rho_+ \le t\} \cap \{|z| \le 1\} \subseteq A_{k+1} \tag{4.27}$$

and

$$\big(\{\rho_+ \le t\} \cup A_{k+1}\big) \cap \big\{|z| \le 1\big\} = \big\{\rho_- \le t\big\} \cap \big\{|z| \le 1\big\}.$$

Together with (4.24), the latter yields

$$\big\{\rho_+ \le t\big\} \cup A_{k+1} = \big\{\rho_- \le t\big\}. \tag{4.28}$$

From (4.25) it follows that

$$\big\{\rho_+ \le t\big\} \cap \big\{1/2 \le |z| \le 1\big\} = A_{k+1} \cap \big\{1/2 \le |z| \le 1\big\} = \{z \in \Delta_{II}\}. \tag{4.29}$$

By (4.26) - (4.29), $\big(\{\rho_+ \le t\}, A_{k+1}\big)$ is a bump in X (condition (4.5) is satisfied) with $\big\{\rho_+ \le t\big\} \cup A_{k+1} = \big\{\rho_- \le t\big\}$ and such that, by (4.16), $A_{k+1} \cap \{\rho \le t - \delta - 1\} = \emptyset$ and A_{k+1} is contained in at least one set of $\mathcal{U}$. Together with (a) and (b) it follows that the m-tuple $(A_1, \dots, A_m)$ has the required properties. □

5. Z-adapted pairs of compact sets on 1-dimensional complex spaces

DEFINITION 5.1. Let X be a 1-dimensional complex space, and let Z be a discrete and closed subset of X such that all points of $X \setminus Z$ are smooth. A pair (Γ_1, Γ_2) will be called a **Z-adapted pair of compact sets in** X if $\Gamma_1 \cap \Gamma_2 = \Gamma^o \cup \Gamma$, where

- $\Gamma^o \cap \Gamma = \emptyset$;
- $\Gamma^o \subseteq Z$;
- $\Gamma \cap Z = \emptyset$ and, if $\Gamma \neq \emptyset$, then Γ consists of a finite number[3] of connected components each of which has a basis of contractible open neighborhoods.

LEMMA 5.2. *Let X be a 1-dimensional complex space, and let Z be a discrete and closed subset of X such that all points of $X \setminus Z$ are smooth. Let $\pi : \widetilde{X} \to X$ be the normalization of X (see, e.g.,* [**20**, Ch. VI, §4]*), and let (B_1, B_2) be a bump in $\widetilde{X}$ (Def. 4.1). Then there exists a Z-adapted pair of compact sets in X, (Γ_1, Γ_2), such that*

$$\Gamma_1 \subseteq \pi(B_1), \tag{5.1}$$

$$\Gamma_2 \subseteq \pi(B_2), \tag{5.2}$$

$$\Gamma_1 \cup \Gamma_2 = \pi(B_1 \cup B_2). \tag{5.3}$$

PROOF. First let $B_1 \cap B_2 = \emptyset$. Since all points of $X \setminus Z$ are smooth and, hence, π is bijective from $\widetilde{X} \setminus \pi^{-1}(Z)$ onto $X \setminus Z$, then $\pi(B_1) \cap \pi(B_2) \subseteq Z$. Set $\Gamma_1 = \pi(B_1)$ and $\Gamma_2 = \pi(B_2)$. Then (Γ_1, Γ_2) a Z-adapted pair of compact sets in X (we can take $\Gamma^o = \pi(B_1) \cap \pi(B_2)$ and $\Gamma = \emptyset$ in Def. 5.1), which trivially satisfies conditions (5.1)-(5.3).

---

[3]In the applications below, this 'finite number' will be one or two.

Now let $B_1 \cap B_2 \neq \emptyset$. Then (by Definition 4.1) we have a neighborhood U of B_2 and a diffeomorphic map, z, from U onto a neighborhood of Δ satisfying (4.2) and (4.3), and one of the relations (4.4) or (4.5).

Since Z is discrete and closed in X, $\pi^{-1}(Z)$ is discrete and closed in $\widetilde{X}$. Therefore, we can find $1/2 < r < R < 1$ such that $\pi^{-1}(Z) \cap \{r \leq |z| \leq R\} = \emptyset$ and, hence,

$$Z \cap \pi\big(\{r \leq |z| \leq R\}\big) = \emptyset. \tag{5.4}$$

Set

$$\begin{aligned}
&K = B_1 \cap B_2 \cap \{r \leq |z| \leq R\},\\
&K_1 = B_1 \cap (\widetilde{X} \setminus \{|z| < R\}),\\
&K_2 = B_2 \cap \{|z| \leq r\},\\
&\Gamma = \pi(K), \quad \Gamma_1 = \pi(K \cup K_1), \quad \Gamma_2 = \pi(K \cup K_2), \quad \Gamma^{\mathrm{o}} = \pi(K_1) \cap \pi(K_2).
\end{aligned}$$

Then

$$\begin{aligned}
(5.5)\quad \Gamma_1 \cap \Gamma_2 &= \Big(\pi(K) \cup \pi(K_1)\Big) \cap \Big(\pi(K) \cup \pi(K_2)\Big)\\
&= \pi(K) \cup \Big(\pi(K_1) \cap \pi(K)\Big) \cup \Big(\pi(K_1) \cap \pi(K_2)\Big) = \pi(K) \cup \Big(\pi(K_1) \cap \pi(K_2)\Big) = \Gamma \cup \Gamma^0.
\end{aligned}$$

Since all points in $X \setminus Z$ are smooth points of X, π is bijective from $\widetilde{X} \setminus \pi^{-1}(Z)$ onto $X \setminus Z$. As $K_1 \cap K_2 = \emptyset$, this implies that

$$\Gamma^{\mathrm{o}} \subseteq Z. \tag{5.6}$$

By (5.4), $\Gamma \cap Z = \emptyset$. Together with (5.5), this gives

$$\Gamma \cap \Gamma^{\mathrm{o}} = \emptyset. \tag{5.7}$$

From (4.4) resp. (4.5) we get

$$K = \begin{cases} \{z \in \Delta_I\} \cap \{r \leq |z| \leq R\} & \text{in case of (4.4),}\\ \{z \in \Delta_{II}\} \cap \{r \leq |z| \leq R\} & \text{in case of (4.5).}\end{cases}$$

Since z diffeomorphic, this implies that K consists of a finite number of connected components (one in case (4.4) and two in case (4.5)) each of which has a basis of contractible open neighborhoods. Since, by (5.4), π is biholomorphic from an open neighborhood of K onto an open neighborhood of Γ, this further implies that Γ consists of a finite number of connected components each of which has a basis of contractible open neighborhoods. Together with (5.5)-(5.7), this shows that (Γ_1, Γ_2) is a Z-adapted pair of compact sets in X. Moreover, since $K \cup K_1 \subseteq B_1$ and $K \cup K_2 \subseteq B_2$, (5.1), (5.2) and "$\subseteq$" in (5.3) are clear. Therefore it remains to prove "$\supseteq$" in (5.3). For the latter it is sufficient to prove that

$$\pi\big((B_1 \cup B_2) \cap \{|z| \leq r\}\big) \subseteq \Gamma_2, \tag{5.8}$$

$$\pi\big((B_1 \cup B_2) \cap \{r \leq |z| \leq R\}\big) = \Gamma, \tag{5.9}$$

$$\pi\big((B_1 \cup B_2) \cap (\widetilde{X} \setminus \{|z| < R\})\big) \subseteq \Gamma_1. \tag{5.10}$$

From (4.3) we get $B_1 \cap \{|z| \leq r\} \subseteq B_2$. Hence

$$(B_1 \cup B_2) \cap \{|z| \leq r\} \subseteq B_2 \cap \{|z| \leq r\} = K_2 \subseteq K \cup K_2,$$

which yields (5.8) be definition of Γ_2.

Since, by the first equality in (4.4) resp. (4.5),

$$(B_1 \cup B_2) \cap \{r \leq |z| \leq R\} = B_1 \cap B_2 \cap \{r \leq |z| \leq R\} = K,$$

we get (5.9).

By (4.2), $B_2 \subseteq \{|z| \leq 1\}$ and, hence,

$$B_2 \cap \big(\widetilde{X} \setminus \{|z| < R\}\big) = B_2 \cap \{R \leq |z| \leq 1\}.$$

Again using the first equality in (4.4) resp. (4.5), this yields

$$B_2 \cap \big(\widetilde{X} \setminus \{|z| < R\}\big) = B_1 \cap \{R \leq |z| \leq 1\}$$

and, further,

$$B_2 \cap \big(\widetilde{X} \setminus \{|z| < R\}\big) \subseteq B_1 \cap \big(\widetilde{X} \setminus \{|z| < R\}\big).$$

It follows that

$$(B_1 \cup B_2) \cap \big(\widetilde{X} \setminus \{|z| < R\}\big) \subseteq B_1 \cap \big(\widetilde{X} \setminus \{|z| < R\}\big) = K_1,$$

which implies (5.10). □

6. Jordan stable points

In this section, X is a complex space (of arbitrary dimension), and $A : X \to \mathrm{Mat}(n \times n, \mathbb{C})$ is a holomorphic map.

DEFINITION 6.1. A point $\xi \in X$ will be called **Jordan stable** for A if there exists a neighbourhood U of ξ such that the following two conditions are satisfied:

(a) there are holomorphic functions $\lambda_1, \ldots, \lambda_m : U \to \mathbb{C}$ such that, for each $\zeta \in U$, $\lambda_1(\zeta), \ldots, \lambda_m(\zeta)$ are the different eigenvalues of $A(\zeta)$;

(b) there is a holomorphic map $T : U \to \mathrm{GL}(n\mathbb{C})$ such that, for all $\zeta \in U$, $T(\zeta)^{-1}A(\zeta)T(\zeta)$ is in Jordan normal form.[4]

PROPOSITION 6.2. *The points in X which are not Jordan stable for A form a nowhere dense analytic subset of X. (If X is 1-dimensional, this means that this set is discrete and closed in X.)*

This proposition can be found in [**18**, Theorem 5.5]. If X is 1-dimensional and smooth, it was first proved H. Baumgärtel [**1**], [**2**, Kap. 5, §7], [**4**, 5.7]. If X is of arbitrary dimension and smooth, H. Baumgärtel [**3**], [**4**, S 3.4] obtained the slightly weaker statement that the points in X which are not Jordan stable for A are contained in a nowhere dense analytic subset of X.

THEOREM 6.3. *Let $\xi \in X$ be Jordan stable for A. Then there exist a neighborhood U of ξ and a holomorphic map $H : U \to GL(n, \mathbb{C})$ such that (cp. Def. 3.1)*

$$H(\zeta)^{-1}\big(\mathrm{Com}\, A(\zeta)\big)H(\zeta) = \mathrm{Com}\, A(\xi) \quad \textit{for all} \quad \zeta \in U. \tag{6.1}$$

[4] Equivalently, one could define: ξ is Jordan stable for A if and only if there exists a neighborhood U of ξ such that the number of different eigenvalues of $A(\zeta)$ is the same for all $\zeta \in U$ and, for all integers $1 \leq k \leq n$, the number of Jordan blocks in the Jordan normal forms of $A(\zeta)$ is the same for all $\zeta \in U$. This was proved by G. P. A. Thijsse [**23**] (see also [**18**, Lemma 5.3]).

PROOF. Let U, λ_j, $1 \le j \le m$, and T be as in Definition 6.1. Set $J(\zeta) = T(\zeta)^{-1}A(\zeta)T(\zeta)$ and $H(\zeta) = T(\zeta)T(\xi)^{-1}$ for $\zeta \in U$. We may assume that U is connected. Since J is continuous and, for each $\zeta \in U$, $\lambda_1(\zeta), \dots, \lambda_m(\zeta)$ are the different eigenvalues of $J(\zeta)$, and since $J(\zeta)$ is in Jordan normal form, then, after a possible change of the numbering, there are integers $n_1, \dots, n_m \ge 1$ and matrices $M_j \in \mathrm{Mat}(n_j \times n_j, \mathbb{C})$, $1 \le j \le m$, in Jordan normal form and with the only eigenvalue 0, such that, for each $\zeta \in U$, $J(\zeta)$ is the block diagonal matrix with the diagonal $\lambda_1(\zeta)I_{n_1} + M_1, \dots, \lambda_m(\zeta)I_{n_m} + M_m$.

Since the eigenvalues $\lambda_1(\zeta), \dots, \lambda_m(\zeta)$ are pairwise different, then [**10**, Ch. VIII, §1], a matrix $\Theta \in \mathrm{Mat}(n \times n, \mathbb{C})$ belongs to $\mathrm{Com}\, J(\zeta)$ if and only if Θ is a block diagonal matrix with matrices $Z_j \in \mathrm{Com}\big(\lambda_j(\zeta)I_{n_j} + M_j\big)$, $1 \le j \le m$, on the diagonal. Obviously, $\mathrm{Com}\big(\lambda_j(\zeta)I_{k_j} + M_j\big) = \mathrm{Com}\big(\lambda_j(\xi)I_{k_j} + M_j\big)$ for all $\zeta \in U$. Therefore, $\mathrm{Com}\, J(\zeta) = \mathrm{Com}\, J(\xi)$ for all $\zeta \in U$. (6.1) now follows by (3.2). □

REMARK 6.4. Let Z be the set of all points in X which are not Jordan stable for A (which is, by Proposition 6.2, a nowhere dense analytic subset of X). Assume that $X \setminus Z$ is connected, fix some point $\xi \in X$, and denote by N the normalizer of $\mathrm{GCom}\, A(\xi)$ in $\mathrm{GL}(n, \mathbb{C})$, i.e., the complex Lie group of all $\Phi \in \mathrm{GL}(n, \mathbb{C})$ with $\Phi^{-1}\big(\mathrm{GCom}\, A(\xi)\big)\Phi = \mathrm{GCom}\, A(\xi)$. Then Theorem 6.3 implies that the family

$$\big\{\mathrm{GCom}\, A(\zeta)\big\}_{\zeta \in X \setminus Z}$$

is a holomorphic N-principal bundle of complex Lie groups, with the characteristic fiber $\mathrm{GCom}\, A(\xi)$. Note that $\mathrm{GCom}\, A(\xi)$ is connected (as easy to see – cp. [**19**, Lemma 4.2]), whereas N need not be connected (cp. Remark 6.8 below).

Theorem 6.3 immediately yields

COROLLARY 6.5. *For each open set $W \subseteq X$ which contains only Jordan stable points of A, we have $\widehat{\mathcal{O}}^{\mathrm{Com}\, A}(W) = \mathcal{C}^{\mathrm{Com}\, A}(W)$.*

THEOREM 6.6. *Let $W \subseteq X$ be an open set which contains only Jordan stable points of A. Further suppose that W is contractible. Fix some point $\xi \in W$. Then there exists a continuous map $T : W \to \mathrm{GL}(n, \mathbb{C})$ such that*

$$T^{-1}(\zeta)\big(\mathrm{Com}\, A(\zeta)\big)T(\zeta) = \mathrm{Com}\, A(\xi) \quad \textit{for all} \quad \zeta \in W. \tag{6.2}$$

If, moreover, W is Stein, then this T can be chosen holomorphically on W.

PROOF. Since W is connected, by Theorem 6.3, for each $\eta \in W$, we can find a neighborhood $U_\eta \subseteq W$ of η and a holomorphic map $H_\eta : U_\eta \to GL(n, \mathbb{C})$ such that

$$H_\eta^{-1}(\zeta)\big(\mathrm{Com}\, A(\zeta)\big)H_\eta(\zeta) = \mathrm{Com}\, A(\xi) \quad \text{for all} \quad \zeta \in U_\eta. \tag{6.3}$$

Let N the normalizer of $\mathrm{GCom}\, A(\xi)$ in $\mathrm{GL}(n, \mathbb{C})$, i.e., the complex Lie group of all $\Phi \in \mathrm{GL}(n, \mathbb{C})$ with $\Phi^{-1}\big(\mathrm{GCom}\, A(\xi)\big)\Phi = \mathrm{GCom}\, A(\xi)$. Then, by (6.3),

$$H_\eta^{-1}(\zeta)H_\tau(\zeta) \in N \quad \text{for all} \quad \zeta \in U_\eta \cap U_\tau \quad \text{and all} \quad \eta, \tau \in W,$$

i.e., the family $\{H_\eta^{-1}H_\tau\}_{\eta\tau}$ is an $\mathcal{O}^N$-cocycle with the covering $\{U_\eta\}_{\eta \in W}$. Since W is contractible, this cocycle splits as a $\mathcal{C}^N$-cocycle [**22**, Corollary 11.6], i.e., there is a family $T_\eta \in \mathcal{C}^N(U_\eta)$ such that $T_\eta^{-1}T_\tau = H_\eta H_\tau^{-1}$ on $U_\eta \cap U_\tau$. Then

$$H_\eta T_\eta = H_\tau T_\tau \quad \text{on} \quad U_\eta \cap U_\tau, \quad \eta, \tau \in W.$$

Hence, there is a well-defined continuous map $T : X \to \mathrm{GL}(n, \mathbb{C})$ with

$$T = H_\eta T_\eta \quad \text{on} \quad U_\eta, \quad \eta \in W.$$

By (6.3) and since $T_\eta(\zeta) \in N$, this T satisfies (6.2). If W is Stein, by Grauert's Oka principle [**11**, Satz 6] (see also [**9**, Theorem 8.2.1 (i)]), the maps T_η can be chosen holomorphically, which implies that T is holomorphic. $\square$

COROLLARY 6.7. *Under the hypotheses of Theorem* 6.6, $\widehat{\mathcal{O}}^{\mathrm{GCom}\,A}(W)$, *endowed with the topology of uniform convergence on the compact subsets of* W, *is connected.*

PROOF. Let $G := \mathrm{GCom}\,A(\xi)$, and let $\mathcal{C}^G(W)$ be also endowed with the topology of uniform convergence on the compact subsets of W. Since W is contractible, each element of $\mathcal{C}^G(W)$ can be connected by a continuous path in $\mathcal{C}^G(W)$ with a constant map. It is easy to see that G is connected [**19**, Lemma 4.2]. Therefore, this implies that $\mathcal{C}^G(W)$ is connected. Since, by (6.1), $\mathcal{C}^{\mathrm{GCom}\,A}(W)$ and $\mathcal{C}^G(W)$ are isomorphic as topological groups, it follows that $\mathcal{C}^{\mathrm{GCom}\,A}(W)$ is connected. Since $\widehat{\mathcal{O}}^{\mathrm{GCom}\,A}(W) = \mathcal{C}^{\mathrm{GCom}\,A}(W)$ (Corollary 6.5), this completes the proof. $\square$

REMARK 6.8. Assume that X is a domain in the complex plane and all points of X are Jordan stable for A. Let $\xi \in X$. If, moreover, X is simply connected (and, hence, contractible), then, by Theorem 6.6, we can find a holomorphic map $T : X \to \mathrm{GL}(n,\mathbb{C})$ such that $T^{-1}(\zeta)\big(\mathrm{Com}\,A(\zeta)\big)T(\zeta) = \mathrm{Com}\,A(\xi)$ for all $\zeta \in X$.

Question: Is this true also if X is not simply connected?

If the normalizer, N, of $\mathrm{GCom}\,A(\xi)$ in $\mathrm{GL}(n,\mathbb{C})$, which appears in the proof of Theorem 6.6, is connected, this is the case. This follows by the same proof, using [**11**, Satz 7] (see also, [**9**, Theorem 8.2.1 (iii)]) (saying that $H^1(Y,\mathcal{O}^G) = 0$ for each connected complex Lie group G and each noncompact connected Riemann surface Y). However, this is not always the case.

For example, assume that $A(\xi) = \begin{pmatrix} 1 & 0 \\ 0 & 0 \end{pmatrix}$. Then

$$N = \left\{ \begin{pmatrix} a & 0 \\ 0 & d \end{pmatrix} \,\middle|\, a, d \in \mathbb{C} \setminus \{0\} \right\} \cup \left\{ \begin{pmatrix} 0 & b \\ c & 0 \end{pmatrix} \,\middle|\, b, c \in \mathbb{C} \setminus \{0\} \right\}, \tag{6.4}$$

which is not connected.

Proof of (6.4). It is easy to see that

$$\mathrm{GCom} \begin{pmatrix} 1 & 0 \\ 0 & 0 \end{pmatrix} = \left\{ \begin{pmatrix} a & 0 \\ 0 & d \end{pmatrix} \,\middle|\, a, d \in \mathbb{C} \setminus \{0\} \right\}. \tag{6.5}$$

To prove "$\supseteq$" in (6.4), we therefore only have to prove that, for all $b, c \in \mathbb{C} \setminus \{0\}$,

$$\begin{pmatrix} 0 & b \\ c & 0 \end{pmatrix} \in N. \tag{6.6}$$

Let $b, c \in \mathbb{C} \setminus \{0\}$ be given. Then, for all $\alpha, \delta \in \mathbb{C} \setminus \{0\}$,

$$\begin{pmatrix} 0 & b \\ c & 0 \end{pmatrix}^{-1} \begin{pmatrix} \alpha & 0 \\ 0 & \delta \end{pmatrix} \begin{pmatrix} 0 & b \\ c & 0 \end{pmatrix} = \begin{pmatrix} 0 & c^{-1} \\ b^{-1} & 0 \end{pmatrix} \begin{pmatrix} 0 & \alpha b \\ \delta c & 0 \end{pmatrix} = \begin{pmatrix} \delta & 0 \\ 0 & \alpha \end{pmatrix}.$$

In view of (6.5), this implies that

$$\begin{pmatrix} 0 & b \\ c & 0 \end{pmatrix}^{-1} \mathrm{GCom} \begin{pmatrix} 1 & 0 \\ 0 & 0 \end{pmatrix} \begin{pmatrix} 0 & b \\ c & 0 \end{pmatrix} = \mathrm{GCom} \begin{pmatrix} 1 & 0 \\ 0 & 0 \end{pmatrix},$$

which means (6.6), by definition of N.

To prove "$\subseteq$" in (6.4), let $\begin{pmatrix} a & b \\ c & d \end{pmatrix} \in N$ be given. We have to prove that

$$\text{either} \qquad b = c = 0, \qquad \text{or} \qquad a = d = 0. \tag{6.7}$$

By definition of N and by (6.5), we can find $\alpha, \delta \in \mathbb{C} \setminus \{0\}$ with

$$\begin{pmatrix} a & b \\ c & d \end{pmatrix}^{-1} \begin{pmatrix} 2 & 0 \\ 0 & 1 \end{pmatrix} \begin{pmatrix} a & b \\ c & d \end{pmatrix} = \begin{pmatrix} \alpha & 0 \\ 0 & \delta \end{pmatrix},$$

which implies

$$\begin{pmatrix} a & b \\ c & d \end{pmatrix}^{-1} \begin{pmatrix} 1 & 0 \\ 0 & 0 \end{pmatrix} \begin{pmatrix} a & b \\ c & d \end{pmatrix} = \begin{pmatrix} 1+\alpha & 0 \\ 0 & 1+\delta \end{pmatrix} \tag{6.8}$$

and, hence,

$$\begin{pmatrix} a & b \\ 0 & 0 \end{pmatrix} = \begin{pmatrix} a(1+\alpha) & b(1+\delta) \\ c(1+\alpha) & d(1+\delta) \end{pmatrix}. \tag{6.9}$$

By (6.8), $\begin{pmatrix} 1+\alpha & 0 \\ 0 & 1+\delta \end{pmatrix}$ is of rank 1. Therefore

$$\text{either} \quad 1+\alpha \neq 0 \quad \text{and} \quad 1+\delta = 0, \qquad \text{or} \qquad 1+\alpha = 0 \quad \text{and} \quad 1+\delta \neq 0.$$

By (6.9), this proves (6.7). □

7. Proof of Theorem 1.2

In this section we prove Theorem 1.2. Let X be a 1-dimensional Stein space, and $A : X \to \operatorname{Mat}(n \times n, \mathbb{C})$ a holomorphic map. By Corollary 3.5, it is sufficient to prove the following

THEOREM 7.1. $H^1(X, \widehat{\mathcal{O}}^{\mathrm{GCom}\,A}) = 0$.

DEFINITION 7.2. Let Z be a discrete and closed subset of X.

If U is a nonempty open subset of X, then we define:

$\widehat{\mathcal{O}}^{\mathrm{Com}\,A}(U, Z)$ is the subalgebra of all $f \in \widehat{\mathcal{O}}^{\mathrm{Com}\,A}(U)$ such that there exists an open neighborhood U_Z of Z (depending on f) with $f = 0$ on $U_Z \cap U$.

$\widehat{\mathcal{O}}^{\mathrm{GCom}\,A}(U, Z)$ is the subgroup of all $f \in \widehat{\mathcal{O}}^{\mathrm{GCom}\,A}(U)$ such that there exists an open neighborhood U_Z of Z (depending on f) with $f = I$ on $U_Z \cap U$.

The so defined sheaves on X will be denoted by $\widehat{\mathcal{O}}^{\mathrm{Com}\,A}(\cdot, Z)$ and $\widehat{\mathcal{O}}^{\mathrm{GCom}\,A}(\cdot, Z)$, respectively.

DEFINITION 7.3. Let Z be a discrete and closed subset of X. An open covering of X, $\mathcal{U} = \{U_{ij}\}_{i,j \in I}$, will be called **$Z$-adapted** if

$$U_i \cap U_j \cap Z = \emptyset \quad \text{for all } i, j \in I \text{ with } i \neq j. \tag{7.1}$$

LEMMA 7.4. *Let Z be a discrete and closed subset of X. Then each open covering of X admits a Z-adapted refinement.*

PROOF. Let an open covering $\mathcal{U} = \{U_i\}_{i \in I}$ of X be given. Since Z is discrete and closed in X, we can find a family $\{U_\xi^*\}_{\xi \in Z}$ such that, for each $\xi \in Z$, U_ξ^* is an open neighborhood of ξ, and $U_\xi^* \cap U_\eta^* = \emptyset$ for all $\xi, \eta \in Z$ with $\xi \neq \eta$. Further, we set $U_i^* = U_i \cap (X \setminus Z)$ for $i \in I$. Then $\{U_\alpha^*\}_{\alpha \in I \cup Z}$ (we may assume that $I \cap Z = \emptyset$) is a Z-adapted refinement of $\mathcal{U}$. □

REMARK 7.5. To prove Theorem 7.1, it is sufficient to find a discrete and closed subset Z of X such that

$$H^1\big(X, \widehat{\mathcal{O}}^{\mathrm{GCom}\,A}(\cdot, Z)\big) = 0. \tag{7.2}$$

Indeed, assume Z is such a set and f is an $\widehat{\mathcal{O}}^{\mathrm{GCom}\,A}$-cocycle. Let $\mathcal{U}$ be the covering of f. Then, by Lemma 7.4, we can find a Z-adapted open covering of X, $\mathcal{U}^*$, which is a refinement of $\mathcal{U}$. Let f^* be a $(\mathcal{U}^*, \widehat{\mathcal{O}}^{\mathrm{GCom}\,A})$-cocycle, which is induced by f. Since $\mathcal{U}^*$ is Z-adapted, then f^* can be interpreted as a $\big(\mathcal{U}^*, \widehat{\mathcal{O}}^{\mathrm{GCom}\,A}(\cdot, Z)\big)$-cocycle, and it follows from (7.2) that f^* splits as an $\widehat{\mathcal{O}}^{\mathrm{GCom}\,A}(\cdot, Z)$-cocycle. As $\widehat{\mathcal{O}}^{\mathrm{GCom}\,A}(\cdot, Z)$ is a subsheaf of $\widehat{\mathcal{O}}^{\mathrm{GCom}\,A}$, this means in particular that f^* splits as an $\widehat{\mathcal{O}}^{\mathrm{GCom}\,A}$-cocycle.

LEMMA 7.6. *Let Z be a discrete and closed subset of X, which contains all points of X which are not Jordan stable for A (by Proposition 6.2 such Z exist), and let (Γ_1, Γ_2) be a Z-adapted pair of compact sets in X (Def. 5.1). Then:*

(i) *Let U be an open neighborhood of $\Gamma_1 \cap \Gamma_2$ and $f \in \widehat{\mathcal{O}}^{\mathrm{GCom}\,A}(U, Z)$. Then there exist a section $\widetilde{f} \in \widehat{\mathcal{O}}^{\mathrm{GCom}\,A}(X, Z)$ and an open neighborhood $V \subseteq U$ of $\Gamma_1 \cap \Gamma_2$ such that $\widetilde{f} = f$ on V and $\widetilde{f} = I$ on $X \setminus U$.*

(ii) *Let U_1 be an open neighborhood of Γ_1, $f_1 \in \widehat{\mathcal{O}}^{\mathrm{GCom}\,A}(U_1, Z)$, and let V be a neighborhood of $\Gamma_1 \cap \Gamma_2$. Then there exist an open neighborhood W of $\Gamma_1 \cup \Gamma_2$ and a section $f \in \widehat{\mathcal{O}}^{\mathrm{GCom}\,A}(W, Z)$ such that $f = f_1$ on $(U_1 \cap W) \setminus V$.*

(iii) *Let $\mathcal{U} = \{U_i\}_{i \in I}$ be an open covering of X such that Γ_2 is contained in at least one set of $\mathcal{U}$. Let $f \in Z^1\big(\mathcal{U}, \widehat{\mathcal{O}}^{\mathrm{GCom}\,A}(\cdot, Z)\big)$ such that, for some open neighborhood W_1 of Γ_1, $f\big|_{W_1}$ splits. Then there exist an open neighborhood W of $\Gamma_1 \cup \Gamma_2$ such that also $f\big|_W$ splits.*

PROOF. (i) Let Γ° and Γ be as in Definition 5.1. If $\Gamma = \emptyset$, i.e., $\Gamma_1 \cap \Gamma_2 \subseteq Z$, then, by definition of $\widehat{\mathcal{O}}^{\mathrm{GCom}\,A}(U, Z)$, there is a neighborhood $V \subseteq U$ of $\Gamma_1 \cap \Gamma_2$ such that $f = I$ on V, and we can define $\widetilde{f} \equiv I$.

Therefore, we may assume that $\Gamma \neq \emptyset$. Then, shrinking U, we can achieve that $U = W^\circ \cup W$, where W° and W are open neighborhoods of Γ° and Γ, respectively, such that $W^\circ \cap W = \emptyset$, $f = I$ on W°, W consists of a finite number of connected components each of which is contractible, and $W \cap Z = \emptyset$. Then all points of W are Jordan stable for A, and it follows from Corollary 6.7 that $\widehat{\mathcal{O}}^{\mathrm{GCom}\,A}(W)$ is connected. Therefore, we can find a continuous map $\theta : [0,1] \times W \to \mathrm{GL}(n, \mathbb{C})$ such that

$$\begin{aligned} &\theta(t, \cdot) \in \widehat{\mathcal{O}}^{\mathrm{GCom}\,A}(W) \quad \text{for all} \quad 0 \leq t \leq 1, \\ &\theta(0, \cdot) = f\big|_W \quad \text{and} \quad \theta(1, \cdot) = I \quad \text{on} \quad W. \end{aligned}$$

Choose open neighborhoods $W'' \subseteq W' \subseteq W$ of Γ such that W'' is relatively compact in W', and W' is relatively compact in W. Then we can find a partition of $[0,1]$, $0 = t_1 < t_2 < \ldots < t_m = 1$, so fine that the maps $g_j \in \widehat{\mathcal{O}}^{\mathrm{Com}\,A}(W')$, $1 \leq j \leq m-1$, defined by $g_j(\zeta) := \theta(t_j, \zeta)\theta(t_{j+1}, \zeta)^{-1} - I$ satisfy $\|g_j\| \leq 1/2$ on W'. Then $I + g_j \in \widehat{\mathcal{O}}^{\mathrm{GCom}\,A}(W')$ and

$$f = (I + g_1) \cdot \ldots \cdot (I + g_{m-1}) \quad \text{on} \quad W'. \tag{7.3}$$

Choose a continuous function $\chi : X \to [0,1]$ such that $\chi = 1$ in W'' and $\chi = 0$ on a neighborhood of $X \setminus W'$, and define $\widetilde{f} \in \widehat{\mathcal{O}}^{\mathrm{GCom}\,A}(X)$ by

$$\widetilde{f}(\zeta) = \begin{cases} \big(I + \chi(\zeta)g_1(\zeta)\big) \cdot \ldots \cdot \big(I + \chi(\zeta)g_{m-1}(\zeta)\big) & \text{if} \quad \zeta \in W', \\ I & \text{if} \quad \zeta \in X \setminus W'. \end{cases}$$

Since $Z \cap W' \subseteq Z \cap W = \emptyset$, then $\widetilde{f} \in \widehat{\mathcal{O}}^{\mathrm{GCom}\,A}(X, Z)$. Since $W^\circ \cap W'' \subseteq W^\circ \cap W = \emptyset$ and $f = I$ on W°, it follows from (7.3) that $\widetilde{f} = f$ on $V := W^\circ \cup W''$, which is a neighborhood of $\Gamma_1 \cap \Gamma_2$.

(ii) Since $U_1 \cap V$ is a neighborhood of $\Gamma_1 \cap \Gamma_2$, by part (i) of the lemma, we can find a section $\widetilde{f} \in \widehat{\mathcal{O}}^{\mathrm{GCom}\,A}(X, Z)$ and an open neighborhood $V' \subseteq U_1 \cap V$ of $\Gamma_1 \cap \Gamma_2$ such that $\widetilde{f} = f_1$ on V' and $\widetilde{f} = I$ on $X \setminus (U_1 \cap V)$. Choose open neighborhoods U_1' of Γ_1 and an open neighborhood U_2 of Γ_2 such that $U_1' \cap U_2 \subseteq V'$ and $U_1' \subseteq U_1$. Further choose an open neighborhood U_2' of Γ_2 such that $\overline{U_2'} \subseteq U_2$. Then $W := U_1' \cup U_2'$ is an open neighborhood of $\Gamma_1 \cup \Gamma_2$. Consider the open sets

$$W_1 := W \setminus \overline{U_2'} \quad \text{and} \quad W_2 := W \cap U_2.$$

Since $\overline{U_2'} \subseteq U_2$, then $W = W_1 \cup W_2$. Moreover, $W_1 \subseteq U_1'$, $W_2 \subseteq U_2$ and therefore $W_1 \cap W_2 \subseteq U_1' \cap U_2 \subseteq V'$, which implies that $f_1 \widetilde{f}^{-1} = I$ on $W_1 \cap W_2$. Hence, there is a well-defined section $f \in \widehat{\mathcal{O}}^{\mathrm{GCom}\,A}(W, Z)$ such that $f = f_1 \widetilde{f}^{-1}$ on W_1 and $f = I$ on W_2. As $\widetilde{f} = I$ on $X \setminus (U_1 \cap V) \supseteq (U_1 \cap W) \setminus V$, we have $f = f_1$ on $(U_1 \cap W) \setminus V$.

(iii) Since $f\big|_{W_1}$ is trivial, we have a family $f_i \in \widehat{\mathcal{O}}^{\mathrm{GCom}\,A}(U_i \cap W_1, Z)$, $i \in I$, with

$$f_{ij} = f_i f_j^{-1} \quad \text{on} \quad U_i \cap U_j \cap W_1, \qquad i, j \in I. \tag{7.4}$$

By hypothesis, we can find $i_0 \in I$ with $\Gamma_2 \subseteq U_{i_0}$. By (7.4), then

$$f_i^{-1} f_{ii_0} = f_i^{-1} f_{ij} f_{ji_0} = f_j^{-1} f_{ji_0} \quad \text{on} \quad U_i \cap U_j \cap W_1 \cap U_{i_0}, \quad i, j \in I.$$

Therefore, we have a well-defined section $g \in \widehat{\mathcal{O}}^{\mathrm{GCom}\,A}(W_1 \cap U_{i_0}, Z)$ with

$$g = f_i^{-1} f_{ii_0} \quad \text{on} \quad U_i \cap W_1 \cap U_{i_0}, \quad i \in I. \tag{7.5}$$

By part (i) of the lemma, we can find an open neighborhood $V \subseteq W_1 \cap U_{i_0}$ of $\Gamma_1 \cap \Gamma_2$ and a section $\widetilde{g} \in \widehat{\mathcal{O}}^{\mathrm{GCom}\,A}(X, Z))$ with $\widetilde{g} = g$ on V and, hence, by (7.5),

$$\widetilde{g} = f_i^{-1} f_{ii_0} \quad \text{on} \quad U_i \cap V, \quad i \in I. \tag{7.6}$$

Choose an open neighborhood $W_1' \subseteq W_1$ of Γ_1 and an open neighborhood $U_{i_0}' \subseteq U_{i_0}$ of Γ_1 such that $W_1' \cap U_{i_0}' \subseteq V$, and set $W := W_1' \cup U_{i_0}'$. Then W is an open neighborhood of $\Gamma_1 \cup \Gamma_2$, and, by (7.6),

$$\widetilde{g} = f_i^{-1} f_{ii_0} \quad \text{on} \quad U_i \cap W_1' \cap U_{i_0}', \quad i \in I. \tag{7.7}$$

Consider the open covering $\mathcal{U}^* = \{U_i^*\}_{i \in I}$ of W defined by

$$U_i^* = \begin{cases} U_i \cap W_1' & \text{for } i \in I \setminus \{i_0\}. \\ U_{i_0}' & \text{for } i = i_0, \end{cases}$$

Then $\mathcal{U}^*$ is a refinement of $\mathcal{U} \cap W$, and $f^* := \{f_{ij}\big|_{U_i^* \cap U_j^*}\}_{i,j \in I}$ is a $(\mathcal{U}^*, \widehat{\mathcal{O}}^{\mathrm{GCom}\,A}(\cdot, Z))$-cocycle which is induced by $f\big|_W$. Further, let

$$f_i^* := \begin{cases} f_i\big|_{U_i^*} & \text{if} \quad i \in I \setminus \{i_0\}, \\ \widetilde{g}^{-1}\big|_{U_{i_0}'} & \text{if} \quad i = i_0. \end{cases}$$

Then it follows from (7.4) (if $i, j \in I \setminus \{i_0\}$) and from (7.7) (if $i = i_0$ or $j = i_0$) that $f_i^*(f_j^*)^{-1} = f_{ij}^*$ on $U_i^* \cap U_j^*$, for all $i, j \in I$. Hence, f^* is trivial, which means, by Proposition 2.1, that $f\big|_W$ is trivial. $\square$

LEMMA 7.7. *Assume that X is irreducible*[5], *and Z is a discrete and closed subset of X which contains all nonsmooth points of X and all points of X which are not Jordan-stable for A. Then $H^1\big(X, \widehat{\mathcal{O}}^{\mathrm{GCom}\,A}(\cdot, Z)\big) = 0$.*

PROOF. Let an $\widehat{\mathcal{O}}^{\mathrm{GCom}\,A}(\cdot, Z)$-cocycle f on X be given, and let $\mathcal{U} = \{U_\alpha\}_{\alpha \in I}$ be the covering of f.

Let $\pi : \widetilde{X} \to X$ be the normalization of X (see, e.g., [**20**, Ch. VI, §4]). Since X is irreducible, $\widetilde{X}$ is connected (see, e.g., [**20**, Ch. VI, §4.2]). Since X is 1-dimensional, by the Puiseaux theorem (see, e.g., [**20**, Ch. VI, §4.1]), $\widetilde{X}$ is a Riemann surfaces. Since X is Stein and, hence, noncompact, $\widetilde{X}$ is noncompact.

Therefore we can apply Theorem 4.3 to $\widetilde{X}$ and the covering $\widetilde{\mathcal{U}} := \big\{\pi^{-1}(U_i)\big\}_{i \in I}$ of $\widetilde{X}$. This gives a sequence $(B_\mu)_{\mu \in \mathbb{N}}$ of compact subsets of $\widetilde{X}$ such that

(a) for each $\mu \in \mathbb{N}$, B_μ is contained in at least one set of $\widetilde{\mathcal{U}}$;
(b) for each $\mu \in \mathbb{N}^*$, $(B_0 \cup \ldots \cup B_{\mu-1}, B_\mu)$ is a bump in $\widetilde{X}$;
(c) $\widetilde{X} = \bigcup_{\mu \in \mathbb{N}} B_\mu$.
(d) for each compact set $L \subseteq \widetilde{X}$, there exists $N(\Gamma) \in \mathbb{N}$ such that $B_\mu \cap L = \emptyset$ if $\mu \geq N(\Gamma)$.

From (a) we obtain

(a') For each $\mu \in \mathbb{N}$, $\pi(B_\mu)$ is contained in at least one set of $\mathcal{U}$.

We introduce the abbreviation $K_m = \pi(B_0 \cup \ldots \cup B_m)$ and denote by $\mathrm{Int}\, K_m$ the interior of K_m, for $m \in \mathbb{N}$.

Statement 1. Let $m \in \mathbb{N}$. Then there exists a neighborhood W of K_m such that $f|_W$ splits.

Proof of Statement 1. By (a'), we have $\alpha_0 \in I$ such that U_{α_0} is a neighborhood of K_0. By Remark 2.2) $f|_{U_{\alpha_0}}$ splits. Hence, the claim of Statement 1 holds for $m = 0$.

Now, proceeding by induction, let $k \in \mathbb{N}$ such that, we already have an open neighborhood W of K_m such that $f|_W$ splits.

By (b) and Lemma 5.2, we can find a Z-adapted pair of compact sets in X, (Γ_1, Γ_2), such that

$$\Gamma_1 \subseteq K_k, \tag{7.8}$$

$$\Gamma_2 \subseteq \pi(B_{k+1}), \tag{7.9}$$

$$\Gamma_1 \cup \Gamma_2 = K_{k+1}. \tag{7.10}$$

Then, by (7.8), W is an open neighborhood of Γ_1. Since $f|_W$ splits and since, by (7.9) and (a'), Γ_2 is contained in at least one set of $\mathcal{U}$, from Lemma 7.6 (iii) we get an open neighborhood V of $\Gamma_1 \cup \Gamma_2$ such that $f|_V$ splits. By (7.10), this implies the claim of Statement 1 for $m = k + 1$. Statement 1 is proved.

By (c) and (d) we can find a strictly increasing sequence $\{m(j)\}_{j \in \mathbb{N}^*}$ of natural numbers such that

$$\pi(B_\mu) \cap K_{m(j)} = \emptyset \quad \text{if} \quad \mu \geq m(j+1), \quad \text{for all} \quad j \in \mathbb{N}^*, \tag{7.11}$$

$$K_{m(j)} \subseteq \mathrm{Int}\, K_{m(j+1)} \quad \text{for all} \quad j \in \mathbb{N}^*, \tag{7.12}$$

where $\mathrm{Int}\, K_{m(j+1)}$ denotes the interior of $K_{m(j+1)}$.

---

[5] X is called **irreducible** if the set of smooth points of X is connected, see, e.g., [**12**, Ch. 9, §1.2] or [**20**, Ch. V, §4.5, Prop. α].

Statement 2. Let $j, \ell \in \mathbb{N}^*$ with $m(j+1) \leq \ell$, let W be an open neighborhood of K_ℓ and $g \in \widehat{\mathcal{O}}^{\mathrm{GCom}\, A}(W, Z)$. Then there exist an open neighborhood W' of $K_{\ell+1}$ and $h \in \widehat{\mathcal{O}}^{\mathrm{GCom}\, A}(W', Z)$ such that

$$h = g \quad \text{on} \quad K_{m(j)}. \tag{7.13}$$

Proof of Statement 2. By (7.11), $\pi(B_{\ell+1}) \cap K_{m(j)} = \emptyset$. Therefore, we can find an open set V with

$$\pi(B_{\ell+1}) \subseteq V, \tag{7.14}$$

$$K_{m(j)} \cap V = \emptyset. \tag{7.15}$$

By (b) and Lemma 5.2, we can find a Z-adapted pair of compact sets in X, (Γ_1, Γ_2), such that

$$\Gamma_1 \subseteq K_\ell, \tag{7.16}$$

$$\Gamma_2 \subseteq \pi(B_{\ell+1}), \tag{7.17}$$

$$\Gamma_1 \cup \Gamma_2 = K_{\ell+1}. \tag{7.18}$$

Then, by (7.16), W is an open neighborhood of Γ_1, and, by (7.17) and (7.14), V is a neighborhood of Γ_2. Therefore, from Lemma 7.6 (ii) we get an open neighborhood W' of $\Gamma_1 \cup \Gamma_2$, i.e., by (7.18), of $K_{\ell+1}$, and a section $h \in \widehat{\mathcal{O}}^{\mathrm{GCom}\, A}(W', Z)$ such that

$$h = g \quad \text{on} \quad (W \cap W') \setminus V.$$

Since, by (7.15), $K_{m(j)} \subseteq (W \cap W') \setminus V$, in particular we have (7.13). Statement 2 is proved.

Statement 3. Let $j \in \mathbb{N}^*$, let W be an open neighborhood of $K_{m(j+1)}$ and $g \in \widehat{\mathcal{O}}^{\mathrm{GCom}\, A}(W, Z)$. Then there exist an open neighborhood W' of $K_{m(j+2)}$ and $h \in \widehat{\mathcal{O}}^{\mathrm{GCom}\, A}(W', Z)$ such that

$$g = h \quad \text{on} \quad K_{m(j)}. \tag{7.19}$$

Proof of Statement 3. Application of Statement 2 for $\ell = 1, 2, \ldots, m(j+2) - 1$.

To prove the lemma, we have to find a family $f = \{f_\alpha\}_{\alpha \in I}$ of sections $f_\alpha \in \widehat{\mathcal{O}}^{\mathrm{GCom}\, A}(U_\alpha, Z)$ such that

$$f_{\alpha\beta} = f_\alpha f_\beta^{-1} \quad \text{on} \quad U_\alpha \cap U_\beta, \quad \alpha, \beta \in I.$$

Since $\bigcup_{j=1}^\infty \operatorname{Int} K_{m(j)} = X$, for that it is sufficient to find a sequence of families $\{f_\alpha^{(j)}\}_{\alpha \in I}$, $j \in \mathbb{N}^*$, of sections $f_\alpha^{(j)} \in \widehat{\mathcal{O}}^{\mathrm{GCom}\, A}(\operatorname{Int} K_{m(j)} \cap U_\alpha, Z)$ such that, for each $j \in \mathbb{N}^*$, the following conditions S(j) and T(j) are satisfied.

S(j): $f_{\alpha\beta} = f_\alpha^{(j)} (f_\beta^{(j)})^{-1}$ on $\operatorname{Int} K_{m(j)} \cap U_\alpha \cap U_\beta$ for all $\alpha, \beta \in I$.

T(j): If $j > 2$, then $f_\alpha^{(j)} = f_\alpha^{(j-1)}$ on $K_{m(j-2)} \cap U_\alpha$ for all $\alpha \in I$.

We do this by induction.

Beginning of induction: By Statement 1 there is a neighborhood W of $K(m(1))$ such that $f|_W$ splits, i.e., there is a family $\{f_\alpha^{(W)}\}_{\alpha \in I}$ of sections $f_\alpha^{(W)} \in \widehat{\mathcal{O}}^{\mathrm{GCom}\, A}(W \cap U_\alpha, Z)$ such that $f_{\alpha\beta} = f_\alpha^{(W)} (f_\beta^{(W)})^{-1}$ on $W \cap U_\alpha \cap U_\beta$ for all $\alpha, \beta \in I$. Setting $f_\alpha^{(1)} = f_\alpha^{(W)}|_{\operatorname{Int} K_{m(1)} \cap U_\alpha}$, we get a family $\{f_\alpha^{(1)}\}_{\alpha \in I}$ of sections $f_\alpha^{(1)} \in \widehat{\mathcal{O}}^{\mathrm{GCom}\, A}(\operatorname{Int} K_{m(1)} \cap U_\alpha, Z)$ satisfying condition S(1). Condition T(1) is trivially satisfied, because $1 \not> 2$.

Step of induction: Assume, for some $k \in \mathbb{N}^*$, we already have families $\{f_\alpha^{(j)}\}_{\alpha \in I}$, $1 \le j \le k$, of sections $f_\alpha^{(j)} \in \widehat{\mathcal{O}}^{\mathrm{GCom}\,A}(\mathrm{Int}\,K_{m(j)} \cap U_\alpha, Z)$ such that conditions S(j) and T(j) are satisfied for $1 \le j \le k$.

We have to find a family $f_\alpha^{(k+1)} \in \widehat{\mathcal{O}}^{\mathrm{GCom}\,A}(\mathrm{Int}\,K_{m(k+1)} \cap U_\alpha, Z)$, $\alpha \in I$, such that also conditions S($k+1$) and T($k+1$) are satisfied.

Again by Statement 1, we have a family $F_\alpha \in \widehat{\mathcal{O}}^{\mathrm{GCom}\,A}(\mathrm{Int}\,K_{m(k+1)} \cap U_\alpha, Z)$, $\alpha \in I$, such that

$$f_{\alpha\beta} = F_\alpha F_\beta^{-1} \text{ on } \mathrm{Int}\,K_{m(k+1)} \cap U_\alpha \cap U_\beta, \quad \text{for all } \alpha, \beta \in I. \tag{7.20}$$

First consider the case $k = 2$. Then the family $f_\alpha^{(k)} := F_\alpha$, $\alpha \in I$, is a required, because, with this definition, condition S(k) holds by (7.20), and condition T(k) holds trivially, as $k \not> 2$.

Now let $k > 2$. Then it follows from (7.20) and condition S(k) that, for all $\alpha, \beta \in I$, on $\mathrm{Int}\,K_{m(k)} \cap U_\alpha \cap U_\beta$ we have

$$F_\alpha F_\beta^{-1} = f_{\alpha\beta} = f_\alpha^{(k)}(f_\beta^{(k)})^{-1} \quad \text{and, hence,} \quad (f_\alpha^{(k)})^{-1}F_\alpha = (f_\beta^{(k)})^{-1}F_\beta.$$

Therefore, there is a well-defined section $g \in \widehat{\mathcal{O}}^{\mathrm{GCom}\,A}(\mathrm{Int}\,K_{m(k)}, Z)$ with

$$g = (f_\alpha^{(k)})^{-1}F_\alpha \text{ on } \mathrm{Int}\,K_{m(k)} \cap U_\alpha \quad \text{for all } \alpha \in I, \tag{7.21}$$

Ba Statement 3, we can find a section $h \in \widehat{\mathcal{O}}^{\mathrm{GCom}\,A}(\mathrm{Int}\,K_{m(k+1)}, Z)$ with

$$h = g \quad \text{on} \quad \mathrm{Int}\,K_{m(k-1)}. \tag{7.22}$$

Set $f_\alpha^{(k+1)} = F_\alpha h^{-1}$ on $\mathrm{Int}\,K_{m(k+1)} \cap U_\alpha$ for $\alpha \in I$. Then, by (7.20), for all $\alpha, \beta \in I$, on $\mathrm{Int}\,K_{m(k+1)} \cap U_\alpha \cap U_\beta$, we have

$$f_\alpha^{(k+1)}(f_\beta^{(k+1)})^{-1} = F_\alpha h^{-1} h F_\beta^{-1} = F_\alpha F_\beta^{-1} = f_{\alpha\beta}.$$

i.e., condition S($k+1$) is satisfied. From (7.22) and (7.21) it follows that, for all $\alpha \in I$, on $\mathrm{Int}\,K_{m(k-1)}$, we have

$$f_\alpha^{(k+1)} = F_\alpha h^{-1} = F_\alpha g^{-1} = F_\alpha (F_\alpha)^{-1} f_\alpha^{(k)} = f_\alpha^{(k)},$$

i.e., also condition T(k+1) is satisfied. □

Proof of Theorem 7.1. Let Z be the union of the set of nonsmooth points of X and the set of points in X which are not Jordan stable for A (Def. 6.1). By Proposition 6.2, Z is discrete and closed in X. Therefore, by Remark 7.5 it is sufficient to prove that $H^1\big(X, \widehat{\mathcal{O}}^{\mathrm{GCom}\,A}(\cdot, Z)\big) = 0$.

Let an open covering $\mathcal{U} = \{U_i\}_{i \in I}$ of X and a $\big(\mathcal{U}, \widehat{\mathcal{O}}^{\mathrm{GCom}\,A}(\cdot, Z)\big)$-cocycle $f = \{f_{ij}\}_{j \in I}$ be given. We have to find a family $f_i \in \widehat{\mathcal{O}}^{\mathrm{GCom}\,A}(U_i, Z)$, $i \in I$, such that, for all $i, j \in I$ and $\zeta \in U_i \cap U_j$,

$$f_{ij}(\zeta) = f_i(\zeta)(f_j(\zeta))^{-1}. \tag{7.23}$$

Let $\mathfrak{X}$ be the family of irreducible components of X (see, e.g., [**12**, Ch. 9, §9.2] or [**20**, Ch. IV, §2.9, Theorem 4, and Ch. V, §4.5]). Then, for each $Y \in \mathfrak{X}$, $Z \cap Y$ is discrete and closed in Y, contains all nonsmooth points of Y and all points of Y which are not Jordan stable for $A|_Y$. Therefore, by Lemma 7.7, for each $Y \in \mathfrak{X}$, we can find a family $f_i^Y \in \widehat{\mathcal{O}}^{\mathrm{GCom}\,A|_Y}(U_i \cap Y, Z \cap Y)$ such that

$$f_{ij} = f_i^Y (f_j^Y)^{-1} \quad \text{on} \quad Y \cap U_i \cap U_j, \quad i, j \in I. \tag{7.24}$$

Set $U_i' = U_i \setminus Z$ for $i \in I$. Then, for each $i \in I$, the family of sets $\{Y \cap U_i'\}_{Y \in \mathfrak{X}}$ is a pairwise disjoint open covering of U_i'. Therefore, for each $i \in I$, there is a well-defined section $f_i' \in \widehat{\mathcal{O}}^{\mathrm{GCom}\,A|_Y}(U_i', Z)$ $(= \widehat{\mathcal{O}}^{\mathrm{GCom}\,A|_Y}(U_i'))$ such that

$$f_i' = f_i^Y \quad \text{on} \quad Y \cap U_i' \quad \text{for all} \quad Y \in \mathfrak{X}, \tag{7.25}$$

and it follows from (7.24) that

$$f_{ij} = f_i'(f_j')^{-1} \quad \text{on} \quad U_i' \cap U_j', \quad i, j \in I. \tag{7.26}$$

For $\xi \in X$, denote by $\mathfrak{X}_\xi$ the family of all $Y \in \mathfrak{X}$ with $\xi \in Y$. Since $\mathfrak{X}$ is locally finite, for each $\xi \in X$, $\mathfrak{X}_\xi$ is finite and there exists an open neighborhood Ω_ξ of ξ such that $Y \cap \Omega_\xi = \emptyset$ for all $Y \in \mathfrak{X} \setminus \mathfrak{X}_\xi$.

Claim. For each $i \in I$, there is an open neighborhood U_i'' of $Z \cap U_i$ such that

$$f_i' = I \quad \text{on} \quad U_i'' \cap U_i'. \tag{7.27}$$

Proof of the Claim. Let $i \in I$ be given. Since $f_i^Y \in \widehat{\mathcal{O}}^{\mathrm{GCom}\,A|_Y}(U_i \cap Y, Z \cap Y)$ for all $Y \in \mathfrak{X}$, then, for each point $\xi \in Z \cap U_i$ and each $Y \in \mathfrak{X}_\xi$, we can find an open neighborhood $U''_{i,\xi,Y}$ of ξ (in X) such that

$$f_i^Y = I \quad \text{on} \quad Y \cap U''_{i,\xi,Y}.$$

Since $\mathfrak{X}_\xi$ is finite, then, for each $\xi \in Z \cap U_i$,

$$U''_{i,\xi} := \bigcap_{Y \in \mathfrak{X}_\xi} U''_{i,\xi,Y}$$

is an open neighborhood of ξ with

$$f_i^Y = I \quad \text{on} \quad Y \cap U''_{i,\xi} \quad \text{for all} \quad Y \in \mathfrak{X}_\xi.$$

Since $Y \cap \Omega_\xi = \emptyset$ for all $Y \in \mathfrak{X} \setminus \mathfrak{X}_\xi$, this implies that, for all $\xi \in Z \cap U_i$,

$$f_i^Y = I \quad \text{on} \quad Y \cap \Omega_\xi \cap U''_{i,\xi} \quad \text{for all} \quad Y \in \mathfrak{X}.$$

By (7.25), it follows that, for all $\xi \in Z \cap U_i$,

$$f_i = I \quad \text{on} \quad U_i' \cap \Omega_\xi \cap U''_{i,\xi}.$$

To prove the Claim, it remains to set $U_i'' = \bigcup_{\xi \in Z \cap U_i} (\Omega_\xi \cap \widetilde{U}''_{i,\xi})$.

Since $U_i'' \cup U_i' \supseteq U_i$, it follows from (7.27) that there are well-defined sections $f_i \in \widehat{\mathcal{O}}^{\mathrm{GCom}\,A}(U_i, Z)$, $i \in I$, such that

$$f_i = f_i' \quad \text{on} \quad U_i', \quad \text{and} \quad f_i = I \quad \text{on} \quad U_i'' \cap U_i.$$

Then, for $\zeta \in U_i' \cap U_j' = (U_i \cap U_j) \setminus Z$, (7.23) holds by (7.24) and (7.25). For $\zeta \in U_i \cap U_j \cap Z$ this is trivial, because then $f_{ij}(\zeta) = f_i(\zeta) = f_j(\zeta) = I$. This completes the proof of Theorem 7.1 and, hence, that of Theorem 1.2.

References

[1] H. Baumgärtel, *Jordansche Normalform holomorpher Matrizen* (German), Monatsb. Deutsch. Akad. Wiss. Berlin **11** (1969), 23–24. MR0251572

[2] H. Baumgärtel, *Endlichdimensionale analytische Störungstheorie* (German), Akademie-Verlag, Berlin, 1972. Mathematische Lehrbücher und Monographien, II. Abteilung. Mathematische Monographien, Band 28. MR0634965

[3] H. Baumgertel′, *The theory of analytic perturbations for linear operators depending on several complex variables* (Russian), Mat. Issled. **9** (1974), no. 1 (31), 17–39. MR0353019

[4] H. Baumgärtel, *Analytic perturbation theory for matrices and operators*, Operator Theory: Advances and Applications, vol. 15, Birkhäuser Verlag, Basel, 1985. MR878974

[5] H. Cartan, *Espaces fibrés analytiques* (French), Symposium internacional de topología algebraica International symposium on algebraic topology, Universidad Nacional Autónoma de México and UNESCO, Mexico City, 1958, pp. 97–121. MR0098196

[6] O. Forster, *Lectures on Riemann surfaces*, Graduate Texts in Mathematics, vol. 81, Springer-Verlag, New York-Berlin, 1981. Translated from the German by Bruce Gilligan. MR648106

[7] O. Forster and K. J. Ramspott, *Okasche Paare von Garben nicht-abelscher Gruppen* (German), Invent. Math. **1** (1966), 260–286, DOI 10.1007/BF01452245. MR0212211

[8] F. Forstnerič, *The Oka principle for multivalued sections of ramified mappings*, Forum Math. **15** (2003), no. 2, 309–328, DOI 10.1515/form.2003.018. MR1956971

[9] F. Forstnerič, *Stein manifolds and holomorphic mappings*, 2nd ed., Ergebnisse der Mathematik und ihrer Grenzgebiete. 3. Folge. A Series of Modern Surveys in Mathematics [Results in Mathematics and Related Areas. 3rd Series. A Series of Modern Surveys in Mathematics], vol. 56, Springer, Cham, 2017. The homotopy principle in complex analysis. MR3700709

[10] F. R. Gantmacher, *The theory of matrices. Vols. 1, 2*, Translated by K. A. Hirsch, Chelsea Publishing Co., New York, 1959. MR0107649

[11] H. Grauert, *Analytische Faserungen über holomorph-vollständigen Räumen* (German), Math. Ann. **135** (1958), 263–273, DOI 10.1007/BF01351803. MR0098199

[12] H. Grauert and R. Remmert, *Coherent analytic sheaves*, Grundlehren der Mathematischen Wissenschaften [Fundamental Principles of Mathematical Sciences], vol. 265, Springer-Verlag, Berlin, 1984. MR755331

[13] V. Guillemin and A. Pollack, *Differential topology*, Prentice-Hall, Inc., Englewood Cliffs, N.J., 1974. MR0348781

[14] R. M. Guralnick, *Similarity of holomorphic matrices*, Linear Algebra Appl. **99** (1988), 85–96, DOI 10.1016/0024-3795(88)90126-7. MR925151

[15] F. Hirzebruch, *Neue topologische Methoden in der algebraischen Geometrie* (German), Ergebnisse der Mathematik und ihrer Grenzgebiete (N.F.), Heft 9, Springer-Verlag, Berlin-Göttingen-Heidelberg, 1956. MR0082174

[16] L. Hörmander, *An introduction to complex analysis in several variables*, 3rd ed., North-Holland Mathematical Library, vol. 7, North-Holland Publishing Co., Amsterdam, 1990. MR1045639

[17] J. Leiterer, *Local and global similarity of holomorphic matrices*, arXiv:1703.09524.

[18] J. Leiterer, *On the Jordan structure of holomorphic matrices*, arXiv:1703.09535.

[19] J. Leiterer, *On the similarity of holomorphic matrices*, arXiv:1703.09530.

[20] S. Łojasiewicz, *Introduction to complex analytic geometry*, Birkhäuser Verlag, Basel, 1991. Translated from the Polish by Maciej Klimek. MR1131081

[21] J. Milnor, *Morse theory*, Based on lecture notes by M. Spivak and R. Wells. Annals of Mathematics Studies, No. 51, Princeton University Press, Princeton, N.J., 1963. MR0163331

[22] N. Steenrod, *The Topology of Fibre Bundles*, Princeton Mathematical Series, vol. 14, Princeton University Press, Princeton, N. J., 1951. MR0039258

[23] G. P. A. Thijsse, *Global holomorphic similarity to a Jordan form*, Results Math. **8** (1985), no. 1, 78–87, DOI 10.1007/BF03322659. MR812247

Institut für Mathematik, Humboldt-Universität zu Berlin, Rudower Chaussee 25, D-12489 Berlin , Germany

Email address: `leiterer@mathematik.hu-berlin.de`

Contemporary Mathematics
Volume **733**, 2019
https://doi.org/10.1090/conm/733/14745

Weighted geometric means of convex bodies

Vitali Milman and Liran Rotem

ABSTRACT. Given two convex bodies K and T and a number $0 < \lambda < 1$, we present two constructions for the geometric mean of K and T with weights $1-\lambda$ and λ. If K and T are the unit balls of two norms, one may think of the weighted geometric means as the unit balls of a "geometric interpolation" between the two normed spaces.

Our first construction is explicit, but lacks the so called "duality property" – a natural property one expects from the geometric mean. The second construction has this property, but is based on the choice of an ultrafilter. Both definitions extend previous constructions by the authors for the power K^{λ}, and the results proved here also prove some previously announced results.

1. Introduction

We begin by quickly recalling some basic definitions from convex geometry. More background information can be found, e.g. in [**13**]. By a convex body in $\mathbb{R}^n$ we mean a set $K \subseteq \mathbb{R}^n$ which is convex, compact and has non-empty interior. We will always assume our convex bodies are centrally symmetric, i.e. $x \in K$ implies $-x \in K$. We denote the class of such convex bodies by $\mathcal{K}_s^n$.

For convex bodies $K, T \in \mathcal{K}_s^n$ their Minkowski sum is defined by

$$K + T = \{x + y : \ x \in K, \ y \in T\}.$$

For $\lambda > 0$ the homothety λK is defined in the obvious way as $\lambda K = \{\lambda x : \ x \in K\}$. The polar body of K is

$$K^\circ = \{y \in \mathbb{R}^n : \ \langle x, y \rangle \leq 1 \text{ for all } x \in K\},$$

where $\langle \cdot, \cdot \rangle$ denotes the standard inner product on $\mathbb{R}^n$. Finally, the Hausdorff metric on $\mathcal{K}_s^n$ is

$$d_H(K,T) = \min\{t > 0 : \ K \subseteq T + tB_2^n \text{ and } T \subseteq K + tB_2^n\},$$

where B_2^n denotes the unit Euclidean ball.

2010 *Mathematics Subject Classification.* Primary 52A20; Secondary 15B48.

Key words and phrases. Convex geometry, geometric mean, power, positive-definite matrices.

The first author is supported by ISF grant number 519/17.

Both authors are supported by BSF grant number 2016050. Part of the research was conducted while the authors were in residence at the Mathematical Sciences Research Institute in Berkeley, California, supported by NSF grant DMS-1440140. The authors also thank the reviewers for their helpful comments.

In [**9**] the authors presented a new construction which we will call the pre-geometric mean. Given $K, T \in \mathcal{K}_s^n$ one defines two sequences of convex bodies $\{A_m\}_{m=0}^{\infty}$, $\{H_m\}_{m=0}^{\infty}$ by

$$\begin{aligned} A_0 &= K & H_0 &= T \\ A_{m+1} &= \frac{A_m + H_m}{2} & H_{m+1} &= \left(\frac{A_m^\circ + H_m^\circ}{2}\right)^\circ . \end{aligned}$$

Note that A_{m+1} is the arithmetic mean of A_m and H_m. If we think of K° as the inverse "K^{-1}", then H_{n+1} is the harmonic mean of A_m and H_m. Using these definitions Firey proved an arithmetic mean – harmonic mean inequality for convex bodies ([**4**]). We then define

$$g(K,T) = \lim_{m\to\infty} A_m = \lim_{m\to\infty} H_m,$$

where the limits are taken in the Hausdorff sense.

The pre-mean $g(K,T)$ has many of the natural properties we would expect a geometric mean to have. It is continuous and monotone, it satisfies the harmonic mean – geometric mean – arithmetic mean inequality

$$\left(\frac{K^\circ + T^\circ}{2}\right)^\circ \subseteq g(K,T) \subseteq \frac{K+T}{2},$$

and it commutes with polarity in sense that $g(K,T)^\circ = g(K^\circ, T^\circ)$. This last property is the natural extension of the fact that the geometric mean of $\frac{1}{x}$ and $\frac{1}{y}$ is the inverse of the geometric mean of x and y. In particular, we have $g(K, K^\circ) = B_2^n$.

Of particular interest for this paper is the geometric mean of (origin-symmetric) ellipsoids. Recall that the support function $h_K : \mathbb{R}^n \to (0,\infty)$ of a convex body $K \in \mathcal{K}_s^n$ is defined by $h_K(y) = \sup_{x\in K} \langle x, y\rangle$. Ellipsoids are characterized by the fact that their support function is of the form $h_E(x) = \sqrt{\langle ux, x\rangle}$, where u is a positive-definite matrix. The matrix u is uniquely defined given E, and we will sometimes write u_E for the matrix that corresponds to the ellipsoid E.

It is proved in [**9**] that if E and F are ellipsoids, so is $g(E,F)$. Moreover, the matrix $u_{g(E,F)}$ is the geometric mean of u_E and u_F, a notion first introduced by Pusz and Woronowicz ([**11**]). If u and v commute then their product is also positive definite, and the geometric mean can be defined as $u\#v = (uv)^{\frac{1}{2}}$. In the general case the formula for the matrix geometric mean is more complicated, and is given by

$$u\#v = u^{\frac{1}{2}}\left(u^{-\frac{1}{2}} v u^{-\frac{1}{2}}\right)^{\frac{1}{2}} u^{\frac{1}{2}}.$$

The reader may consult the survey paper [**6**] for several equivalent definitions of the matrix geometric mean and some of its nice properties. More generally, one may also construct the weighted geometric mean with weights $1-\lambda$ and λ. The formula for such a weighted mean is

$$u\#_\lambda v = u^{\frac{1}{2}}\left(u^{-\frac{1}{2}} v u^{-\frac{1}{2}}\right)^{\lambda} u^{\frac{1}{2}}, \tag{1.1}$$

which reduces to $u^{1-\lambda}v^\lambda$ in the commutative case.

Despite its many good properties, the pre-mean $g(K,T)$ lacks the important scaling property: in general it does not satisfy $g(\alpha K, \beta T) = \sqrt{\alpha\beta}\, g(K,T)$ (see the appendix to [**9**] by Magazinov). Therefore the definition needs to be modified. One such modification appeared in [**12**], but we will not discuss it here. Instead we will discuss an alternative approach from [8] which uses ellipsoids. In this approach we define the ellipsoidal envelopes of the geometric mean as

$$\overline{g}(K,T)=\bigcap\left\{g(E,F): \begin{array}{c}E,F \text{ are ellipsoids}\\ K\subseteq E \text{ and } T\subseteq F\end{array}\right\}$$

and

$$\underline{g}(K,T)=\operatorname{conv}\left(\bigcup\left\{g(E,F): \begin{array}{c}E,F \text{ are ellipsoids}\\ K\supseteq E \text{ and } T\supseteq F\end{array}\right\}\right),$$

where conv denotes the convex hull. Finally we set

$$G(K,T)=g\left(\overline{g}(K,T),\underline{g}(K,T)\right).$$

It is proved in [8] that $G(K,T)$ satisfies many natural properties expected of the geometric mean, including the scaling property.

Another goal of the paper [8] was to introduce powers of convex bodies. For an ellipsoid E, one may define the power E^λ for every $\lambda\in\mathbb{R}$ by setting $u_{E^\lambda}=(u_E)^\lambda$. In geometric terms, the power E^λ has the following interpretation:

PROPOSITION 1.1. *Assume the principal axes of E have lengths $a_1, a_2, \dots, a_n$. Then the principal axes of E^λ have the same direction as the axes of E, and have lengths $a_1^\lambda, a_2^\lambda, \dots, a_n^\lambda$.*

PROOF. Since u_E is symmetric there exists an orthogonal matrix v such that vu_Ev^t is diagonal. Let $b_1, b_2, \dots, b_n$ be the diagonal entries of this matrix. Since

$$h_{vE}(x)=h_E(v^tx)=\sqrt{\langle u_Ev^tx,v^tx\rangle}=\sqrt{\langle vu_Ev^tx,x\rangle}=\sqrt{\sum_{i=1}^n b_ix_i^2},$$

we see that vE is an axes-aligned ellipsoid with axes lengths $b_1^{1/2}, b_2^{1/2}, \dots, b_n^{1/2}$. Since $vu_{E^\lambda}v^t=(vu_Ev^t)^\lambda$ is a diagonal matrix with entries $b_1^\lambda, b_2^\lambda, \dots, b_n^\lambda$, we conclude similarly that vE^λ is an axes-aligned ellipsoid with axes lengths $b_1^{\lambda/2}, b_2^{\lambda/2}, \dots, b_n^{\lambda/2}$. The result follows. □

It turns out that there is an important difference between the case $0\le\lambda\le 1$ and the case $\lambda>1$. To explain it we need:

DEFINITION 1.2 (Löwner [7]). A function $f:(0,\infty)\to\mathbb{R}$ is operator monotone if for every positive definite matrices u and v (of any size) $u\succeq v$ implies $f(u)\succeq f(v)$.

Recall that $u\succeq v$ in the matrix order means that $u-v$ is positive semi-definite. It is well known that the function $f(x)=x^\lambda$ is operator monotone for $0\le\lambda\le 1$, but not for $\lambda>1$ (see, e.g. [1], Section V.1). Since $E\subseteq F$ if and only if $u_E\preceq u_F$, we deduce that if $E\subseteq F$ and $0<\lambda<1$ then $E^\lambda\subseteq F^\lambda$. Therefore it makes sense to define the upper pre-power by:

$$\overline{P}_\lambda(K)=\bigcap\left\{E^\lambda:\ E \text{ is an ellipsoid and } K\subseteq E\right\}.$$

The map $K\mapsto\overline{P}_\lambda(K)$ satisfies some natural properties, such as monotonicity, but it does not satisfy the power law $\overline{P}_{\lambda\mu}(K)=\overline{P}_\lambda\left(\overline{P}_\mu(K)\right)$, which is the analogue of the identity $x^{\lambda\mu}=(x^\mu)^\lambda$. Instead, the pre-powers only satisfy the inclusion $\overline{P}_{\lambda\mu}(K)\supseteq\overline{P}_\lambda\left(\overline{P}_\mu(K)\right)$ (see Proposition 2.5).

A better construction was defined in [8]: For a fixed finite partition of $[\lambda,1]$,

$$\Pi:\ \lambda=t_0<t_1<\dots<t_m=1,$$

we set $s_i = t_{i-1}/t_i$ for $i = 1, 2, ..m$ and define

$$\overline{P}_\Pi(K) = \left(\overline{P}_{s_1} \circ \overline{P}_{s_2} \circ \cdots \circ \overline{P}_{s_m}\right)(K).$$

We then have the following theorem:

THEOREM 1.3. *Fix $\mathcal{K}_s^n$ and $0 < \lambda < 1$. Then the limit*

$$K^{\overline{\lambda}} := \lim_{\|\Pi\| \to 0} P_\Pi(K) \in \mathcal{K}_s^n \tag{1.2}$$

exists in the Hausdorff sense. Here $\|\Pi\| = \max|t_{i+1} - t_i|$ denotes the length of the longest interval in Π. Furthermore, the maps $K \mapsto K^{\overline{\lambda}}$ have the following properties:

(1) *For every $0 < \lambda < 1$, if $K \subseteq T$ then $K^{\overline{\lambda}} \subseteq T^{\overline{\lambda}}$.*
(2) *For every $0 < \lambda < 1$, every $K \in \mathcal{K}_s^n$ and every $c > 0$ we have $(cK)^{\overline{\lambda}} = c^\lambda K^{\overline{\lambda}}$.*
(3) *For every $0 < \lambda, \mu < 1$ and every $K \in \mathcal{K}_s^n$ we have $\left(K^{\overline{\mu}}\right)^{\overline{\lambda}} = K^{\overline{\lambda\mu}}$.*
(4) *If $E \in \mathcal{K}_s^n$ is an ellipsoid, then $E^{\overline{\lambda}}$ agrees with its linear algebra definition for every $0 < \lambda < 1$.*
(5) *If $K \supseteq B_2^n$ then $K^{\overline{\lambda}} \subseteq K$ for all $0 < \lambda < 1$.*

This theorem was announced in [**10**], after a slightly weaker version was proved in [**8**]. It is an immediate corollary of Theorems 2.7 and 2.9 of this paper.

We call the map $K \mapsto K^{\overline{\lambda}}$ the upper power. Naturally, we can also define the lower power by taking

$$\underline{P}_\lambda(K) = \operatorname{conv} \bigcup \left\{E^\lambda : \ E \text{ is an ellipsoid and } K \supseteq E\right\}$$

and carrying out the rest of the construction in the same way. The lower powers $K^{\underline{\lambda}}$ will have the same good properties as the upper powers, and of course $K^{\underline{\lambda}} \subseteq K^{\overline{\lambda}}$ for all $K \in \mathcal{K}_s^n$ and $0 < \lambda < 1$.

However, the upper and lower powers still don't satisfy all properties one would like to see in the power function. Like for the geometric mean, we expect our power to commute with duality, in the sense that $\left(K^\lambda\right)^\circ = (K^\circ)^\lambda$, as both sides should be equal to "$K^{-\lambda}$". There is no reason for the upper and lower powers to satisfy this property.

The paper [**10**] announced an alternative construction which does commute with polarity. We fix a sequence $\{\gamma_m\}_{m=1}^\infty \subseteq (0, 1)$ such that $\gamma_m \to 1$ as $m \to \infty$. We also fix a free ultrafilter $\mathcal{U}$ on the natural numbers (see Section 3 for information about ultrafilters and ultralimits). For every $K \in \mathcal{K}_s^n$ and every $0 < \lambda < 1$ we set

$$Q_m(K, \lambda) = \left(\underline{P}_{\gamma_m} \circ \overline{P}_{\gamma_m}\right)^\ell (K),$$

where $\ell = \ell(m)$ is the biggest integer such that $\gamma_m^{2\ell} \geq \lambda$. Finally, we define

$$K^\lambda = \lim_{\mathcal{U}} Q_m(K, \lambda). \tag{1.3}$$

We then have the following result, announced in [**10**]:

THEOREM 1.4. *The maps $K \mapsto K^\lambda$ which send convex bodies to convex bodies satisfy properties 1-5 from Theorem 1.3, together with the additional property:*

6. *For every $0 < \lambda < 1$ and every $K \in \mathcal{K}_s^n$ we have $\left(K^\lambda\right)^\circ = (K^\circ)^\lambda$.*

Furthermore, $K^{\underline{\lambda}} \subseteq K^{\lambda} \subseteq K^{\overline{\lambda}}$ *for all* $K \in \mathcal{K}_s^n$ *and all* $0 < \lambda < 1$.

Theorem 1.4 will follow immediately from Theorem 3.4.

So far we described two separate constructions: The geometric mean $G(K,T)$ (with equal weights) and the power K^{λ}. There is one case that is common to both definitions: we expect to have $G(B_2^n, K) = K^{\frac{1}{2}}$. However, this is not true with the definitions given above. The main goal of this paper is to present a weighted version of the geometric mean $G_\lambda(K,T)$ which satisfies all the natural properties and is also related to the power by $G_\lambda(B_2^n, K) = K^{\lambda}$. In Section 2 we will construct the upper mean $\overline{G}_\lambda(K,T)$ which generalizes the upper power $K^{\overline{\lambda}}$. It has the advantage of being constructive, as it does not depend on an arbitrary choice of an ultrafilter. However, it does not satisfy the duality relation $\overline{G}_\lambda(K,T)^\circ = \overline{G}_\lambda(K^\circ, T^\circ)$. In Section 2 we will construct a weighted mean that does satisfy this relation. Our theorems will generalize Theorems 1.3 and 1.4 and will provide proofs for these announced results.

Our expectation is that geometric means $G_\lambda(K,T)$ should play the role of "geometric interpolation" between the bodies K and T. However, we did not (yet) develop this direction.

2. Weighted geometric means

We now turn our attention to defining the geometric mean of two convex bodies K and T with weighted $1-\lambda$ and λ. We begin by collecting some useful properties of the matrix geometric mean, defined in equation (1.1):

PROPOSITION 2.1. *For every positive-definite matrices* u *and* v *we have:*

(1) $u\#_\lambda v = v\#_{1-\lambda} u$
(2) *If* $u_1 \preceq u_2$ *and* $v_1 \preceq v_2$ *then* $u_1\#_\lambda v_1 \preceq u_2\#_\lambda v_2$.
(3) $u\#_\lambda (u\#_\mu v) = u\#_{\lambda\mu} v$.
(4) $(\alpha u)\#_\lambda (\beta v) = \alpha^{1-\lambda}\beta^{\lambda} \cdot u\#_\lambda v$ *for all* $\alpha, \beta > 0$.
(5) *For every invertible matrix* w, *not necessarily positive-definite, we have* $(wuw^t)\#_\lambda(wvw^t) = w(u\#_\lambda v)w^t$.

Properties (3) and (4) are a direct computation. Properties (1) and (5) follow e.g. from Theorem 6.1.6 of [**2**]. In Property (2) the monotonicity in v follows from the fact that the maps $v \mapsto u^{-\frac{1}{2}}vu^{-\frac{1}{2}}$, $v \mapsto v^{\lambda}$, and $v \mapsto u^{\frac{1}{2}}vu^{\frac{1}{2}}$ are operator monotone, so their composition is also operator monotone. Finally, the monotonicity in u is deduced from the monotonicity in v and Property (1).

We are now ready to define the geometric mean of two ellipsoids:

DEFINITION 2.2. Let $E_1, E_2 \subseteq \mathbb{R}^n$ be ellipsoids and fix $0 < \lambda < 1$. The geometric mean $G_\lambda(E_1, E_2)$ is the unique ellipsoid E_3 such that $u_{E_3} = u_{E_1}\#_\lambda u_{E_2}$.

The weighted geometric mean for ellipsoids is monotone: If $E_1 \subseteq E_2$ and $F_1 \subseteq F_2$ then $G_\lambda(E_1, F_1) \subseteq G_\lambda(E_2, F_2)$. This follows from Proposition 2.1(2). Therefore it makes sense to define:

DEFINITION 2.3. Fix $K, T \in \mathcal{K}_s^n$ and $0 < \lambda < 1$. We define the upper ellipsoidal envelope $\overline{G}_\lambda(K,T)$ by

$$\overline{G}_\lambda(K,T) = \bigcap \left\{ G_\lambda(E,F) : \begin{array}{c} E, F \text{ are ellipsoids} \\ K \subseteq E \text{ and } T \subseteq F \end{array} \right\}.$$

Similarly, the lower ellipsoidal envelope $\underline{G}_\lambda(K,T)$ is defined by

$$\underline{G}_\lambda(K,T)=\operatorname{conv}\left(\bigcup\left\{G_\lambda(E,F):\begin{array}{l}E,F\text{ are ellipsoids}\\ K\supseteq E\text{ and }T\supseteq F\end{array}\right\}\right)$$

where conv denotes the convex hull.

The monotonicity of G_λ for ellipsoids implies that $\underline{G}_\lambda(K,T)\subseteq\overline{G}_\lambda(K,T)$. Furthermore, the two envelopes are related to each other by duality:

$$\overline{G}_\lambda(K,T)^\circ=\underline{G}_\lambda(K^\circ,T^\circ).$$

To avoid repeating every result twice we will mostly discuss the upper envelope, but every result we prove in this section also has an analogue for the lower envelope which can be deduced by duality.

Many of the properties of the upper ellipsoidal envelope are straightforward. For example, it is immediate from the definition that $\overline{G}_\lambda$ is monotone:

$$K_1\subseteq K_2\text{ and }T_1\subseteq T_2\text{ implies }\overline{G}_\lambda(K_1,T_1)\subseteq\overline{G}_\lambda(K_2,T_2).\tag{2.1}$$

Furthermore, by Proposition 2.1(4) the weighted geometric mean for ellipsoids has the scaling property

$$G_\lambda(\alpha E,\beta F)=\alpha^{1-\lambda}\beta^\lambda G_\lambda(E,F),$$

so we also have

$$\overline{G}_\lambda(\alpha K,\beta T)=\alpha^{1-\lambda}\beta^\lambda\overline{G}_\lambda(K,T).\tag{2.2}$$

Similarly, from Proposition 2.1(5) we deduce that

$$\overline{G}_\lambda(wK,wT)=w\left(\overline{G}_\lambda(K,T)\right)\tag{2.3}$$

for all linear maps w.

The next property is slightly more delicate. It essentially states that for $\lambda\approx 1$ we have $\overline{G}_\lambda(K,T)\approx T$:

PROPOSITION 2.4. *Fix $K,T\in\mathcal{K}_s^n$ and $0<\lambda<1$. If $\frac{1}{R}\cdot B_2^n\subseteq K,T\subseteq R\cdot B_2^n$ for some $R>0$ then*

$$R^{-2(1-\lambda)}\cdot T\subseteq\overline{G}_\lambda(K,T)\subseteq R^{2(1-\lambda)}\cdot T.$$

PROOF. By (2.1) and (2.2) we have

$$\overline{G}_\lambda(K,T)\subseteq\overline{G}_\lambda(R\cdot B_2^n,T)=R^{1-\lambda}\cdot\overline{G}_\lambda(B_2^n,T).$$

Let E be any ellipsoid such that $E\supseteq T$. If the principal axes of E have lengths $a_1,a_2,\ldots,a_n$ then $G_\lambda\left(B_2^n,E\right)$ has axes of length $a_1^\lambda,a_2^\lambda,\ldots,a_n^\lambda$ by Proposition 1.1. Since $E\supseteq T\supseteq\frac{1}{R}\cdot B_2^n$ we have

$$a_i^\lambda=a_i\cdot\left(\frac{1}{a_i}\right)^{1-\lambda}\le a_i\cdot R^{1-\lambda}$$

so $G_\lambda\left(B_2^n,E\right)\subseteq R^{1-\lambda}\cdot E$. Intersecting over all $E\supseteq T$ we obtain $\overline{G}_\lambda\left(B_2^n,T\right)\subseteq R^{1-\lambda}\cdot T$, so $\overline{G}_\lambda(K,T)\subseteq R^{2(1-\lambda)}\cdot T$.

The left inequality follows from the right one by duality: We also have $\frac{1}{R}\cdot B_2^n\subseteq K^\circ,T^\circ\subseteq R\cdot B_2^n$, so

$$\overline{G}_\lambda(K,T)^\circ\subseteq\underline{G}_\lambda(K,T)^\circ=\overline{G}_\lambda\left(K^\circ,T^\circ\right)\subseteq R^{2(1-\lambda)}T^\circ,$$

and the result follows. □

Finally, we will need the following result relating different ellipsoidal envelopes for different values of λ:

PROPOSITION 2.5. *Fix $K, T \in \mathcal{K}_s^n$. For every $0 < \lambda, \mu < 1$ one has*

$$\overline{G}_\lambda\left(K, \overline{G}_\mu(K,T)\right) \subseteq \overline{G}_{\lambda\mu}(K,T).$$

PROOF. It follows from Proposition 2.1(3) that for every ellipsoids $E, F \subseteq \mathbb{R}^n$ we have

$$G_\lambda\left(E, G_\mu(E,F)\right) = G_{\lambda\mu}(E,F).$$

Now if $E, F \subseteq \mathbb{R}^n$ are any ellipsoids such that $K \subseteq E$ and $T \subseteq F$, then by definition $G_\mu(E,F)$ is an ellipsoid and $\overline{G}_\mu(K,T) \subseteq G_\mu(E,F)$. Hence

$$\overline{G}_\lambda\left(K, \overline{G}_\mu(K,T)\right) \subseteq G_\lambda\left(E, G_\mu(E,F)\right) = G_{\lambda\mu}(E,F).$$

Intersecting over all such E and F we obtain the result. □

One may consider $\overline{G}_\lambda\left(K, \overline{G}_\mu(K,T)\right)$ as an "iterated mean" of length 2. We would like to consider such iterated means of an arbitrary size m. To handle such expressions, let us introduce a notation: Given $K \in \mathcal{K}_s^n$ and $0 < \lambda < 1$ we define the operator $\overline{G}_{\lambda,K} : \mathcal{K}_s^n \to \mathcal{K}_s^n$ by $\overline{G}_{\lambda,K}(T) = \overline{G}_\lambda(K,T)$. This will allow us write expressions like $\overline{G}_\lambda\left(K, \overline{G}_\mu(K,T)\right)$ as $\left(\overline{G}_{\lambda,K} \circ \overline{G}_{\mu,K}\right)(T)$. Using this notation we may now define:

DEFINITION 2.6. Fix $0 < \lambda < 1$, and let Π be a finite partition of $[\lambda, 1]$:

$$\Pi : \ \lambda = t_0 < t_1 < \cdots < t_m = 1.$$

We define

$$\overline{G}_\Pi(K,T) = \overline{G}_{\Pi,K}(T) = \left(\overline{G}_{s_1,K} \circ \overline{G}_{s_2,K} \circ \cdots \circ \overline{G}_{s_m,K}\right)(T),$$

where $s_i = t_{i-1}/t_i$ for $i = 1, 2, ..m$.

Note that for every partition Π of $[\lambda, 1]$ and every $b > 0$ we have $\overline{G}_{\Pi,K}(bT) = b^\lambda \overline{G}_{\Pi,K}(T)$. For convenience we will also use the simpler inclusions

$$\begin{aligned} \overline{G}_{\Pi,K}(bT) &\supseteq b\overline{G}_{\Pi,K}(T) \text{ if } b \le 1 \\ \overline{G}_{\Pi,K}(bT) &\subseteq b\overline{G}_{\Pi,K}(T) \text{ if } b \ge 1. \end{aligned} \tag{2.4}$$

We are ready to state the first main result of this section. Recall that $\|\Pi\| = \max_i |t_i - t_{i-1}|$ denotes the length of the longest interval of Π. We have:

THEOREM 2.7. *For every $K, T \in \mathcal{K}_s^n$ and $0 < \lambda < 1$ the limit*

$$\lim_{\|\Pi\| \to 0} \overline{G}_\Pi(K,T)$$

exists in the Hausdorff sense and is equal to the intersection $\bigcap_\Pi \overline{G}_\Pi(K,T)$. We call this body the upper λ-mean of K and T and denote it by $G_{\overline{\lambda}}(K,T)$.

In order to prove the theorem we need the following lemma:

LEMMA 2.8. *Fix $K, T \in \mathcal{K}_s^n$ and $0 < \lambda < 1$. Then:*

(1) *If Π and $\widetilde{\Pi}$ are two partitions of $[\lambda, 1]$ and $\Pi \supseteq \widetilde{\Pi}$ then $\overline{G}_\Pi(K,T) \subseteq \overline{G}_{\widetilde{\Pi}}(K,T)$.*
(2) *For every $\epsilon > 0$ and every integer m there exists $\delta > 0$ with the following property: If $\left\|\widetilde{\Pi}\right\| < \delta$ and Π is obtained from $\widetilde{\Pi}$ by adding at most m points, then $\overline{G}_{\widetilde{\Pi}}(K,T) \subseteq (1+\epsilon) \cdot \overline{G}_\Pi(K,T)$.*

PROOF. The proof of (1) appeared in [**8**] in the case of the power, and the general case of the geometric mean is almost identical. We may assume that $|\Pi| = \left|\widetilde{\Pi}\right| + 1$. Write

$$\Pi : \ \lambda = t_0 < t_1 < \cdots < t_m = 1$$

and assume $\widetilde{\Pi}$ is obtained from Π by removing the point t_k. Let Π_1 be the partition $\{t_0, t_1, \ldots, t_{k-1}\}$ and Π_2 be the partition $\{t_{k+1}, t_{k+2}, \ldots, t_m\}$. Then

$$\overline{G}_{\Pi,K} = \overline{G}_{\Pi_1,K} \circ \overline{G}_{t_{k-1}/t_k,K} \circ \overline{G}_{t_k/t_{k+1},K} \circ \overline{G}_{\Pi_2,K}$$

and

$$\overline{G}_{\widetilde{\Pi},K} = \overline{G}_{\Pi_1,K} \circ \overline{G}_{t_{k-1}/t_{k+1},K} \circ \overline{G}_{\Pi_2,K}.$$

By Proposition 2.5 we have

$$\left(\overline{G}_{t_{k-1}/t_k,K} \circ \overline{G}_{t_k/t_{k+1},K}\right)(A) \subseteq \overline{G}_{t_{k-1}/t_{k+1},K}\,(A)$$

for any body A. Taking $A = \overline{G}_{\Pi_2,K}(T)$ and using the fact that $\overline{G}_{\Pi_1,K}$ is monotone we complete the proof of (1).

The proof of (2) is more delicate. First we prove the case $m = 1$. Fix a number $R > 0$ such that $\frac{1}{R} \cdot B_2^n \subseteq K, T \subseteq R \cdot B_2^n$, and choose $\delta > 0$ such that $R^{6\delta/\lambda} < 1 + \epsilon$. Let Π and $\widetilde{\Pi}$ be partitions of $[\lambda, 1]$ such that $\left\|\widetilde{\Pi}\right\| < \delta$ and Π is obtained from $\widetilde{\Pi}$ by adding one point. Keeping the notation from the first part of the proof we have

$$1 - \frac{t_{k-1}}{t_{k+1}} = \frac{t_{k+1} - t_{k-1}}{t_{k+1}} \leq \frac{\left\|\widetilde{\Pi}\right\|}{\lambda} < \frac{\delta}{\lambda}$$

and the same bound holds for t_{k-1}/t_k and t_k/t_{k+1}. Furthermore, all bodies constructed during the computation of $\overline{G}_{\Pi,K}(T)$ and $\overline{G}_{\widetilde{\Pi},K}(T)$ are always between $\frac{1}{R} \cdot B_2^n$ and $R \cdot B_2^n$. By Proposition 2.4, the monotonicity of $\overline{G}_{\Pi_1,K}$ and (2.4) we have

$$\overline{G}_{\Pi,K}(T) = \left(\overline{G}_{\Pi_1,K} \circ \overline{G}_{t_{k-1}/t_k,K} \circ \overline{G}_{t_k/t_{k+1},K} \circ \overline{G}_{\Pi_2,K}\right)(T)$$
$$\supseteq \overline{G}_{\Pi_1,K}\left(R^{-2\delta/\lambda} \cdot R^{-2\delta/\lambda} \cdot \overline{G}_{\Pi_2,K}(T)\right) \supseteq R^{-4\delta/\lambda} \cdot \left(\overline{G}_{\Pi_1,K} \circ \overline{G}_{\Pi_2,K}\right)(T).$$

Similarly, we have

$$\overline{G}_{\widetilde{\Pi},K}(T) \subseteq \overline{G}_{\Pi_1,K}\left(R^{2\delta/\lambda} \cdot \overline{G}_{\Pi_2,K}(T)\right) \subseteq R^{2\delta/\lambda} \cdot \left(\overline{G}_{\Pi_1,K} \circ \overline{G}_{\Pi_2,K}\right)(T).$$

Combining the two inclusions we deduce that $\overline{G}_{\widetilde{\Pi}}(K,T) \subseteq R^{6\delta/\lambda} \cdot \overline{G}_{\Pi}(K,T) \subseteq (1+\epsilon)\overline{G}_{\Pi}(K,T)$ like we wanted.

Finally, the case of general m follows from the case $m = 1$: Given $\epsilon > 0$, choose $\widetilde{\epsilon} > 0$ such that $(1 + \widetilde{\epsilon})^m < 1 + \epsilon$. The $\delta > 0$ that works for $\widetilde{\epsilon}$ and $m = 1$ will also work for ϵ and m. □

Now we are ready to prove Theorem 2.7:

PROOF OF THEOREM 2.7. Write $G = \bigcap_{\Pi} \overline{G}_{\Pi}(K,T)$. In order to prove that $\lim_{\|\Pi\| \to 0} \overline{G}_{\Pi}(K,T) = G$ it is enough to prove that for $\epsilon > 0$ there exists $\delta > 0$ such that if Π is a partition of $[\lambda, 1]$ with $\|\Pi\| < \delta$ then $\overline{G}_{\Pi}(K,T) \subseteq (1+\epsilon)^2\, G$.

Fix $\epsilon > 0$. First we show that there exists a single partition Π_0 such that $\overline{G}_{\Pi_0}(K,T) \subseteq (1+\epsilon)\,G$. The proof of this fact essentially appeared in [**8**], and we repeat it here. Assume by contradiction that $\overline{G}_{\Pi}(K,T) \not\subseteq (1+\epsilon)\,G$ for all Π. Set

$$A_\Pi = \overline{G}_\Pi(K,T) \setminus \operatorname{int}\left((1+\epsilon)\,G\right) \neq \emptyset,$$

where int denotes the interior of a set. For every finite family of partitions $\Pi_1, \Pi_2, \ldots, \Pi_n$ we have by Lemma 2.8(1)

$$\bigcap_{i=1}^{n} A_{\Pi_i} \supseteq A_{\Pi_1 \cup \Pi_2 \cup \cdots \cup \Pi_n} \neq \emptyset,$$

so the family $\{A_\Pi\}_\Pi$ has the finite intersection property. Since this is a family of compact sets, it follows that $\bigcap_\Pi A_\Pi \neq \emptyset$.

However, every point $a \in \bigcap_\Pi A_\Pi$ satisfies both $a \in \bigcap_\Pi \overline{G}_\Pi(K,T) = G$ and $a \notin \operatorname{int}\left((1+\epsilon)\,G\right)$, which is clearly impossible. We arrived at a contradiction, so the desired Π_0 exists.

To conclude the proof, denote the number of points in Π_0 by m. Let $\delta > 0$ be the number whose existence was guaranteed by Lemma 2.8(2). Let Π be any partition with $\|\Pi\| < \delta$. Since $\widetilde{\Pi} = \Pi \cup \Pi_0$ can be created from Π by adding at most m points we have

$$\overline{G}_\Pi(K,T) \subseteq (1+\epsilon)\overline{G}_{\widetilde{\Pi}}(K,T) \subseteq (1+\epsilon)\overline{G}_{\Pi_0}(K,T) \subseteq (1+\epsilon)^2\,G$$

which proves the result. □

The upper λ-mean has many of the properties we would expect from a weighted geometric mean. These properties are summarized in the following theorem:

THEOREM 2.9. *The upper λ-mean satisfies the following properties:*

(1) $G_{\overline{\lambda}}(K,K) = K$.
(2) *If $K_1 \subseteq T_1$ and $K_2 \subseteq T_2$ then $G_{\overline{\lambda}}(K_1,K_2) \subseteq G_{\overline{\lambda}}(T_1,T_2)$.*
(3) $G_{\overline{\lambda}}(\alpha K, \beta T) = \alpha^{1-\lambda}\beta^{\lambda} G_{\overline{\lambda}}(K,T)$ *for all $\alpha, \beta > 0$.*
(4) *$G_{\overline{\lambda}}$ is a continuous function of K, T and λ with respect to the Hausdorff metric on $\mathcal{K}_s^n$.*
(5) *$G_{\overline{\lambda}}$ satisfies the harmonic mean – geometric mean – arithmetic mean inequality*

$$\left((1-\lambda)\,K^\circ + \lambda T^\circ\right)^\circ \subseteq G_{\overline{\lambda}}(K,T) \subseteq (1-\lambda)K + \lambda T.$$

(6) $G_{\overline{\lambda}}(uK, uT) = u\left(G_{\overline{\lambda}}(K,T)\right)$ *for all invertible linear maps u.*
(7) $G_{\overline{\lambda}}\left(K, G_{\overline{\mu}}(K,T)\right) = G_{\overline{\lambda\mu}}(K,T)$.
(8) $G_{\overline{\lambda}}(B_2^n, K) = K^{\overline{\lambda}}$, *where the upper power $K^{\overline{\lambda}}$ was defined in* (1.2).
(9) *For ellipsoids E and F we have $G_{\overline{\lambda}}(E,F) = G_\lambda(E,F)$, where $G_\lambda(E,F)$ was defined in Definition* 2.2.

Unfortunately, as our construction of $G_{\overline{\lambda}}$ is not symmetric in K and T, we don't know if we have the property $G_{1/2}(K,T) = G_{1/2}(T,K)$.

PROOF. Properties (2), (3), (6) and (9) are straightforward: First we note that the upper envelope $\overline{G}_\lambda$ satisfies these properties by (2.1), (2.2) and (2.3). It follows immediately that the same is true for the operators $\overline{G}_\Pi$, and by taking a limit $\|\Pi\| \to 0$ we conclude that $G_{\overline{\lambda}}$ has the same properties. Property (8) is immediate from comparing the two definitions.

Continuity of $G_{\overline{\lambda}}(K,T)$ with respect to K and T is a corollary of properties (2) and (3). Indeed, if $K_m \to K$ and $T_m \to T$ then for every $\epsilon > 0$ we have

$$(1-\epsilon)K \subseteq K_m \subseteq (1+\epsilon)K$$
$$(1-\epsilon)T \subseteq T_m \subseteq (1+\epsilon)T$$

for all large enough m. But then we have

$$G_{\overline{\lambda}}(K_m,T_m) \subseteq G_{\overline{\lambda}}((1+\epsilon)K,(1+\epsilon)T) = (1+\epsilon)G_{\overline{\lambda}}(K,T),$$

and similarly $G_{\overline{\lambda}}(K_m,T_m) \supseteq (1-\epsilon)G_{\overline{\lambda}}(K,T)$. This implies that $G_{\overline{\lambda}}(K_m,T_m) \to G_{\overline{\lambda}}(K,T)$.

To prove (5) let us first show that for two ellipsoids E and F one has

$$G_\lambda(E,F) \subseteq (1-\lambda)E + \lambda F \tag{2.5}$$

Indeed, by applying a suitable linear map we may assume that $E = B_2^n$ and $h_F(x) = \left(\sum_{i=1}^n a_i x_i^2\right)^{1/2}$ for some constants $a_i > 0$. By Hölder's inequality we have for every $x \in \mathbb{R}^n$

$$h^2_{G_\lambda(E,F)}(x) = \sum_{i=1}^n a_i^\lambda x_i^2 = \sum_{i=1}^n \left(a_i x_i^2\right)^\lambda \cdot x_i^{2(1-\lambda)} \le \left(\sum_{i=1}^n a_i x_i^2\right)^\lambda \cdot \left(\sum_{i=1}^n x_i^2\right)^{1-\lambda},$$

so $h_{G_\lambda(E,F)}(x) \le h_E(x)^{1-\lambda} \cdot h_F(x)^\lambda$. By the geometric mean – arithmetic mean inequality for numbers we deduce that $h_{G_\lambda(E,F)}(x) \le (1-\lambda)h_E(x) + \lambda h_F(x)$ and (2.5) is proved.

Next, we fix convex bodies K and T. Intersecting (2.5) over all ellipsoids E and F such that $E \supseteq K$ and $F \supseteq T$ we conclude that $\overline{G}_\lambda(K,T) \subseteq (1-\lambda)K + \lambda T$. From here it is immediate to deduce that $\overline{G}_\Pi$ and $G_{\overline{\lambda}}$ have the same property, finishing the proof of the right inequality. The left inequality follows from the right one by duality since $G_{\overline{\lambda}}(K^\circ,T^\circ)^\circ \subseteq G_{\overline{\lambda}}(K,T)$ (we already observed this property for $\overline{G}_\lambda$, and it extends to $G_{\overline{\lambda}}$ in the standard way). Property (1) follows immediately from (5).

Now we prove property (7). Fix a partition Π_λ of $[\lambda,1]$ and a partition Π_μ of $[\mu,1]$. If

$$\Pi_\lambda = \{\lambda = t_0, t_1, t_2, \ldots, t_m = 1\},$$

let us write $\mu\Pi_\lambda$ for the partition $\{\mu t_0, \mu t_1, \ldots, \mu t_m\}$ of the interval $[\lambda\mu,\mu]$. Then $\Pi = \mu\Pi_\lambda \cup \Pi_\mu$ is a partition of $[\lambda\mu,1]$, and we have

$$\overline{G}_\Pi(K,T) = \overline{G}_{\mu\Pi_\lambda}\left(K, \overline{G}_{\Pi_\mu}(K,T)\right) = \overline{G}_{\Pi_\lambda}\left(K, \overline{G}_{\Pi_\mu}(K,T)\right).$$

Letting $\|\Pi_\lambda\|, \|\Pi_\mu\| \to 0$ we also have $\|\Pi\| \to 0$ and (7) follows.

Finally, we complete the proof of (4) by showing continuity with respect to λ (a similar argument shows joint continuity with respect to K,T and λ). Fix $K,T \in \mathcal{K}_s^n$ and $0 < \lambda < 1$. Also fix a number $R > 0$ such that $\frac{1}{R}B_2^n \subseteq K,T \subseteq R\cdot B_2^n$, and fix a sequence $\{\lambda_m\}$ such that $\lambda_m \to \lambda$. We will assume that $\lambda_m < \lambda$ for all m, but the proof in the case $\lambda_m > \lambda$ is almost identical.

Using property (7), Proposition 2.4 and Property (3) we have

$$\begin{aligned} G_{\overline{\lambda_m}}(K,T) &= G_{\overline{\lambda}}\left(K, G_{\overline{\lambda_m/\lambda}}(K,T)\right) \subseteq G_{\overline{\lambda}}\left(K, R^{2(1-\lambda_m/\lambda)}T\right) \\ &= R^{2(\lambda-\lambda_m)} \cdot G_{\overline{\lambda}}(K,T). \end{aligned}$$

In the same way we also have

$$G_{\overline{\lambda_m}}(K,T) \supseteq R^{-2(\lambda-\lambda_m)}G_{\overline{\lambda}}(K,T).$$

Letting $m \to \infty$ we conclude that $G_{\overline{\lambda_m}}(K,T) \to G_{\overline{\lambda}}(K,T)$ and the proof is complete. □

Unfortunately, unlike the pre-mean $g(K,T)$ introduced in Section 1, the upper λ-mean $G_{\overline{\lambda}}(K,T)$ does not commute with polarity. In general we only have

$$G_{\overline{\lambda}}(K,T)^\circ = G_{\underline{\lambda}}(K^\circ,T^\circ) \subseteq G_{\overline{\lambda}}(K^\circ,T^\circ),$$

where the lower λ-mean $G_{\underline{\lambda}}(K,T)$ is defined in the obvious way using lower ellipsoidal envelopes. In the next section we will fix this problem and construct a weighted geometric mean $G_\lambda(K,T)$ which satisfies all properties of Theorem 2.9, together with the duality property $G_\lambda(K,T)^\circ = G_\lambda(K^\circ,T^\circ)$.

3. A self-dual geometric mean

The goal of this section is to construct a weighted geometric mean for convex bodies which satisfies the duality property $G_\lambda(K,T)^\circ = G_\lambda(K^\circ,T^\circ)$. We will do so by properly "mixing" upper and lower ellipsoidal envelopes. We keep the notation $\overline{G}_{\lambda,K}$ from the previous section, and define its lower analogue $\underline{G}_{\lambda,K}$ in the obvious way. We also fix a sequence $\{\gamma_m\}_{m=1}^\infty \subseteq (0,1)$ such that $\gamma_m \to 1$ as $m \to \infty$. The sequence $\{\gamma_m\}$ will remain fixed throughout this section.

Definition 3.1. For $K,T \in \mathcal{K}_s^n$ and $0<\lambda<1$ we define a sequence $\{Q_m(K,T,\lambda)\}_{m=1}^\infty$ of convex bodies by

$$Q_m(K,T,\lambda) = \left(\underline{G}_{\gamma_m,K} \circ \overline{G}_{\gamma_m,K}\right)^\ell (T).$$

Here $\ell = \ell(m)$ is the largest integer such that $\gamma_m^{2\ell} \ge \lambda$.

We would like to define $G_\lambda(K,T) = \lim_{m\to\infty} Q_m(K,T,\lambda)$, but it is not clear that this limit exists. Instead, we will use the notion of an ultrafilter limit. For the basic definition of a limit along an ultrafilter, we refer the reader for example to Chapter 3 of [**5**]. In the following proposition we summarize basic information about ultralimits for sequences of convex bodies:

Proposition 3.2. *Fix an ultrafilter $\mathcal{U}$ on the natural numbers $\mathbb{N}$. Then:*

(1) *Fix a sequence $\{K_m\}_{m=1}^\infty \subseteq \mathcal{K}_s^n$ such that $\frac{1}{R}\cdot B_2^n \subseteq K_m \subseteq R\cdot B_2^n$ for all $m\in\mathbb{N}$, where $R>0$ is some fixed number. Then $\lim_{\mathcal{U}} K_m$ exists and belongs to $\mathcal{K}_s^n$.*
(2) *If $K_m \subseteq T_m$ for all $m\in\mathbb{N}$, then $\lim_{\mathcal{U}} K_m \subseteq \lim_{\mathcal{U}} T_m$.*
(3) *$\lim_{\mathcal{U}}(K_m+T_m) = \lim_{\mathcal{U}} K_m + \lim_{\mathcal{U}} T_m$ for every two sequences $\{K_m\}_{m=1}^\infty, \{T_m\}_{m=1}^\infty \subseteq \mathcal{K}_s^n$.*
(4) *If $\{\lambda_m\}_{m=1}^\infty$ is a sequence of positive real numbers such that $\lambda_m \to \lambda$, then $\lim_{\mathcal{U}}(\lambda_m K_m) = \lambda\cdot\lim_{\mathcal{U}} K_m$ for every sequence $\{K_m\}_{m=1}^\infty \subseteq \mathcal{K}_s^n$.*
(5) *$\lim_{\mathcal{U}}(K_m^\circ) = (\lim_{\mathcal{U}} K_m)^\circ$ for every sequence $\{K_m\}_{m=1}^\infty \subseteq \mathcal{K}_s^n$*
(6) *If u is an invertible linear map on $\mathbb{R}^n$ then $\lim_{\mathcal{U}}(uK_m) = u(\lim_{\mathcal{U}} K_m)$ for every sequence $\{K_m\}_{m=1}^\infty \subseteq \mathcal{K}_s^n$.*

Proof. The set

$$\mathcal{K}_R = \left\{K\in\mathcal{K}_s^n : \frac{1}{R}\cdot B_2^n \subseteq K \subseteq R\cdot B_2^n\right\} \subseteq \mathcal{K}_s^n \tag{3.1}$$

is compact with respect to the Hausdorff metric, by the Blaschke selection theorem (see [**13**]). Property (1) now follows using the standard theory of ultrafilters – see e.g. Theorem 3.48 of [**5**] . Properties (2)–(6) are easy corollaries of the definitions, using the fact that the four maps $(K,T) \mapsto K+T$, $(\lambda, K) \mapsto \lambda K$, $K \mapsto K^\circ$, $K \mapsto uK$ are all continuous with respect to the Hausdorff metric – see Theorem 3.49 of [**5**]. □

From now on we fix the ultrafilter $\mathcal{U}$. We then define:

DEFINITION 3.3. For $K, T \in \mathcal{K}_s^n$ and $0<\lambda<1$ we define

$$G_\lambda(K,T) = \lim_{\mathcal{U}} Q_m(K,T,\lambda).$$

Notice that if $\frac{1}{R}\cdot B_2^n \subseteq K, T \subseteq R\cdot B_2^n$ then $\frac{1}{R}\cdot B_2^n \subseteq Q_m(K,T,\lambda) \subseteq R\cdot B_2^n$, so the ultralimit in Definition 3.3 exists.

We summarize the important properties of $G_\lambda(K,T)$ in the following theorem:

THEOREM 3.4. *The weighted geometric mean satisfies the following properties:*

(1) $G_\lambda(K,K) = K$.
(2) *If* $K_1 \subseteq T_1$ *and* $K_2 \subseteq T_2$ *then* $G_\lambda(K_1,K_2) \subseteq G_\lambda(T_1,T_2)$.
(3) $G_\lambda(\alpha K, \beta T) = \alpha^{1-\lambda}\beta^\lambda G_\lambda(K,T)$ *for all* $\alpha, \beta > 0$.
(4) G_λ *is a continuous function of* K, T *and* λ *with respect to the Hausdorff metric on* $\mathcal{K}_s^n$.
(5) G_λ *satisfies the harmonic mean – geometric mean – arithmetic mean inequality*

$$((1-\lambda)K^\circ + \lambda T^\circ)^\circ \subseteq G_\lambda(K,T) \subseteq (1-\lambda)K + \lambda T.$$

In fact, G_λ *is bounded between the upper and lower* λ*-means:*

$$G_{\underline{\lambda}}(K,T) \subseteq G_\lambda(K,T) \subseteq G_{\overline{\lambda}}(K,T).$$

(6) $G_\lambda(K,T)^\circ = G_\lambda(K^\circ, T^\circ)$
(7) $G_\lambda(uK,uT) = u(G_\lambda(K,T))$ *for all invertible linear maps* u.
(8) $G_\lambda(K, G_\mu(K,T)) = G_{\lambda\mu}(K,T)$.
(9) $G_\lambda(B_2^n, K) = K^\lambda$, *where the power* K^λ *was defined in* (1.3).
(10) *For ellipsoids* E *and* F *the mean* $G_\lambda(E,F)$ *agrees with the one defined in Definition* 2.2.

For the proof of property (8) we will need the following technical lemma:

LEMMA 3.5. *Fix* $K, T \in \mathcal{K}_s^n$ *and* $0 < \lambda, \mu < 1$. *Then*

$$\lim_{\mathcal{U}} Q_m(K, Q_m(K,T,\mu), \lambda) = G_\lambda(K, G_\mu(K,T)).$$

PROOF. The main idea is that the maps $T \mapsto Q_m(K,T,\mu)$ are uniformly Lipschitz. More concretely, if K, T_1 and T_2 all belong to the set $\mathcal{K}_R$ from (3.1) then

$$d_H(Q_m(K,T_1,\mu), Q_m(K,T_2,\mu)) \le R^2 \cdot d_H(T_1,T_2).$$

To prove this fact, let us write $d = d_H(T_1,T_2)$. Then $T_1 \subseteq T_2 + d\cdot B_2^n \subseteq (1+R\cdot d)T_2$, and so

$$\begin{aligned} Q_m(K,T_1,\mu) &\subseteq Q_m(K,(1+R\cdot d)T_2,\mu) \subseteq (1+Rd)\cdot Q_m(K,T_2,\mu) \\ &\subseteq Q_m(K,T_2,\mu) + R^2 d\cdot B_2^n. \end{aligned}$$

Exchanging the roles of T_1 and T_2 we conclude that

$$d_H\left(Q_m(K,T_1,\mu),Q_m(K,T_2,\mu)\right) < R^2 \cdot d,$$

so the maps are indeed uniformly Lipschitz.

The rest of the proof is a standard computation with ultrafilters. We fix a number R such that $K,T \in \mathcal{K}_R$. For every fixed $\epsilon > 0$ we have by definition

$$I = \left\{m:\ d_H\left(Q_m(K,T,\mu),G_\mu(K,T)\right) < \frac{\epsilon}{2R^2}\right\} \in \mathcal{U}$$

and

$$J = \left\{m:\ d_H\left(Q_m\left(K,G_\mu\left(K,T\right),\lambda\right),G_\lambda\left(K,G_\mu\left(K,T\right)\right)\right) < \frac{\epsilon}{2}\right\} \in \mathcal{U}.$$

Hence $I \cap J \in \mathcal{U}$, and for every $m \in I \cap J$ we have

$$d_H\left(Q_m\left(K,Q_m(K,T,\mu),\lambda\right),G_\lambda\left(K,G_\mu\left(K,T\right)\right)\right) \leq R^2 \cdot \frac{\epsilon}{2R^2} + \frac{\epsilon}{2} = \epsilon,$$

completing the proof. □

We can now prove Theorem 3.4:

Proof of Theorem 3.4. Properties (2), (7) and (10) are obvious from the corresponding properties of the ellipsoidal envelopes. Property (9) is immediate from comparing the two definitions.

For property (3), notice that

$$Q_m(\alpha K,\beta T,\lambda) = \alpha^{1-\gamma_m^{2\ell}} \cdot \beta^{\gamma_m^{2\ell}} \cdot Q_m\left(K,T,\lambda\right).$$

Since $\gamma_m^{2\ell} \to \lambda$ as $m \to \infty$ we can apply $\lim_{\mathcal{U}}$ to both sides and obtain (3).

For property (5) we obviously have

$$Q_m(K,T,\lambda) \subseteq \left(\overline{G}_{\gamma_m,K} \circ \overline{G}_{\gamma_m,K}\right)^{\ell}(T) = \overline{G}_{\Pi,K}(T),$$

where $\Pi = \left\{\gamma_m^{2\ell},\gamma_m^{2\ell-1},\ldots,\gamma_m,1\right\}$. If we write $\widetilde{\Pi} = \{\lambda\} \cup \Pi$ then $\widetilde{\Pi}$ is a partition of $[\lambda,1]$, and by Proposition 2.4 we have

$$\overline{G}_{\widetilde{\Pi},K}(T) = \overline{G}_{\lambda\gamma_m^{-2\ell},K}\left(\overline{G}_{\Pi,K}(T)\right) \supseteq R^{-2\left(1-\lambda\gamma_m^{-2\ell}\right)} \cdot \overline{G}_{\Pi,K}(T),$$

where R is some number such that $\frac{1}{R} \cdot B_2^n \subseteq K,T \subseteq R \cdot B_2^n$. Hence

$$Q_m(K,T,\lambda) \subseteq R^{2\left(1-\lambda\gamma_m^{-2\ell}\right)} \cdot \overline{G}_{\widetilde{\Pi},K}(T).$$

As $m \to \infty$ we have $\lambda\gamma_m^{-2\ell} \to 1$ and $\left\|\widetilde{\Pi}\right\| \to 0$, so by applying $\lim_{\mathcal{U}}$ to both sides we conclude that $G_\lambda(K,T) \subseteq G_{\overline{\lambda}}(K,T)$. The inequality $G_\lambda(K,T) \supseteq G_{\underline{\lambda}}(K,T)$ is similar, and the harmonic mean – geometric mean – arithmetic mean inequality is now an immediate corollary of Theorem 2.9(5). Property (1) is immediate from Property (5).

Next we prove property (6). Since the upper and lower envelopes are dual to each other we have

$$\begin{aligned} Q_m\left(K,T,\lambda\right)^\circ &= \left[\left(\underline{G}_{\gamma_m,K} \circ \overline{G}_{\gamma_m,K}\right)^{\ell}(T)\right]^\circ = \left(\overline{G}_{\gamma_m,K^\circ} \circ \underline{G}_{\gamma_m,K^\circ}\right)^{\ell}(T^\circ). \\ &= \left[\overline{G}_{\gamma_m,K^\circ} \circ \left(\underline{G}_{\gamma_m,K^\circ} \circ \overline{G}_{\gamma_m,K^\circ}\right)^{\ell-1} \circ \underline{G}_{\gamma_m,K^\circ}\right](T^\circ). \end{aligned}$$

If we now fix a number R as before then again all intermediate bodies created in the computation of $Q_m(K,T,\lambda)$ and $Q_m(K^\circ,T^\circ,\lambda)$ are also between $\frac{1}{R}\cdot B_2^n$ and $R\cdot B_2^n$. Hence we may apply Proposition 2.4 to conclude that

$$Q_m(K,T,\lambda)^\circ \subseteq R^{2(1-\gamma_m)}\cdot\left[\left(\underline{G}_{\gamma_m,K^\circ}\circ\overline{G}_{\gamma_m,K^\circ}\right)^{\ell-1}\circ\underline{G}_{\gamma_m,K^\circ}\right](T^\circ).$$

By the same proposition we also have $T^\circ\subseteq R^{2(1-\gamma_m)}\cdot\overline{G}_{\gamma_m,K^\circ}(T^\circ)$, so

$$Q_m(K,T,\lambda)^\circ\subseteq R^{4(1-\gamma_m)}\cdot\left(\underline{G}_{\gamma_m,K^\circ}\circ\overline{G}_{\gamma_m,K^\circ}\right)^{\ell}(T^\circ)=R^{4(1-\gamma_m)}\cdot Q_m(K^\circ,T^\circ,\lambda).$$

Applying $\lim_{\mathcal{U}}$ on both sides we conclude that $G_\lambda(K,T)^\circ\subseteq G_\lambda(K^\circ,T^\circ)$. The proof of the opposite inclusion is similar.

Now we prove property (8). By definition we have

$$Q_m(K,T,\lambda\mu)=\left(\underline{G}_{\gamma_m,K}\circ\overline{G}_{\gamma_m,K}\right)^{\ell}(T)$$

where ℓ is the largest integer satisfying $\gamma_m^{2\ell}\geq\lambda\mu$. Similarly we have

$$\begin{aligned}Q_m\left(K,Q_m(K,T,\mu),\lambda\right)&=\left(\underline{G}_{\gamma_m,K}\circ\overline{G}_{\gamma_m,K}\right)^{\ell_1}\left(Q_m(K,T,\mu)\right)\\&=\left(\underline{G}_{\gamma_m,K}\circ\overline{G}_{\gamma_m,K}\right)^{\ell_1}\circ\left(\underline{G}_{\gamma_m,K}\circ\overline{G}_{\gamma_m,K}\right)^{\ell_2}(T)\\&=\left(\underline{G}_{\gamma_m,K}\circ\overline{G}_{\gamma_m,K}\right)^{\ell_1+\ell_2}(T),\end{aligned}$$

where ℓ_1 is the largest integer satisfying $\gamma_m^{2\ell_1}\geq\lambda$ and ℓ_2 is the largest integer satisfying $\gamma_m^{2\ell_2}\geq\mu$.

Since $\gamma_m^{2(\ell_1+\ell_2)}\geq\lambda\mu$, we must have $\ell\geq\ell_1+\ell_2$. Conversely

$$\gamma_m^{2(\ell_1+\ell_2+2)}=\gamma_m^{2(\ell_1+1)}\cdot\gamma_m^{2(\ell_2+1)}<\lambda\mu,$$

so $\ell<\ell_1+\ell_2+2$. It follows that either $\ell=\ell_1+\ell_2$ or $\ell=\ell_1+\ell_2+1$.

If $\ell=\ell_1+\ell_2+1$ then

$$Q_m(K,T,\lambda\mu)=\left(\underline{G}_{\gamma_m,K}\circ\overline{G}_{\gamma_m,K}\right)\left(Q_m\left(K,Q_m(K,T,\mu),\lambda\right)\right),$$

and using Proposition 2.4 again we conclude that

$$\begin{aligned}(3.2)\qquad R^{-4(1-\gamma_m)}Q_m\left(K,Q_m(K,T,\mu),\lambda\right)&\subseteq Q_m(K,T,\lambda\mu)\\&\subseteq R^{4(1-\gamma_m)}Q_m\left(K,Q_m(K,T,\mu),\lambda\right),\end{aligned}$$

where R is as before. Of course these inclusions also hold when $\ell=\ell_1+\ell_2$, since in this case both bodies are equal. Hence (3.2) holds for every m. Applying $\lim_{\mathcal{U}}$ and using Lemma 3.5 we get (8).

Finally, property (4) is proved just like in Theorem 2.9, and the proof of the theorem is complete. □

Let us remark that we do not know if the definition of $G_\lambda(K,T)$ depends on the choice of an ultrafilter $\mathcal{U}$ (or on the choice of the sequence $\{\gamma_m\}$). It is known that $\lim_{\mathcal{U}}K_m=K$ for every ultrafilter $\mathcal{U}$ if and only if $\lim_{m\to\infty}K_m=K$ in the classic sense. Hence the above question is equivalent to the following: Does the limit $\lim_{m\to\infty}Q_m(K,T,\lambda)$ exist in the Hausdorff sense?

4. Additional remarks

The constructions of Sections 2 and 3 were based on ellipsoids. This is an especially nice class of bodies to consider, since we have a simple definition for the geometric mean $G_\lambda(E,F)$ of two ellipsoids $E,F\subseteq\mathbb{R}^n$ that satisfies all properties one expects. Furthermore, this class is closed under polarity, so we can define both the upper envelope $\overline{G}_\lambda(K,T)$ and the lower envelope $\underline{G}_\lambda(K,T)$ in terms of ellipsoids and have the polarity relation $\overline{G}_\lambda(K,T)^\circ=\underline{G}_\lambda(K^\circ,T^\circ)$. However, this is not the only possible choice.

Another possibility is to define the upper envelope using only a subfamily of ellipsoids, or even of degenerated ellipsoids. Of course, we need this subfamily to be closed with respect to the weighted geometric mean. One important example of such a family is the family of strips. For a direction θ on the unit sphere S^{n-1} and a number $c>0$, we define

$$S_{\theta,c}=\{x:\ |\langle x,\theta\rangle|\le c\}.$$

Obviously $S_{\theta,c}\notin\mathcal{K}_s^n$, as it is not compact. Still, we may define the geometric mean of two strips by

$$G_\lambda\left(S_{\theta,c},S_{\eta,d}\right)=\begin{cases}\mathbb{R}^n & \text{if } \theta\neq\eta\\ S_{\theta,c^{1-\lambda}d^\lambda} & \text{if } \theta=\eta.\end{cases}\tag{4.1}$$

This definition is consistent with our usual definition for the geometric mean of ellipsoids. More concretely, we may approximate every strip $S_{\theta,c}$ with ellipsoids of the form

$$\left\{x:\ \langle x,\theta\rangle^2+\frac{1}{R^2}\left|\mathrm{Proj}_{\theta^\perp}x\right|^2\le c^2\right\},$$

where $|\mathrm{Proj}_{\theta^\perp}x|$ is the length of the orthogonal projection of x to the hyperplane perpendicular to θ. Computing the geometric mean of such ellipsoids, and letting $R\to\infty$, we obtain (4.1).

We now define a variant of the upper envelope that only uses strips:

$$\overline{L}_\lambda(K,T)=\bigcap\left\{G_\lambda\left(S_1,S_2\right):\ \begin{array}{c}S_1,S_2 \text{ are strips}\\ K\subseteq S_1 \text{ and } T\subseteq S_2\end{array}\right\}.$$

One may check that $L=\overline{L}_\lambda(K,T)$ is the largest convex body such that $h_L(x)\le h_K(x)^{1-\lambda}h_T(x)^\lambda$ for all $x\in\mathbb{R}^n$. This body was studied by Böröczky, Lutwak, Yang and Zhang in [**3**] and is known as the logarithmic mean of K and T.

The lower envelope is now defined as $\underline{L}_\lambda(K,T)=\overline{L}_\lambda(K^\circ,T^\circ)^\circ$. The lower envelope will not be defined in terms of strips, but in terms of segments. We have $S_{\theta,c}^\circ=I_{\theta,c^{-1}}$, where $I_{\theta,d}=[-d\theta,d\theta]$ is the interval with endpoints $(-d\theta)$ and $d\theta$. If we set

$$G_\lambda(I_{\theta,c},I_{\eta,d})=\begin{cases}\{0\} & \text{if } \theta\neq\eta\\ I_{\theta,c^{1-\lambda}d^\lambda} & \text{if } \theta=\eta,\end{cases}$$

we may write

$$\underline{L}_\lambda(K,T)=\mathrm{conv}\left(\bigcup\left\{G_\lambda\left(I_1,I_2\right):\ \begin{array}{c}I_1,I_2 \text{ are intervals}\\ K\supseteq I_1 \text{ and } T\supseteq I_2\end{array}\right\}\right).$$

Since every strip can be approximated by ellipsoids, we have

$$\underline{L}_\lambda(K,T)\subseteq\underline{G}_\lambda(K,T)\subseteq\overline{G}_\lambda(K,T)\subseteq\overline{L}_\lambda(K,T).$$

In particular the lower envelope is contained in the upper envelope, as expected.

Once the upper and lower envelopes are defined, we proceed in exactly the same way as in Sections 2 and 3. We may define the operators $\overline{L}_{\lambda,K}$ and $\underline{L}_{\lambda,K}$, their analogues for partitions $\overline{L}_{\Pi,K}$ and $\underline{L}_{\Pi,K}$, the upper and lower means $L_{\overline{\lambda}}(K,T)$ and $L_{\underline{\lambda}}(K,T)$, and finally the self-dual mean $L_{\lambda}(K,T)$. Very similar proofs will give:

THEOREM 4.1.

(1) *The upper mean $L_{\overline{\lambda}}(K,T)$ satisfies the all properties of Theorem 2.9, except properties (8) and (9). Moreover,*

$$L_{\underline{\lambda}}(K,T) \subseteq G_{\underline{\lambda}}(K,T) \subseteq G_{\overline{\lambda}}(K,T) \subseteq L_{\overline{\lambda}}(K,T)$$

for every $K,T \in \mathcal{K}_s^n$ and $0 < \lambda < 1$.

(2) *The self-polar mean $L_{\lambda}(K,T)$ satisfies all properties of Theorem 3.4, except properties (9) and (10).*

Of course, one may also define the upper power function using strips instead of ellipsoids, by taking $K^{\overline{\lambda}} = L_{\overline{\lambda}}(B_2^n, K)$. Under this definition the analogue of Theorem 2.9(8) will trivially hold. The same remark applies to the power K^{λ} and Theorem 3.4(9).

As a final comment, when one works with strips and intervals instead of ellipsoids it is no longer necessary to assume that our bodies are centrally symmetric. Indeed, one may replace the strips $S_{\theta,c}$ with half-spaces

$$H_{\theta,c} = \{x : \langle x, \theta\rangle \leq c\},$$

and arrive at constructions that make sense for all compact convex bodies with the origin in their interior. That way one obtains an exact analogue of Theorem 4.1 that holds for all such bodies.

References

[1] R. Bhatia, *Matrix analysis*, Graduate Texts in Mathematics, vol. 169, Springer-Verlag, New York, 1997. MR1477662

[2] R. Bhatia, *Positive definite matrices*, Princeton Series in Applied Mathematics, Princeton University Press, Princeton, NJ, 2007. MR2284176

[3] K. J. Böröczky, E. Lutwak, D. Yang, and G. Zhang, *The log-Brunn-Minkowski inequality*, Adv. Math. **231** (2012), no. 3-4, 1974–1997, DOI 10.1016/j.aim.2012.07.015. MR2964630

[4] W. J. Firey, *Polar means of convex bodies and a dual to the Brunn-Minkowski theorem*, Canad. J. Math. **13** (1961), 444–453, DOI 10.4153/CJM-1961-037-0. MR0177350

[5] N. Hindman and D. Strauss, *Algebra in the Stone-Čech compactification: Theory and applications*, De Gruyter Textbook, Walter de Gruyter & Co., Berlin, 2012. Second revised and extended edition [of MR1642231]. MR2893605

[6] J. D. Lawson and Y. Lim, *The geometric mean, matrices, metrics, and more*, Amer. Math. Monthly **108** (2001), no. 9, 797–812, DOI 10.2307/2695553. MR1864051

[7] K. Löwner, *Über monotone Matrixfunktionen* (German), Math. Z. **38** (1934), no. 1, 177–216, DOI 10.1007/BF01170633. MR1545446

[8] V. Milman and L. Rotem, *"Irrational" constructions in convex geometry*, Algebra i Analiz **29** (2017), no. 1, 222–236, DOI 10.1090/spmj/1487. Reprinted in St. Petersburg Math. J. **29** (2018), no. 1, 165–175. MR3660690

[9] V. Milman and L. Rotem, *Non-standard constructions in convex geometry: geometric means of convex bodies*, Convexity and concentration, IMA Vol. Math. Appl., vol. 161, Springer, New York, 2017, pp. 361–390. MR3837277

[10] V. Milman and L. Rotem, *Powers and logarithms of convex bodies* (English, with English and French summaries), C. R. Math. Acad. Sci. Paris **355** (2017), no. 9, 981–986, DOI 10.1016/j.crma.2017.09.002. MR3709538

[11] W. Pusz and S. L. Woronowicz, *Functional calculus for sesquilinear forms and the purification map*, Rep. Mathematical Phys. **8** (1975), no. 2, 159–170, DOI 10.1016/0034-4877(75)90061-0. MR0420302

[12] L. Rotem, *Banach limit in convexity and geometric means for convex bodies*, Electron. Res. Announc. Math. Sci. **23** (2016), 41–51, DOI 10.3934/era.2016.23.005. MR3591943

[13] R. Schneider, *Convex bodies: the Brunn-Minkowski theory*, Second expanded edition, Encyclopedia of Mathematics and its Applications, vol. 151, Cambridge University Press, Cambridge, 2014. MR3155183

School of Mathematical Sciences, Tel Aviv University, Tel Aviv 69978, Israel
Email address: milman@post.tau.ac.il

School of Mathematics, University of Minnesota, Minneapolis, MN 55455, USA
Email address: lrotem@umn.edu

Contemporary Mathematics
Volume **733**, 2019
https://doi.org/10.1090/conm/733/14746

Sobolev, Besov and Paley-Wiener vectors in Banach and Hilbert spaces

Isaac Z. Pesenson

Dedicated to the 100th birthday of my teacher S.G. Krein.

Abstract. We consider Banach spaces equipped with a set of strongly continuous bounded semigroups satisfying certain conditions. Using these semigroups we introduce an analog of a modulus of continuity and define analogs of Besov norms. A generalization of a classical interpolation theorem is proven in which the role of Sobolev spaces is played by subspaces defined in terms of infinitesimal operators of these semigroups. We show that our assumptions about a given set of semigroups are satisfied in the case of a strongly continuous bounded representation of a Lie group. In the case of a unitary representation in a Hilbert space we consider an analog of the Laplace operator and use it to define Paley-Wiener vectors. It allows us to develop a generalization of the Shannon-type sampling in Paley-Wiener subspaces and to construct Paley-Wiener nearly Parseval frames in the entire Hilbert space. It is shown that Besov spaces defined previously in terms of the modulus of continuity can be described in terms of approximation by Paley-Wiener vectors and also in terms of the frame coefficients. Throughout the paper we extensively use theory of interpolation and approximation spaces. The paper ends with applications of our results to function spaces on homogeneous manifolds.

1. Introduction and main results

I am honored to have had Selim Grigorievich Krein as my academic advisor and co-author. Without his steady, strong interest in my research and his support I would not have been able to have the privilege of becoming a mathematician. His inspiring encouragement shaped not only my professional trajectory but my life in general.

The first five sections of this paper are devoted to Sobolev and Besov subspaces and to relevant moduli of continuity in Banach spaces. This theory is rooted in results obtained in 70s [**9**, **19**–**21**, **23**]. It is a far going generalization of the one-dimensional theory by J. Lions [**11**] and J. Lions-J. Peetre [**12**], which the author learned from the nice book by P. Butzer and H. Berens [**3**]. The Paley-Wiener vectors and corresponding approximation theory in abstract Hilbert spaces were

2010 *Mathematics Subject Classification.* Primary 43A85, 41A17.
Key words and phrases. Groups and semigroups of operators, Sobolev, Besov, and Paley-Wiener vectors, Interpolation and approximation spaces.
The author thanks Dr. Meyer Pesenson for a number of useful suggestions.

introduced in 80s in [**9**, **22**, **24**, **25**] (see sections 6 and 7 below). In papers [**26**]-[**34**] the notion of Paley-Wiener vectors is used to prove Shannon-type sampling theorems on manifolds and in general Hilbert spaces. The construction of frames and description of Besov subspaces on manifolds and in Hilbert spaces in terms of frame coefficients is rather recent development and can be found in [**7**, **39**]. Some of these results were summarized in [**6**].

We consider a Banach space $\mathbf{E}$ and operators $D_1, D_2, ..., D_d$ which generate strongly continuous uniformly bounded semigroups $T_1(t), T_2(t), ..., T_d(t)$, where $\|T(t)\| \leq 1,\ t \geq 0$. An analog of a Sobolev space is introduced as the space $\mathbf{E}^r$ of vectors in $\mathbf{E}$ for which the following norm is finite

$$|||f|||_{\mathbf{E}^r} = \|f\|_{\mathbf{E}} + \sum_{k=1}^{r} \sum_{1 \leq j_1, ... j_k \leq d} \|D_{j_1}...D_{j_k} f\|_{\mathbf{E}},$$

where $r \in \mathbb{N},\ f \in \mathbf{E}$. By using the closed graph theorem and the fact that each D_i is a closed operator in $\mathbf{E}$, one can show that this norm is equivalent to the norm

$$\|f\|_r = \|f\|_{\mathbf{E}} + \sum_{1 \leq i_1, ..., i_r \leq d} \|D_{i_1}...D_{i_k} f\|_{\mathbf{E}},\quad r \in \mathbb{N}. \tag{1.1}$$

The mixed modulus of continuity is introduced as

$$\Omega^r(s, f) = \sum_{1 \leq j_1, ..., j_r \leq d} \sup_{0 \leq \tau_{j_1} \leq s} ... \sup_{0 \leq \tau_{j_r} \leq s} \|\left(T_{j_1}(\tau_{j_1}) - I\right) ... \left(T_{j_r}(\tau_{j_r}) - I\right) f\|_{\mathbf{E}}, \tag{1.2}$$

where $f \in \mathbf{E}$, $r \in \mathbb{N}$, and I is the identity operator in $\mathbf{E}$. Let $\mathcal{D}(D_i)$ be the domain of the operator D_i. For every $f \in \mathbf{E}$ we introduce a vector-valued function $Tf : \mathbb{R}^d \longmapsto \mathbf{E}$ defined as $Tf(t_1, t_2, ..., t_d) = T_1(t_1)T_2(t_2)...T_d(t_d)f$.

ASSUMPTION 1. We assume that the following properties hold.

(1) The set $\mathbf{E}^1 = \bigcap_{i=1}^{d} \mathcal{D}(D_i)$ is dense in $\mathbf{E}$.
(2) The set $\mathbf{E}^1$ is invariant with respect to all $T_i(t),\ 1 \leq i \leq d,\ t \geq 0$.
(3) For every $1 \leq i \leq d$, every $f \in \mathbf{E}^1$ and all $\mathbf{t} = (t_1, ..., t_d)$ in the standard open unit ball U in $\mathbb{R}^d$

$$D_i Tf(t_1, ..., t_d) = \sum_{k=1}^{d} \zeta_i^k(\mathbf{t}) \left(\partial_k Tf\right)(t_1, ..., t_d), \tag{1.3}$$

where $\zeta_i^k(\mathbf{t})$ belong to $C^\infty(U)$, $\partial_k = \frac{\partial}{\partial t_k}$.

REMARK 1.1. Since $\mathbf{E}^1$ is invariant with respect to all bounded operators $T_i(t_i),\ 1 \leq i \leq d$, we obtain for every $f \in \mathbf{E}^1$

$$T_1(t_1)T_2(t_2)...D_k T_k(t_k)...T_d(t_d)f = \tag{1.4}$$

$$T_1(t_1)T_2(t_2)...T_{k-1}(t_{k-1}) \lim_{s \to 0} \frac{1}{s} \left(T_k(t_k + s) - T_k(t_k)\right) T_{k+1}(t_{k+1})...T_d(t_d)f =$$

$$\lim_{s \to 0} \frac{1}{s} T_1(t_1)T_2(t_2)...T_{k-1}(t_{k-1}) \left(T_k(t_k + s) - T_k(t_k)\right) T_{k+1}(t_{k+1})...T_d(t_d)f =$$

$$\frac{\partial}{\partial t_k} T_1(t_1)T_2(t_2)...T_d(t_d)f = \left(\frac{\partial}{\partial t_k} Tf\right)(t_1, ..., t_d).$$

Thus the formula (1.3) can be rewritten as

$$D_iT_1(t_1)T_2(t_2)...T_d(t_d)f=\sum_{k=1}^{d}\zeta_i^k(\mathbf{t})T_1(t_1)T_2(t_2)...D_kT_k(t_k)...T_d(t_d)f, \tag{1.5}$$

where $\zeta_i^k(\mathbf{t})$ belong to $C^\infty(U)$, $\mathbf{t}=(t_1,...,t_d)\in U$.

REMARK 1.2. When semigroups commute with each other

$$T_i(t_i)T_j(t_j)=T_j(t_j)T_i(t_i),\ 1\le i,j\le d,\ t_i,t_j\ge 0,$$

then $\zeta_i^k=\delta_i^k$.

It is well known that most of remarkable properties of the so-called Besov functional spaces follow from the fact that they are interpolation spaces (see section 4 below) between two Sobolev spaces [**3**, **10**]. For this reason we define Besov spaces by the formula

$$\mathbf{E}^{\alpha,q}=(\mathbf{E},\ \mathbf{E}^r)^K_{\alpha/r,\,q},\ \ 0<\alpha<r\in\mathbb{N},\ 1\le p,\ q\le\infty, \tag{1.6}$$

where K is the so-called Peetre's interpolation functor (see section 4 below). The main result is the following.

THEOREM 1.3. *If Assumption 1 is satisfied, then the following holds true.*

(1) *The functionals $\Omega_r(s,f)$ and $K(s^r,f,\mathbf{E},\mathbf{E}^r)$ are equivalent. Namely, there exist constants $c>0$, $C>0$, such that for all $f\in\mathbf{E}$, $t\ge0$*

$$c\,\Omega^r(s,f)\le K(s^r,f,\mathbf{E},\mathbf{E}^r)\le C\left(\Omega^r(s,f)+\min(s^r,1)\|f\|\right). \tag{1.7}$$

(2) *The norm of the Besov space $\mathbf{E}^{\alpha,q}=(\mathbf{E},\ \mathbf{E}^r)^K_{\alpha/r,q}$, $0<\alpha<r\in\mathbb{N}$, $1\le p,q\le\infty$, is equivalent to the norm*

$$\|f\|_{\mathbf{E}}+\left(\int_0^\infty(s^{-\alpha}\Omega^r(s,f))^q\frac{ds}{s}\right)^{1/q},1\le q<\infty, \tag{1.8}$$

with the usual modifications for $q=\infty$.

(3) *The following isomorphism holds*

$$(\mathbf{E},\ \mathbf{E}^r)^K_{\alpha/r,q}=\left(\mathbf{E}^{k_1},\ \mathbf{E}^{k_2}\right)^K_{(\alpha-k_1)/(k_2-k_1),q},$$

where $0\le k_1<\alpha<k_2\le r\in\mathbb{N}$, $1\le q\le\infty$.

(4) *If α is not integer then the norm (1.8) is equivalent to the norm*

$$\|f\|_{\mathbf{E}^{[\alpha]}}+\sum_{1\le j_1,\dots,j_{[\alpha]}\le d}\left(\int_0^\infty\left(s^{[\alpha]-\alpha}\Omega^1(s,D_{j_1}...D_{j_{[\alpha]}}f)\right)^q\frac{ds}{s}\right)^{1/q} \tag{1.9}$$

where $[\alpha]$ is the integer part of α.

(5) *If $\alpha=k\in\mathbb{N}$ is an integer then the norm (1.8) is equivalent to the norm (Zygmund condition)*

$$\|f\|_{\mathbf{E}^{k-1}}+\sum_{1\le j_1,\dots,j_{k-1}\le d}\left(\int_0^\infty\left(s^{-1}\Omega^2(s,D_{j_1}...D_{j_{k-1}}f)\right)^q\frac{ds}{s}\right)^{1/q}. \tag{1.10}$$

Next we make another assumption.

ASSUMPTION 2. In addition to Assumption 1, we assume that $\mathbf{E}=\mathbf{H}$ is a Hilbert space and the following properties hold.

(1) The operator $L = D_1^2 + ... + D_d^2$ is a non-negative self-adjoint operator in $\mathbf{H}$.
(2) The domain $\mathcal{D}(L^{k/2})$, $k \in \mathbb{N}$, of the non-negative square root $L^{k/2}$ coincides with the space $\mathbf{H}^k$ and the norms (1.1) and $\|f\|_{\mathbf{H}} + \|L^{k/2}f\|_{\mathbf{H}}$ are equivalent.

This assumption allows us to introduce notion of Paley-Wiener vectors (bandlimited vectors) (Definition 6.2) and to prove an abstract version of the Paley-Wiener Theorem (Theorem 6.4).

To formulate an analog of the Shannon-type sampling in abstract Paley-Wiener spaces (Theorem 6.6) we consider the following assumptions.

ASSUMPTION 3. We assume that there exist $C, c > 0$ and $m_0 \geq 0$ such that for any $0 < \rho < 1$ there exists a set of functionals $\mathcal{A}^{(\rho)} = \left\{\mathcal{A}_k^{(\rho)}\right\}_{k \in \mathcal{K}}$, defined on $\mathbf{H}^{m_0}$, for which

$$c \sum_k \left|\mathcal{A}_k^{(\rho)}(f)\right|^2 \leq \|f\|_{\mathbf{H}}^2 \leq C \left(\sum_{k \in \mathcal{K}} \left|\mathcal{A}_k^{(\rho)}(f)\right|^2 + \rho^{2m} \|L^{m/2} f\|_{\mathbf{H}}^2\right), \tag{1.11}$$

for all $f \in \mathbf{H}^m$, $m > m_0$.

This Abstract Sampling Theorem 6.6 is used to construct bandlimited frames in $\mathbf{E}$ (Theorem 6.7). By exploring relations between Interpolation and Approximation spaces we develop an approximation theory by Paley-Wiener vectors in Theorems 7.5 and 7.6. Finally, in Theorem 7.7 we obtain description of Besov spaces in terms of frame coefficients.

In section 2 we show that our Assumptions 1 are satisfied for strongly continuous representations of Lie groups in Banach spaces and Assumptions 2 are satisfied for unitary representations of Lie groups in Hilbert spaces. Concerning Assumptions 3 we note that the results of [**26**]-[**33**] imply that at least when one is considering a so-called regular or quasi-regular representation of a Lie group in a function space on a homogeneous manifold M, some specific sets of Dirac measures (or even more general functionals [**29**]) "uniformly" distributed over M can serve as functionals $\mathcal{A}^{(\rho)} = \left\{\mathcal{A}_k^{(\rho)}\right\}_{k \in \mathcal{K}_\rho}$, where $\rho \in \mathbb{R}_+$ represents a specific "spacing" of these Dirac measures.

It should be noted that due to importance of the theory of function spaces, there is constant interest in extending classical constructions and results from Euclidean to non-Euclidean settings. It is impossible to list even the most significant publications on this subject that appeared in recent years. Here we mention just a very few papers [**5, 15, 17**], which are relevant to our work.

2. Lie groups and their representations

2.1. Lie groups and their representations in Banach spaces. Lie algebra $\mathbf{g}$ of a Lie group G can be identified with the tangent space $T_e(G)$ of G at the identity $e \in G$. Let $\exp(tX) : T_e(G) \to G$, $t \in \mathbb{R}$, $X \in T_e(G)$, be the exponential geodesic map i. e. $\exp(tX) = \gamma(1)$, where $\gamma(t)$ is a geodesic of a fixed left-invariant metric on G which starts at e with the initial vector $tX \in T_e(G)$: $\gamma(0) = e$, $\frac{d\gamma(0)}{dt} = tX$. It is known that exp is an analytic homomorphism of $\mathbb{R}$ onto one parameter subgroup $\exp tX$ of G: $\exp((s+t)X) = \exp(sX)\exp(tX)$, $s, t \in \mathbb{R}$.

Let $X_1, ..., X_d,\ d = dim\, G$ form a basis in the Lie algebra of G, then one can consider the following coordinate system in a neighborhood of identity e

$$(t_1, ..., t_d) \mapsto \exp(t_1 X_1 + ... + t_d X_d). \tag{2.1}$$

If $Y_1 = s_1 X_1 + ... + s_d X_d$ and $Y_2 = t_1 X_1 + ... + t_d X_d$ then $\exp Y_1 \exp Y_2 = \exp Z$, where Z is given by the Campbell-Hausdorff formula

(2.2)

$$Z = Y_1 + Y_2 + \frac{1}{2}[Y_1, Y_2] + \frac{1}{12}[Y_1, [Y_1, Y_2]] - \frac{1}{12}[Y_2, [Y_1, Y_2]] - \frac{1}{24}[Y_2, [Y_1, [Y_1, Y_2]]] + ...$$

It implies that $Z = \zeta_1 X_1 + ... + \zeta_d X_d$, where

$$\zeta_j = s_j + t_j + O(\epsilon^2), \quad |t_j|, |s_j| \leq \epsilon, \quad 1 \leq j \leq d. \tag{2.3}$$

One can also consider another local coordinate system around e which is given by the formula

$$(t_1, ..., t_d) \mapsto \varphi\left(\sum_{j=1}^{d} t_j X_j\right) = \exp(t_1 X_1) ... \exp(t_d X_d). \tag{2.4}$$

Let us recall that a strongly continuous representation of a Lie group G in a Banach space $\mathbf{E}$ is a homomorphism $g \mapsto T(g),\ g \in G,\ T(g) \in GL(\mathbf{E})$, of G into the group $GL(\mathbf{E})$ of linear bounded invertible operators in $\mathbf{E}$ such that trajectory $T(g)f,\ g \in G,\ f \in \mathbf{E}$, is continuous with respect to g for every $f \in \mathbf{E}$. We will consider only uniformly bounded representations. In this case one can introduce a new norm $\|f\|_{\mathbf{E}}^{'} = \sup_{g \in G} \|T(g)f\|_{\mathbf{E}},\ f \in \mathbf{E}$, in which $\|T(g)f\|_{\mathbf{E}}^{'} \leq \|f\|_{\mathbf{E}}^{'}$. Thus, without any restriction we will assume that the last inequality is satisfied in the original norm $\|\cdot\|_{\mathbf{E}}$. Let $X_1, ..., X_d$ be a basis in $\mathbf{g}$. With every $X_j,\ 1 \leq j \leq d$, one associates a strongly continuous one-parameter group of isometries $t \mapsto T(\exp tX_j),\ t \in \mathbb{R}$, whose generator is denoted as $D_j,\ 1 \leq j \leq d$.

LEMMA 2.1. *If $T : G \mapsto GL(\mathbf{E})$ is a strongly continuous bounded representation of G in a Banach space $\mathbf{E}$ and $T_j(t) = T(\exp tX_j)$, where $\{X_1, ..., X_d\}$ is a basis in $\mathbf{g}$ then the Assumption 1 is satisfied for groups T_j and their infinitesimal operators $D_j,\ 1 \leq j \leq d$.*

PROOF. The fact that $\mathbf{E}^1 = \bigcap_{j=1}^{d} \mathcal{D}(D_j)$ is dense in $\mathbf{E}$ and invariant with respect to T is well known [**13, 14**]. Since exp and φ are diffeomorphisms in a neighborhood of zero in $\mathbf{g}$ the map $\exp^{-1} \circ \varphi : \mathbf{g} \mapsto \mathbf{g}$ is also a diffeomorphism. The formulas (2.2) and (2.3) give connection between (2.1) and (2.4)

$$\varphi\left(\sum_{j=1}^{d} t_j X_j\right) = \exp\left(\sum_{k=1}^{d} \alpha_k(t) X_k\right), \quad t = (t_1, , , t_d), \tag{2.5}$$

where

$$\alpha_j(t) = t_j + O(\epsilon^2), \quad |t_j| \leq \epsilon, \quad 1 \leq j \leq d. \tag{2.6}$$

In particular, (2.2) implies $\exp \tau X_j \exp \sum_{i=1}^{d} t_i X_i = \exp \sum_{k=1}^{d} \gamma_k^j(t, \tau) X_k,\quad t = (t_1, ..., t_d)$, where

$$\gamma_k^j(t, \tau) = t_k + \tau \zeta_k^j(t) + \tau^2 R_k^j(t, \tau), \tag{2.7}$$

and $\zeta_k^j(t) = \delta_k^j + Q_k^j(t)$, where δ_k^j is the Kronecker symbol and $Q_k^j(t)$ and $R_k^j(t, \tau)$ are convergent series in $t_1, ..., t_d$ and $t_1, ..., t_d, \tau$ respectively.

Since for $f \in \mathbf{E}^1$ one has $D_jT(g)f = \frac{d}{d\tau}T(\exp \tau X_j)T(g)f|_{\tau=0}$ we obtain for $f \in \mathbf{E}^1$ the following

$$D_jT_1(t_1)...T_d(t_d)f = D_jT\left(\varphi\left(\sum_{i=1}^{d} t_iX_i\right)\right) = D_jT\left(\exp\sum_{i=1}^{d}\alpha_i(t)X_i\right)f =$$

$$\frac{d}{d\tau}T(\exp \tau X_j)T\left(\varphi\left(\sum_{i=1}^{d} t_iX_i\right)\right)f|_{\tau=0} = \frac{d}{d\tau}T\left(\exp\sum_{i=1}^{d}\gamma_i^j(\alpha,\tau)X_i\right)f|_{\tau=0},$$

where $\alpha = (\alpha_1(t), ..., \alpha_d(t))$ and according to (2.7) $\gamma_i^j(\alpha,\tau) = \alpha_i(t) + \tau\zeta_i^j(\alpha(t)) + \tau^2R_i^j(\alpha,\tau)$. By using the Chain Rule and (2.5) we finally obtain the formula (1.3):

$$D_jT_1(t_1)...T_d(t_d)f = \frac{d}{d\tau}T\left(\exp\sum_{i=1}^{d}\gamma_i^j(\alpha,\tau)X_i\right)f|_{\tau=0} =$$

$$\sum_{k=1}^{d}\left(\frac{d}{d\tau}\gamma_k^j(\alpha,\tau)|_{\tau=0}\right)\partial_kT\left(\exp\sum_{i=1}^{d}\gamma_i^j(\alpha,\tau)X_i\right)f|_{\tau=0}$$

$$= \sum_{k=1}^{d}\zeta_k^j(t)\partial_kT_1(t_1)...T_d(t_d)f.$$

Lemma is proved. □

2.2. Unitary representations in Hilbert spaces. A strongly continuous unitary representation of a Lie group G in a Hilbert space $\mathbf{H}$ is a homomorphism $T : G \mapsto U(\mathbf{H})$ where $U(\mathbf{E})$ is the group of unitary operators of $\mathbf{H}$ such that $T(g)f,\ g \in G$, is continuous on G for any $f \in \mathbf{H}$. The Gårding space $\mathcal{G}$ is defined as the set of vectors h in $\mathbf{H}$ that have the representation $h = \int_G \varphi(g)T(g)fdg$, where $f \in \mathbf{H},\ \ \varphi \in C_0^\infty(G),\ \ dg$ is a left-invariant measure on G. If $X \in \mathbf{g}$ is identified with a right-invariant vector field

$$X\varphi(g) = \lim_{t\to 0}\frac{\varphi(\exp tX \cdot g) - \varphi(g)}{t},$$

then one has a representation $D(X)$ of $\mathbf{g}$ by operators which act on $\mathcal{G}$ by the formula $D(X)h = -\int_G X\varphi(g)T(g)fdg$. It is known that $\mathcal{G} \subset \bigcap_{r\in N}\mathbf{H}^r = \mathbf{H}^\infty$ is invariant with respect to all operators $D(X),\ X \in \mathbf{g}$, and dense in every $\mathbf{H}^r$. If $X_1, ..., X_d$ is a basis in $\mathbf{g}$ and $D_i = D(X_i),\ 1 \le i \le d$, we consider the operator $L_\mathcal{G} = -\sum_{i=1}^d D_i^2$ defined on $\mathcal{G}$.

Since $L_\mathcal{G}$ is symmetric and the differential operator $-\sum_{i=1}^d X_i^2$ is elliptic on the group G the Theorem 2.2 in [**14**] implies that $L_\mathcal{G}$ is essentially self-adjoint, which means $\overline{L}_\mathcal{G} = L_\mathcal{G}^*$. In other words, the closure $\overline{L}_\mathcal{G} = L$ of $L_\mathcal{G}$ from $\mathcal{G}$ is a self-adjoint operator. Obviously, $L \ge 0$. We introduce the self-adjoint operator $\Lambda = I + L \ge 0$.

THEOREM 2.2. *The space* $\mathbf{H}^r$ *with the norm* (1.1) *is isomorphic to the domain of* $\Lambda^{r/2}$ *with the norm* $\|\Lambda^{r/2}f\|_\mathbf{H}$.

PROOF. In the case $r = 2k$, the inequality

$$\|f\|_{\mathbf{H}^{2k}} \le C(k)\|\Lambda^k f\|_\mathbf{H} \tag{2.8}$$

is shown in [**13**], Lemma 6.3. The reverse inequality is obvious. We consider now the case $r = 2k+1$. If $f \in \mathbf{H}^2 = \mathcal{D}(\Lambda)$, then since $\mathcal{D}(\Lambda) \subset \mathcal{D}(\Lambda^{1/2})$ we have

(2.9)

$$\|f\|^2_{\mathbf{H}} + \sum_j \|D_j f\|^2_{\mathbf{H}} = \langle f, f\rangle + \sum_j \langle D_j f, D_j f\rangle = \langle f, f\rangle + \left\langle -\sum_j D_j^2 f, f\right\rangle =$$
$$\left\langle f - \sum_j D_j^2 f, f\right\rangle = \langle \Lambda f, f\rangle = \|\Lambda^{1/2} f\|^2_{\mathbf{H}}.$$

These equalities imply that $\mathbf{H}^1$ is isomorphic to $\mathcal{D}(\Lambda^{1/2})$. Our goal is to prove existence of an isomorphism between $\mathbf{H}^{2k+1}$ and $\mathcal{D}(\Lambda^{k+1/2})$. It is enough to establish equivalence of the corresponding norms on the set $\mathbf{H}^{4k+2} = \mathcal{D}(\Lambda^{2k+1})$ since the latest is dense in $\mathbf{H}^{2k+1}$. If $f \in \mathbf{H}^{4k+2} \subset \mathbf{H}^{2k}$ then $D_j f \in \mathbf{H}^{4k+1} \subset \mathbf{H}^{2k}$ and $\Lambda^k f = \sum_{m\le k} \sum D^2_{j_1} ... D^2_{j_m} f$. Thus if $f \in \mathbf{H}^{4k+2}$ then

$$\left\|D_{j_1}...D_{j_{2k+1}} f\right\|_{\mathbf{H}} \le C \left\|\Lambda^k D_{j_{2k+1}} f\right\|_{\mathbf{H}} = \left\|\sum_{m\le k}\sum D^2_{j_1}...D^2_{j_m} D_{j_{2k+1}} f\right\|_{\mathbf{H}}.$$

Multiple applications of the identity $D_iD_j - D_jD_i = \sum_k c^k_{i,j} D_k$ which holds on $\mathbf{H}^2$ lead to the inequality $\left\|D_{j_1}...D_{j_{2k+1}} f\right\|_{\mathbf{H}} \le C\left(\|D_{j_{2k+1}}\Lambda^k f\|_{\mathbf{H}} + \|Rf\|_{\mathbf{H}}\right)$, where R is a polynomial in $D_1, ..., D_d$ whose degree $\le 2k$. According to (2.8) and (2.9) we have that

$$\left\|D_{j_{2k+1}}\Lambda^k f\right\|_{\mathbf{H}} \le \left\|\Lambda^{1/2}\Lambda^k f\right\|_{\mathbf{H}} = \left\|\Lambda^{k+1/2} f\right\|_{\mathbf{H}}$$

and also $\|Rf\|_{\mathbf{H}} \le \|f\|_{\mathbf{H}^{2k}} \le C(k)\left\|\Lambda^k f\right\|_{\mathbf{H}}$. Since $\|\Lambda^k f\|_{\mathbf{H}}$ is not decreasing with k we get the following estimate

$$\|D_{j_1}...D_{j_{2k+1}} f\|_{\mathbf{H}} \le C(k)\|\Lambda^{k+1/2} f\|_{\mathbf{H}}, \quad f \in \mathbf{H}^{4k+2}.$$

Now, since for $f \in \mathbf{H}^{4k+2}$ we have $D_{j_1}...D_{j_{2k}} f \in \mathbf{H}^{2k+2} \subset \mathbf{H}^1 = \mathcal{D}(\Lambda^{1/2})$, and the equality $\Lambda^k f = \sum_{m\le k}\sum D^2_{j_1}...D^2_{j_m} f$, holds we obtain, by using (2.9)

$$\|\Lambda^{k+1/2} f\|_{\mathbf{H}} = \|\Lambda^{1/2}\sum_{m\le k}\sum D^2_{j_1}...D^2_{j_m} f\|_{\mathbf{H}} \le C\|f\|_{\mathbf{H}^{2k+1}}, \quad C = C(k).$$

Theorem is proved. □

COROLLARY 2.1. *If T is a strongly continuous unitary representation of a Lie group in a Hilbert space $\mathbf{H}$ and $X_1, ..., X_d$ is a basis in the corresponding algebra Lie $\mathbf{g}$ then for $T_j(t) = T(\exp tX_j)$, $1 \le j \le d$, and their generators D_j, $1 \le j \le d$, the Assumption 2 is satisfied.*

3. Hardy-Steklov operator associated with operators $D_1, D_2, ..., D_d$.

We return to the general set up. Namely, we consider a Banach space $\mathbf{E}$ and operators $D_1, D_2, ..., D_d$ that generate strongly continuous uniformly bounded semigroups $T_1(t), T_2(t), ..., T_d(t)$, $\|T(t)\| \le 1$, $t \ge 0$. It is assumed that the Assumption 1 holds.

3.1. Preliminaries. If D generates in $\mathbf{E}$ a strongly continuous bounded semigroup $T_D(t),\ t \geq 0$, and

$$\Omega^r(s,f) = \sup_{0\leq\tau\leq s} \| (T_D(\tau) - I)^r f\|_{\mathbf{E}},$$

where I is the identity operator, then one can easily prove the following two inequalities

$$\Omega^m(f,s) \leq s^k \Omega^{m-k}(D^k f, s), \tag{3.1}$$

and

$$\Omega^m(f,as) \leq (1+a)^m \Omega^m(f,s), \quad a \in \mathbb{R}. \tag{3.2}$$

Let $F(x_1, x_2, ..., x_N)$ be a function on $\mathbb{R}^N$ that takes values in the Banach space $\mathbf{E}$

$$F : \mathbb{R}^N \longmapsto \mathbf{E}.$$

For $1 \leq i \leq N$ we introduce difference operator by the formula

$$(\Delta_i(s)F)(x_1, x_2, ..., x_N) = F(x_1, x_2, ..., x_{i-1}, s, x_{i+1}, ..., x_N) - \\ F(x_1, x_2, ..., x_{i-1}, 0, x_{i+1}, ..., x_N). \tag{3.3}$$

For a scalar differentiable function $\theta : \mathbb{R}^N \longmapsto \mathbb{R}$ and for $1 \leq i_1, ..., i_l \leq N,\ i_m \neq i_k$, the following inequality holds

$$\max_{0\leq x_j\leq s_j} |\Delta_{i_1}(s_{i_1})...\Delta_{i_l}(s_{i_l})\theta(x_1,...,x_N)| \leq s_{i_1}...s_{i_l} \max_{0\leq x_j\leq s_j} \left| \frac{\partial^k}{\partial_{i_1}...\partial_{i_k}} \theta(x_1,...,x_N) \right|.$$

One has for $1 \leq i \leq N$

$$\Delta_i(s)(\theta F)(x_1,...,x_N) = \Delta_i(s)\theta(x_1,...,x_N)F(x_1,...,x_{i-1},s,x_{i+1},...,x_N) + \\ \theta(x_1,...,x_{i-1},0,x_{i+1},...,x_N)\Delta_i(s)F(x_1,...,x_N), \tag{3.4}$$

and then if $i_m \neq i_k$ one has

$$\Delta_{i_1}(s_{i_1})...\Delta_{i_l}(s_{i_l})(\theta F)(x_1,...,x_N) = \\ \sum \Delta_{i_1^{(1)}}(s_{i_1^{(1)}})...\Delta_{i_l^{(1)}}(s_{i_l^{(1)}})\theta(x^{(0)}_{i_1^{(2)},...,i_l^{(2)}})\Delta_{i_1^{(2)}}(s_{i_1^{(2)}})...\Delta_{i_l^{(2)}}(s_{i_l^{(2)}})F(x^{(s)}_{i_1^{(1)},...,i_l^{(1)}}), \tag{3.5}$$

where the sum is taken over all possible partitions of the natural vector $(i_1, ..., i_l)$ in the sum of nonnegative integer vectors $(i_1^{(1)}, ..., i_l^{(1)})$ and $(i_1^{(2)}, ..., i_l^{(2)})$. The $x^{(0)}_{i_1^{(2)},...,i_l^{(2)}}$ denotes a vector obtained from the vector $(x_1, ..., x_N)$ by replacing the coordinates $x_{i_1^{(2)},...,i_l^{(2)}}$ by 0 and $x^{(s)}_{i_1^{(1)},...,i_l^{(1)}}$, denotes the vector obtained by replacing $i_1^{(1)}, ..., i_l^{(1)}$ by $s_{i_1^{(1)}}, ..., s_{i_l^{(1)}}$, respectively.

If F is a differentiable function and $m \neq i$ then we have

$$\Delta_i(s)\frac{\partial}{\partial x_m}F(x_1, x_2, ..., x_N) = \frac{\partial}{\partial x_m}\Delta_i(s)F(x_1, x_2, ..., x_N), \quad m \neq i. \tag{3.6}$$

In particular, for any $r \geq 2$ for $\tau_j = \tau_{j,1} + ... + \tau_{j,r},\ j = 1, ..., d,\ 1 \leq i, i' \leq d,\ 1 \leq k, k' \leq r$, if $(i', k') \neq (i, k)$ we have

$$\frac{\partial}{\partial \tau_{i',k'}}\Delta_{i,k}(s)\prod_{j=1}^{d} T_j(\tau_j) f = \Delta_{i,k}(s)\frac{\partial}{\partial \tau_{i',k'}}\prod_{j=1}^{d} T_j(\tau_j) f, \quad f \in \mathbf{E}^1,$$

or, equivalently,

$$\frac{\partial}{\partial \tau_{i'}} \Delta_{i,k}(s) \prod_{j=1}^{d} T_j\left(\tau_j\right) f = \Delta_{i,k}(s) \frac{\partial}{\partial \tau_{i'}} \prod_{j=1}^{d} T_j\left(\tau_j\right) f, \quad f \in \mathbf{E}^1,$$

where the right-hand side and $\Delta_{i,k}(s) \prod_{j=1}^{d} T_j\left(\tau_j\right) f$ are defined according to (3.3). In the same notations $\tau_j = \tau_{j,1} + ... + \tau_{j,r}$, $1 \leq m, i \leq d$, $1 \leq k \leq r$, we clearly have the identity

$$D_m \Delta_{i,k}(s) \prod_{j=1}^{d} T_j\left(\tau_j\right) f = \Delta_{i,k}(s) D_m \prod_{j=1}^{d} T_j\left(\tau_j\right) f, \quad f \in \mathbf{E}^1,$$

which implies for every $f \in \mathbf{E}^1$ the next formula

$$D_m \Delta_{i_1,k_1}(s)..\Delta_{i_l,k_l}(s) \prod_{j=1}^{d} T_j\left(\tau_j\right) f = \Delta_{i_1,k_1}(s)..\Delta_{i_l,k_l}(s) D_m \prod_{j=1}^{d} T_j\left(\tau_j\right) f. \tag{3.7}$$

Using (3.7), (1.3), (3.5) and (3.6) we obtain for $f \in \mathbf{E}^1$, $\tau_j = \tau_{j,1} + ... + \tau_{j,r}$, $j = 1, ..., d$, $r \geq l+1$, $1 \leq m \leq d$,

$$D_m \Delta_{i_1,k_1}(s) ... \Delta_{i_l,k_l}(s) \prod_{j=1}^{d} T_j\left(\tau_j\right) f = \tag{3.8}$$

$$\sum \sum_{i_{l+1}=1}^{d} \Delta_{i_1^{(1)},k_1^{(1)}}(s) ... \Delta_{i_l^{(1)},k_l^{(1)}}(s) \zeta_m^{i_{l+1}}(...) \frac{\partial}{\partial \tau_{i_{l+1}}} \Delta_{i_1^{(2)},k_1^{(2)}}(s) ... \Delta_{i_l^{(2)},k_l^{(2)}}(s) \prod_{j=1}^{d} T_j\left(...\right) f,$$

where outer summation $\sum$ and arguments of $\zeta_m^{i_{l+1}}(...)$ and $\prod_{j=1}^{d} T_j\left(...\right) f$ the same as in (3.5).

3.2. The Hardy-Steklov operator. We introduce a generalization of the classical Hardy-Steklov operator. For a positive small s, natural r and $1 \leq j \leq d$ we set

$$H_{j.r}(s) f = (s/r)^{-r} \int_0^{s/r} ... \int_0^{s/r} \sum_{k=1}^{r} (-1)^k C_r^k T_j(k(\tau_{j,1} + ... + \tau_{j,r}) f d\tau_{j,1} ... d\tau_{j,r},$$

where C_r^k are the binomial coefficients and then define the Hardy-Steklov operator: $H_r(s) f = \prod_{j=1}^{d} H_{j,r}(s) f = H_{1,r}(s) H_{2,r}(s) ... H_{d,r}(s) f$, $f \in \mathbf{E}$. For every fixed $f \in \mathbf{E}$ the function $H_r(s) f$ is an abstract valued function from $\mathbb{R}$ to $\mathbf{E}$ and it is a linear combination of some abstract valued functions of the form

$$(s/r)^{-rd} \int_0^{s/r} ... \int_0^{s/r} T f(\tau) d\tau_{1,1} ... d\tau_{d,r}, \tag{3.9}$$

where $\tau = (k_1 \tau_1, k_2 \tau_2, ..., k_d \tau_d)$, $1 \leq k_j \leq r$, $\tau_j = (\tau_{j,1} + \tau_{j,2} + ... + \tau_{j,r})$, $1 \leq j \leq d$, and

$$T f(\tau) = T_1(k_1 \tau_1) T_2(k_2 \tau_2) ... T_d(k_d \tau_d) f. \tag{3.10}$$

LEMMA 3.1. *The following holds:*

(1) *For every $f \in \mathbf{E}$ the function $H_r(s) f$ maps $\mathbb{R}$ to $\mathbf{E}^r$.*

(2) *For every* $0 \le q \le r$ *the "mixed derivative"* $D_{j_1}...D_{j_q}H_r(s)f, \ \ 1 \le j_k \le d$ *is another abstract valued function with values in* $\mathbf{E}^{r-q}$ *and it is a linear combination of abstract valued functions (with values in* $\mathbf{E}^r$*) of the form*

$$(s/r)^{-rd} \underbrace{\int_0^{s/r} ..... \int_0^{s/r}}_{rd-m} \mu_{j_1,...,j_q}^{i_1,...,i_l;m}(\cdot)\Delta_{i_1,k_1}(s/r)....\Delta_{i_l,k_l}(s/r)Tf(\cdot)d\cdot, \tag{3.11}$$

where

$$\max_{0 \le \tau_{i,j} \le s} |\partial^p \mu_{j_1,...,j_q}^{i_1,...,i_l;m}(\cdot)| \le cs^{m-l}, \quad p \in \mathbb{N} \cup \{0\}, \tag{3.12}$$

and $0 < s < 1, \ \ 0 \le m \le rd, \ \ 0 \le l \le m,$ *where* $l = 0$ *corresponds to the case when the set of indices* $\{i_1, ..., i_l\}$ *is empty.*

PROOF. The proof proceeds by induction in q. We shall show that $H_r(s)f$ is in $\mathbf{E}^1$ for every $f \in \mathbf{E}$. Let's assume first that $f \in \mathbf{E}^1$. Since every $D_j, \ 1 \le j \le d$, is a closed operator to show that the term (3.9) belongs to $\mathcal{D}(D_j)$ it is sufficient to show existence of the integral (see notations (5.1), (5.2))

$$\left(\frac{s}{r}\right)^{-rd} \int_0^{s/r} ... \int_0^{s/r} D_jTf(\tau)d\tau_{1,1}...d\tau_{d,r}. \tag{3.13}$$

According to (1.3) the last integral equals to

$$\sum_{i=1}^{d} \left(\frac{s}{r}\right)^{-rd} \int_0^{s/r} ... \int_0^{s/r} \zeta_j^i(\tau)\partial_i Tf(k_1\tau_1, ..., k_d\tau_d)d\tau_{1,1}...d\tau_{d,r}, \tag{3.14}$$

where τ and τ_j described in (5.1) and (5.2). Since derivative ∂_i is the same as the derivative $\frac{\partial}{\partial \tau_{i_1,1}}$ the integration by parts formula and (3.4) allow to continue (3.14) as follows

$$(s/r)^{-rd} \sum_{i=1}^{d} \left(\underbrace{\int_0^{s/r} ... \int_0^{s/r}}_{rd-1} \zeta_j^i(\tau_i^{(s/r)})\Delta_i(s/r)Tf(\tau)(d\tau)_i \right) + \tag{3.15}$$

$$(s/r)^{-rd} \sum_{i=1}^{d} \left(\underbrace{\int_0^{s/r} ... \int_0^{s/r}}_{rd-1} (\Delta_i(s/r)\zeta_j^i(\tau))Tf(\tau_i^{(0)})(d\tau)_i \right) -$$

$$(s/r)^{-rd} \sum_{i=1}^{d} \left(\underbrace{\int_0^{s/r} ... \int_0^{s/r}}_{rd} (\partial_i\zeta_j^i)(\tau)Tf(\tau)d\tau \right) = B_j(s)f,$$

where $\tau = (\tau_{1,1} + ... + \tau_{1,r}, \ ..., \tau_{n,1} + ... + \tau_{n,r}), \ \ \tau_i^{s/r} = (\tau_{1,1} + ... + \tau_{1,r}, \ ..., \tau_{i,1} + ... + \tau_{i,r-1} + s/r, ..., \tau_{n,1} + ... + \tau_{n,r}), \ \ \tau_i^0 = (\tau_{1,1} + ... + \tau_{1,r}, \ ..., \tau_{i,1} + ... + \tau_{i,r-1} + 0, ..., \tau_{n,1} + ... + \tau_{n,r}), \ \ d\tau = d\tau_{1,1}....d\tau_{n,r}$, and $(d\tau)_i = d\tau_{1,1}....d\tau_{i,r-1}\widehat{d\tau_{i,r}}d\tau_{i+1,1}...d\tau_{n,r}$, where the term $d\tau_{i,r}$ is missing.

Since integrand of each of the three integrals is bounded it implies existence of (3.13) which, in turn, shows that (3.9) is an element of $\mathcal{D}(D_j)$ for every $1 \le j \le d$.

Thus if f belongs to $\mathbf{E}^1$ then $H_r(s)f$ takes values in $\mathbf{E}^1$ and the formula $D_jH_r(s)f = B_j(s)f$ holds where the operator $B_m(s)$ is bounded. This fact along with the fact that $\mathbf{E}^1$ is dense in $\mathbf{E}$ implies the formula $D_jH_r(s)f = B_j(s)f$ for every $f \in \mathbf{E}$. Thus we proved the first part of the Lemma for $q = 1$.

To verify the second claim of the Lemma it is sufficient to note that in the first line of (3.15) one has $m = l = q = 1,\ j = j_1,\ \mu_{j_1}^{1,1} = \zeta_j^i$, in the second line $m = 1,\ l = 0,\ q = 1,\ j = j_1,\ \mu_{j_1}^{0,1} = \Delta_i(s/r)\zeta_j^i$, in the third line $m = l = 0,\ q = 1,\ \mu_{j_1}^{0,0} = \partial_i\zeta_j^i$. It is easy to verify that in all of these cases the estimates (3.12) hold.

We now assume that theorem is proved for all $q < r$ and we shall prove it for $q + 1 \leq r$. Suppose $f \in \mathbf{E}^1$. We consider

$$\int_0^{s/r} \dots\dots \int_0^{s/r} \mu_{j_1,\dots,j_q}^{i_1,\dots,i_l;m}(\cdot) D_{j_{q+1}} \Delta_{i_1,k_1}(s/r) \dots\dots \Delta_{i_l,k_l}(s/r) T(\cdot) f d\cdot\,. \tag{3.16}$$

By (3.8) this expression is decomposed into a sum of terms of the form

$$\int_0^{s/r} \dots \int_0^{s/r} \nu_{j_1,\dots,j_{q+1}}^{i_1,\dots,i_{l+1};m}(\cdot) \partial_{i_{l+1}} \Delta_{i_1^{(2)},k_1^{(2)}}(s/r) \dots \Delta_{i_l^{(2)},k_l^{(2)}}(s/r) T(\tau) f d\tau, \tag{3.17}$$

where $\nu_{j_1,\dots,j_{q+1}}^{i_1,\dots,i_{l+1};m}$ are constructed from $\mu_{j_1,\dots,j_q}^{i_1,\dots,i_l;m}$ in accordance with (3.8). Integrating by parts as in the first step of induction, we obtain that the integral (3.17) exists and is equal to the value of some bounded operator at an element $f \in \mathbf{E}^1$. As a result we obtain that $H_r(s)$ belongs to the space $\mathbf{E}^{q+1}$ and also that $D_{j_1} \dots D_{j_{q+1}} H_r(s) f$ is a linear combination of terms of the form (3.11) . The rest of the theorem fellows from the induction hypothesis. Theorem is proved. □

4. Interpolation spaces

The goal of the section is to introduce basic notions of the theory of interpolation spaces [**1**, **3**, **10**]. Later in section 7 we will also introduce the so-called approximation spaces [**4**, **18**]. It is important to realize that the relations between interpolation and approximation spaces cannot be described in the language of normed spaces. We have to make use of quasi-normed linear spaces in order to treat them simultaneously.

A quasi-norm $\|\cdot\|_{\mathbf{E}}$ on linear space $\mathbf{E}$ is a real-valued function on $\mathbf{E}$ such that for any $f, f_1, f_2 \in \mathbf{E}$ the following holds true: (1)$\|f\|_{\mathbf{E}} \geq 0$; (2) $\|f\|_{\mathbf{E}} = 0 \Longleftrightarrow f = 0$; (3)$\|-f\|_{\mathbf{E}} = \|f\|_{\mathbf{E}}$; (4) there exists some $C_{\mathbf{E}} \geq 1$ such that $\|f_1 + f_2\|_{\mathbf{E}} \leq C_{\mathbf{E}}(\|f_1\|_{\mathbf{E}} + \|f_2\|_{\mathbf{E}})$. Two quasi-normed linear spaces $\mathbf{E}$ and $\mathbf{F}$ form a pair if they are linear subspaces of a common linear space $\mathbf{A}$ and the conditions $\|f_k - g\|_{\mathbf{E}} \to 0$, and $\|f_k - h\|_{\mathbf{F}} \to 0$ imply equality $g = h$ (in $\mathbf{A}$). For any such pair $\mathbf{E}, \mathbf{F}$ one can construct the space $\mathbf{E} \cap \mathbf{F}$ with quasi-norm $\|f\|_{\mathbf{E}\cap\mathbf{F}} = \max\left(\|f\|_{\mathbf{E}}, \|f\|_{\mathbf{F}}\right)$ and the sum of the spaces, $\mathbf{E} + \mathbf{F}$ consisting of all sums $f_0 + f_1$ with $f_0 \in \mathbf{E}, f_1 \in \mathbf{F}$, and endowed with the quasi-norm $\|f\|_{\mathbf{E}+\mathbf{F}} = \inf_{f=f_0+f_1, f_0\in\mathbf{E}, f_1\in\mathbf{F}}\left(\|f_0\|_{\mathbf{E}} + \|f_1\|_{\mathbf{F}}\right)$.

Quasi-normed spaces $\mathbf{H}$ with $\mathbf{E} \cap \mathbf{F} \subset \mathbf{H} \subset \mathbf{E} + \mathbf{F}$ are called intermediate between $\mathbf{E}$ and $\mathbf{F}$. If both E and F are complete the inclusion mappings are automatically continuous. An additive homomorphism $T : \mathbf{E} \to \mathbf{F}$ is called bounded if $\|T\| = \sup_{f\in\mathbf{E}, f\neq 0} \|Tf\|_{\mathbf{F}}/\|f\|_{\mathbf{E}} < \infty$. An intermediate quasi-normed linear space $\mathbf{H}$ interpolates between $\mathbf{E}$ and $\mathbf{F}$ if every bounded homomorphism $T : \mathbf{E}+\mathbf{F} \to \mathbf{E}+\mathbf{F}$

which is a bounded homomorphism of $\mathbf{E}$ into $\mathbf{E}$ and a bounded homomorphism of $\mathbf{F}$ into $\mathbf{F}$ is also a bounded homomorphism of $\mathbf{H}$ into $\mathbf{H}$. On $\mathbf{E}+\mathbf{F}$ one considers the so-called Peetre's K-functional

$$K(f,t)=K(f,t,\mathbf{E},\mathbf{F})=\inf_{f=f_0+f_1,f_0\in\mathbf{E},f_1\in\mathbf{F}}\left(\|f_0\|_{\mathbf{E}}+t\|f_1\|_{\mathbf{F}}\right).$$

The quasi-normed linear space $(\mathbf{E},\mathbf{F})^K_{\theta,q}$, with parameters $0<\theta<1$, $0<q\leq\infty$, or $0\leq\theta\leq 1$, $q=\infty$, is introduced as the set of elements f in $\mathbf{E}+\mathbf{F}$ for which

$$\|f\|_{\theta,q}=\left(\int_0^\infty\left(t^{-\theta}K(f,t)\right)^q\frac{dt}{t}\right)^{1/q}<\infty. \tag{4.1}$$

It turns out that $(\mathbf{E},\mathbf{F})^K_{\theta,q}$ with the quasi-norm (4.1) interpolates between $\mathbf{E}$ and $\mathbf{F}$.

5. Approximation by Hardy-Steklov averages, K-functor and modulus of continuity

We are going to prove items (1) and (2) of Theorem 1.3.

PROOF. First we prove the right-hand side of the inequality (1.7). The following simple lemma plays an important role in the roof.

LEMMA 5.1. *In any ring $\mathcal{R}$ with multiplicative identity 1 the following formulas hold for $a_1,a_2,....,a_n\in\mathcal{R}$,*

$$a_1a_2...a_n-1=a_1(a_2-1)+...+a_1a_2...a_{n-1}(a_n-1), \tag{5.1}$$

$$(a_1-1)a_2...a_n=(a_1-1)+a_1(a_2-1)+...+a_1a_2...a_{n-1}(a_n-1). \tag{5.2}$$

By using the first formula we obtain

$$\left\|(-1)^{n(r+1)}H_r(s)f-f\right\|_{\mathbf{E}}=\left\|\prod_{j=1}^n(-1)^{n+1}H_{j,r}(s)f-f\right\|_{\mathbf{E}}\leq \tag{5.3}$$

$$c\sum_{j=1}^n\sup_{0\leq\tau_j\leq s/r}\|(T_j(\tau_j)-I)^rf\|_{\mathbf{E}}\leq C\Omega^r(s,f).$$

To estimate $s^r\|H_r(s)f\|_{\mathbf{E}}$ we note that $\|H_r(s)f\|_{\mathbf{E}}\leq C\|f\|_{\mathbf{E}}$. According to Lemma 3.1 the quantity $\|D_{j_1}...D_{j_r}H_r(s)f\|_{\mathbf{E}}$ is estimated for $0\leq s\leq 1$ by

$$s^{-l}\sup_{0\leq\tau_{j,k}\leq s}\|\Delta_{j_1,k_1}(s/r)...\Delta_{j_l,k_l}(s/r)T(\cdot)f\|_{\mathbf{E}}, \tag{5.4}$$

where $T(\cdot)=\prod_{j=1}^n\prod_{k=1}^rT_j(\tau_{j,k})$. By the definition of $\Delta_{j,k}(s/r)$ the expression $\Delta_{j,k}(s/r)T(\cdot)$ differs from $T(\cdot)$ only in that in place of the factor $T_j(\tau_{j,k})$ the factor $T_j(s/r)-I$ appears. Multiple applications of the identity (5.2) to the operator $\Delta_{j_1,k_1}(s/r)...\Delta_{j_l,k_l}(s/r)T(\cdot)$ allow its expansion into a sum of operators each of which is a product of not less than $l\leq r$ of operators $T_i(\sigma_i)-I$, $\sigma_i\in(0,s/r)$, $1\leq i\leq n$. Consequently, (5.4) is dominated by a multiple of $s^{-l}\Omega^l(s,f)$. By summing the estimates obtained above we arrive at the inequality

$$K(s^r,f,\mathbf{E},\mathbf{E}^r)\leq C\left(\sum_{l=1}^n s^{r-l}\Omega^l(s,f)+s^r\|f\|_{\mathbf{E}}\right),\quad 0\leq s\leq 1. \tag{5.5}$$

Note, that by repeating the known proof for the classical modulus of continuity one can prove the inequality

$$\Omega^l(s,f) \leq C\left(s^l\|f\|_{\mathbf{E}} + s^l\int_s^1 \sigma^{-1-l}\Omega^{k+r}(\sigma,f)d\sigma\right),$$

which implies $s^{r-l}\Omega^l(s,f) \leq C\left(s^r\|f\|_{\mathbf{E}} + \Omega^r(s,f)\right)$. By applying this inequality to (5.5) and taking into account the inequality $K(s^r,f,\mathbf{E},\mathbf{E}^r) \leq \|f\|$, we obtain the right-hand side of the estimate (1.7). To prove the left-hand side of (1.7) we first notice that the following inequality holds

$$\Omega^r(s,g) \leq Cs^k \sum_{1\leq j_1,\ldots j_k\leq n} \Omega^{r-k}\left(s, D_{j_1}...D_{j_k}g\right),\ g\in\mathbf{E}^k,\ C=C(k,r),\ k\leq r, \tag{5.6}$$

which is an easy consequence of the identity (5.2) and the identity $(T_j(t)-I)\,g = \int_0^t T_j(\tau)D_jg d\tau,\ g\in\mathcal{D}(D_j)$. From here, for any $f\in\mathbf{E},\ g\in\mathbf{E}^r$ we obtain $\Omega^r(s,f) \leq \Omega^r(s,f-g)+\Omega^r(s,g) \leq C\left(\|f-g\|_{\mathbf{E}} + s^r\|g\|_{\mathbf{E}^r}\right)$. The first item of Theorem 1.3 is proven and it obviously implies second item of the same Theorem. □

REMARK 5.2. The proof shows that the left-hand side of (1.7) holds true for any finite set of one-parameter strongly continuous bounded semigroups.

Below is the proof of items (3)-(5) of Theorem 1.3.

PROOF. We will need the following lemma.

LEMMA 5.3. *The following inequalities hold*

$$\|f\|_{\mathbf{E}^k} \leq C\|f\|_{\mathbf{E}}^{1-k/r}\|f\|_{\mathbf{E}^r}^{k/r},\quad f\in\mathbf{E}^r,\quad C=C(k,r), \tag{5.7}$$

$$K(s^r,f,\mathbf{E},\mathbf{E}^r) \leq Cs^k\|f\|_{\mathbf{E}^k},\quad f\in\mathbf{E}^k,\quad C=C(k,r). \tag{5.8}$$

PROOF. The first inequality follows from its well-known one-dimensional version (see also [**33**]). The second one follows from the right-hand estimate of (1.7) and (5.6). □

This lemma shows that one can use the Reiteration Theorem (see [**3, 10**]), which immediately implies item (3) of Theorem 1.3. Next, let $\alpha>0,\ [\alpha]$ be a non-integer and its integer part respectively. According to item (3) of Theorem 1.3 we have $(\mathbf{E},\mathbf{E}^r)^K_{\alpha/r,q} = \left(\mathbf{E}^{[\alpha]},\mathbf{E}^r\right)^K_{(\alpha-[\alpha])/(r-[\alpha]),q}$ and $\left(\mathbf{E},\mathbf{E}^1\right)^K_{\alpha-[\alpha],q} = \left(\mathbf{E},\mathbf{E}^{r-[\alpha]}\right)^K_{(\alpha-[\alpha])/(r-[\alpha]),q}$. Note, that $D_{j_1}D_{j_2}...D_{j_{[\alpha]}}$ is a continuous map from $\left(\mathbf{E}^{[\alpha]},\mathbf{E}^r\right)^K_{(\alpha-[\alpha])/(r-[\alpha]),q}$ to $\left(\mathbf{E},\mathbf{E}^{r-[\alpha]}\right)^K_{(\alpha-[\alpha])/(r-[\alpha]),q}$. All together it shows that if $f\in(\mathbf{E},\mathbf{E}^r)^K_{\alpha/r,q}$ then $D_{j_1}D_{j_2}...D_{j_{[\alpha]}}f\subset\left(\mathbf{E},\mathbf{E}^1\right)^K_{\alpha-[\alpha],q}$ and

$$\left\|D_{j_1}D_{j_2}...D_{j_{[\alpha]}}f\right\|_{(\mathbf{E},\mathbf{E}^1)^K_{\alpha-[\alpha],q}} \leq C\|f\|_{(\mathbf{E},\mathbf{E}^r)^K_{\alpha/r,q}}. \tag{5.9}$$

Conversely, let $D_{j_1}D_{j_2}...D_{j_{[\alpha]}}f\in\left(\mathbf{E},\mathbf{E}^1\right)^K_{\alpha-[\alpha],q} = \left(\mathbf{E}^{[\alpha]},\mathbf{E}^r\right)^K_{(\alpha-[\alpha])/(r-[\alpha]),q}$. Then the right-hand estimate of (1.7) and (5.6) imply

$$\|f\|_{(\mathbf{E},\mathbf{E}^r)^K_{\alpha/r,q}} \leq C\sum_{j_1,\ldots,j_{[\alpha]}=1}^{n}\left\|D_{j_1}D_{j_2}...D_{j_{[\alpha]}}f\right\|_{(\mathbf{E},\mathbf{E}^1)^K_{\alpha-[\alpha],q}}. \tag{5.10}$$

Inequalities (5.9) and (5.10) imply item (4) of Theorem 1.3. Proof of item (5) is similar. Theorem 1.3 is completely proved. □

6. Shannon sampling, Paley-Wiener frames and abstract Besov subspaces

6.1. Paley-Wiener vectors in Hilbert spaces. Consider a self-adjoint positive definite operator L in a Hilbert space $\mathbf{H}$. Let $\sqrt{L}$ be the positive square root of L. According to the spectral theory for such operators [**2**] there exists a direct integral of Hilbert spaces $X = \int X(\lambda)dm(\lambda)$ and a unitary operator $\mathcal{F}$ from $\mathbf{H}$ onto X, which transforms the domains of $L^{k/2}, k \in \mathbb{N}$, onto the sets $X_k = \{x \in X | \lambda^k x \in X\}$ with the norm

$$\|x(\lambda)\|_{X_k} = \langle x(\lambda), x(\lambda)\rangle_{X(\lambda)}^{1/2} = \left(\int_0^\infty \lambda^{2k}\|x(\lambda)\|^2_{X(\lambda)}dm(\lambda)\right)^{1/2}. \tag{6.1}$$

and satisfies the identity $\mathcal{F}(L^{k/2}f)(\lambda) = \lambda^k(\mathcal{F}f)(\lambda)$, if f belongs to the domain of $L^{k/2}$. We call the operator $\mathcal{F}$ the Spectral Fourier Transform [**25, 26**]. As known, X is the set of all m-measurable functions $\lambda \mapsto x(\lambda) \in X(\lambda)$, for which the following norm is finite:

$$\|x\|_X = \left(\int_0^\infty \|x(\lambda)\|^2_{X(\lambda)}dm(\lambda)\right)^{1/2}$$

For a function F on $[0, \infty)$ which is bounded and measurable with respect to dm one can introduce the operator $F(\sqrt{L})$ by using the formula

$$F(\sqrt{L})f = \mathcal{F}^{-1}F(\lambda)\mathcal{F}f, \quad f \in \mathbf{H}. \tag{6.2}$$

If F is real-valued the operator $F(\sqrt{L})$ is self-adjoint.

REMARK 6.1. In many applications L is a second-order differential operator and then $\sqrt{L}$ is a first-order pseudo-differential operator.

DEFINITION 6.2. For $\sqrt{L}$ as above we will say that a vector $f \in \mathbf{H}$ belongs to the *Paley-Wiener space* $\mathbf{PW}_\omega(\sqrt{L})$ if the support of the Spectral Fourier Transform $\mathcal{F}f$ is contained in $[0, \omega]$.

The next two facts are obvious.

THEOREM 6.3. *The spaces* $\mathbf{PW}_\omega(\sqrt{L})$ *have the following properties:*

(1) *the space* $\mathbf{PW}_\omega(\sqrt{L})$ *is a linear closed subspace in* $\mathbf{H}$.
(2) *the space* $\bigcup_{\omega>0}\mathbf{PW}_\omega(\sqrt{L})$ *is dense in* $\mathbf{H}$.

Next we denote by $\mathbf{H}^k$ the domain of $L^{k/2}$. It is a Banach space, equipped with the graph norm $\|f\|_k = \|f\|_{\mathbf{H}} + \|L^{k/2}f\|_{\mathbf{H}}$. The next theorem contains generalizations of several results from classical harmonic analysis (in particular the Paley-Wiener theorem). It follows from our results in [**26**].

THEOREM 6.4. *The following statements hold:*

(1) *(Bernstein inequality)* $f \in \mathbf{PW}_\omega(\sqrt{L})$ *if and only if* $f \in \mathbf{H}^\infty = \bigcap_{k=1}^\infty \mathbf{H}^k$, *and the following Bernstein inequalities holds true*

$$\|L^{s/2}f\|_{\mathbf{H}} \leq \omega^s\|f\|_{\mathbf{H}} \quad \textit{for all } s \in \mathbb{R}_+; \tag{6.3}$$

(2) *(Paley-Wiener theorem)* $f \in \mathbf{PW}_\omega(\sqrt{L})$ *if and only if for every* $g \in \mathbf{H}$ *the scalar-valued function of the real variable* $t \mapsto \langle e^{it\sqrt{L}}f, g\rangle$ *is bounded on the real line and has an extension to the complex plane as an entire function of the exponential type* ω;

(3) *(Riesz-Boas interpolation formula)* $f \in \mathbf{PW}_\omega(\sqrt{L})$ *if and only if* $f \in \mathbf{H}^\infty$ *and the following Riesz-Boas interpolation formula holds for all* $\omega > 0$*:*

$$i\sqrt{L}f = \frac{\omega}{\pi^2}\sum_{k\in\mathbb{Z}} \frac{(-1)^{k-1}}{(k-1/2)^2} e^{i\left(\frac{\pi}{\omega}(k-1/2)\right)\sqrt{L}} f. \tag{6.4}$$

PROOF. (1) follows immediately from the definition and representation (6.1). To prove (2) it is sufficient to apply the classical Bernstein inequality [**16**] in the uniform norm on $\mathbb{R}$ to every function $\langle e^{it\sqrt{L}}f, g\rangle$, $g \in \mathbf{H}$. To prove (3) one has to apply the classical Riesz-Boas interpolation formula on $\mathbb{R}$, [**16, 36**] to a function $\langle e^{it\sqrt{L}}f, g\rangle$. □

6.2. Frames in Hilbert spaces. A family of vectors $\{\theta_v\}$ in a Hilbert space $\mathbf{H}$ is called a frame if there exist constants $A, B > 0$ such that

$$A\|f\|_{\mathbf{H}}^2 \leq \sum_v |\langle f, \theta_v\rangle|^2 \leq B\|f\|_{\mathbf{H}}^2 \quad \text{for all} \quad f \in \mathbf{H}. \tag{6.5}$$

The largest A and smallest B are called lower and upper frame bounds.

The family of scalars $\{\langle f, \theta_v\rangle\}$ represents a set of measurements of a vector f. In order to resynthesize the vector f from this collection of measurements in a linear way one has to find another (dual) frame $\{\Theta_v\}$. Then a reconstruction formula is $f = \sum_v \langle f, \theta_v\rangle \Theta_v$. Dual frames are not unique in general. Moreover it may be difficult to find a dual frame in concrete situations. If $A = B = 1$ the frame is said to be tight or Parseval. Parseval frames are similar in many respects to orthonormal wavelet bases. For example, if in addition all vectors θ_v are unit vectors, then the frame is an orthonormal basis. The main feature of Parseval frames is that decomposing and synthesizing a vector from known data are tasks carried out with the same family of functions, i.e., the Parseval frame is its own dual frame.

6.3. Sampling in abstract Paley-Wiener spaces. We now assume that the **Assumption 3** is satisfied. It meant that there exists a $C > 0$ and $m_0 \geq 0$ such that for any $0 < \rho < 1$ there exists a set of functionals $\mathcal{A}^{(\rho)} = \left\{\mathcal{A}_k^{(\rho)}\right\}$, defined on $\mathbf{H}^{m_0}$, for which the inequalities (1.11) hold true.

REMARK 6.5. Following [**27, 29**], we call inequality (1.11) a *Poincaré-type inequality* since it is an estimate of the norm of f through the norm of its "derivative" $L^{m/2}f$.

Let us introduce vectors $\mu_k \in \mathbf{H}$ such that $\langle f, \mu_k\rangle = \mathcal{A}_k^{(\rho)}(f)$, $f \in \mathbf{H}^m$, $m > m_0$. Let $\mathcal{P}_\omega$ be the orthogonal projection of $\mathbf{H}$ onto $\mathbf{PW}_\omega(\sqrt{L})$ and put

$$\phi_k^\omega = \mathcal{P}_\omega \mu_k. \tag{6.6}$$

Using the Bernstein inequality (6.3) we obtain the following statement.

THEOREM 6.6. *(Sampling Theorem) Assume that inequality* (1.11) *holds and for a given* $\omega > 0$ *and* $\delta \in (0,1)$ *pick a* ρ *such that*

$$\rho^{2m} = C^{-1}\omega^{-2m}\delta. \tag{6.7}$$

Then the family of vectors $\{\phi_k^\omega\}$ *is a frame for the Hilbert space* $\mathbf{PW}_\omega(\sqrt{L})$ *and*

$$(1-\delta)\|f\|_{\mathbf{H}}^2 \leq \sum_k |\langle f, \phi_k^\omega\rangle|^2 \leq \|f\|_{\mathbf{H}}^2, \quad f \in \mathbf{PW}_\omega(\sqrt{L}). \tag{6.8}$$

The canonical dual frame $\{\Theta_k^\omega\}$ has the property $\Theta_k^\omega \in \mathbf{PW}_\omega(\sqrt{L})$ and provides the following reconstruction formulas

$$f = \sum_k \langle f, \phi_k^\omega \rangle \, \Theta_k^\omega = \sum_k \langle f, \Theta_k^\omega \rangle \, \phi_k^\omega, \quad f \in \mathbf{PW}_\omega(\sqrt{L}). \tag{6.9}$$

6.4. Partitions of unity on the frequency side. The construction of frequency-localized frames is achieved via spectral calculus. The idea is to start from a partition of unity on the positive real axis. In the following, we will be considering two different types of such partitions, whose construction we now describe in some detail.

Let $g \in C^\infty(\mathbb{R}_+)$ be a non-increasing function such that $supp(g) \subset [0,\ 2]$, and $g(\lambda) = 1$ for $\lambda \in [0,\ 1]$, $0 \le g(\lambda) \le 1$, $\lambda > 0$. We now let $h(\lambda) = g(\lambda) - g(2\lambda)$, which entails $supp(h) \subset [2^{-1}, 2]$, and use this to define $F_0(\lambda) = \sqrt{g(\lambda)}$, $F_j(\lambda) = \sqrt{h(2^{-j}\lambda)}$, $j \ge 1$, as well as $G_j(\lambda) = [F_j(\lambda)]^2 = F_j^2(\lambda)$, $j \ge 0$. As a result of the definitions, we get for all $\lambda \ge 0$ the equations $\sum_{j=0}^n G_j(\lambda) = \sum_{j=0}^n F_j^2(\lambda) = g(2^{-n}\lambda)$, and as a consequence $\sum_{j\ge 0} G_j(\lambda) = \sum_{j \ge 0} F_j^2(\lambda) = 1$, $\lambda \ge 0$, with finitely many nonzero terms occurring in the sums for each fixed λ. We call the sequence $(G_j)_{j\ge 0}$ a **(dyadic) partition of unity**, and $(F_j)_{j\ge 0}$ a **quadratic (dyadic) partition of unity**. As will become soon apparent, quadratic partitions are useful for the construction of frames. Using the spectral theorem one has $F_j^2(\sqrt{L})f = \mathcal{F}^{-1}\left(F_j^2(\lambda)\mathcal{F}f(\lambda)\right)$, $j \ge 1$, and thus

$$f = \mathcal{F}^{-1}\mathcal{F}f(\lambda) = \mathcal{F}^{-1}\left(\sum_{j\ge 0} F_j^2(\lambda)\mathcal{F}f(\lambda)\right) = \sum_{j \ge 0} F_j^2(\sqrt{L})f. \tag{6.10}$$

Taking inner product with f gives $\|F_j(\sqrt{L})f\|_{\mathbf{H}}^2 = \langle F_j^2(\sqrt{L})f, f\rangle$ and $\|f\|_{\mathbf{H}}^2 = \sum_{j\ge 0}\langle F_j^2(\sqrt{L})f, f\rangle = \sum_{j\ge 0}\|F_j(\sqrt{L})f\|_{\mathbf{H}}^2$. Similarly, we get the identity

$$\sum_{j\ge 0} G_j(\sqrt{L})f = f\ .$$

Moreover, since the functions G_j, F_j, have their supports in $[2^{j-1},\ 2^{j+1}]$, the elements $F_j(\sqrt{L})f$ and $G_j(\sqrt{L})f$ are bandlimited to $[2^{j-1},\ 2^{j+1}]$, whenever $j \ge 1$, and to $[0, 2]$ for $j = 0$.

6.5. Paley-Wiener frames in Hilbert spaces. Using the notation from above and Theorem 6.6, one can describe the following Paley-Wiener frame in an abstract Hilbert space $\mathbf{H}$.

THEOREM 6.7. *(Paley-Wiener nearly Parseval frame in* $\mathbf{H}$*)*

For a fixed $\delta \in (0,1)$ and $j \in \mathbb{N}$ let $\{\phi_k^j\}$ be a set of vectors described in Theorem 6.6 *that correspond to $\omega = 2^{j+1}$. Then for functions F_j the family of Paley-Wiener vectors $\Phi_k^j = F_j(\sqrt{L})\phi_k^j$ has the following properties:*

(1) *Each vector Φ_k^j belongs to $\mathbf{PW}_{[2^{j-1},\,2^{j+1}]}(\sqrt{L})$, $j \in N$, $k = 1, \ldots$.*
(2) *The family $\left\{\Phi_k^j\right\}$ is a frame in $\mathbf{H}$ with constants $1-\delta$ and 1:*

$$(1-\delta)\|f\|_{\mathbf{H}}^2 \le \sum_{j\ge 0}\sum_k \left|\left\langle f, \Phi_k^j\right\rangle\right|^2 \le \|f\|_{\mathbf{H}}^2, \quad f \in \mathbf{H}. \tag{6.11}$$

(3) *The canonical dual frame* $\{\Psi_k^j\}$ *also consists of bandlimited vectors* $\Psi_k^j \in \mathbf{PW}_{[2^{j-1},\, 2^{j+1}]}(\sqrt{L})$, $j \in [0, \infty)$, $k = 1, ...,$ *and has frame bounds* $A = 1$, $B = (1-\delta)^{-1}$.

(4) *The reconstruction formulas hold for every* $f \in \mathbf{H}$

$$f = \sum_j \sum_k \left\langle f, \Phi_k^j \right\rangle \Psi_k^j = \sum_j \sum_k \left\langle f, \Psi_k^j \right\rangle \Phi_k^j.$$

The last two items here follow from the first two and general properties of frames. We also note that for reconstruction of a Paley-Wiener vector from a set of samples one can use, besides dual frames, the variational (polyharmonic) splines in Hilbert spaces developed in [**27**].

7. Besov subspaces in Hilbert spaces

7.1. Approximation spaces. Let us introduce another functional on $\mathbf{E}+\mathbf{F}$, where $\mathbf{E}$ and $\mathbf{F}$ form a pair of quasi-normed linear spaces $\mathcal{E}(f,t) = \mathcal{E}(f,t,\mathbf{E},\mathbf{F}) = \inf_{g\in\mathbf{F}, \|g\|_{\mathbf{F}}\le t} \|f-g\|_{\mathbf{E}}$.

DEFINITION 7.1. The approximation space $\mathcal{E}_{\alpha,q}(\mathbf{E},\mathbf{F}), 0<\alpha<\infty, 0<q\le\infty$ is the quasi-normed linear spaces of all $f\in\mathbf{E}+\mathbf{F}$ for which the quasi-norm

$$\|f\|_{\mathcal{E}_{\alpha,q}(\mathbf{E},\mathbf{F})} = \left(\int_0^\infty (t^\alpha \mathcal{E}(f,t))^q \frac{dt}{t}\right)^{1/q} \tag{7.1}$$

is finite.

The next theorem represents a very abstract version of what is known as an Equivalence Approximation Theorem [**2, 18**]. In the form it is stated below it was proved in [**9**].

THEOREM 7.2. *Suppose that* $\mathcal{T}\subset\mathbf{F}\subset\mathbf{E}$ *are quasi-normed linear spaces and* $\mathbf{E}$ *and* $\mathbf{F}$ *are complete. If there exist* $C>0$ *and* $\beta>0$ *such that the following Jackson-type inequality is satisfied* $t^\beta\mathcal{E}(t,f,\mathcal{T},\mathbf{E})\le C\|f\|_{\mathbf{F}}$, $t>0$, $f\in\mathbf{F}$, *then the following embedding holds true*

$$(\mathbf{E},\mathbf{F})^K_{\theta,q} \subset \mathcal{E}_{\theta\beta,q}(\mathbf{E},\mathcal{T}), \quad 0<\theta<1,\ 0<q\le\infty. \tag{7.2}$$

If there exist $C>0$ *and* $\beta>0$ *such that the following Bernstein-type inequality holds* $\|f\|_{\mathbf{F}}\le C\|f\|^\beta_{\mathcal{T}}\|f\|_{\mathbf{E}}$, $f\in\mathcal{T}$, *then the following embedding holds true*

$$\mathcal{E}_{\theta\beta,q}(\mathbf{E},\mathcal{T}) \subset (\mathbf{E},\mathbf{F})^K_{\theta,q}, \quad 0<\theta<1,\ 0<q\le\infty. \tag{7.3}$$

7.2. Besov subspaces in Hilbert spaces. According to (1.6) we introduce $\mathcal{B}^\alpha_{\mathbf{H},q}(\sqrt{L}) = (\mathbf{H},\mathbf{H}^r)^K_{\theta,q}$, $0<\theta=\alpha/r<1$, $1\le q\le\infty$. We also introduce a notion of best approximation: $\mathcal{E}(f,\omega) = \inf_{g\in\mathbf{PW}_\omega(\sqrt{L})}\|f-g\|_{\mathbf{H}}$. Our goal is to apply Theorem 7.2 in the situation where $\mathbf{E}=\mathbf{H}$, $\mathbf{F}=\mathbf{H}^r$ and $\mathcal{T}=\mathbf{PW}_\omega(\sqrt{L})$ is a natural abelian group as the additive group of a vector space, with the quasi-norm $\|f\|_{\mathcal{T}} = \inf\left\{\omega'>0\ : f\in\mathbf{PW}_{\omega'}\left(\sqrt{L}\right)\right\}$. To be more precise it is the space of finite sequences of Fourier coefficients $\mathbf{c}=(c_1,...c_m)\in\mathbf{PW}_\omega(\sqrt{L})$ where m is the greatest index such that the eigenvalue $\lambda_m\le\omega$. For a $\mathbf{c}=(c_1,...c_m)\in\mathbf{PW}_\omega(\sqrt{L})$ the quasi-norm is defined as $\|\mathbf{c}\|_{E_\omega(L)} = \max\left\{\sqrt{\lambda_j} : c_j\ne 0,\ c_{j+1}=...=c_m=0\right\}$.

REMARK 7.3. Let us emphasize that the reason we need the language of quasi-normed spaces is because $\|\cdot\|_{\mathcal{T}}$ is clearly not a norm, only a quasi-norm on $\mathbf{PW}_\omega(\sqrt{L})$.

The Plancherel Theorem allows us to verify a generalization of the Bernstein inequality for bandlimited functions in $f \in \mathbf{PW}_{\omega}(\sqrt{L})$. One can prove the following statement (see [**22, 26**]).

LEMMA 7.4. *A vector f belongs to the space $\mathbf{PW}_{\omega}(\sqrt{L})$ if and only if the following Bernstein inequality holds* $\|L^{r/2}f\|_{\mathbf{H}} \leq \omega^r\|f\|_{\mathbf{H}}, \quad r \in \mathbb{R}_+$.

One also has an analogue of the Jackson inequality (see [**22, 26**]) $\mathcal{E}(f,\omega) \leq \omega^{-r}\|f\|_{\mathbf{H}^r}, \quad f \in \mathbf{H}^r$. These two inequalities and Theorem 7.2 imply the following result (compare to [**31, 34**]).

THEOREM 7.5. *For $\alpha > 0, 1 \leq q \leq \infty$ the norm of $\mathcal{B}^{\alpha}_{\mathbf{H},q}(\sqrt{L})$, is equivalent to*

$$\|f\|_{\mathbf{H}} + \left(\sum_{j=0}^{\infty}\left(2^{j\alpha}\mathcal{E}(f,2^j)\right)^q\right)^{1/q}. \tag{7.4}$$

Let the functions F_j be as in Subsection 6.4.

THEOREM 7.6. *For $\alpha > 0, 1 \leq q \leq \infty$ the norm of $\mathcal{B}^{\alpha}_{\mathbf{H},q}(\sqrt{L})$, is equivalent to*

$$f \mapsto \left(\sum_{j=0}^{\infty}\left(2^{j\alpha}\left\|F_j(\sqrt{L})f\right\|_{\mathbf{H}}\right)^q\right)^{1/q}, \tag{7.5}$$

with the standard modifications for $q = \infty$.

PROOF. We obviously have $\mathcal{E}(f,2^l) \leq \sum_{j>l}\left\|F_j(\sqrt{L})f\right\|_{\mathbf{H}}$. By using a discrete version of Hardy's inequality [**3**] we obtain the estimate

$$\|f\| + \left(\sum_{l=0}^{\infty}\left(2^{l\alpha}\mathcal{E}(f,2^l)\right)^q\right)^{1/q} \leq C\left(\sum_{j=0}^{\infty}\left(2^{j\alpha}\left\|F_j(\sqrt{L})f\right\|_{\mathbf{H}}\right)^q\right)^{1/q}. \tag{7.6}$$

Conversely, for any $g \in \mathbf{PW}_{2^{j-1}}(\sqrt{L})$ we have $\left\|F_j(\sqrt{L})f\right\|_{\mathbf{H}} = \left\|F_j(\sqrt{L})(f-g)\right\|_{\mathbf{H}}$ $\leq \|f-g\|_{\mathbf{H}}$. This implies the estimate $\left\|F_j(\sqrt{L})f\right\|_{\mathbf{H}} \leq \mathcal{E}(f,\, 2^{j-1})$, which shows that the inequality opposite to (7.6) holds. The proof is complete. □

THEOREM 7.7. *For $\alpha > 0, 1 \leq q \leq \infty$ the norm of $\mathcal{B}^{\alpha}_{\mathbf{H},q}(\sqrt{L})$ is equivalent to*

$$\left(\sum_{j=0}^{\infty}2^{j\alpha q}\left(\sum_{k}\left|\left\langle f,\Phi^j_k\right\rangle\right|^2\right)^{q/2}\right)^{1/q} \asymp \|f\|_{B^{\alpha}_q}, \tag{7.7}$$

with the standard modifications for $q = \infty$.

PROOF. For $f \in \mathbf{H}$ and operator $F_j(\sqrt{L})$ we apply (6.8) to $F_j(\sqrt{L})f \in \mathbf{PW}_{2^{j+1}}(\sqrt{L})$ to obtain

$$(1-\delta)\left\|F_j(\sqrt{L})f\right\|^2_{\mathbf{H}} \leq \sum_k\left|\left\langle F_j(\sqrt{L})f,\phi^j_k\right\rangle\right|^2 \leq \left\|F_j(\sqrt{L})f\right\|^2_{\mathbf{H}}. \tag{7.8}$$

Since $\Phi_k^j = F_j(\sqrt{L})\phi_k^j$ we obtain the following inequality

$$\sum_k \left|\left\langle f, \Phi_k^j\right\rangle\right|^2 \leq \left\|F_j(\sqrt{L})f\right\|_{\mathbf{H}}^2 \leq \frac{1}{1-\delta}\sum_k \left|\left\langle f, \Phi_k^j\right\rangle\right|^2 \quad \text{for all } f \in \mathbf{H}.$$

Our statement follows now from Theorem 7.6. □

8. Applications

8.1. Analysis on $\mathbb{S}^d$. We will specify the general setup in the case of standard unit sphere. Let $\mathbb{S}^d = \left\{x \in \mathbb{R}^{d+1} : \|x\| = 1\right\}$. Let $\mathcal{P}_n$ denote the space of spherical harmonics of degree n, which are restrictions to $\mathbb{S}^d$ of harmonic homogeneous polynomials of degree n in $\mathbb{R}^d$. The Laplace-Beltrami operator $\Delta_{\mathbb{S}^d}$ on $\mathbb{S}^d$ is a restriction of the regular Laplace operator Δ in $\mathbb{R}^d$. Namely, $\Delta_{\mathbb{S}^d} f(x) = \Delta\widetilde{f}(x)$, $x \in \mathbb{S}^d$, where $\widetilde{f}(x)$ is the homogeneous extension of f: $\widetilde{f}(x) = f\left(x/\|x\|\right)$. Another way to compute $\Delta_{\mathbb{S}^d} f(x)$ is to express both $\Delta_{\mathbb{S}^d}$ and f in a spherical coordinate system. Each $\mathcal{P}_n$ is the eigenspace of $\Delta_{\mathbb{S}^d}$ that corresponds to the eigenvalue $-n(n+d-1)$. Let $Y_{n,l}$, $l = 1, ..., l_n$ be an orthonormal basis in $\mathcal{P}_n$.

Let $e_1, ..., e_{d+1}$ be the standard orthonormal basis in $\mathbb{R}^{d+1}$. If $SO(d+1)$ and $SO(d)$ are the groups of rotations of $\mathbb{R}^{d+1}$ and $\mathbb{R}^d$ respectively then $\mathbb{S}^d = SO(d+1)/SO(d)$. On $\mathbb{S}^d$ we consider vector fields $X_{i,j} = x_j\partial_{x_i} - x_i\partial_{x_j}$ which are generators of one-parameter groups of rotations $\exp tX_{i,j} \in SO(d+1)$ in the plane (x_i, x_j). These groups are defined by the formulas for $\tau \in \mathbb{R}$,

$$\exp\tau X_{i,j}\cdot(x_1, ..., x_{d+1}) = (x_1, ..., x_i\cos\tau - x_j\sin\tau, ..., x_i\sin\tau + x_j\cos\tau, ..., x_{d+1})$$

Let $T_{i,j}(\tau)$ be a one-parameter group which is a representation of $\exp\tau X_{i,j}$ in the space $L_p(\mathbb{S}^d)$. It acts on $f \in L_p(\mathbb{S}^d)$ by the following formula

$$T_{i,j}(\tau)f(x_1, ..., x_{d+1}) = f(x_1, ..., x_i\cos\tau - x_j\sin\tau, ..., x_i\sin\tau + x_j\cos\tau, ..., x_{d+1}).$$

Let $D_{i,j}$ be a generator of $T_{i,j}$ in $L_p(\mathbb{S}^d)$. The Laplace-Beltrami operator $\Delta_{\mathbb{S}^d}$ can be identified with the operator $L = \sum_{i<j} D_{i,j}^2$. One can easily illustrate our results by describing norms in Sobolev and Besov spaces on $\mathbb{S}^d$ in terms of operators $T_{i,j}$ and $D_{i,j}$. In this situation role of Paley-Wiener subspaces is played by subspaces $\mathcal{P}_n$. A set of functionals $\left\{\mathcal{A}_k^{(\rho)}\right\}_{k\in\mathcal{K}_\rho}$ described in (1.11), (6.7) can be represented by a set of Dirac measures at nodes $\{x_k^{(\rho)}\}$ "nearly uniformly" distributed over the sphere $\mathbb{S}^d$.

8.2. Compact homogeneous manifolds. It should be noted, that a similar situation holds on any compact homogeneous manifolds $M = G/K$ where G is a compact Lie group and K is its closed subgroup. Moreover, in this case description of Besov spaces in terms of approximation by Paley-Wiener vectors (eigenfunctions of a corresponding Laplace-Beltrami operator) and in terms of frame coefficients can be extended to any $1 \leq p \leq \infty$ [**6, 7, 38**]. A set of functionals $\left\{\mathcal{A}_k^{(\rho)}\right\}_{k\in\mathcal{K}_\rho}$ can be represented by a set of Dirac measures (or some other functionals [**29**]) at a set of nodes $\{x_k^{(\rho)}\}$ "nearly uniformly" distributed over the manifold M with the spacing comparable to $\rho > 0$. The Weyl's asymptotic formula [**8**] implies [**28**] that a rate of sampling which is given by (6.7) is essentially optimal.

8.3. Non-compact symmetric spaces. Our framework also holds on non-compact symmetric spaces [**30**]-[**35**]. Besov spaces can be characterized either using corresponding modulus of continuity [**32**] or by approximation by Paley-Wiener vectors [**33**] or in terms of frames [**35**]. In this situation Paley-Wiener functions which admit explicit description in terms of the Helgason-Fourier transform [**33**] can be represented by a set of Dirac measures (or even more general functionals [**29**]) at a set of nodes $\{x_k^{(\rho)}\}$ "nearly uniformly" distributed over the manifold M with the spacing comparable to $\rho > 0$.

REMARK 8.1. It should be noted that in the case of non-compact symmetric spaces our approach leads to Sobolev and Besov spaces which are different from the conventional ones generated by the Laplace-Beltrami operator associated with the natural metric.

References

[1] J. Bergh, J. Lofstrom, *Interpolation spaces*, Springer-Verlag, 1976.

[2] M. Sh. Birman and M. Z. Solomjak, *Spectral theory of selfadjoint operators in Hilbert space*, Mathematics and its Applications (Soviet Series), D. Reidel Publishing Co., Dordrecht, 1987. Translated from the 1980 Russian original by S. Khrushchëv and V. Peller. MR1192782

[3] P. L. Butzer and H. Berens, *Semi-groups of operators and approximation*, Die Grundlehren der mathematischen Wissenschaften, Band 145, Springer-Verlag New York Inc., New York, 1967. MR0230022

[4] P. L. Butzer and K. Scherer, *Jackson and Bernstein-type inequalities for families of commutative operators in Banach spaces*, J. Approximation Theory **5** (1972), 308–342, DOI 10.1016/0021-9045(72)90019-6. Collection of articles dedicated to J. L. Walsh on his 75th birthday, III. MR0346396

[5] F. Dai and Y. Xu, *Approximation theory and harmonic analysis on spheres and balls*, Springer Monographs in Mathematics, Springer, New York, 2013. MR3060033

[6] H. G. Feichtinger, H. Führ, and I. Z. Pesenson, *Geometric space-frequency analysis on manifolds*, J. Fourier Anal. Appl. **22** (2016), no. 6, 1294–1355, DOI 10.1007/s00041-015-9457-3. MR3572904

[7] D. Geller and I. Z. Pesenson, *Band-limited localized Parseval frames and Besov spaces on compact homogeneous manifolds*, J. Geom. Anal. **21** (2011), no. 2, 334–371, DOI 10.1007/s12220-010-9150-3. MR2772076

[8] L. Hörmander, *The analysis of linear partial differential operators. III: Pseudo-differential operators*, Classics in Mathematics, Springer, Berlin, 2007. Reprint of the 1994 edition. MR2304165

[9] S. Krein, I. Pesenson, *Interpolation Spaces and Approximation on Lie Groups*, The Voronezh State University, Voronezh, 1990.

[10] S. G. Kreĭn, Yu. Ī. Petunīn, and E. M. Semënov, *Interpolation of linear operators*, Translations of Mathematical Monographs, vol. 54, American Mathematical Society, Providence, R.I., 1982. Translated from the Russian by J. Szűcs. MR649411

[11] J.-L. Lions, *Théorèmes de trace et d'interpolation. I*, Ann. Scuola Norm. Sup. Pisa (3) **13** (1959), 389–403. MR0119092

[12] J.-L. Lions and J. Peetre, *Sur une classe d'espaces d'interpolation* (French), Inst. Hautes Études Sci. Publ. Math. **19** (1964), 5–68. MR0165343

[13] E. Nelson, *Analytic vectors*, Ann. of Math. (2) **70** (1959), 572–615, DOI 10.2307/1970331. MR0107176

[14] E. Nelson and W. F. Stinespring, *Representation of elliptic operators in an enveloping algebra*, Amer. J. Math. **81** (1959), 547–560, DOI 10.2307/2372913. MR0110024

[15] D. Müller and D. Yang, *A difference characterization of Besov and Triebel-Lizorkin spaces on RD-spaces*, Forum Math. **21** (2009), no. 2, 259–298, DOI 10.1515/FORUM.2009.013. MR2503306

[16] S. M. Nikol′skiĭ, *Approximation of functions of several variables and imbedding theorems*, Springer-Verlag, New York-Heidelberg., 1975. Translated from the Russian by John M. Danskin, Jr.; Die Grundlehren der Mathematischen Wissenschaften, Band 205. MR0374877

[17] E. Nursultanov, M. Ruzhansky, and S. Tikhonov, *Nikolskii inequality and Besov, Triebel-Lizorkin, Wiener and Beurling spaces on compact homogeneous manifolds*, Ann. Sc. Norm. Super. Pisa Cl. Sci. (5) **16** (2016), no. 3, 981–1017. MR3618082

[18] J. Peetre and G. Sparr, *Interpolation of normed abelian groups*, Ann. Mat. Pura Appl. (4) **92** (1972), 217–262, DOI 10.1007/BF02417949. MR0322529

[19] I. Z. Pesenson, *Interpolation of noncommuting operators* (Russian), Uspehi Mat. Nauk **33** (1978), no. 3(201), 183–184. MR0493415

[20] I. Z. Pesenson, *Interpolation spaces on Lie groups* (Russian), Dokl. Akad. Nauk SSSR **246** (1979), no. 6, 1298–1303. MR546322

[21] I. Z. Pesenson, *Nikol′skiĭ-Besov spaces connected with representations of Lie groups* (Russian), Dokl. Akad. Nauk SSSR **273** (1983), no. 1, 45–49. MR722886

[22] I. Z. Pesenson, *Best approximations in a space of the representation of a Lie group* (Russian), Dokl. Akad. Nauk SSSR **302** (1988), no. 5, 1055–1058; English transl., Soviet Math. Dokl. **38** (1989), no. 2, 384–388. MR981052

[23] I. Z. Pesenson, *On the abstract theory of Nikol′skiĭ-Besov spaces* (Russian), Izv. Vyssh. Uchebn. Zaved. Mat. **6** (1988), 59–68; English transl., Soviet Math. (Iz. VUZ) **32** (1988), no. 6, 80–92. MR960795

[24] I. Z. Pesenson, *Approximations in the representation space of a Lie group* (Russian), Izv. Vyssh. Uchebn. Zaved. Mat. **7** (1990), 43–50; English transl., Soviet Math. (Iz. VUZ) **34** (1990), no. 7, 49–57. MR1104147

[25] I. Z. Pesenson, *The Bernstein inequality in representations of Lie groups* (Russian), Dokl. Akad. Nauk SSSR **313** (1990), no. 4, 803–806; English transl., Soviet Math. Dokl. **42** (1991), no. 1, 87–90. MR1080635

[26] I. Pesenson, *A sampling theorem on homogeneous manifolds*, Trans. Amer. Math. Soc. **352** (2000), no. 9, 4257–4269, DOI 10.1090/S0002-9947-00-02592-7. MR1707201

[27] I. Pesenson, *Sampling of band-limited vectors*, J. Fourier Anal. Appl. **7** (2001), no. 1, 93–100, DOI 10.1007/s00041-001-0007-9. MR1812998

[28] I. Pesenson, *An approach to spectral problems on Riemannian manifolds*, Pacific J. Math. **215** (2004), no. 1, 183–199, DOI 10.2140/pjm.2004.215.183. MR2060498

[29] I. Pesenson, *Poincaré-type inequalities and reconstruction of Paley-Wiener functions on manifolds*, J. Geom. Anal. **14** (2004), no. 1, 101–121, DOI 10.1007/BF02921868. MR2030577

[30] I. Pesenson, *Bernstein-Nikolskii inequalities and Riesz interpolation formula on compact homogeneous manifolds*, J. Approx. Theory **150** (2008), no. 2, 175–198, DOI 10.1016/j.jat.2007.06.001. MR2388855

[31] I. Pesenson, *Paley-Wiener approximations and multiscale approximations in Sobolev and Besov spaces on manifolds*, J. Geom. Anal. **19** (2009), no. 2, 390–419, DOI 10.1007/s12220-008-9059-2. MR2481967

[32] I. Pesenson, *Bernstein-Nikolskii and Plancherel-Polya inequalities in L_p-norms on non-compact symmetric spaces*, Math. Nachr. **282** (2009), no. 2, 253–269, DOI 10.1002/mana.200510736. MR2493515

[33] I. Pesenson, *A discrete Helgason-Fourier transform for Sobolev and Besov functions on noncompact symmetric spaces*, Radon transforms, geometry, and wavelets, Contemp. Math., vol. 464, Amer. Math. Soc., Providence, RI, 2008, pp. 231–247, DOI 10.1090/conm/464/09087. MR2440139

[34] I. Z. Pesenson and M. Z. Pesenson, *Approximation of Besov vectors by Paley-Wiener vectors in Hilbert spaces*, Approximation theory XIII: San Antonio 2010, Springer Proc. Math., vol. 13, Springer, New York, 2012, pp. 249–262, DOI 10.1007/978-1-4614-0772-0_15. MR3026214

[35] I. Z. Pesenson, *Paley-Wiener-Schwartz nearly Parseval frames on noncompact symmetric spaces*, Commutative and noncommutative harmonic analysis and applications, Contemp. Math., vol. 603, Amer. Math. Soc., Providence, RI, 2013, pp. 55–71, DOI 10.1090/conm/603/12044. MR3204026

[36] I. Z. Pesenson, *Boas-type formulas and sampling in Banach spaces with applications to analysis on manifolds*, New perspectives on approximation and sampling theory, Appl. Numer. Harmon. Anal., Birkhäuser/Springer, Cham, 2014, pp. 39–61. MR3363026

[37] I. Z. Pesenson, *Splines and Wavelets on Geophysically Relevant Manifolds*, Handbook of Geomathematics, Springer, Berlin (2014), 1-33.

[38] I. Z. Pesenson, *Approximations in L_p-norms and Besov spaces on compact manifolds*, Trends in harmonic analysis and its applications, Contemp. Math., vol. 650, Amer. Math. Soc., Providence, RI, 2015, pp. 199–209, DOI 10.1090/conm/650/13033. MR3441739

[39] I. Z. Pesenson, *Sampling, splines and frames on compact manifolds*, GEM Int. J. Geomath. **6** (2015), no. 1, 43–81, DOI 10.1007/s13137-015-0069-5. MR3322489

Department of Mathematics, Temple University, Philadelphia, PA 19122
Email address: pesenson@temple.edu

Contemporary Mathematics
Volume **733**, 2019
https://doi.org/10.1090/conm/733/14747

Toeplitz operators in polyanalytic Bergman type spaces

Grigori Rozenblum and Nikolai Vasilevski

In memory of Selim Grigorievich Krein, a great mathematician and a charming person

ABSTRACT. We consider Toeplitz operators in Bergman and Fock type spaces of polyanalytic L^2-functions on the disk or on the half-plane with respect to the Lebesgue measure (resp., on $\mathbb{C}$ with the plane Gaussian measure). The structure involving creation and annihilation operators, similar to the classical one present for the Landau Hamiltonian, enables us to reduce Toeplitz operators in true polyanalytic spaces to the ones in the usual Bergman type spaces, however with distributional symbols. This reduction leads to describing a number of properties of the operators in the title, which may differ from the properties of the usual Bergman-Toeplitz operators.

1. Introduction

The paper is devoted to the study of Toeplitz operators in a relatively unexplored class of Bergman type spaces. While there is a vast literature devoted to operators in the classical Bergman and Fock spaces of analytic functions, considerably less is known about the case of spaces of polyanalytic functions. As it concerns general properties of these spaces, one can mention the books [**3, 28**] and the review papers [**2, 4**]. More specific information concerning properties of these spaces can be found in [**9, 11, 12, 14, 16–18, 25, 26**] and other sources, see the bibliography at the end of this paper. However, only one paper [**6**] was devoted to the study of Toeplitz operators in nonanalytic spaces. Meanwhile, the application fields of polyanalytic spaces keep expanding, including classical and quantum physics, mechanics, signal processing, wavelets and more. As it has been rather recently discovered, the spaces of polyanalytic functions possess an interesting structure, reminding the creation-annihilation operators in certain classical quantum-physics problems. Say, for the Landau Hamiltonian, i.e., the operator describing the motion of a charged quantum particle confined to the plane under the action of the uniform magnetic field orthogonal to the plane, the whole L^2 space splits into the sequence of orthogonal (Landau) subspaces $\mathcal{L}_q$, $q = 0, 1, \dots,$ invariant for the operator. These subspaces are interconnected by creation, Q^+, and annihilation, Q, operators, $Q^+ : \mathcal{L}_q \to \mathcal{L}_{q+1}$, $Q : \mathcal{L}_q \to \mathcal{L}_{q-1}$, which are isometries, up to a numerical coefficient, while $Q : \mathcal{L}_0 \to \{0\}$. These operators, discovered in 1928 by V.Fock [**7**], play an important role in the analysis of various problems concerning two- and

2010 *Mathematics Subject Classification.* Primary 30H20, Secondary 47B35.
Key words and phrases. Polyanalytic functions, Bergman spaces, Fock spaces, Toeplitz operators, creation and annihilation.

three-dimensional quantum models with magnetic fields. The fact important for us is that the creation and annihilation operators enable one to reduce a Toeplitz operator in the Landau subspace $\mathcal{L}_q$, $q > 0$, to the unitary equivalent Toeplitz operator in the lowest subspace $\mathcal{L}_0$, although with a different symbol. Such reduction was first discovered in [**5**]. A few years later, in [**19, 20**], such equivalence, (more exactly, 'almost equivalence', up to a very week error term) was established for an approximate creation-annihilation structure arising for the Landau Hamiltonian with weakly (and not that weakly) perturbed uniform magnetic field.

It turns out that the Bergman type spaces of polyanalytic functions admit such creation-annihilation structure as well. The ambient Hilbert space $L^2(\Omega)$, for a domain $\Omega \subset \mathbb{C}$, splits into the direct orthogonal sum of 'true polyanalytic' subspaces $\mathcal{A}_{(j)}(\Omega) := \mathcal{A}_j(\Omega) \ominus \mathcal{A}_{j-1}(\Omega)$, where $\mathcal{A}_j(\Omega)$ consists of polyanalytic functions, $\mathcal{A}_j(\Omega) = \{f \in L^2, \bar{\partial}^j f = 0\}$, and, for the case of the disk or the half-plane, also of analogous spaces of anti-polyanalytic functions. This property was established by Ramazanov in [**17, 18**] and, in a different way, by Y.Karlovich in [**11**] for the Bergman space on the unit disk, by N.Vasilevski, in [**26**] and Y.Karlovich–L.Pessoa in [**10**], for the Bergman space on the upper half-plane and by N.Vasilevski in [**25**] for the Fock space. In these papers, it was found that the creation-annihilation operators, acting as partial isometries, can be represented as two-dimensional singular integral operators.

In the present paper we follow the pattern of the analysis of the (perturbed) Landau Hamiltonian and establish reduction theorems that associate to a Toeplitz operator in a polyanalytic Bergman type space another Toeplitz operator, but now acting in the corresponding usual analytic Bergman type space. The symbol of the new operator is obtained from the symbol of the initial operator by applying an elliptic differential operator. Thus, if the initial symbol was not sufficiently smooth, the reduced symbol may turn out to be not a function but a distribution. The new operator turns out to be unitarily equivalent (or, for the case of the Bergman spaces on the disk, cosimilar, see Definition 4.4) to the initial one.

Thus, an alternative arises. Should one, for further analysis, consider operators with nice symbols in 'bad' spaces (the polyanalytic spaces are 'bad' due to the absence of many useful properties and structures present for analytic spaces), or, in the opposite, consider operators in nice spaces with 'bad' symbols. The second option seems to be more productive. Fortunately, the analysis of Toeplitz operators with distributional (and, more generally, singular) symbols is now available. For operators in the Bergman space in the disk and in the Fock space, such analysis was done by the authors in [**21, 22**]; the case of the Bergman space in the half-plane will be presented in [**23**]. Using this analysis, we find conditions for boundedness and compactness of Toeplitz operators in true polyanalytic Bergman type spaces, which turn out to be similar to the conditions in the standard Bergman spaces, and discuss the questions of uniqueness of the symbol and the finite rank problem, which differ somewhat from the classical ones.

The results are presented for the case of Bergman type spaces on domains in the complex plane. For higher dimensions the approach works as well, with certain modifications. These results will be presented elsewhere.

2. The structure of poly-Bergman and poly-Fock spaces

2.1. Polyanalytic and true polyanalytic spaces. A Bergman type Hilbert space $\mathbf{B}^2(\Omega)$ is the Hilbert space of solutions of some elliptic equation or system in a domain Ω in $\mathbb{R}^d$ or $\mathbb{C}^d$, belonging to $L^2(\Omega)$ with respect to some measure μ (Banach spaces, involving L^p theory, are also considered in the literature, but we do not discuss them here). Specific for Bergman type spaces is the existence of the reproducing kernel $\kappa(z,w)$: for any $z \in \Omega$, the linear evaluation functional $\boldsymbol{ev}_z : \mathbf{B}^2(\Omega) \ni u \mapsto u(z)$ is continuous and therefore admits the representation $\boldsymbol{ev}_z(u) = \langle u(\cdot), \kappa(z,\cdot)\rangle$. A classical Toeplitz operator in $\mathbf{B}^2(\Omega)$ with symbol $F \in L^\infty(\Omega)$ is the operator acting as $\mathbf{T}_F \equiv \mathbf{T}_F(\mathbf{B}^2(\Omega)) : \mathbf{B}^2(\Omega) \ni u \mapsto \mathbf{P}Fu \in \mathbf{B}^2(\Omega)$, where $\mathbf{P}$ is the orthogonal (Bergman) projection $\mathbf{P} : L^2(\Omega) \to \mathbf{B}^2(\Omega)$. In [**21, 22**] an approach was developed for defining Toeplitz operators in Bergman type spaces with much more general symbols, including F being a distribution in Ω or $\Omega \times \Omega$ or even a hyper-function in Ω. This approach is based upon defining operators by means of sesquilinear forms with a special structure and further using the reproducing kernel to describe the action of the operator.

The typical and best studied examples here are the classical Bergman space $\mathcal{A}^2(\mathbb{D})$ of analytic functions on the unit disk $\mathbb{D}$, a similar space $\mathcal{A}^2(\Pi)$ of analytic functions on the upper half-plane Π, both considered as subspaces of L^2 with respect to the corresponding Lebesgue area measure $dA(z)$, and the Fock (Bargmann-Segal) space $\mathcal{F}^2$ of entire analytic functions on $\mathbb{C} \equiv \mathbb{R}^2$, belonging to L^2 with respect to the plane Gaussian measure $d\mu = d\rho \equiv \pi^{-1}e^{-|z|^2}dA(z)$. Since we deal with Hilbert spaces only, we suppress further on the superscript 2 in the notation of spaces and write simply $\mathcal{A}(\mathbb{D})$, $\mathcal{A}(\Pi)$, $\mathcal{F}$, and so on.

In the present paper we study Toeplitz operators in analogous Hilbert spaces of polyanalytic functions, i.e., spaces of solutions of the iterated Cauchy-Riemann equation $\overline{\partial}^j u = 0, \quad j = 2, 3, \dots$, where, as usual, $\overline{\partial} = \frac{1}{2}(\partial_x + i\partial_y)$. The polyanalytic Bergman spaces (or, shorter, poly-Bergman spaces) $\mathcal{A}_j(\mathbb{D})$, $\mathcal{A}_j(\Pi)$, are the spaces of such functions in L^2 in the disk, resp., half-plane, and the poly-Fock space $\mathcal{F}_j$ is the space of entire $j-$analytic functions in $L^2(\mathbb{C}, d\rho)$. On properties of these spaces and applications, see the monographs [**3, 28**], the review articles [**2, 4**] and references therein (note that in the monograph [**28**], for the case of the spaces on the disk, the poly-Bergman spaces are defined in a slightly different way, as functions satisfying $(\overline{z}\overline{\partial})^j u = 0$; our results essentially carry over to this case as well). Polyanalytic functions find now applications in various problems in quantum and classical physics and mechanics, signal processing, wavelet theory etc.

Of course, $\mathcal{A}_{j-1}(\Omega) \subset \mathcal{A}_j(\Omega)$, therefore, to exclude poly-analytic functions of lower order, we define *true* poly-Bergman and *true* poly-Fock spaces as

$$\mathcal{A}_{(j)}(\mathbb{D}) = \mathcal{A}_j(\mathbb{D}) \ominus \mathcal{A}_{j-1}(\mathbb{D}),\ \mathcal{A}_{(j)}(\Pi) = \mathcal{A}_j(\Pi) \ominus \mathcal{A}_{j-1}(\Pi), \mathcal{F}_{(j)} = \mathcal{F}_j \ominus \mathcal{F}_{j-1}.$$

These spaces have been introduced and studied in [**17, 25, 26**]. Along with poly-Bergman type spaces $\mathcal{A}_j$, one can introduce anti-poly-Bergman type spaces $\tilde{\mathcal{A}}_j$, i.e., spaces of solutions of the iterated adjoint Cauchy-Riemann equation $\partial^j u = 0$, $\partial = \frac{1}{2}(\partial_x - i\partial_y)$ and the corresponding true anti-poly-Bergman type spaces $\tilde{\mathcal{A}}_{(j)}$. Here, the properties for Bergman spaces and Fock spaces differ essentially. It is established (see [**25**]) that true poly-Fock spaces form a complete orthogonal system in $L^2(\mathbb{C}, \rho)$ in the sense that

$$\overline{\bigcup_{j=1}^{\infty} \mathcal{F}_j} = \bigoplus_{j=1}^{\infty} \mathcal{F}_{(j)} = L^2(\mathbb{C}, d\rho).$$

On the other hand, any true poly-Bergman space on the upper half-plane is orthogonal to any true anti-poly-Bergman space; and therefore

$$\left(\bigoplus_{j=1}^{\infty} \mathcal{A}_{(j)}(\Pi)\right) \oplus \left(\bigoplus_{j=1}^{\infty} \tilde{\mathcal{A}}_{(j)}(\Pi)\right) = L^2(\Pi, dA),$$

with a similar relation for the unit disk. Since the iterated Cauchy-Riemann operator is elliptic, all the above spaces are reproducing kernel ones.

2.2. The structure of the Fock spaces. Along with true poly-Fock spaces $\mathcal{F}_{(j)}$, we consider the Landau subspaces $\mathcal{L}_q \in L^2(\mathbb{C}, dA)$, $q = 0, 1, \dots$, the eigenspaces of the magnetic Schrödinger operator $\mathbf{H} = -(\nabla + i\mathbf{A}(z))^2$, where $\mathbf{A} = \frac{1}{2}(y, -x)$, is the potential of the uniform magnetic field, $z = x + iy$. As it was found by V.Fock [**7**] in 1928, the eigenspace $\mathcal{L}_0$ consists of functions of the form $g = \sigma u \in L^2(\mathbb{C}, dA)$, $\sigma(z) = \pi^{-\frac{1}{2}} e^{-|z|^2/2} = \omega(z)^{\frac{1}{2}}$, with u being an entire analytic function on $\mathbb{C}$. Thus the unitary mapping $\mathbf{V} : L^2(\mathbb{C}, d\rho) \to L^2(\mathbb{C}, dA)$, $\mathbf{V} : u(z) \mapsto \sigma(z)u(z)$ maps isometrically the Fock space $\mathcal{F} = \mathcal{F}_{(1)}$ onto $\mathcal{L}_0$. Further on, the 'creation operator' $\mathbf{Q}^+ = (-\partial + \frac{1}{2}\bar{z})$ maps the Landau subspace $\mathcal{L}_q$ onto $\mathcal{L}_{q+1}$, isometrically, up to a constant factor depending on q, while the 'annihilation operator' $\mathbf{Q} = \bar{\partial} + \frac{1}{2}z$ maps $\mathcal{L}_q$ onto $\mathcal{L}_{q-1}$, $q > 0$, $\mathbf{Q}\mathcal{L}_0 = \{0\}$. Under the unitary transformation $\mathbf{V}$, the operators $\mathbf{Q}^+, \mathbf{Q}$ are transformed to the operators

$$\mathbf{S}^+ = \mathbf{V}^{-1}\mathbf{Q}^+\mathbf{V} = -\partial + \bar{z}, \quad \mathbf{S} = \mathbf{V}^{-1}\mathbf{Q}\mathbf{V} = \bar{\partial}. \tag{2.1}$$

Note that for the space $\mathbf{V}^{-1}\mathcal{L}_{q-1}$,

$$\bar{\partial}^q(\mathbf{V}^{-1}\mathcal{L}_{q-1}) = \mathbf{S}^q(\mathbf{V}^{-1}\mathcal{L}_{q-1}) = \mathbf{V}(\mathbf{Q}^q\mathcal{L}_{q-1})) = 0,$$

thus the space $\mathbf{V}^{-1}\mathcal{L}_{q-1} = \mathbf{V}^{-1}((\mathbf{Q}^+)^{q-1}\mathcal{L}_0) = (\mathbf{S}^+)^{q-1}\mathcal{F}_{(1)} = (\mathbf{S}^+)^{q-1}\mathcal{F}$ consists of q-analytic functions. On the other hand, since the subspaces $\mathcal{L}_{q-1} = \mathbf{Q}^{q-1}\mathcal{L}_0$ are mutually orthogonal, the subspace

$$(\mathbf{S}^+)^{q-1}\mathcal{F} = \mathbf{V}^{-1}(\mathbf{Q}^+)^{q-1}\mathcal{L}_0$$

is orthogonal to the previous subspaces $(\mathbf{S}^+)^l\mathcal{F}$, $l = 0, \dots, q-2$. Therefore,

$$(\mathbf{S}^+)^{q-1}\mathcal{F} = \mathcal{F}_{(q)} \tag{2.2}$$

is the space of true q-analytic functions, and the creation and annihilation operators $\mathbf{S}^+$ and $\mathbf{S}$ connect isometrically, up to multiplicative constants, the true poly-Fock spaces $\mathcal{F}_{(j)}$.

Another approach to these properties, based upon the Fourier representation of the true poly-Fock spaces is elaborated in [**25**], where, in particular, an exact isometry between these spaces is described.

THEOREM 2.1. *Given natural numbers $k < n$, the operator*

$$\sqrt{\frac{(k-1)!}{(n-1)!}}(\mathbf{S}^+)^{n-k}|_{\mathcal{F}_{(k)}} : \mathcal{F}_{(k)} \longrightarrow \mathcal{F}_{(n)}$$

is an isometric isomorphism, together with its inverse

$$\sqrt{\frac{(k-1)!}{(n-1)!}}\mathbf{S}^{n-k}|_{\mathcal{F}_{(n)}} : \mathcal{F}_{(n)} \longrightarrow \mathcal{F}_{(k)}.$$

2.3. The structure of the Bergman spaces on the half-plane and on the disk. Although the Bergman spaces on the disk and on the half-plane are related by the Cayley transform, the properties of the corresponding true poly-Bergman spaces differ somewhat.

For the case of the Bergman space on the upper half-plane, with the *standard orthonormal basis* $\sqrt{\frac{k+1}{\pi}}\frac{(z-i)^k}{(z+i)^{k+1}}$, $k = 0, 1, \ldots$ there exists a system of creation and annihilation operators, described in [**10**, **26**, **27**]. These operators are two-dimensional singular integral operators,

$$(\mathbf{S}_\Pi u)(w) = -\frac{1}{\pi}\int_\Pi \frac{u(z)dA(z)}{(z-w)^2} \qquad \text{and} \qquad (\mathbf{S}_\Pi^* u)(w) = -\frac{1}{\pi}\int_\Pi \frac{u(z)dA(z)}{(\bar{z}-\bar{w})^2}.$$

Understood in the principal value sense, they are bounded in $L^2(\Pi)$ and adjoint to each other. They are, in fact, the Beurling–Ahlfors operators compressed to the half-plane, and are surjective isometries,

$$\mathbf{S}_\Pi : \mathcal{A}_{(j)}(\Pi) \to \mathcal{A}_{(j+1)}(\Pi), \quad \mathbf{S}_\Pi : \tilde{\mathcal{A}}_{(j)}(\Pi) \to \tilde{\mathcal{A}}_{(j-1)}(\Pi), \quad j > 1, \tag{2.3}$$

and

$$\mathbf{S}_\Pi^* : \mathcal{A}_{(j+1)}(\Pi) \to \mathcal{A}_{(j)}(\Pi), \quad \mathbf{S}_\Pi^* : \tilde{\mathcal{A}}_{(j-1)}(\Pi) \to \tilde{\mathcal{A}}_{(j)}(\Pi), \quad j > 1, \tag{2.4}$$

while

$$\mathbf{S}_\Pi^* : \mathcal{A}_1(\Pi) \to \{0\}, \quad \mathbf{S}_\Pi : \tilde{\mathcal{A}}_1(\Pi) \to \{0\}.$$

Thus, we have surjective isometries

$$(\mathbf{S}_\Pi)^j \mathcal{A}(\Pi) \equiv (\mathbf{S}_\Pi)^j \mathcal{A}_{(1)}(\Pi) = \mathcal{A}_{(j+1)}(\Pi)$$

and

$$(\mathbf{S}_\Pi^*)^j \tilde{\mathcal{A}}(\Pi) \equiv (\mathbf{S}_\Pi^*)^j \tilde{\mathcal{A}}_{(1)}(\Pi) = \tilde{\mathcal{A}}_{(j+1)}(\Pi).$$

Formulas (2.3), (2.4), similar to (2.2), justify calling $\mathbf{S}_\Pi, \mathbf{S}_\Pi^*$ creation and annihilation operators.

REMARK 2.2. Here one can notice a certain discrepancy in notations: $\mathbf{S}$ denotes the annihilation operator in the poly-Fock spaces while $\mathbf{S}_\Pi$ denotes the creation operator in the poly-Bergman spaces – however, this is the tradition we do not want to break.

The operators $\mathbf{S}_\Pi^j$, restricted to $\mathcal{A}(\Pi)$, admit a representation, found in [**15**], which is much more convenient for using in further reductions.

THEOREM 2.3 ([**15**, Theorem 3.3]). *For* $u \in \mathcal{A}(\Pi)$,

$$(\mathbf{S}_\Pi^j u)(z) = \frac{\partial^j[(z-\bar{z})^j u(z)]}{j!}, j \geq 0. \tag{2.5}$$

Now we pass on to the case of the Bergman space on the unit disk $\mathbb{D}$, with the *standard orthonormal basis* $\sqrt{\frac{k+1}{\pi}}z^k$, $k = 0, 1, \ldots$. Similar to the case of the half plane we define the operators

$$(\mathbf{S}_\mathbb{D} u)(w) = -\frac{1}{\pi}\int_\mathbb{D} \frac{u(z)dA(z)}{(z-w)^2} \qquad \text{and} \qquad (\mathbf{S}_\mathbb{D}^* u)(w) = -\frac{1}{\pi}\int_\mathbb{D} \frac{u(z)dA(z)}{(\bar{z}-\bar{w})^2}.$$

These operator are however not isometries of true poly-Bergman spaces but only partial isometries. Let L_j denote the one-dimensional space $\{\tau\bar{z}^{j-1}, \tau \in \mathbb{C}\}$, $\widetilde{\mathrm{L}}_j =$

$\{\tau z^{j-1}, \tau \in \mathbb{C}\}$. As shown in [**11**, Theorem 3.5], the operator $S_{\mathbb{D}}$ acts surjective isometrically:

$$\mathbf{S}_{\mathbb{D}} : \mathcal{A}_{(j)}(\mathbb{D}) \ominus \mathrm{L}_j \to \mathcal{A}_{(j+1)}(\mathbb{D}),$$

while $\mathbf{S}_{\mathbb{D}}\mathrm{L}_j = \{0\}$ (with a natural modification for $\mathbf{S}^*_{\mathbb{D}}$.)

Our aim now is to present differential operators which can replace $\mathbf{S}^j_{\mathbb{D}}$, similarly to (2.5). We apply the results of A. Ramazanov [**17**, Theorem 1], see also [**18**, Theorem 1, Corollary 2], and [**14**, Theorem 4.1]. As it has been proved there (in the notations of the present paper), the operator

$$\Sigma_j : \mathcal{A}_{(1)}(\mathbb{D}) \ni u \mapsto \partial^j((1-|z|^2)^j u) \tag{2.6}$$

is a bounded and boundedly invertible operator from $\mathcal{A}_{(1)}(\mathbb{D}) \equiv \mathcal{A}(\mathbb{D})$ onto $\mathcal{A}_{(j+1)}(\mathbb{D})$. This means that there exist constants C_j such that

$$C_j^{-1}\|u\|_{L^2} \leq \|\Sigma_j u\|_{L^2} \leq C_j\|u\|_{L^2}, u \in \mathcal{A}(\mathbb{D}).$$

In other words, the mapping Σ_j is a Banach isomorphism, but not an isometry, of the standard Bergman space $\mathcal{A}(\mathbb{D})$ onto the true poly-Bergman space $\mathcal{A}_{(j+1)}(\mathbb{D})$.

REMARK 2.4. It might be tempting to use the isometry of Bergman spaces $W : u(z) \mapsto (1-z)^{-1}u(i\frac{z+1}{1-z})$, $W : \mathcal{A}(\mathbb{D}) \to \mathcal{A}(\Pi)$, generated by the standard Möbius transform, to derive unitary isomorphisms in the differential form between true poly-Bergman spaces on the disk from the ones found for the half plane, instead of Banach isomorphisms. This idea is, unfortunately, doomed since true poly-Bergman spaces of order higher than 1 are not invariant under conformal mappings, as simple examples (say, $u(z) = |z|^2$) show (one more drawback of polyanalytic spaces).

3. Toeplitz operators with distributional symbols

The standard definition of a Toeplitz operator in a Bergman type space $\mathbf{B}^2(\Omega) \subset L^2(\Omega)$ as $u \mapsto \mathbf{P}Fu$, where $\mathbf{P}$ is the orthogonal projection from $L^2(\Omega)$ to $\mathbf{B}^2(\Omega)$, works nicely for bounded functions F, since Fu may be outside $L^2(\Omega)$ for an unbounded F. This restriction can be somewhat relaxed, by admitting *some* classes of functions $F \in L^2(\Omega)$, since functions in Bergman type spaces are infinitely differentiable and, therefore, locally bounded. However further extension of *this* definition to more singular objects F acting as symbols encounters serious obstacles.

In [**21**, **22**], an approach was developed to define Toeplitz operators with fairly singular symbols. This approach is based upon a representation of usual Toeplitz operators by means of sesquilinear forms. If, initially, F is a bounded function on Ω, the sesquilinear form of the Toeplitz operator $\mathbf{T}_F = \mathbf{P}F$ in $\mathbf{B}^2$ can be transformed to

$$\langle \mathbf{T}_F u, v\rangle = \langle \mathbf{P}Fu, v\rangle = \langle Fu, \mathbf{P}v\rangle = \langle Fu, v\rangle, \quad u, v \in \mathbf{B}^2. \tag{3.1}$$

The right-hand side in (3.1) may make sense and define a bounded sesquilinear form for rather general objects acting as F. Many examples of such objects are presented in [**21**–**23**].

3.1. Toeplitz operators in Bergman spaces on the disk and on the half-plane with symbols being derivatives of k–C measures. In $L_2(\Omega, dA)$, $\Omega = \mathbb{D}$ or $\Omega = \Pi$, the L^2 scalar product is consistent, up to the complex conjugation, with the standard action of a distribution-function or distribution-measure on a smooth function on Ω. Therefore, the Toeplitz operator in $\mathcal{A}(\Omega)$ with distributional symbol $F \in \mathcal{D}'(\Omega)$ is determined by the sesquilinear form on the right-hand side in

$$\mathbf{f}_F[u, v] = \langle Fu, v \rangle_{L^2(\Omega, dA)} := (F, u\bar{v}) \tag{3.2}$$

(the expression on the right-hand side in (3.2) is understood as the action of the distribution F on the smooth and, actually, real-analytic, function $u(z)\bar{v}(z)$). This sesquilinear form is defined initially on the linear space of such functions $u, v \in \mathcal{A}$, for which the expression in (3.2) is finite. In particular, in the cases we consider here, the distributions in question admit a natural extension from $\mathcal{D}(\Omega)$ (there are no nontrivial real analytic functions in $\mathcal{D}(\Omega)$) to finite linear combinations of functions in the standard bases. Further, *if* the sesquilinear form (3.2) turns out to be bounded in the norm of $\mathcal{A}(\Omega)$ for such functions, it is extended to the whole of $\mathcal{A}(\Omega)$ by L^2- continuity. If it is the case, the action of the Toeplitz operator with this distributional symbol is

$$(\mathbf{T}_F u)(z) = \mathbf{f}_F[u, \kappa(z, \cdot)],$$

where, recall, $\kappa(z, \cdot)$ is the reproducing kernel for the space $\mathcal{A}(\Omega)$.

Here we describe briefly the realization of this approach as applied to a wide class of symbols-distributions, being derivatives of k–Carleson measures on the disk or on the half-plane. The details of this realization differ somewhat from the case of the Fock space which will be explained later on.

DEFINITION 3.1. A measure ν on Ω is called k–Carleson measure for derivatives (shortly, k–C measure) *for the Bergman space* $\mathcal{A}(\Omega)$, Ω being $\mathbb{D}$ or Π, if

$$\mathbf{f}_{\nu,\kappa}[u] = \int_\Omega |\partial^k u(z)|^2 d\nu(z) \le C\|u\|^2_{L^2(\Omega, dA)}$$

for any function $u \in \mathcal{A}(\Omega)$.

In [**22, 23**], for the Bergman spaces on $\mathbb{D}$ and Π, conditions, sufficient for a measure to be a k–C measure, have been found. (These conditions are even necessary for the case of a positive measure.) Under these conditions, extended in a natural way to half-integer values of k, the sesquilinear form

$$\mathbf{f}_{\nu,\alpha,\beta}[u, v] = \int_\Omega \partial^\alpha u \overline{\partial^\beta v} d\nu \equiv (-1)^{\alpha+\beta}(\partial^\alpha \bar{\partial}^\beta \nu, u\bar{v}), \quad u, v \in \mathcal{A}(\Omega),\ \alpha + \beta = 2k, \tag{3.3}$$

is bounded in the corresponding Bergman type space $\mathcal{A}(\Omega)$, where the derivative of a measure is understood in the sense of distributions.

With this in view, for $\alpha, \beta \ge 0$, we define the Toeplitz operator in $\mathcal{A}(\Omega)$ with symbol $\partial^\alpha \bar{\partial}^\beta \nu$ as determined by the sesquilinear form $(-1)^{\alpha+\beta}\mathbf{f}_{\nu,\alpha,\beta}[u, v]$ in (3.3), provided that the latter form is bounded in $\mathcal{A}(\Omega)$.

In the above cited papers, some sharp estimates have been established, involving, in particular, the dependence of the norm of the corresponding operator on the differentiation orders α, β. Here we do not need such details, so we present simplified versions. Below, $B(z,r)$ denotes the disk with center z and radius r (this r can be chosen arbitrarily) and $|\nu|$ denotes the variation of the measure ν.

THEOREM 3.2 ([**22**, Theorem 6.1]). *Let ν be a measure on $\mathbb{D}$ and suppose that for some integer or half-integer $k \geq 0$,*

$$C_k(\nu, \mathbb{D}) = \sup_{z\in\mathbb{D}} \left\{(1-|z|)^{-2(k+1)}|\nu|(B(z, \tfrac{1}{2}(1-|z|)))\right\} < \infty. \tag{3.4}$$

Then, for any $u, v \in \mathcal{A}(\mathbb{D})$ and any $\alpha, \beta : \alpha + \beta = 2k$, the inequality

$$|\mathbf{f}_{\nu,\alpha,\beta}[u,v]| \equiv |(\partial^\alpha \bar{\partial}^\beta \nu, u\bar{v})| \leq c_k C_k(\nu, \mathbb{D}) \|u\|_{\mathcal{A}(\mathbb{D})} \|v\|_{\mathcal{A}(\mathbb{D})}$$

holds, with constant c_k depending on k only, i.e., the measure ν is a k-C measure on $\mathbb{D}$

The result for the Bergman space on Π is the following (see [**23**]).

THEOREM 3.3. *Let ν be a measure on Π and suppose that for some integer or half-integer $k \geq 0$,*

$$C_k(\nu, \Pi) = \sup_{z\in\Pi} \left\{(\operatorname{Im} z)^{-2(k+1)}|\nu|(B(z, \tfrac{1}{2}\operatorname{Im} z))\right\} < \infty. \tag{3.5}$$

Then, for any $u, v \in \mathcal{A}(\Pi)$ and any $\alpha, \beta : \alpha + \beta = 2k$, the inequality

$$|\mathbf{f}_{\nu,\alpha,\beta}[u,v]| \equiv |(\partial^\alpha \bar{\partial}^\beta \nu, u\bar{v})| \leq c_k C_k(\nu, \Pi) \|u\|_{\mathcal{A}(\Pi)} \|v\|_{\mathcal{A}(\Pi)}.$$

holds, with constant c_k depending on k only, i.e., the measure ν is a k-C measure on Π.

A bounded sesquilinear form in a Hilbert space generates a bounded operator.

THEOREM 3.4. *Under conditions* (3.4), *resp.,* (3.5), *the Toeplitz operator with distributional symbol $\partial^\alpha \bar{\partial}^\beta \nu$, $\alpha + \beta = 2k$ is a bounded operator in $\mathcal{A}(\mathbb{D})$, resp., $\mathcal{A}(\Pi)$.*

As usual, the results on the boundedness of certain classes of Toeplitz operators are accompanied by analogous results on the compactness conditions, proved in a standard way.

THEOREM 3.5. *a) Let ν be a measure on $\mathbb{D}$ and*

$$\lim_{\varepsilon\to 0} \sup_{z\in\mathbb{D}, |z|>1-\varepsilon} \left\{(1-|z|)^{-2(k+1)}|\nu|(B(z, \tfrac{1}{2}(1-|z|)))\right\} = 0. \tag{3.6}$$

Then the Toeplitz operator in $\mathcal{A}(\mathbb{D})$ with distributional symbol $\partial^\alpha \bar{\partial}^\beta \nu$, $\alpha + \beta = 2k$, is compact.
b) Let ν be a measure on Π and

$$\lim_{\varepsilon\to 0} \sup_{z\in\Pi, \operatorname{Im} z<\varepsilon} \left\{(\operatorname{Im} z)^{-2(k+1)}|\nu|(B(z, \tfrac{1}{2}\operatorname{Im} z))\right\} = 0. \tag{3.7}$$

Then the Toeplitz operator in $\mathcal{A}(\Pi)$ with distributional symbol $\partial^\alpha \bar{\partial}^\beta \nu$, $\alpha + \beta = 2k$, is compact.

Measures satisfying (3.6), (3.7) are called vanishing k–C measures in [**22, 23**].

3.2. Toeplitz operators in $\mathcal{F}$ with symbols being coderivatives of k–FC measures.

DEFINITION 3.6. Let ν be a measure on $\mathbb{C} \equiv \mathbb{R}^2$. The measure ν is called a Fock-Carleson measure for derivatives of order k (shortly, a k-FC measure) if

$$\int_{\mathbb{C}} |\partial^k u|^2 \omega(z) d|\nu|(z) \leq \|u\|_{\mathcal{F}}^2, u \in \mathcal{F}.$$

In [**21**, Theorem 5.4] sufficient conditions for a measure on $\mathbb{C}$ to be a k-FC measure, which are also necessary for positive measures, have been found. These conditions enable one to extend the notion of k-C measure to half-integer values of k.

We define a *coderivative* $\boldsymbol{\partial} F$ of a distribution $F \in \mathcal{D}'(\mathbb{C})$ as

$$\boldsymbol{\partial} F = \omega(z)^{-1} \partial(\omega(z) F) = \left(\partial - \tfrac{1}{2}\bar{z}\right) F. \tag{3.8}$$

With this definition,

$$(\partial(\omega F), \phi) = \langle \boldsymbol{\partial} F, \bar{\phi} \rangle_{L^2(\mathbb{C}, d\rho)}, \tag{3.9}$$

in the case both sides of (3.9) make sense. Thus, the coderivative of a distribution defined in (3.8) is consistent with the usual scalar product in the weighted space.

The Toeplitz operator with coderivative of a k-FC measure ν is defined by the sesquilinear form

$$\begin{aligned} \mathbf{f}_{\alpha,\beta,\nu}[u,v] &\equiv (-1)^{\alpha+\beta} (\omega \boldsymbol{\partial}^\alpha \bar{\boldsymbol{\partial}}^\beta \nu, u\bar{v}) \\ \equiv (\omega\nu, \partial^\alpha u \overline{\partial^\beta v}) &\equiv \int_{\mathbb{C}} \partial^\alpha u(z) \overline{\partial^\beta v(z)} \omega(z) d\nu(z). \end{aligned} \tag{3.10}$$

A sufficient condition for boundedness of the sesquilinear form (3.10) in $\mathcal{F}$ was found in [**21**]:

THEOREM 3.7 ([**21**, Theorem 5.4]). *Let, for some integer or half-integer $k \geq 0$ the measure ν on $\mathbb{C} \equiv \mathbb{R}^2$ satisfy the condition*

$$C_k(\nu, \mathbb{C}) = \sup_{z \in \mathbb{C}} \{(1 + |z^2|)^k |\nu|(B(z,r))\} < \infty. \tag{3.11}$$

Then for any $u, v \in \mathcal{F}$ and $\alpha, \beta : \alpha + \beta = 2k$, the inequality

$$|\mathbf{f}_{\nu,\alpha,\beta}[u,v]| \leq c_k C_k(\nu) \|u\|_{\mathcal{F}} \|v\|_{\mathcal{F}}$$

holds, with constant c_k depending on k only.

Thus, under the condition (3.11), the Toeplitz operator

$$\mathbf{T} : u \mapsto \mathbf{f}_{\nu,\alpha,\beta}[u, \kappa(z,\cdot)]$$

is bounded in $\mathcal{F}$. In a similar way to Theorem 3.5 the condition for compactness is found:

THEOREM 3.8. *Suppose that the measure ν on $\mathbb{C}$ satisfies*

$$\lim_{R \to \infty} \sup_{z \in \mathbb{C}, |z| > R} \{(1 + |z^2|)^k |\nu|(B(z,r))\} = 0.$$

Then the operator $\mathbf{T}_{\nu,\alpha,\beta}$ is compact.

The iterated coderivative $\boldsymbol{\partial}$ in (3.8) can be expressed via the usual derivative and vice versa.

PROPOSITION 3.9. *Let $F \in \mathcal{S}'(\mathbb{C})$ be a distribution. Then, for some polynomials $\mathbf{q}_k(s,t)$ and $\tilde{\mathbf{q}}_k(s,t)$ of degree k,*

$$\boldsymbol{\partial}^k = \mathbf{q}_k(\partial, z) \tag{3.12}$$

and

$$\partial^k = \tilde{\mathbf{q}}_k(\boldsymbol{\partial}, z). \tag{3.13}$$

PROOF. The relation (3.12) follows by iteration of (3.8); the inverse, (3.13), is obtained by solving the triangular system of equations $\boldsymbol{\partial}^j = \mathbf{q}_j(\partial, z)$, $j = 1, \dots, \kappa$. □

We always mean that (this can be achieved by means of commutation) in each term in the polynomials $\mathbf{q}_k(\partial, z), \tilde{\mathbf{q}}_k(\boldsymbol{\partial}, z)$, first the multiplication by a power of z and then the differentiation are applied.

4. Transformations of Toeplitz operators

Having an operator $\mathbf{T}$ in the Hilbert space $\mathcal{H}$ and an isometry (or a Banach isomorphism) $\mathbf{Z} : \mathcal{H} \to \mathcal{K}$ of $\mathcal{H}$ onto another Hilbert space $\mathcal{K}$, one can consider the operator $\mathbf{T}^{\mathbf{Z}} = \mathbf{Z}\mathbf{T}\mathbf{Z}^*$ in the space $\mathcal{K}$ on the proper domain. Of course, a lot of properties of $\mathbf{T}$ and $\mathbf{T}^{\mathbf{Z}}$ coincide. This simple idea can be successfully applied to Toeplitz operators in true Bergman type spaces described above. We will show in this Section that Toeplitz operators with symbol-function in true poly-Fock and in true poly-Bergman spaces on the half-plane and on the disk are unitary equivalent (or, in the latter case cosimilar) to certain Toeplitz operators in the corresponding standard spaces with symbols obtained by applying proper elliptic differential operators to the initial symbols.

4.1. Transformation of Toeplitz operators in the true poly-Bergman space on the half-plane. Let F be a distribution on Π, defined at least on $C^\infty(\bar{\Pi}) \cap L^1(\Pi)$. We consider the sesquilinear form

$$\mathbf{t}_F[u, v] = \langle Fu, v\rangle_{L^2(\Pi, dA)}, \quad u = \mathbf{S}_\Pi^j f \in \mathcal{A}_{(j+1)}(\Pi), \quad v = \mathbf{S}_\Pi^j g \in \mathcal{A}_{(j+1)}(\Pi),$$

with f, g being elements of the standard orthonormal basis (further on in this subsection we will only use the Hilbert space $L^2(\Pi, dA)$ and therefore we will suppress the notation of the space in the scalar product.)

The following result leads to establishing a relation between Toeplitz operators in the true poly-Bergman space $\mathcal{A}_{(j)}(\Pi)$ and the Bergman space $\mathcal{A}(\Pi)$.

PROPOSITION 4.1. *Let F be a distribution in the half-plane Π. Then*

$$\langle F\mathbf{S}_\Pi^j f, \mathbf{S}_\Pi^j g\rangle = \langle (\mathbf{K}F) f, g\rangle, \tag{4.1}$$

with $\mathbf{S}$ defined in (2.5) where $\mathbf{K}$ is a differential operator of order $2j$ having the form

$$\mathbf{K} = \underline{\mathbf{K}}(\Delta(y^2\cdot), \bar{\partial}(y\cdot), \partial(y\cdot)), \tag{4.2}$$

with $\underline{\mathbf{K}}$ being a polynomial of degree j. Moreover, if we assign the weight -1 to the differentiation and the weight 1 to the multiplication by y, with weights adding under the multiplication, then all monomials in $\underline{\mathbf{K}}$ have weight 0.

PROOF. We demonstrate the reasoning for the case $j = 1$. The general case uses the same machinery with some tedious bookkeeping. We consider the sesquilinear form $\mathbf{f}_F[f,g] = \langle F\partial(yf), \partial(yg)\rangle \equiv (F, \partial(yf)\overline{\partial(yg)})$ for f, g being some elements in the standard orthonormal basis in $\mathcal{A}(\Pi)$. Due to $\partial^* = -\bar{\partial}$:

$$(4.3)\qquad \begin{aligned}(F, \partial(yf)\overline{\partial(yg)}) &= (F, (-if + y\partial f)\overline{(-ig + y\partial g)})\\ &= (F, f\bar{g}) + (F, -ify\bar{\partial}\bar{g}) + (F, i\bar{g}y\partial f) + (F, y^2(\partial f)\bar{\partial}\bar{g})\end{aligned}$$

(the last transformation uses $\bar{\partial}g = 0$). For the second term on the right-hand side in (4.3), by the general rules of manipulation with distributions, we have

$$(F, -ify\bar{\partial}\bar{g}) = (-iyF, f\overline{\partial g}) = (-iyF, \overline{\partial(\bar{f}g)}) - (-iyF, (\bar{\partial}f)g) = (\bar{\partial}(iyF), f\bar{g}),$$

because $\bar{\partial}f = 0$. The third term on the right in (4.3) is transformed in a similar way, and for the last one,

$$\begin{aligned}(F, y^2\partial f\bar{\partial}\bar{g}) &= (y^2F, \bar{\partial}(\partial f\bar{g}) - \bar{\partial}(\partial f)\bar{g}) = -(\bar{\partial}(y^2F), (\partial f)\bar{g})\\ &= -(\bar{\partial}(y^2F), \partial(f\bar{g}) - \{f\partial\bar{g})\}) = (\partial\bar{\partial}(y^2F), f\bar{g}),\end{aligned}$$

again, the terms in curly bracket vanishing due to $\bar{\partial}g = 0$. Collecting the terms in (4.3), after simple transformations, we obtain the required relation.

Note that we would arrive at the same result, and even somewhat faster, by making formal commutations of terms in the expression $\langle F\partial(yf), \partial(yg)\rangle$, using the commutation relations $[F, \partial] = -(\partial F)$, and $[[F, \partial], \bar{\partial}]F = \frac{1}{4}\Delta F$, obtain the representation (4.2).

For higher order, the procedure of transformation is similar, by means of formally commuting F and factors in the creation operators $\mathbf{S}^j_\Pi$ in the expression $\langle F\mathbf{S}^j_\Pi f, \mathbf{S}^j_\Pi g\rangle \equiv (F, \mathbf{S}^j_\Pi f\overline{\mathbf{S}^j_\Pi g})$, so that the Cauchy-Riemann operator falls on the functions f, g, while any commutation with F produces a derivative of F. It remains to notice that by the commuting operation of the terms in the expression on the left-hand side in (4.1) the weight of the terms does not change. Alternatively, one can make the calculations similar to the ones shown above, again by moving the Cauchy-Riemann operator to the functions f, g and on F. □

As before, the equality (4.1) extends to the whole of $\mathcal{A}(\Pi)$, as soon as we know that the right-hand side or on the left-hand side is a bounded sesquilinear form in $\mathcal{A}(\Pi)$. Thus, the statement of Proposition 4.1 can be formulated as the following theorem.

THEOREM 4.2. *The operators $\mathbf{T}_F(\mathcal{A}_{(j+1)}(\Pi))$ and $\mathbf{T}_{\mathbf{K}F}(\mathcal{A}(\Pi))$ are unitarily equivalent (up to a numerical factor) as soon as one of them is bounded. In this case, if one of these operators is compact, or belongs to a Schatten class, or is finite rank, or zero, then the same holds for the other one.*

4.2. Transformation of Toeplitz operators in the true poly-Bergman space on the disk. The case of true poly-Bergman spaces on the disk is considered in the same way, with a single essential difference consisting in replacing the unitary equivalence of Toeplitz operators by their cosimilarity (see Definition 4.4). The proposition to follow is analogous to Proposition 4.1.

PROPOSITION 4.3. *Let F be a distribution in the unit circle $\mathbb{D}$, defined at least on the functions in $C^\infty(\bar{\mathbb{D}})$ Then for functions f, g, being linear combinations of* *the standard basis functions in* $\mathcal{A}(\mathbb{D})$,

$$\mathbf{f}_F[f,g] \equiv (F\Sigma_j f, \overline{\Sigma_j g}) \equiv (F, \Sigma_j f\overline{\Sigma_j g}) = ((\mathbf{M}_j F)f, g), \tag{4.4}$$

where Σ_j *is the operator* (2.6), $\mathbf{M}$ *is an elliptic differential operator of degree* $2j-2$, $\mathbf{M} = \underline{\mathbf{M}}(\Delta(1-|z|^2)^2\cdot, \partial z(1-|z|^2)\cdot, \bar{z}\bar{\partial}(1-|z|^2)\cdot)$, *with a polynomial* $\underline{\mathbf{M}}$ *of degree* j. *In this polynomial, if we assign the weight* -1 *to differentiation, the weight* 1 *to* $(1-|z|^2)$ *and the weight* 0 *to* $z, \bar{z}$, *with weights adding under multiplication, all monomials have weight* 0.

PROOF. We will suppress $L^2(\mathbb{D}, dA)$ in the notation of the scalar product. Again, we demonstrate (4.4) for $j=2$ and then indicate how the general case is handled. We perform calculations, similar to the ones in (4.3):

$$\begin{aligned} \langle F\Sigma_1 f, \Sigma_1 g\rangle &= C(F, \partial(1-|z|^2)f\overline{\partial(1-|z|^2)g}) \\ = C(F, \partial((1-|z|^2)f\overline{\partial(1-|z|^2)g})) &- C(F, (1-|z|^2)f\partial\overline{\partial(1-|z|^2)g}) = \\ -C((1-|z|^2)\bar{\partial}F, f\overline{\partial(1-|z|^2)g}) &- ((1-|z|^2)F, f\bar{\partial}\partial(1-|z|^2)\bar{g}). \end{aligned} \tag{4.5}$$

For the first term in (4.5), we have

$$\begin{aligned} &-((1-|z|^2)\partial F, f\bar{\partial}((1-|z|^2)\bar{g})) \\ &= -((1-|z|^2)\partial F, \bar{\partial}(f(1-|z|^2)\bar{g}) + (\bar{\partial}f)(1-|z|^2)\bar{g}) \\ &= ((1-|z|^2)\bar{\partial}(1-|z|^2)\partial F, f\bar{g}) = \tfrac{1}{4}(1-|z|^2)^2\Delta F - (1-|z|^2)z\bar{\partial}F, f\bar{g}). \end{aligned} \tag{4.6}$$

A similar transformation takes care of the second term in (4.5). After collecting all terms, and moving all differentiations to the left, we obtain (4.4). For a higher order, one should perform analogous commutations of the terms in the operator Σ_j, noticing that the weight of terms does not change under these commutation. □

Again, the equality (4.4) extends to all functions $f, g \in \mathcal{A}(\mathbb{D})$ as soon as we know that the right-hand side or the left-hand side is a bounded sesquilinear form in $\mathcal{A}(\mathbb{D})$.

We introduce here the notion of *cosimilar* operators.

DEFINITION 4.4. Let $\mathcal{H}, \mathcal{K}$ be Hilbert spaces. Operators $\mathbf{T}_1$ in $\mathcal{H}$ and $\mathbf{T}_2$ in $\mathcal{K}$ are called *cosimilar* if there exists a bounded and invertible operator $\mathbf{Z} : \mathcal{H} \to \mathcal{K}$ such that $\mathbf{T}_1 = \mathbf{Z}^*\mathbf{T}_2\mathbf{Z}$.

Thus, Proposition 4.3 leads to the following relation between operators.

THEOREM 4.5. *The operators* $\mathbf{T}_F(\mathcal{A}_{(j+1)}(\mathbb{D}))$ *and* $\mathbf{T}_{\mathbf{M}F}(\mathcal{A}(\mathbb{D}))$ *are cosimilar as soon as one of them is bounded. In this case, if one of these operators is compact, or belongs to some Schatten class, or is finite rank, or is the zero operator, then the same holds for the other one.*

PROOF. According to the definition of Toeplitz operators by means of their sesquilinear forms, the relation (4.4) can be written as

$$\langle \mathbf{T}_F(\mathcal{A}_{(j+1)}(\mathbb{D}))\Sigma_j f, \Sigma_j g\rangle = \langle \mathbf{T}_{\mathbf{M}F}(\mathcal{A}(\mathbb{D}))f, g\rangle,$$

which means

$$\Sigma_j^*\mathbf{T}_F(\mathcal{A}_{(j+1)}(\mathbb{D}))\Sigma_j = \mathbf{T}_{\mathbf{M}F}(\mathcal{A}(\mathbb{D})),\ \mathbf{T}_F(\mathcal{A}_{(j+1)}(\mathbb{D})) = (\Sigma_j^{-1})^*\mathbf{T}_{\mathbf{M}F}(\mathcal{A}(\mathbb{D}))\Sigma_j^{-1}.$$

The operator Σ_j^{-1} is a Banach isomorphism, therefore, the operators in question are cosimilar, with Σ_j^{-1} playing the role of $\mathbf{Z}$. The remaining statements of Theorem follow automatically, since the properties in question are invariant under the multiplication by a bounded invertible operator. □

4.3. Transformation of Toeplitz operators in the true poly-Fock space. We start by recalling Lemma 9.2 in [**5**].

THEOREM 4.6. *Consider the orthonormal in $L^2(\mathbb{C})$ basis in the lowest Landau subspaces $\mathcal{L}_0$, $\varphi_k = \frac{z^k}{\sqrt{k!}}\sigma(z)$ (recall that $\sigma(z) = \pi^{-1/2}e^{-|z|^2/2}$). Let F be a distribution in $\mathcal{S}'(\mathbb{C})$. Then for any basis functions, φ_k, φ_l*

$$\langle F(Q^+)^q\varphi_k, (Q^+)^q\varphi_l\rangle_{L^2(\mathbb{C},dA)} = \langle(\mathbf{D}_qF)\varphi_k, \varphi_l\rangle_{L^2(\mathbb{C},dA)}, \tag{4.7}$$

where $\mathbf{D}_q = \mathbf{D}_q(\Delta)$ is a constant coefficients differential operator (understood as acting in $\mathcal{S}'$), being a polynomial of degree q with positive coefficients of the Laplacian in $\mathbb{C} \equiv \mathbb{R}^2$.

Note that in [**5**], the objects in (4.7) are defined consistently with our definition in Section 3; say,

$$\langle F(Q^+)^q\varphi_k, (Q^+)^q\varphi_l\rangle_{L^2(\mathbb{C},dA)} := (F, (Q^+)^q\varphi_k\overline{(Q^+)^q\varphi_l}).$$

In [**5**] explicit formulas for coefficients of the operator $\mathbf{D}_q(\Delta)$ have been derived, however they are not needed in our study. A simple extension of this result is, however, important. The polynomial $\mathbf{D}_q$ does not depend on the choice of the basis functions in (4.7). Therefore, (4.7) extends by linearity to finite linear combinations of basis functions and further, if by some reason, for a certain F, we happen to know that

$$|\langle(\mathbf{D}_qF)\phi, \psi\rangle_{L^2(\mathbb{C},dA)}| \le C\|\phi\|_{L^2(\mathbb{C},dA)}\|\psi\|_{L^2(\mathbb{C},dA)}$$

for ϕ, ψ being such linear combinations, the equality (4.7) extends by continuity to the whole of $\mathcal{L}_0$.

Now we use the surjective isometry $\mathbf{V} : L^2(\mathbb{C}, d\rho) \to L^2(\mathbb{C}, dA)$, $\mathbf{V}f = \sigma(z)f$, which, as it is explained in Section 2.2, relates the creation-annihilation structure in true poly-Fock spaces with the one in Landau subspaces, see (2.1). Setting $\phi = \mathbf{V}f \in \mathcal{L}_0$, $\psi = \mathbf{V}g \in \mathcal{L}_0$ for $f, g \in \mathcal{F}$, we obtain

$$\begin{aligned}\langle F(\mathbf{S}^+)^qf, (\mathbf{S}^+)^qg\rangle_{L^2(\mathbb{C},d\rho)} &= \langle\mathbf{V}F\mathbf{V}^{-1}\mathbf{V}(\mathbf{S}^+)^qf, \mathbf{V}(\mathbf{S}^+)^qg\rangle_{L^2(\mathbb{C},dA)}\\ &= (F, \mathbf{V}(\mathbf{S}^+)^qf\overline{\mathbf{V}(\mathbf{S}^+)^qg}) = (F, \mathbf{V}(\mathbf{S}^+)^q\mathbf{V}^{-1}\phi\overline{\mathbf{V}(\mathbf{S}^+)^q\mathbf{V}^{-1}\psi})\\ &= (F, (Q^+)^q\phi\overline{(Q^+)^q\psi}) = ((\mathcal{D}_qF), \mathbf{V}f\bar{\mathbf{V}}g).\end{aligned}$$

For $F \in \mathcal{S}'$, the above equality holds for every $\phi = \mathbf{V}f$, $\psi = \mathbf{V}g$ with f, g being elements in the standard monomial basis in $\mathcal{F}$, and, therefore it extends to the finite linear combinations of such elements. Suppose that we know that the inequality

$$|((\mathbf{D}_qF), \mathbf{V}f\overline{\mathbf{V}g})| \le C\|f\|_{\mathcal{F}}\|g\|_{\mathcal{F}} \tag{4.8}$$

is satisfied for elements in this subspace. Since $\mathbf{Q}^+$ is (up to a constant) a surjective isometry of Landau subspaces, (4.8) means that

$$\begin{aligned}|\mathbf{f}_F[u, v]| \equiv |\langle Fu, v\rangle_{L^2(\mathbb{C},d\rho)}| &\le C\|u\|_{L^2(\mathbb{C},d\rho)}\|v\|_{L^2(\mathbb{C},d\rho)},\\ u = (\mathbf{Q}^+)^qf \in \mathcal{F}_{(q+1)},\ & v = (\mathbf{Q}^+)^qg \in \mathcal{F}_{(q+1)}\end{aligned} \tag{4.9}$$

for a dense linear subspace of f, g in $\mathcal{F}$, and, therefore, extends to the whole of $\mathcal{F}$. Thus we arrive to the following transformation result.

PROPOSITION 4.7. *Let for some $F \in \mathcal{S}'(\mathbb{C})$ the inequality* (4.9) *be satisfied for all u, v in a dense linear subset of linear combinations of basic functions in the true poly-Fock space $\mathcal{F}_{(q+1)}$. Then*

$$\langle F(\mathbf{S}^+)^q f, (\mathbf{S}^+)^q g\rangle_{\mathcal{F}_{(q+1)}} = \langle \mathbf{V}^{-1}(\mathbf{D}_q F)\mathbf{V} f, g\rangle_{\mathcal{F}} = (\mathbf{D}_q F, \mathbf{V} f \mathbf{V}\bar{g}),$$

where $f = \mathbf{Q}^q u \in \mathcal{F}$ and $g = \mathbf{Q}^q v \in \mathcal{F}$.

5. Properties of Toeplitz operators in true poly-Bergman type spaces

Now we use our transformation results to investigate some properties of operators in true spaces of polyanalytic functions. Since the formulations and explanations are mostly the same for all three types of spaces we consider in this paper, we will refer to all of them as to true poly-Bergman type spaces (*tpBt* spaces.)

5.1. Compactness and degeneracy.

5.1.1. *Symbols with compact support and generalizations.* We start with the most simple results. It was established in [**21**–**23**] that a Toeplitz operator with symbol-distribution having compact support in $\mathbb{C}$, $\mathbb{D}$, Π is compact in the corresponding Bergman type space. Of course, if the symbol F belongs to $\mathrm{E}'(\Omega)$, $\Omega = \mathbb{C}, \mathbb{D}$ or Π, the result of applying a differential operator to F has compact support as well. This leads to the first, quite simple, result.

THEOREM 5.1. *If the symbol F has compact support in Ω then the Toeplitz operator with symbol F in the corresponding* tpBt*-space is compact.*

Now, as it will happen repeatedly further on, it is quite possible that the symbol F, probably, does not have compact support, but, after applying to F a differential operator $\mathbf{E}$ (this is $\mathbf{K}, \mathbf{M}, \mathbf{D}$, depending on the space), we obtain the symbol $\mathbf{E}F$ with compact support. Thus, we have the following theorem.

THEOREM 5.2. *Suppose that for a symbol F the symbol $\mathbf{E}F$ has compact support. Then the Toeplitz operator with symbol F in the corresponding* tpBt *space is compact.*

REMARK 5.3. As explained in [**22**, Section 5], a Toeplitz operator with a compactly supported distributional symbol in $\mathcal{A}(\mathbb{D})$ has eigenvalues (and s-numbers, if the operator is not self-adjoint) tending to zero at least exponentially. The same holds for operators in the space $\mathcal{A}(\Pi)$; for the Fock space, the eigenvalues (s-numbers) decay superexponentially. By our transformation theorems, the same takes place for Toeplitz operators with compactly supported symbol, or even symbol with compactly supported $\mathbf{E}F$, in the corresponding *tpBt* space. We are going to explore this topic in later publications.

5.1.2. *Uniqueness.* It is well known that a Toeplitz operator in Bergman type spaces with more or less non-crazy symbol cannot be zero unless the symbol is zero. For example, if a radial symbol in the Fock space grows at infinity not faster than $\exp a|z^2|$, $a < 1$, such uniqueness takes place (examples of non-uniqueness, therefore, involve symbols growing at infinity *very* rapidly, see [**8**]). The same is correct for operators in Bergman spaces on the disk and on the half-plane. To break uniqueness, the symbol should grow rapidly and oscillate very fast near the boundary. For operators in *tpBt* spaces the situation is completely different.

Suppose that F is a symbol in Ω satisfying in Ω the equation $\mathbf{E}F = 0$ in the sense of distributions, where, as before $\mathbf{E}$ is $\mathbf{K}, \mathbf{M}$ or $\mathbf{D}$, depending on the space in question. Then, by the results of Section 2.2, the Toeplitz operator with symbol F in the corresponding *tpBt* space, will be the zero operator. We delay to another publication the detailed analysis of the properties of such equations. We note a simple, however not quite trivial, special case.

THEOREM 5.4. *Let F be a bounded function (of course, this restriction can be relaxed) with compact support in Ω. Suppose that the Toeplitz operator in a tpBt space with symbol F is zero. Then F is zero.*

Not giving a formal proof, we just explain that the operator with symbol $\mathbf{E}F$ in the usual Bergman space is zero, therefore, $\mathbf{E}f = 0$. A solution of this equation is a real-analytic function and therefore cannot have compact support unless it is zero everywhere.

5.1.3. *The finite rank property.* Recently there has been certain research activity concerning the problem of Toeplitz operators in Bergman type spaces having finite rank. Some final results on this topic have been obtained in [**1**, **24**] (see also the bibliography there.) Due to these results, a Toeplitz operator in $\mathcal{A}(\mathbb{D})$ and $\mathcal{F}$ with distributional symbol G, subject to some restrictions concerning 'growth at infinity', may have finite rank if only if the symbol G is, in fact, a finite sum of point masses and their derivatives. In particular, this characterization holds for symbols with compact support. More generally, if a distribution G satisfies these growth conditions then any derivative of G satisfies them (see [**24**] for details). From these results, it follows, that no function with a moderate growth (and, for sure, no function with compact support) may be the symbol of a finite rank Toeplitz operator in a Bergman type space.

Now we touch upon this problem for operators in a *tpBt* spaces. Suppose that F is a function satisfying the conditions mentioned above. Suppose that the Toeplitz operator in the *tpBt* space has finite rank. Then, by our transformation results, the Toeplitz operator in the corresponding usual Bergman type space with *distributional* symbol $G = \mathbf{E}F$ has finite rank as well, therefore, G must be a finite linear combination of point masses and their derivatives. Thus any solution F of the elliptic equation $\mathbf{E}F = G$ produces a Toeplitz operator with finite rank. In particular, if the order $j+1$ of the *tpBt* space is larger than the singularity order of the distribution G then, by the elliptic regularity, any solution of the equation $\mathbf{E}F = G$ is a continuous function. Thus, in *tpBt* spaces there can exist symbols-functions producing nontrivial finite rank Toeplitz operators. Note, however, that if we suppose that $F = 0$ on some open set in Ω (in particular, if it has compact support in Ω), then the function F, being a solution of a real-analytic elliptic equation $\mathbf{E}F = G$, must be real-analytic everywhere outside the singularities of G, and this means, outside some finite set. Therefore, F must be zero outside this set, where F must be a finite linear combination of point masses and their derivatives.

5.2. Boundedness. Here we combine the results on transformation of Section 4 and the boundedness conditions described in Section 3. We have everything at hand already.

THEOREM 5.5. *Let F be a function in $\mathbb{D}$ such that*

$$\sup_{z\in\mathbb{D}}\left\{(1-|z|)^{-2}\int_{B(z,\frac{1}{2}(1-|z|))}|F(w)|dA(w)\right\}<\infty, \tag{5.1}$$

then the Toeplitz operator with symbol F in $\mathcal{A}_{(k+1)}$ is bounded. Under the additional condition

$$\lim_{\varepsilon\to 0}\sup_{|z|>1-\varepsilon}\left\{(1-|z|)^{-2}\int_{B(z,\frac{1}{2}(1-|w|))}|F(w)|dA(w)\right\}=0 \tag{5.2}$$

this operator is compact.

Thus, strangely enough, the boundedness conditions for Toeplitz operators in $\mathcal{A}_{(j+1)}(\mathbb{D})$ are the same as in $\mathcal{A}(\mathbb{D})$.

PROOF. By Theorem 4.5, it is sufficient to prove that the Toeplitz operator in the Bergman space $\mathcal{A}(\mathbb{D})$ with distributional symbol $\mathbf{M}_jF$ is bounded. We look closer at the structure of the operator $\mathbf{M}_j$. By Proposition 4.3, it consists of the sum of terms of weight zero. These terms can be transformed to the ones $h_\alpha(z)\ell_\alpha(\partial,\bar\partial)(1-|z|^2)^{|\alpha|}$, with $|\alpha|\le 2j$. Here, $\ell_\alpha(\partial,\bar\partial)$ is a constant coefficients differential operator of order $|\alpha|$ and h_α is a smooth bounded function on $\mathbb{D}$. Each such term generates a bounded Toeplitz operator in $\mathcal{A}(\mathbb{D})$, by Theorem 3.2, applied to the measure ν_α, $\nu_a(E)=\int_E(1-|w|^2)^{|\alpha|}|F(w)|dA(w)$ (the factors depending on the distance to the boundary cancel). The statement about compactness follows in the same way from Theorem 3.5. □

The boundedness and compactness theorem for Toeplitz operators in $\mathcal{A}_{(j+1)}(\Pi)$ follows completely analogously from Theorem 4.2, Proposition 4.1 and Theorem 3.3.

THEOREM 5.6. *Let F be a function on Π If*

$$\sup_{z\in\Pi}\left\{(\operatorname{Im} z)^{-2}\int_{B(z,\frac{1}{2}\operatorname{Im} z)}|F(w)|dA(w)\right\}<\infty \tag{5.3}$$

then the Toeplitz operator in $\mathcal{A}_{(j)}$ with symbol F is bounded. If, additionally,

$$\lim_{\varepsilon\to 0}\sup_{\operatorname{Im} z<\varepsilon}\left\{(\operatorname{Im} z)^{-2}\int_{B(z,\frac{1}{2}\operatorname{Im} z)}|F(w)|dA(w)\right\}=0 \tag{5.4}$$

then this operator is compact.

Finally, we present the result for true poly-Fock spaces. It follows in the same way from Proposition 4.7 and Theorem 3.7. Here, unlike two previous cases, the boundedness and compactness conditions are different for poly-Fock spaces of different order. This is explained by the absence of cancelation of weight factors in the transformed symbol with the weight factors in the boundedness conditions.

THEOREM 5.7. *Suppose that the function F satisfies*

$$\sup_{z\in\mathbb{C}}\left\{(1+|z|^2)^{(j-1)}\int_{B(z,r)}|F(w)|dA(w)\right\}<\infty.$$

Then the Toeplitz operator in $\mathcal{F}_{(j)}$ with symbol F is bounded in $\mathcal{F}_{(j)}$. If, additionally

$$\lim_{R\to\infty}\sup_{|z|>R}\left\{(1+|z|^2)^{(j-1)}\int_{B(z,r)}|F(w)|dA(w)\right\}=0,$$

then this operator is compact.

PROOF. As it is shown in Proposition 3.9, the derivatives entering into the operator $\mathbf{D}$ can be expressed in a linear way through the coderivatives. To each of the corresponding terms in $\mathbf{D}F$, we can, therefore, apply Theorem 3.7, which gives the required estimate. The compactness follows automatically. □

References

[1] A. Alexandrov and G. Rozenblum, *Finite rank Toeplitz operators: some extensions of D. Luecking's theorem*, J. Funct. Anal. **256** (2009), no. 7, 2291–2303, DOI 10.1016/j.jfa.2008.11.006. MR2498766

[2] L. D. Abreu and H. G. Feichtinger, *Function spaces of polyanalytic functions*, Harmonic and complex analysis and its applications, Trends Math., Birkhäuser/Springer, Cham, 2014, pp. 1–38, DOI 10.1007/978-3-319-01806-5_1. MR3203099

[3] M. B. Balk, *Polyanalytic functions*, Mathematical Research, vol. 63, Akademie-Verlag, Berlin, 1991. MR1184141

[4] M. B. Balk, *Polyanalytic functions and their generalizations* (Russian), Complex analysis. One variable, 1 (Russian), Itogi Nauki i Tekhniki, Akad. Nauk SSSR, Vsesoyuz. Inst. Nauchn. i Tekhn. Inform., Moscow, 1991, pp. 187–246, 256. MR1155418

[5] V. Bruneau, A. Pushnitski, and G. Raikov, *Spectral shift function in strong magnetic fields*, Algebra i Analiz **16** (2004), no. 1, 207–238, DOI 10.1090/S1061-0022-04-00847-7; English transl., St. Petersburg Math. J. **16** (2005), no. 1, 181–209. MR2069004

[6] Ž. Čučković and T. Le, *Toeplitz operators on Bergman spaces of polyanalytic functions*, Bull. Lond. Math. Soc. **44** (2012), no. 5, 961–973, DOI 10.1112/blms/bds024. MR2975154

[7] V. Fock, *Bemerkung zur Quantelung des harmonischen oscillators im magnetfeld* [Remarks on the quantization of the harmonic oscilator in a magnetic field.] Zeit. für Fys. A. 47 (1928) 446–448.

[8] S. M. Grudsky and N. L. Vasilevski, *Toeplitz operators on the Fock space: radial component effects*, Integral Equations Operator Theory **44** (2002), no. 1, 10–37, DOI 10.1007/BF01197858. MR1913421

[9] A. Haimi and H. Hedenmalm, *Asymptotic expansion of polyanalytic Bergman kernels*, J. Funct. Anal. **267** (2014), no. 12, 4667–4731, DOI 10.1016/j.jfa.2014.09.002. MR3275106

[10] Yu. I. Karlovich and L. V. Pessoa, *C^*-algebras of Bergman type operators with piecewise continuous coefficients*, Integral Equations Operator Theory **57** (2007), no. 4, 521–565, DOI 10.1007/s00020-006-1473-x. MR2313284

[11] Y. I. Karlovich and L. V. Pessoa, *Poly-Bergman projections and orthogonal decompositions of L^2-spaces over bounded domains*, Operator algebras, operator theory and applications, Oper. Theory Adv. Appl., vol. 181, Birkhäuser Verlag, Basel, 2008, pp. 263–282, DOI 10.1007/978-3-7643-8684-9_12. MR2681891

[12] M. Pavlović, *Decompositions of L^p and Hardy spaces of polyharmonic functions*, J. Math. Anal. Appl. **216** (1997), no. 2, 499–509, DOI 10.1006/jmaa.1997.5675. MR1489593

[13] G. Ren and H. R. Malonek, *Decomposing kernels of iterated operators—a unified approach*, Math. Methods Appl. Sci. **30** (2007), no. 9, 1037–1047, DOI 10.1002/mma.823. MR2321941

[14] L. V. Pessoa, *On the structure of polyharmonic Bergman-type spaces over the unit disk*, Complex Var. Elliptic Equ. **60** (2015), no. 12, 1668–1684, DOI 10.1080/17476933.2015.1041111. MR3399305

[15] L. V. Pessoa and A. M. Santos, *Theorems of Paley-Wiener type for spaces of polyanalytic functions*, Current trends in analysis and its applications, Trends Math., Birkhäuser/Springer, Cham, 2015, pp. 605–613. MR3496792

[16] L. V. Pessoa and A. M. Santos, *Polyharmonic Bergman spaces and Bargmann type transforms*, J. Math. Anal. Appl. **449** (2017), no. 1, 619–641, DOI 10.1016/j.jmaa.2016.12.034. MR3595223

[17] A. K. Ramazanov, *Representation of the space of polyanalytic functions as the direct sum of orthogonal subspaces. Application to rational approximations* (Russian, with Russian summary), Mat. Zametki **66** (1999), no. 5, 741–759, DOI 10.1007/BF02674203; English transl., Math. Notes **66** (1999), no. 5-6, 613–627 (2000). MR1754474

[18] A. K. Ramazanov, *On the structure of spaces of polyanalytic functions* (Russian, with Russian summary), Mat. Zametki **72** (2002), no. 5, 750–764, DOI 10.1023/A:1021469308636; English transl., Math. Notes **72** (2002), no. 5-6, 692–704. MR1963139

[19] G. Rozenblum and G. Tashchiyan, *On the spectral properties of the perturbed Landau Hamiltonian*, Comm. Partial Differential Equations **33** (2008), no. 4-6, 1048–1081, DOI 10.1080/03605300701741099. MR2424389

[20] G. Rozenblum and G. Tashchiyan, *On the spectral properties of the Landau Hamiltonian perturbed by a moderately decaying magnetic field*, Spectral and scattering theory for quantum magnetic systems, Contemp. Math., vol. 500, Amer. Math. Soc., Providence, RI, 2009, pp. 169–186, DOI 10.1090/conm/500/09829. MR2655151

[21] G. Rozenblum and N. Vasilevski, *Toeplitz operators defined by sesquilinear forms: Fock space case*, J. Funct. Anal. **267** (2014), no. 11, 4399–4430, DOI 10.1016/j.jfa.2014.10.001. MR3269881

[22] G. Rozenblum and N. Vasilevski, *Toeplitz operators defined by sesquilinear forms: Bergman space case*, J. Math. Sci. (N.Y.) **213** (2016), no. 4, Problems in mathematical analysis. No. 83 (Russian), 582–609, DOI 10.1007/s10958-016-2726-0. MR3464836

[23] G. Rozenblum, N. Vasilevski, *Toeplitz operators defined by sesquilinear forms: Bergman spaces in the half-plane.* (to appear)

[24] G. Rozenblum and N. Shirokov, *Some weighted estimates for the $\overline{\partial}$-equation and a finite rank theorem for Toeplitz operators in the Fock space*, Proc. Lond. Math. Soc. (3) **109** (2014), no. 5, 1281–1303, DOI 10.1112/plms/pdu038. MR3283618

[25] N. L. Vasilevski, *Poly-Fock spaces*, Differential operators and related topics, Vol. I (Odessa, 1997), Oper. Theory Adv. Appl., vol. 117, Birkhäuser, Basel, 2000, pp. 371–386. MR1764974

[26] N. L. Vasilevski, *On the structure of Bergman and poly-Bergman spaces*, Integral Equations Operator Theory **33** (1999), no. 4, 471–488, DOI 10.1007/BF01291838. MR1682807

[27] N. Vasilevski, *Poly-Bergman spaces and two-dimensional singular integral operators*, The extended field of operator theory, Oper. Theory Adv. Appl., vol. 171, Birkhäuser, Basel, 2007, pp. 349–359, DOI 10.1007/978-3-7643-7980-3_16. MR2308569

[28] N. L. Vasilevski, *Commutative algebras of Toeplitz operators on the Bergman space*, Operator Theory: Advances and Applications, vol. 185, Birkhäuser Verlag, Basel, 2008. MR2441227

Chalmers University of Technology and The University of Gothenburg, Gothenburg, Sweden AND Department of Mathematical Physics, St. Petersburg State University, St. Petersburg, Russia.

Email address: grigori@chalmers.se

Cinvestav, Mexico City, Mexico

Email address: nvasilev@cinvestav.mx

Contemporary Mathematics
Volume **733**, 2019
https://doi.org/10.1090/conm/733/14748

Complete metric space of Riemann integrable functions and differential calculus in it

Serguei Samborski

ABSTRACT. In the set of equivalence classes of Riemann integrable functions on a compact X , where equivalence means coincidence almost everywhere, a metric is introduced that turns this set into a complete metric space.The elements of this space have "values" at the points of X , these values are convex subsets. The operator of classical differentiation D, defined initially on smooth functions, as well as the operators of partial differentiation D_{x_i} with respect to coordinates $(x_1, \dots, x_n)$ in X are pre-closed in the spaces under consideration. Their extensions by closure $\mathbf{D}$, $\mathbf{D}_i$ are linear operators and their domains are stable with respect to operations like maximum. The value of $\mathbf{D}F$ at any point $x \in X$ coincides with the value of Clarke's differential at x (as convex subsets). However, unlike Clarke's partial differentials, the equality $\mathbf{D} = (\mathbf{D}_1, \dots, \mathbf{D}_n)$ holds.

1. Introduction

The aim of this paper is to draw attention to functional spaces where the topological structure (a metric in our case) is associated with algebraic structures in a non-traditional way. Namely, algebraic operations are not continuous, but are subject to a less stringent requirement of closure. For example, concerning the addition it means :

$$\text{if } a_i \to a\,, b_i \to b,\; a_i + b_i \to c\,, \text{ then } \; c = a + b.$$

Analysis in such spaces is much more complicated than in the spaces with the continuity of algebraic operations (topological vector spaces, K-spaces etc.). For example, the extension by closure of a pre-closed linear operator may not be a linear mapping. However certain properties desired for applications can be realized only if the continuity of operations is abandoned in favour of their closure. Here is an example. In the set C of continuous functions on a compact X with uniform convergence we want to introduce a metric so that the metric completion simultaneously is the completion of C in the sense of lattices. Recalling the equivalence of the definition of $\mathbb{R}$ by Cantor and by Dedekind, I would like to say that the desired space plays the same role with respect to C that $\mathbb{R}$ plays with respect to $\mathbb{Q}$. This is of course an exaggeration, but precisely such connections which ensure in $\mathbb{R}$ the property of intermediate values for continuous functions (invalid in $\mathbb{Q}$), give also the properties of intermediate values in the resulting space for differential operators

2010 *Mathematics Subject Classification.* Primary 46E99, Secondary 90C.

(invalid in C). Such a metric exists and, in the resulting completion, the algebraic operations are not continuous, but they are closed [**1**, **2**].

Let us return to the space considered in this article. In the set of equivalence classes of Riemann integrable maps on $X \subset \mathbb{R}^n$ with values in $\mathbb{R}^m$, where equivalence means coincidence almost everywhere, a metric is introduced that turns this set into a complete metric space. Elements of this space (that is, equivalence classes) have "values": to each class F and to each $x \in X$ corresponds a convex subset $F(x) \subset \mathbb{R}^m$ denoted "the value of F in x". The resulting multivalued function uniquely determines the class F (note that this is impossible in the space of classes of equivalent Lebesgue-integrable functions). We say that convergence has the property of preserving values if for any $F_i \to F$, $x_i \to x$, $\zeta_i \in F_i(x_i)$ the convergence $\zeta_i \to \zeta$ implies $\zeta \in F(x)$.

Beyond the set of continuous functions, such a property of non-vanishing of limit values is not feasible in the framework of spaces with continuous algebraic operations. At the same time, it is true in the space considered in this paper. Smooth functions are dense in the considered space and the classical differentiation operator D, defined initially on smooth functions, is pre-closed. Let $\mathbf{D}$ be its extension by closure. We show that it is a linear operator and its domain consists of Lipschitz mappings (although not all of them). Let us recall that every Lipschitz map f is almost everywhere differentiable (Theorem of Rademacher) and that Clarke's differential of f at a point x is the following convex subset [**3**]:

$$D_{Cl}f(x) = Co\{limDf(y)\}, \text{ where} y \to x,$$

where x, y are points of differentiability of f and Co means a convex hull.

The above-mentioned values of $\mathbf{D}$ coincide with the values of Clarke's differential. At the same time, unlike Clarke's construction, the presence of "classical" algebraic structures in the space of the image of $\mathbf{D}$ makes it possible to establish the linearity of $\mathbf{D}$ and other rules connecting the differentials of various combinations of functions. This eliminates the analysis of the regularity of points (in Clarke's terminology [**3**]) to deal with equalities instead of inclusions when applying formal differential calculus. However, the principal difference with Clarke's construction is manifested at the level of partial derivatives. Namely, if $(x_1, \dots, x_n)$ are coordinates in X and $\mathbf{D}_i$ is the closure of the operator of partial differentiation D_{x_i}, then the coordinate expression $\mathbf{D} = (\mathbf{D}_1, \dots, \mathbf{D}_n)$ holds. For Clarke's construction there are not even inclusions relating complete differential and partial differentials.

2. Space C_{ae}

Let X be a compact subset in $\mathbb{R}^n$ that is the closure of its interior and E be a normed finite dimensional space over $\mathbb{R}$. We denote by $C_{ae}(X, E)$ the set of equivalence classes of bounded continuous almost everywhere maps from X to E, equivalence means that maps coincide almost everywhere.

According to Lebesgue Criterion if $F \in C_{ae}$ then each representative $f \in F$ is a Riemann-integrable function and each (bounded) Riemann-integrable function defines an element in C_{ae}. Set C_{ae} has the following algebraic structures:

- $C_{ae}(X, E)$ is a vector space over $\mathbb{R}$: if $F, G \in C_{ae}, \lambda \in \mathbb{R}$, then $F + \lambda G$ is the class in C_{ae} with representative $f + \lambda g$, where $f \in F$, $g \in G$ are any representatives;
- $C_{ae}(X, \mathbb{R})$ is a commutative ring;

- $C_{ae}(X, E)$ is a module over $C_{ae}(X, \mathbb{R})$;
- $C_{ae}(X, \mathbb{R})$ is a lattice : $F \leq G$ if for some (and hence for any) representatives $f \in F$, $g \in G$ the inequality $f \leq g$ holds almost everywhere.

Let $F \in C_{ae}(X, \mathbb{R})$, $x \in X$. We define the following interval in $\mathbb{R}$

$$F(x) = [EssInf_x f, EssSup_x f]$$

where f is any representative from F. (Recall, that $EssInf_x(EssSup_x)$ is the $Sup(Inf)$ of local minorants (majorants) almost everywhere in neighborhoods of x). It is easy to see that

$F(x) = Co\{limf(y)\}$, where $y \to x, y$ are points of continuity of f, $f \in F$.

A point x is said to be point of continuity of F if the set $F(x)$ is a singleton. If all points of X are points of continuity of F, than the class F is called a continuous map from C_{ae}. In the same way we speak of Lipschitz or smooth (below, this means continuously differentiable) functions as elements of C_{ae}.

The correspondence $x \to F(x)$ for every $F \in C_{ae}$ determines the function from X to the set of intervals in $\mathbb{R}$. We continue to denote this function by F. If $F \neq G$ then the multivalued functions F and G are different. Remark that if F, $G \in C_{ae}(X, \mathbb{R})$, $F \neq G$,, then there exists a neighborhood $U \subseteq X$ and a real $\alpha > O$ such that

$$\forall x \in U, \forall y \in U, \forall \zeta \in F(x), \forall \eta \in G(y) \text{ we have } |\zeta - \eta| > \alpha.$$

Let $F \in C_{ae}(X, \mathbb{R})$, $n \in \mathbb{N}$. We define two Lipschitz functions

$$F_n^- = sup\{h \in Lip_n(X, \mathbb{R}) \,|\, h \leq F\}$$

,

$$F_n^+ = inf\{h \in Lip_n(X, \mathbb{R}) \,|\, h \geq F\}$$

(recall that the lattice $Lip_n(X, \mathbb{R}$ is conditionally complete and that comparison of functions is performed in the lattice C_{ae} i.e. almost everywhere). On the set of Lipschitz functions on X the following distance is well defined

$$r'(f, g) = h(grf, grg) + \int_X |f - g| dx$$

where gr denote the graph of a function and $h(\cdot, \cdot)$ is the Hausdorff distance between compact subsets.

DEFINITION 2.1 (The distance in $C_{ae}(X, \mathbb{R})$). Let F, G be two elements from $C_{ae}(X, \mathbb{R})$). Put

$$r(F, G) = sup_n\{r'(F_n^-, G_n^-), r'(F_n^+, G_n^+)\}$$

.

It follows from the remark made above, that r is a metric.

THEOREM 2.2. *The set $C_{ae}(X, \mathbb{R})$ equipped with the metric r is a complete metric space. The set of smooth functions is dense in $C_{ae}(X, \mathbb{R})$*

PROOF. Let $(F_k)_{k \in \mathbb{N}}$ be a fundamental sequence in $C_{ae}(X, \mathbb{R})$. Consider the double sequence $((F_k)_n^-, (F_k)_n^+)_{k,n \in \mathbb{N}}$. For each fixed n the sequences $((F_k)_n^-)_{k \in \mathbb{N}}$ and $((F_k)_n^+)_{k \in \mathbb{N}}$ are fundamental in the metric space (Lip_n, r') and, by virtue of the completeness of this space, converge to Lipschitz functions Φ_n^- and Φ_n^+ respectively. Moreover, for each $\varepsilon > 0$ there exists a number $N(\varepsilon) \in \mathbb{N}$, which is independent of n and such that for all $k > N(\varepsilon)$ the inequalities

$$r'((F_k)_n^-, \Phi_n^-) < \varepsilon \text{ and } r'((F_k)_n^-, \Phi_n^-) < \varepsilon$$

are valid. From the continuity of the projectors $F \to F_j^{\pm}$ one obtains the equalities

$$(\Phi_n^+)_j^+ = \Phi_j^+ , \ (\Phi_n^-)_j^- = \Phi_j^- .$$

Further,

$$r'(\Phi_n^-, \Phi_n^+) \leq r'((F_k)_n^-, \Phi_n^-) + r'((F_k)_n^+, \Phi_n^+) + r'((F_k)_n^-, (F_k)_n^+).$$

For each fixed k,r' $((F_k)_n^-, (F_k)_n^+) \to 0$ for $n \to \infty$. Therefore, for $k > N(\varepsilon/3)$ and sufficiently large n, the right-hand side of the last inequality is less than ε. Consequently, if

$$\Phi^-(x) = \underline{lim}\Phi_n^-(x), \Phi^+(x) = \overline{lim}\Phi_n^+(x),$$

then the multi-valued function $x \to [\Phi_n^-(x), \Phi_n^+(x)]$ corresponds to the element $\Phi \in C_{ae}(X, \mathbb{R})$ and $\Phi = limF_k$.

Let $F = (F_n^-, F_n^+) \in C_{ae}(X, \mathbb{R})$. We want to show that the sequence of Lipschitz functions $(F_k^-)_k$ converges to F in the metric r. We have

$$r(F_k^-, F) = Sup(r(F_1^-, F_1^-), \ldots, r(F_k^-, F_k^-), r(F_k^-, F_{k+1}^-), r(F_k^-, F_{k+2}^-), \ldots,$$

$$r((F_k^-)_1^+, F_1^+), r((F_k^-)_2^+, F_2^+), \ldots, r(F_k^-, F_k^+), r(F_k^-, F_{k+1}^+), \ldots).$$

Let $\varepsilon > 0$. In the metric space (Lip_n, r'), every monotone bounded sequence is fundamental, in particular, the sequence $(F_k^-)_k$ is also such. Therefore, all terms of the form (F_k^-, F_{k+l}^-) in the preceding formula do not exceed ε for sufficiently large k. We estimate the terms of the double sequence $W_{s,k} = r((F_k^-)_s^+, F_s^+)$. The monotonicity of the sequence F_s^+ and the inequality $F_k^- \leq F_{k+1}^- \leq F_s^+$ show that $W_{s,k+1} \leq W_{s,k}$ for all s and $k \geq s$. In particular, $W_{s,k} \leq W_{s,s}$ for all k. The sequence $(W_{s,s})_{s \in \mathbb{N}}$ tends to zero for $s \to \infty$. We show that there exists $N \in \mathbb{N}$ such that

$$r'((F_k^-)_s^+, F_s^+) < \varepsilon \text{ for any } k > N \text{ and } k \geq s.$$

In view of the preceding it suffices to show that this inequality holds for $k > N$ and $s < s_0$, where s_0 is some fixed number. The projectors $F \to F_s^+$ are uniformly continuous, hence there exists $\delta > 0$ such that

$$W_{s,k} = r'((F_k^-)_s^+, F_k^+)_s^+) < \varepsilon \text{ if } r'(F_k^-, F_k^+) < \delta \text{ and } s \leq s_0, s \leq k.$$

Therefore $W_{s,k} < \varepsilon$ is true for sufficiently large k, whence the inequality x follows. Since in each space (Lip_n, r') the subset of smooth functions is dense, then such functions are dense in $C_{ae}(X, \mathbb{R})$. □

PROPOSITION 2.3. *Let $F_k \to F$ in $C_{ae}(X, \mathbb{R})$). Then*

(1) *the following property of preservation of values holds: $\forall x_k \to x, \forall \zeta_k \in F_k(x_k)$ the convergence $\zeta_k \to \zeta$ implies $\zeta \in F(x)$.*
(2) *$F_k(x) \to F(x)$ for any x from the full-measure subset in X of common points of continuity of F_k, F*
(3) *If F is a continuous function, then the convergence $F_k \to F$ is uniform: $\forall \varepsilon > 0$ there exists $N \in \mathbb{N}$ such that $\forall x \in X, \forall n > N, \forall \zeta \in F_k(x)$ $|\zeta - F(x)\eta| < \varepsilon$.*

PROOF. Let $F \in C_{ae}(X, \mathbb{R})$. A subset $grF = \{(x, \zeta) \,|\, x \in X, \zeta \in F(x)\}$ in $X \times \mathbb{R}$ is a compact subset. If $F, G \in C_{ae}(X, \mathbb{R})$ we define $h(F, G)$ as the Hausdorff distance between grF and grG. If $F_k \to F$ in $C_{ae}(X, \mathbb{R})$ then the convergences in the distances r and h are equivalent. Hence the formulated properties easily follow. □

PROPOSITION 2.4 (Convergence and algebraic operations).

(1) *If* $F_i \to F$, $G_i \to G$, $\lambda_i \to \lambda$, $F_i + \lambda_i G_i \to H$, then $H = F + \lambda G\,(\lambda_i\,,\,\lambda \in \mathbb{R})$;
(2) *If* $F_i \to F$, $G_i \to G$, $F_i \cdot G_i \to H$, *, then* $H = F \cdot G$;
(3) *If* $F_i \to F$, $G_i \to G$, $max(F_i, G_i) \to H$, *then* $H = max(F, G)$.

Let E be a finite-dimensional vector normed space over $\mathbb{R}$ and let $(e_1, e_2, \ldots, e_m)$ be a basis in E. Then each $F \in C_{ae}(X, E)$ is identified in the usual way with $(F_{(1)}, F_{(2)}, \ldots, F_{(m)})$ where $F_{(i)} \in C_{ae}(X, \mathbb{R})$. Setting

$$r(F, G) = Sup_i\{\, r(F_{(i)}\,,\, G_{(i)}\,)\}$$

we obtain the metric that transforms $C_{ae}(X, E)$ into a complete metric space in which smooth mappings form a dense subset.

3. Differentiation

Recall that a mapping of metric spaces is said to be *closed* if its graph is a closed subset. A mapping is said to be *pre-closed* if the closure of its graph remains a graph of a mapping. Such an extension of a pre-closed mapping by the closure of its graph is called *extension by closure*, it is minimal among possible closed extensions.

THEOREM 3.1. *The operator of classical differentiation D defined on the set of smooth mappings is a pre-closed operator from $C_{ae}(X, \mathbb{R}^m)$ to $C_{ae}(X, (\mathbb{R}^n)^* \otimes \mathbb{R}^m)$. Let $\mathbf{D}$ be its extension by closure and Δ be the domain of this extension. Then Δ is a linear subspace in $C_{ae}(X, \mathbb{R}^m)$ and $\mathbf{D}$ is a linear operator.*

Let $(x_1, x_2, ..., x_n)$ be coordinates in X and D_i be the operator of classical differentiation with respect to the coordinate x_i. Then D_i defined on the set of smooth mappings is a pre-closed operator from $C_{ae}(X, \mathbb{R}^m)$ to $C_{ae}(X, \mathbb{R}^* \otimes \mathbb{R}^m)$. Let $\mathbf{D}_i$ be its extension by closure and Δ_i be the domain of this extension. Then Δ_i is a linear subspace in $C_{ae}(X, \mathbb{R}^m)$ and $\mathbf{D}_i$ is a linear operator.

Also we have $\Delta = \cap_i \Delta_i$ and $\mathbf{D} = (\mathbf{D}_1, \mathbf{D}_2, ..., \mathbf{D}_n)$.

PROOF. It suffices to prove the assertion of the theorem for the operator $\mathbf{D}_i$ and $m = 1$. Let $i = 1$. Suppose by contradiction that $\mathbf{D}_1$ is not pre-closed. Then there exist two sequences of smooth functions $(F_i)_{i \in \mathbb{N}}$ and $(G_i)_{i \in \mathbb{N}}$ such that:

(1) $lim_i F_i = lim_i G_i = F$ in the space $C_{ae}(X, \mathbb{R})$;
(2) The sequences $(\mathbf{D}_1(F_i))_{i \in \mathbb{N}}$ and $(\mathbf{D}_1(G_i))_{i \in \mathbb{N}}$ converge in $C_{ae}(X, \mathbb{R})$ to limits Φ and, respectively, Ψ.
(3) $\Phi \neq \Psi$.

It follows from the last inequality that there exists an open subset $U \subset X$ and real numbers $\alpha, \beta\,(\alpha < \beta)$ such that $\forall x \in U, \forall \zeta \in \Phi(x)\,\forall \eta \in \Psi(x)$ we have $\zeta < \alpha\,,\, \eta > \beta$ (or $\zeta > \beta$ and $\eta < \alpha$, which does not change the course of reasoning). Consequently, there exist reels α', β' $(\alpha < \alpha' < \beta' < \beta)$ and $n \in \mathbb{N}$ such that are true

$$\forall x \in U \ \mathbf{D}_1(F_i(x)) < \alpha' \text{ and } \forall x \in U \ \mathbf{D}_1(G_i(x)) > \beta'.$$

Let $(x_1', x_2, ..., x_n)$ and $(x_1'', x_2, ..., x_n)$ be two points of continuity in U of the mapping F. Using the property of preserving values and passing, if necessary, to subsequences, we can assume that

$$\begin{array}{l} lim_{i\to\infty} F_i(x_1', x_2, \dots, x_n) = lim_{i\to\infty} G_i(x_1', x_2, \dots, x_n) = F(x_1', x_2, \dots, x_n), \\ lim_{i\to\infty} F_i(x_1'', x_2, \dots, x_n) = lim_{i\to\infty} G_i((x_1', x_2, \dots, x_n) = F(x_1'', x_2, \dots, x_n) \end{array}$$

On the other hand, from Main Value Theorem it follows that for $i > N$

$$\begin{array}{l} F_i((x_1', x_2, \dots, x_n) - F_i((x_1'', x_2, \dots, x_n) < \alpha'|x_1' - x_1''| \\ G_i((x_1', x_2, ..., x_n) - G_i((x_1'', x_2, \dots, x_n) > \beta'|x_1' - x_1''|. \end{array}$$

The contradiction proves the pre-closure of the operator $\mathbf{D}_1$.

Let $F, G \in \Delta_1$ and $\Phi = \mathbf{D}_1(F), \Psi = \mathbf{D}_1(G)$. Then for almost all $y = (x_2, \dots, x_n)$ for which $\Phi(\cdot, y)$ and $\Psi(\cdot, y)$ are continuous almost everywhere, we have, after the limit transition

$$F(x, y) = \int_a^x \Phi(t, y,)dt + F(a, y)$$

$$G(x, y) = \int_a^x \Psi(t, y,)dt + G(a, y)$$

where (a, y) is the common point of continuity of F and G. Therefore, for almost all y we have

$$(F + \lambda G)(x, y) = \int_a^x (\Phi + \lambda\Psi(t, y))dt + F(a, y) + \lambda G(a, y).$$

Hence it follows that $F + \lambda G \in \Delta_1$ and $\mathbf{D}_1(F + \lambda G) = \mathbf{D}_1 F + \lambda \mathbf{D}_1 G$. □

For a convex subset $A \subset (\mathbb{R}^n)^* \otimes \mathbb{R}^m$ and a vector $b \in \mathbb{R}^n$ we denote by $A \cdot b$ the convex subset of $\mathbb{R}^m$ equal to $\{\alpha \cdot b \mid \alpha \in A\}$. We note that A can be uniquely reconstructed from the collection of subsets $\{A \cdot b \mid b \in \mathbb{R}^n, \|b\| = 1\}$.

COROLLARY 3.2. *Let $F \in \Delta$, $\Theta = (\theta_1, \theta_2, \dots, \theta_n)$ be a unit vector in $\mathbb{R}^n$. Then for any $x \in X$ the equality of convex subsets in $\mathbb{R}^m$*

$$\mathbf{D}F(x) \cdot \Theta = (\sum_i \theta_i \mathbf{D}_i F)(x)$$

holds.

Below, if the map F is denoted as an element of the space C_{ae}, then we say that F is *differentiable* if $F \in \Delta$. To avoid ambiguity, the classical differentiability will be denoted every time.

PROPOSITION 3.3. *Every differentiable map $F \in C_{ae}(X, \mathbb{R}^m)$ is Lipschitz. A Lipschitz map $F \in C_{ae}(X, \mathbb{R}^m)$ is differentiable if and only if the map $x \to DF(x)$ defined at all points of (classical) differentiability of F is continuous almost everywhere. Moreover, $\mathbf{D}F$ is the class in $C_{ae}(X, (\mathbb{R}^n)^* \otimes \mathbb{R}^m)$ containing the map $x \to DF(x)$ and for any x the equality $\mathbf{D}F(x) = D_{Cl}F(x)$ of convex subsets holds in $(\mathbb{R}^n)^* \otimes \mathbb{R}^m$.*

PROOF. We first show that $F \in \Delta$ implies $F \in Lip_K$ for some K. Recall that for a lower semi-continuous function $f : X \to \mathbb{R}$, the sub-differential $d^- f(x)$ at a point x denotes the subset in $(\mathbb{R}^n)^*$ consisting of all λ such that there is a smooth function $\varphi : X \to \mathbb{R}$, $\varphi \leq f$, $\varphi(x) = f(x)$, $\lambda = D\varphi(x)$. The set $d^- f(x)$ is non empty for x from a dense subset of X. It is known that if there exists a constant K such that for all $x \in X$ $\|d^- f(x)\| \leq K$ then $f \in Lip_K$[**4**].

Let $F \in \Delta$, $F(x) = (F_-(x), F_+(x))$. Then F_- is a lower semi-continuous function. Let us show that $\|d^- F_-(x)\| \leq K$ for every $x \in X$. First we note that for any $\varepsilon > 0$, $F \in C_{ae}(X, \mathbb{R})$ and for $\lambda \in d^- F_-(x)$ there exists $\delta > 0$ such that for every smooth function f satisfying the inequality $h(grf, grF) < \varepsilon$ there is x' verifying

$$\|x - x'\| < \varepsilon, \ |f(x') - F(x))| \, , \ \|\lambda - Df(x')\| < \varepsilon.$$

Let $F_i \to F$ be a sequence of smooth functions, $DF_i \to \mathbf{D}F$. Since $\mathbf{D}F$ is bounded, for sufficiently large i also DF_i are bounded in norm by some constant. Consequently, $\|d^- f(x)\| \leq K$ and F is Lipschitz. Suppose that $\zeta \in D_{Cl}F_-(x)$. Then there are a sequence $x_i \to x$ such that $DF(x_i) \to \zeta$. It follows from the remark made above that there is a sequence $x'_i \to x$ such that $DF_i(x'_i) \to \zeta$. It means that $D_{Cl}F(x) \subseteq \mathbf{D}F(x)$. If x is a continuity point for $\mathbf{D}F$ (that is $\mathbf{D}F(x)$ is a singleton), then $D_{Cl}F(x)$ is also a singleton and from the Proposition 3.2 it follows that F is continuously differentiable at the point x. Therefore $\mathbf{D}F(x) \subseteq Co\{limD(F(y)\}$ where $y \to x, y$ be points of differentiability of $F\} = D_{Cl}F(x)$. □

EXAMPLE 3.4 (Lipschitz function that is not C_{ae}-differentiable). Let $K = [0,1] \setminus \cup I_i$ be a Cantor set of positive measure, where $\{I_i\}$ is a countable system of mutually disjoint open intervals such that K is nowhere dense. We put $\Phi(x) = \int_{[0,x]} \chi_K d\mu$ where χ_K is the characteristic function of the set K. Then Φ is a Lipschitz function and $D_{Cl}\Phi(x) = [0,1]$ at each point x of K

EXAMPLE 3.5. Let us compare the partial differentials $\mathbf{D}_i$ and Clarke's partial differentials using Example 2.5.2 from [**3**], where

$$f(x,y) = max\{min\{x, -y\}, y - x\}.$$

We have, under the standard identification of $\mathbb{R}$ and $\mathbb{R}^*$:

$$D_{1Cl}f(0,0) = [-1,0], D_{2Cl}f(0,0) = [0,1], D_{Cl}f(0,0) = Co\{(1,0),(0,-1),(-1,1)\}.$$

However,

$$\mathbf{D}_1 f(0,0) - \mathbf{D}_2 f(0,0) = [-1,1]$$

and

$$Pr_i \mathbf{D}f(0,0) = \mathbf{D}_i f(0,0).$$

PROPOSITION 3.6. *Let $F, G \in C_{ae}(X, \mathbb{R})$ be differentiable. Then*

(1) *the linear combination $F + \lambda G$ is differentiable ($\lambda \in \mathbb{R}$) and $\mathbf{D}(F + \lambda G) = \mathbf{D}F + \lambda \mathbf{D}G$;*
(2) *the product $F \cdot G$ is differentiable and $\mathbf{D}(F \cdot G) = F \cdot \mathbf{D}G + G \cdot \mathbf{D}F$;*
(3) *the quotient F / G is differentiable (if $G(x) \neq 0$ for any $x \in X$) and $\mathbf{D}(F/G) = (G \cdot \mathbf{D}F - F \cdot \mathbf{D}G)/G^2$;*
(4) *$Max(F, G)$ and $Min(F, G)$ are differentiable.*

PROOF. 1) is proved in Theorem 3.1. 2) and 3) follow from 1) and from the obvious assertion that the differentiability of $F \geq \alpha > 0$ (in C_{ae}) implies differentiability of lnF and vice versa. 4) Since $max\{F, G\} = max\{F - G, 0\} + G$, then in view of 1) we can put $G = 0$. From differentiability of F (in C_{ae}) it follows that the measure of $\{x \in X \,|\, F(x) = 0\}$ is equal to the measure of $\{x \in X \,|\, F(x) = 0, \mathbf{D}F(x) = 0\}$. Therefore, almost everywhere in X either $F(x) \neq G(x)$ or $\mathbf{D}F(x) = \mathbf{D}G(x)$, and the function $max\{F, G\}$ is continuously differentiable almost everywhere. □

4. Compositions

Let $F \in C_{ae}(X, E)$. A representative φ of class F is said to be *regular* if the set of points of continuity of φ coincides with the set of points of continuity of F. For example, if $E = \mathbb{R}$ and $F(x) = [F_-(x), F_+(x)]$, then the functions F_- and F_+ are regular representatives of F.

DEFINITION 4.1. Suppose that $F \in C_{ae}(X, \mathbb{R}^m), G \in C_{ae}(Y, \mathbb{R}^k), F(x) \subset Y$ for any $x \in X$. A class $H \in C_{ae}(X, \mathbb{R}^k)$ is said to be the composition of F and G and is denoted by $G \circ F$ if for any regular $\varphi \in F$ and for any regular $\psi \in G$ the mapping $x \to \psi(\varphi(x))$ belongs to the class H.

If G is a continuous mapping, then $G \circ F$ always exists.

EXAMPLE 4.2. Let $X = [-1, 1], Y = [-1, 1]^2, F(x) = (x, ax), B(x, y)$ be the class containing the representative $(x, y) \to \{1 \text{ if } y > 0, -1 \text{ if } y < 0\}$. Then $B \circ F$ is defined if and only if $a \neq 0$.

If $\alpha = (\alpha_1 \otimes a_2) \in (\mathbb{R}^n)^* \otimes \mathbb{R}^m, \beta \in (\mathbb{R}^m)^*)$, then the correspondence $(\alpha, \beta) \to \beta(\alpha_2)\alpha_1$ extends to a mapping $\bullet : (\mathbb{R}^n)^* \otimes \mathbb{R}^m \times (\mathbb{R}^m)^* \to \mathbb{R}^n)^*$ (the composition of linear operators under the identification of $E^* \otimes F$ with the space of linear bounded operators from E to F). We now consider more complicated compositions of the form $(B \circ F(\cdot)) \bullet A(\cdot)$, where $A \in C_{ae}(X, (\mathbb{R}^n)^* \otimes \mathbb{R}^m), B \in C_{ae}(Y, (\mathbb{R}^m)^*)), F : X \to Y$. In accordance with the general concept, we say that such a composition defines an element $H \in C_{ae}(X, (\mathbb{R}^n)^*)$ if for any regular $\varphi \in A$ and for any regular $\psi \in B$ the mapping $x \to \psi(F(x)) \bullet \varphi(x)$ is continuous almost everywhere and is a representative in H. Note that in the previous expression, $B \circ F(\cdot)$ can not determine an element of $C_{ae}(X, (C)^*))$, as in the case B and F from Example 4.2 and $A(x) = (1, 0)$.

From the definitions given above, the following proposition easily follows.

PROPOSITION 4.3. *Let $F : X \to Y \subset \mathbb{R}^m$, $G : Y \to \mathbb{R}$ and $G \circ F$ be differentiable (in the spaces C_{ae}). Then the composition $(\mathbf{D}G \circ F(\cdot)) \bullet \mathbf{D}F(\cdot)$ is well defined element of the space $C_{ae}(X, (\mathbb{R}^n)^*)$ and it coincides with $\mathbf{D}(G \circ F)$.*

PROPOSITION 4.4. *Let $Y \subset \mathbb{R}$ Let $F \in C_{ae}(X, \mathbb{R}), G \in C_{ae}(Y, \mathbb{R})$ be differentiable. Then the composition $G \circ F$ is differentiable.*

PROOF. If for almost all $x \in X$ the image $F(x)$ belongs to the subset of points of continuous differentiability of G, then the proposition is proved. Suppose that F maps a subset A of positive measure into a subset of measure zero in Y. From the continuous differentiability of F almost everywhere in A it follows that for almost all x in A the differential $DF(x)$ is equal to zero. This means that for almost all $x \in X$, $\mathbf{D}G(F(x)) \bullet \mathbf{D}F(x)$ is a singleton, and the function $G \circ F$ is differentiable. □

EXAMPLE 4.5. Let us return to example 3.4 and let x_i be center of the interval I_i and $2r_i$ be its length. We put $Y = [0,1] \times [-1,1]$ and

$$L = Y \setminus \cup_i B((x_i, 0), r_i)$$

where $B(z,r)$ denote the open ball in Y of radius r and center z.

The function

$$G : (x,y) \to \int_{[0,x]} \chi_L d\mu$$

is continuously differentiable (classically) almost everywhere in Y, and its differential defines an element of $C_{ae}(Y, \mathbb{R})$(here χ denotes the characteristic function). If $F(x) = (x, 0)$, then F and G are differentiable, while $G \circ F$ is not differentiable.

THEOREM 4.6. *Let $X \subset \mathbb{R}^n$,a mapping $F : X \to \mathbb{R}^m$ be obtained from real-analytic functions $F_1, F_2, \ldots, F_k$ by a finite number of operations $\lambda\cdot, +, \cdot, \div, max$, min and compositions $\circ$. Then F is differentiable (in C_{ae}) and in accordance with Propositions 3.6 and 4.3 $\mathbf{D}F$ is expressed in terms of $DF_1, DF_2, \ldots, DF_k$.*

PROOF. Let $\mathcal{A}$ a be the set of mappings obtained from real analytic functions by a finite number of operations indicated in the formulation of the proposition. Only the application of the operations max and min leads to the appearance of points in the domain where the classical differentiability is violated, while at the remaining points the differential remains continuous. Therefore, in view of Proposition 2.3, it suffices to show that $F : X \to Y, G : Y \to \mathbb{R}$, and F, $G \in \mathcal{A}$ implies the differentiability of $G \circ F$ (in the space C_{ae}). Let $Sing(F)$ be the subset of points in X, where $\mathbf{D}F(x)$ is not an singleton. Then in a neighborhood of each point $x \in Sing(F)$ the subset $Sing(F)$ is a finite union of analytic submanifolds. Since F is assumed to be differentiable, then $X \setminus Sing(F)$ is an open dense subset of full measure in X.

Let U be an open subset in $X \setminus Sing(F)$, $x \in U$. The image of U in Y is an analytic submanifold. The subset $Sing(G)$ is also locally a union of a finite number of analytic submanifolds $Sing(G) = \cup_i \Lambda_i$. Note by $T_y\Gamma$ the tangent subspace to a submanifold Γ in $y \in Y$. If for some $x \in X$ and i we have $T_{F(x)}F(U) \subseteq T_{F(x)}\Lambda_i$ then the value of the expression $M = \Theta \bullet DF(x)$ is not dependent on the choice of $\Theta \in \mathbf{D}G(F(x))$ and in this case x is a point of continuity for $D(G \circ F)$. If $T_{F(x)}F(U) \cap T_{F(x)}\Lambda_i$ has dimension less than $dimT_{F(x)}F(U)$, then the set $F^{-1}(SingG) \cap U$ has measure zero in U. So that for a subset of x of full measure in U the differential $x \to D(G \circ F)(x)$ exists (and is continuous). Hence the application of the composition does not output maps from the set $\mathcal{A}$. $\square$

Acknowledgement

The author uses this opportunity to express his admiration for the scale of the scientific and human personality of S. G. Krein, to whom he owes so much.

References

[1] S. N. Samborskiĭ, *On metric completeness and lattice completeness* (Russian, with Russian summary), Tr. Mat. Inst. Steklova **247** (2004), no. Geom. Topol. i Teor. Mnozh., 228–236; English transl., Proc. Steklov Inst. Math. **4(247)** (2004), 209–216. MR2168173

[2] S. N. Samborskiĭ, *On nonsmooth solutions of nonlinear partial differential equations* (Russian), Dokl. Akad. Nauk **424** (2009), no. 1, 16–18, DOI 10.1134/S1064562409010049; English transl., Dokl. Math. **79** (2009), no. 1, 10–12. MR2513136

[3] F. H. Clarke, *Optimization and nonsmooth analysis*, Canadian Mathematical Society Series of Monographs and Advanced Texts, John Wiley & Sons, Inc., New York, 1983. A Wiley-Interscience Publication. MR709590

[4] R. Redheffer and W. Walter, *The subgradient in* $\mathbf{R}^n$, Nonlinear Anal. **20** (1993), no. 11, 1345–1348, DOI 10.1016/0362-546X(93)90135-F. MR1220839

42, rue Notre-Dame de Nazareth, 75003 Paris, France
Email address: `serguei.samborski@unicaen.fr`